Space Sciences Series of ISSI

Volume 50

For further volumes:
www.springer.com/series/6592

Karoly Szego · Nicholas Achilleos · Chris Arridge ·
Sarah Badman · Peter Delamere · Denis Grodent ·
Margaret G. Kivelson · Philippe Louarn
Editors

The Magnetodiscs and Aurorae of Giant Planets

Previously published in *Space Science Reviews* Volume 187,
Issues 1–4, 2015

Editors
Karoly Szego
KFKI Research Institute for Particle
and Nuclear Physics
Budapest, Hungary

Nicholas Achilleos
University College London
London, United Kingdom

Chris Arridge
Department of Physics
University of Lancaster
Lancaster, United Kingdom

Sarah Badman
Department of Physics
University of Lancaster
Lancaster, United Kingdom

Peter Delamere
Geophysical Institute
University of Alaska
Fairbanks, Alaska, USA

Denis Grodent
Universite de Liege
Liege, Belgium

Margaret G. Kivelson
University of California Los Angeles
Los Angeles, California, USA

Philippe Louarn
IRAP
Toulouse, France

ISSN 1385-7525   Space Sciences Series of ISSI
ISBN 978-1-4939-3394-5   ISBN 978-1-4939-3395-2 (eBook)
DOI 10.1007/978-1-4939-3395-2

Library of Congress Control Number: 2015953232

Springer New York Heidelberg Dordrecht London

Cover Image: Sketch of the key magnetospheric regions in the Jovian magnetosphere. Credit: Max Planck Institute for Solar System Research

Printed on acid-free paper

Springer is part of Springer Science+Business Media (www.springer.com)

Contents

Giant Planet Magnetodiscs and Aurorae—An Introduction
K. Szego · N. Achilleos · C. Arridge · S.V. Badman · P. Delamere · D. Grodent ·
M.G. Kivelson · P. Louarn **1**

Planetary Magnetodiscs: Some Unanswered Questions
M.G. Kivelson **5**

A Brief Review of Ultraviolet Auroral Emissions on Giant Planets
D. Grodent **23**

**Solar Wind and Internally Driven Dynamics: Influences on Magnetodiscs and
Auroral Responses**
P.A. Delamere · F. Bagenal · C. Paranicas · A. Masters · A. Radioti · B. Bonfond ·
L. Ray · X. Jia · J. Nichols · C. Arridge **51**

**Auroral Processes at the Giant Planets: Energy Deposition, Emission Mechanisms,
Morphology and Spectra**
S.V. Badman · G. Branduardi-Raymont · M. Galand · S.L.G. Hess · N. Krupp ·
L. Lamy · H. Melin · C. Tao **99**

**Magnetic Reconnection and Associated Transient Phenomena Within the
Magnetospheres of Jupiter and Saturn**
P. Louarn · N. Andre · C.M. Jackman · S. Kasahara · E.A. Kronberg · M.F. Vogt **181**

Transport of Mass, Momentum and Energy in Planetary Magnetodisc Regions
N. Achilleos · N. André · X. Blanco-Cano · P.C. Brandt · P.A. Delamere ·
R. Winglee **229**

Sources of Local Time Asymmetries in Magnetodiscs
C.S. Arridge · M. Kane · N. Sergis · K.K. Khurana · C.M. Jackman **301**

DOI 10.1007/978-1-4939-3395-2_1
Reprinted from *Space Science Reviews* Journal, DOI 10.1007/s11214-014-0131-x

Giant Planet Magnetodiscs and Aurorae— An Introduction

Karoly Szego · Nicholas Achilleos · Chris Arridge ·
Sarah V. Badman · Peter Delamere · Denis Grodent ·
Margaret G. Kivelson · Philippe Louarn

Published online: 6 January 2015
© Springer Science+Business Media Dordrecht 2015

This volume contains the reports discussed during the Workshop "Giant Planet Magnetodiscs and Aurorae" held 26–30 November 2012, at the International Space Science Institute, organised together with the Europlanet project, supported by FP7 (Grant No. 228319).

Magnetodiscs are large current sheets surrounding Jupiter and Saturn (also Uranus and Neptune) that are filled with plasma principally originating in the natural satellites of these worlds. They are also solar system analogues for astrophysical discs. Magnetodiscs are special features of the fast rotating giant planets, a special feature of rotationally driven magnetospheres. Their structure is modified by variability in their plasma sources and by the solar wind. Auroral signatures in the optical and radio wavebands allow a diagnostic of these dynamical processes and enable the visualisation of these large plasma and field structures.

K. Szego (✉)
Wigner Research Centre for Physics, Budapest, Hungary
e-mail: szego.karoly@wigner.mta.hu

N. Achilleos
University College London, London, UK

C. Arridge · S.V. Badman
Lancaster University, Lancaster, UK

P. Delamere
University of Alaska Fairbanks, Fairbanks, USA

D. Grodent
Université de Liège, Liège, Belgium

M.G. Kivelson
Department of Earth, Planetary, and Space Sciences, UCLA, Los Angeles, USA

P. Louarn
IRAP/CNRS, Toulouse, France

The objective of this workshop was to address outstanding issues in the structure and dynamics of magnetodiscs using a comparative approach (see details under topics). More specifically, we aimed to review current understanding of magnetodiscs and auroral responses to magnetodisc dynamics; characterise and understand radial plasma transport in magnetodiscs; determine how magnetic reconnection works in magnetodiscs, and describe the effects on plasma transport; describe the associated auroral responses to internal and external magnetospheric processes; characterise how the solar wind influences magnetodiscs and the auroral responses to solar wind-driven dynamics; characterise the spectral and spatial properties of auroral emissions produced by magnetodisc dynamics; answer the question of whether there are significant differences between solar wind- and internally-driven dynamics; and determine the sources of local-time asymmetries in magnetodiscs.

This volume is a unique synthesis of all aspects of the giant magnetospheres and their aurorae; it provides an interdisciplinary approach to understanding the coupled system from the solar wind to the atmosphere; it combines the latest observations with current theory and models; and it also contains sufficient breadth for students of magnetospheric and space physics to use as a reference for future research.

A few topics in detail:

Radial plasma transport: How does plasma get from its (primary) sources near Io at Jupiter and Enceladus at Saturn into the magnetodisc and out of the magnetosphere? Of particular interest are the timescales for these transport processes, how they might vary with position, the physics of the transport process in the magnetodisc, and how radial transport varies with magnetospheric activity. To address this topic we will exploit the latest data, models and theory together with auroral observations.

Reconnection: Reconnection is a major process by which mass is lost from the magnetosphere and, as such, it is important to characterise reconnection in the magnetodisc. An important unanswered question is how reconnection is triggered in the magnetodisc and the interconnection between the Dungey and Vasyliunas cycles. In terms of remote diagnosis of reconnection, can specific details of the reconnection process (e.g., reconnection of closed or open flux) be identified in auroral observations?

Dynamics: Plasma production, radial transport, reconnection and solar wind influences are sources of dynamics in the magnetodiscs at Jupiter and Saturn (Uranus and Neptune as well). These and other dynamical events, such as injections, produce optical/radio auroral emissions. Here we will examine dynamical events in magnetodiscs, comparing and contrasting Jupiter and Saturn, and use auroral imaging and radio emissions as remote monitors of dynamics. Can the spectra and spatial distributions of various auroral emissions be used to diagnose different types of dynamic event? Can we develop an understanding of Space Weather at the giant planets using knowledge of variability in plasma production, radial transport, instability, and solar wind influences in very large systems?

Solar wind influences: Evidence for solar wind influences on the magnetodisc of Jupiter is substantial; Saturn's magnetodisc appears to respond even more strongly to the solar wind. The mechanisms behind these solar wind effects are not fully understood but involve a combination of Dungey cycle driving, angular momentum conservation and solar wind pressure effects. The dusk flank magnetosphere of Saturn has been studied in far more detail than the corresponding region of Jupiter's magnetosphere and provides an excellent and unique dataset for the study of asymmetries. What can be learned about the solar wind influence by in situ observations and monitoring the location, spectra and strengths of auroral emissions? How much are the magnetospheric structure, magnetospheric dynamics and the aurora of an outer planet influenced by the solar wind?

Sources of local time asymmetries: Magnetodiscs and aurorae at Jupiter and Saturn are known to have structure which is asymmetric in local time. There are asymmetries in magnetodisc location, thickness, field structure, and presumably stress balance. It is not clear what generates these asymmetries. Is it purely driven by the solar wind or do internal processes such as mass loss play a significant role?

DOI 10.1007/978-1-4939-3395-2_2
Reprinted from *Space Science Reviews* Journal, DOI 10.1007/s11214-014-0046-6

Planetary Magnetodiscs: Some Unanswered Questions

Margaret Galland Kivelson

Received: 14 April 2014 / Accepted: 19 April 2014 / Published online: 10 May 2014
© Springer Science+Business Media Dordrecht 2014

Abstract Characteristic of giant planet magnetospheres is a near equatorial region in which a radially stretched magnetic field confines a region of high density plasma. The structure, referred to as a magnetodisc, is present over a large range of local time. This introductory chapter describes some of the physics relevant to understanding the formation of this type of structure. Although many features of the magnetodisc are well understood, some puzzles remain. For example, Jupiter's magnetodisc moves north-south as the planet rotates. The displacement has been attributed to the motion of the dipole equator, but at Saturn the dipole equator does not change its location. This chapter argues that the reasons for flapping may be similar at the two planets and suggests a role for compressional waves in producing the displacement. The development of thermal plasma anisotropy and its role in the structure of Jupiter's magnetodisc are explored. Finally, localized plasma enhancements encountered by the New Horizons spacecraft at large downtail distances in Jupiter's nightside magnetodisc are noted and a firehose instability of stretched flux tubes is proposed as a possible interpretation of the observations.

Keywords Planetary magnetospheres · Plasmas · Compressional perturbations · Firehose instability

1 Introduction

The fundamental physical principles that account for the behavior of a planetary magnetosphere apply to all the magnetospheres of the solar system, but the outcome of the constraints imposed by physical laws is sensitive to critical dimensionless parameters. Special to the giant planet magnetospheres is the composition of the plasma trapped within them:

M.G. Kivelson (✉)
Department of Earth, Planetary, and Space Sciences, UCLA, Los Angeles, CA, USA
e-mail: mkivelson@igpp.ucla.edu

M.G. Kivelson
Department of Atmospheric, Oceanic, and Space Sciences, University of Michigan, Ann Arbor, MI, USA

Table 1 Radii, Sidereal Rotation Periods, and Dominant Magnetospheric Ions of Selected Planets*

Planet/Property	Equatorial radius (km)	Rotation period (h)	Dominant ions
Earth	6,378	23.934	H^+
Jupiter	71,492	9.925	$O^+, O^{++}, S^+, S^{++}, S^{+++}$
Saturn	60,268	10.543**	Water group ions

*Radii from Davies et al. (1996), rotation periods from de Pater and Lissauer (2010)

**The internal rotation period of Saturn is not known. This value is based on indirect evidence

the mass of typical plasma ions is an order of magnitude larger than the mass of the protons that normally dominate terrestrial plasmas. Consequently inertial effects are far more important at Jupiter and Saturn than at Earth. Furthermore the large spatial dimensions and rapid rotation of the giant planets (see Table 1) mean that the solar wind flows past only a portion of these magnetospheres in one planetary rotation period. Interaction with the solar wind does not dominate their dynamics. Magnetospheric and ionospheric plasmas interact through signals carried through the system by magnetohydrodynamic (MHD) waves, and the large travel distances through the giant planet magnetospheres introduce phase delays that are not readily recognizable at Earth. Furthermore, the plasma density drops rapidly as one follows a flux tube from the equator to the ionosphere, inhibiting the coupling of different parts of the system. Indeed, the outer parts of these magnetospheres may be unable to communicate with their ionospheres. Rapid rotation of the heavy ion plasma modifies the geometry and dynamics of the entire magnetosphere and controls aspects of plasma heating and loss through mechanisms that differ from processes significant at Earth.

Schematic illustrations of magnetospheres typically depict the field and plasma in a noon-midnight cut such as shown for Jupiter in Fig. 1. To a considerable degree, such a cartoon image can be viewed as generic, an approximate representation of any planetary magnetosphere. There is an upstream shock, a magnetosheath in which the diverted solar wind flows around a boundary that confines most of the field lines that emerge from the planet. A few of those field lines do not close back on the planet but link directly to the solar wind. In the anti-solar direction, there is a stretched magnetotail centered on a region of high plasma and current density, the plasma sheet. Field lines threading the low density region are probably connected to the solar wind. Of fundamental significance to the discussion of this paper is a feature absent at Earth. At Jupiter and at Saturn field lines are stretched at the equator not only on the night side, as at Earth, but also on the day side. This type of field distortion implies the presence of radially extended azimuthal currents flowing on both day and night sides of the planet. The current-carrying region embedded in relatively dense plasma that is confined near the equatorial plane at a large range of local times is what we here refer to as a magnetodisc.

This book is dedicated to describing the properties of the magnetodiscs of Jupiter and Saturn and interpreting their interactions with remote parts of the magnetosphere/ionosphere system. To set the stage for the discussion, this chapter discusses the origin of the disk-like structure and identifies some problems in the most basic interpretations of their structure.

2 Magnetodisc Formation

Magnetodiscs form in the rotating magnetized plasmas of planetary magnetospheres. Close to a planet, the currents that generate the magnetic field (dominated by the dipolar con-

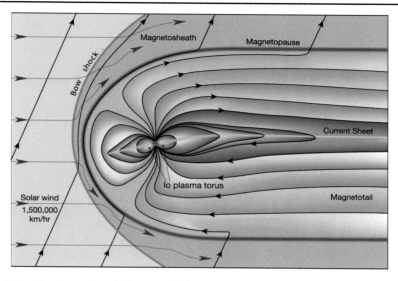

Fig. 1 Schematic of a cut through the noon-midnight meridian of the jovian magnetosphere (Bagenal and Bartlett 2013). Field lines are *black*. Solar wind flows along *red contours*. Plasmas dominated by the solar wind are in *green* and plasma dominated by sources internal to the magnetosphere are shown in *lavender*. The regions marked in *peach* contain the current that flows azimuthally and causes the field lines to stretch radially

tribution) flow within the planet. With increasing distance from the planet, field perturbations arising from currents on the magnetopause or generated in the trapped magnetospheric plasma begin to dominate the contributions from internal sources, and the configuration departs markedly from dipolar.

Field lines stretched radially outward near the equator at both Jupiter and Saturn imply that azimuthal magnetodisc currents extend through much of the equatorial magnetosphere. The regions in which magnetodisc current flows in Jupiter's magnetosphere are indicated schematically by the orange lines in Fig. 1. Within those regions, currents also flow radially outward across the field and link to field-aligned currents that couple to the ionosphere. The resulting current system imposes at least partial corotation on the plasma.

MHD theory identifies the sources of the currents responsible for the configuration of the magnetic field. In a steady state magnetosphere, force balance (with gravitational force negligible) requires

$$\mathbf{j} = \mathbf{B} \times (\nabla \cdot \mathbf{P} + \rho \mathbf{u} \cdot \nabla \mathbf{u}) / B^2 \tag{1}$$

Here $\mathbf{j}$ is the current density, ρ is the mass density, $\mathbf{u} = (u_r, u_\vartheta, u_\varphi)$ is the flow velocity, $\mathbf{B}$ is the magnetic field. Because in some circumstances pressure differs along and across the field, the pressure is taken to be a tensor, $\mathbf{P}$. With $\mathbf{\Omega}$ as the angular velocity, the centripetal force contribution (from the last term on the right) is well approximated as $-\hat{\mathbf{B}} \times \hat{\boldsymbol{\varphi}} \rho u_\varphi^2 / r = -\hat{\mathbf{r}} \rho r \Omega^2$. In Earth's magnetotail, the most significant azimuthal current arises from the north-south pressure gradient across the central plasma sheet and the pressure tensor is well approximated by an isotropic pressure. The centripetal force contribution is negligible. Jupiter and Saturn are different. Even relatively close to the planet, the centripetal force term is extremely large because of the immense spatial scale and the rapid rotation of the central planet and the comparatively high mass density of the plasma. The plasma pressure is not everywhere isotropic. Azimuthal currents can be significant at all local times.

An azimuthal current (whether caused by a pressure gradient or the inertia of the rotating plasma) stretches a field line near the equator and thus reduces its radius of curvature, R_c. The effect of changing the radius of curvature becomes clear in the familiar form of the force balance equation that eliminates $\mathbf{j}$ in terms of $\mathbf{B}$:

$$\nabla \cdot \left(\mathbf{P} + \frac{\mathbf{B}\,\mathbf{B}}{2\mu_o}\right) - \hat{\mathbf{n}}\frac{B^2}{\mu_o R_C} + \rho\mathbf{u}\cdot\nabla\mathbf{u} = 0 \tag{2}$$

Here $\hat{\mathbf{n}}$ is a unit vector normal to the field line, in the outward radial direction near the equator. Near the equator forces other than the curvature force, $-\hat{\mathbf{r}}B^2/\mu_o R_C$, are normally outward directed and as they increase, the inward curvature force must increase to balance them, thus requiring the field lines to stretch and R_C to become smaller. It is of interest to understand which of the outward forces dominates. At Jupiter, Mauk and Krimigis (1987) showed that on the day side, pressure balance requires pressure anisotropy beyond some critical distance. Paranicas et al. (1991) showed that on the night side at distances of 18 R$_J$ and beyond, $p_{\parallel}/p_{\perp} > 1$ and the anisotropy force exceeded other particle forces that have been measured (pressure gradients and the centrifugal force). However, little attention has been directed to the mechanism for generating the pressure anisotropy. This is a question to which I shall return in Sect. 5.

At both Jupiter and Saturn, the magnetodisc is not a rigid object but moves north-south at the period of planetary rotation. At Jupiter, the earliest interpretations of the motion were framed in terms of wave perturbations imposed by the rocking dipole moment and carried outward along the current sheet at the Alfvén velocity (e.g., Northrop et al. 1974; Kivelson et al. 1978). At the time these ideas were first broached, our knowledge of the plasma properties of the magnetosphere was incomplete and quantitative tests of the concept were not possible. Section 4 subjects the implicit assumption of wave propagation through the plasma sheet to quantitative criticism and proposes a different model for how the plasma sheet motion is imposed.

3 Magnetodisc Motion

In all magnetospheres, the plasma sheet is neither rigid nor fixed in its position in the magnetotail. To first order its position is controlled by the tilt of the planetary dipole moment. A cross section of the terrestrial magnetotail reveals that the plasma sheet lies near the magnetic equator, with the center of the plasma sheet bowed northward when magnetic north tilts towards the sun (e.g., Hammond et al. 1994) and bowed southward when the magnetic north tilts away from the sun. At Jupiter, the rotation axis is tilted by 3° relative to the orbit, and the dipole tilted by roughly 10° relative to the rotation axis. Thus, as the dipole axis rotates about the spin axis, the changing tilt of the magnetic equator imposes a north-south displacement on a warped plasma sheet. A spacecraft near the equator (there have been seven to this date) observes the plasma sheet passing up and down over it every ~10 hours as seen in Fig. 2a which shows magnetometer measurements from Voyager 1 as it moved away from Jupiter just a few degrees off the equator near 0200 LT (Khurana and Schwarzl 2005). The dominant component of the magnetic field is the radial component, which is positive north of the equator and reverses sign south of the equator. The periodic sign changes of the radial component of the field indicate that the plasma sheet moved up and down over the

Fig. 2 From Khurana and Schwarzl (2005). (**a**) Three components of the magnetic field (nT) in System III (right-handed) coordinates and the field magnitude vs. UT from measurements of the Voyager 1 magnetometer outbound in the magnetotail of Jupiter. Labels below the plot indicate radial distance in R_J, latitude in degrees, and LT in hours. (**b**) Longitude in degrees of the crossing of the current sheet plotted vs. radial distance in R_J for both N-S crossings (*squares*) and S-N crossings (*diamonds*). The *curves* are predictions from the model of Khurana (1992)

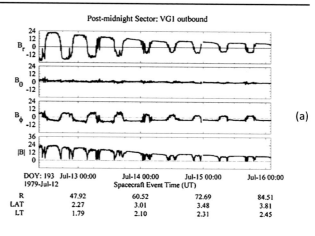

(a)

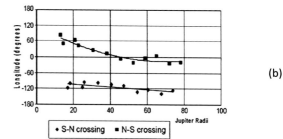

(b)

spacecraft. Khurana and Schwarzl recorded the longitudes at which the radial component reversed sign, finding a strong dependence on distance from Jupiter (Fig. 2b). They attribute the observed phase delay to two processes. Near the planet, the internal magnetic field is strong enough that the plasma sheet remains very close to the magnetic equator. Displacement of the distant plasma sheet launches a signal carried by MHD waves that propagate down tail and impose north-south motion on the plasma with a delay that increases with distance. To this delay Khurana and Schwarzl add a phase delay related to the field configuration. As plasma moves out conserving angular momentum, its angular velocity decreases. This "corotation lag" twists the magnetic field out of the meridian plane beyond $\sim$20 R_J, and produces an additional lag in the crossing time and it is a combination of these two processes that is responsible for the delays shown in Fig. 2b. The phase delay evident in Fig. 2b corresponds to an Alfvén speed of $\sim$39 R_J/hr or 780 km/s. The interpretation is straightforward but puzzles remain, as I discuss in the next section.

4 MHD Waves and Plasma Sheet Flapping?

The previous section introduced the concept that the rocking of Jupiter's magnetic equator at the planetary rotation period launches MHD waves that propagate outward from the inner magnetosphere and set the plasma sheet into north-south motion. It is interesting to ask which wave mode can carry the required signal radially at 780 km/s. In the complex field configuration of the magnetotail, the shear Alfvén mode and the fast mode are coupled. Characteristic values of these wave velocities in the distant magnetotail are provided in

Table 2 Characteristics of Jupiter's distant magnetotail*

Jupiter plasma sheet ($r > 50$ R$_J$)	n (cm^{-3})	T (K)	B (nT)	V_A (km/s)	c_s (km/s)	$V_{f,\max}$
Current sheet	~0.1	$<6 \times 10^7$	~1 nT	17	<226	<226
Lobe	≤0.006	6×10^7	~8 nT	>2250	900	>2420

*Typical values of n, T, and B are estimated from Fig. 13 of Frank et al. (2002). The wave velocities (v_A the Alfvén speed and c_s the sound speed) assume an ion mass of 16 AMU in the plasma sheet. Although the "lobe" entries in the table are intended to apply to Jupiter's magnetotail, the lobe plasma parameters (1 AMU ion mass and electron temperature small compared with ion temperature) are inferred by analogy from a study of lobe composition at Saturn near Titan's orbit (Szego et al. 2011)

Table 2 from data in Frank et al. (2002). The Alfvén wave propagates at V_A but only along the field. The fast mode speed falls between the larger of V_A and c_s and $V_{f,\max} = (V_A^2 + c_s^2)$ depending on the direction of propagation. These basic wave speeds vary greatly across the north-south cross-section of the plasma sheet from hundreds to thousands of km/s as seen in Table 2, which lists values typical of the center of the plasma sheet and the lobes. The actual range of wave speeds may be even greater. Given that the motion must displace the plasma of the central plasma sheet, it seems appropriate to focus on the speeds of propagation near its center, the current sheet. There the fastest wave speed is <226 km/s for the parameters that we have used. This raises the question: how can we account for the observed phase velocity, which is far larger than can be related to plasma and field properties? If no MHD wave can carry the information required to the equatorial plasma, is the model used to interpret the plasma sheet flapping at Jupiter correct?

Before considering further the dilemma posed by the motion of Jupiter's plasma sheet, it is useful to reflect on what happens in Saturn's magnetosphere and extract a lesson. At Saturn, the magnetic axis is aligned with the spin axis and the magnetic equator remains fixed while the planet rotates, yet the magnetotail plasma sheet oscillates north-south at the rotation period, much as observed at Jupiter. Again, as at Jupiter, there is a phase delay that increases with distance. The effective radial propagation speed is a very slow ~8.4 R$_S$/h or ~140 km/s.

If the motion of the magnetic equation close to the planet is not causing the north-south displacement of the plasma sheet, what is? Several possible explanations for the plasma sheet flapping at Saturn have been advanced (see, for example, Khurana et al. 2009). Here we focus on inferences from an MHD simulation (Jia et al. 2012; Jia and Kivelson 2012) that reproduces quantitatively many of the observed aspects of Saturn's dynamics. The simulation is carried out for fixed solar wind conditions appropriate to Saturn's orbit, with magnetospheric plasma properties that closely mimic the observed values. Flow vortices, centered near 70° latitude, are imposed in the southern and northern ionospheres and these structures rotate around the spin axis at prescribed rates, slightly different north and south, completing one rotation in 10.7 ± 0.1 hours. In order to represent southern summer conditions relevant to the early part of the Cassini mission, the ionospheric conductance is assumed to be a factor of 3 larger in the south than in the north. Vorticity drives field-aligned current (FAC), strongest in the southern hemisphere because of its larger conductance; the current source couples strongly into the magnetosphere and imposes perturbations at the rotation period. Even though the spin axis and the magnetic axis are aligned, the MHD simulation shows that the plasma sheet is found to flap north-south at the rotation period, as observed in Cassini data. Figure 3, a sequence of plots of field lines and plasma density in the noon-midnight meridian plane separated by 3 hours of simulated time, shows the changing structure in the

Fig. 3 From the simulation of Jia and Kivelson (2012) showing the noon-midnight meridian plane at times separated by 3 hours. Although different rotation periods (both close to 10.7 hours) have been imposed in the two ionospheres, the difference is ignorable over the 12 hour period shown here. However, the ionospheric conductance is larger in the south by a factor of 3, so currents from the south dominate the dynamics. At 200 hr, the plasma sheet is centered close to the equator. At 203 hr, it has moved distinctly northward, at 206 hr, it has begun to return to the equator, and by 212 hr, it is still near the equator but has started to move up again. In the bottom panel, the yellow dotted contours superimposed on the Jia and Kivelson image are intended to suggest how fast mode wave fronts emerging from close to the southern high latitude source of FACs might propagate radially along the plasma sheet at a speed far different from the wave speed, V_f

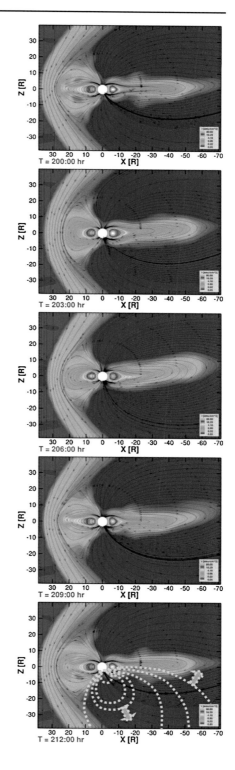

magnetotail through slightly more than one rotation period. The current sheet moves up and back to near the equator, never dropping significantly below. This asymmetry must arise because stronger forces are exerted by currents from the southern source.

It remains to understand what aspect of the simulated system causes the north-south displacement. The flapping cannot be explained in terms of effects of the solar wind because the simulation is run with a steady solar wind flowing perpendicular to the spin axis. The flapping does not relate to reconnection with the solar wind because the interplanetary magnetic field is taken to be aligned with Saturn's equatorial field and there is no evidence of dayside reconnection. It is only the FACs generated by the rotating vortices and dominated by the southern hemisphere source that can account for the periodic motion on the simulated plasma sheet.

In Fig. 3, the density at fixed points varies in successive images, implying that some of the perturbations are carried by compressional waves. The waves propagate across the field, so they must be dominantly fast mode waves. At Saturn, as at Jupiter, the fast mode should propagate at close to the Alfvén speed. The phase delay as a function of down-tail distance can be understood if the wave fronts are skewed relative to the equator, lying closer near Saturn and further at large downtail distances as illustrated by the schematic yellow lines in the bottom panel of Fig. 3. (The wave fronts have not yet been studied in the simulation, but it seems likely that their structure can be identified in future studies.) The skew of the fronts is consistent with a phase delay that increases with distance along the equator.

If Saturn's plasma sheet is driven into north-south motion by compressional waves linked to FACs, rather than by waves propagating outward near its center, is it possible that the same mechanism is at work at Jupiter? That interpretation could explain why the plasma sheet displacements observed at large distances from Jupiter occur with a delay that is short compared with plausible wave propagation speeds down the central plasma sheet. FACs flowing upward from the ionosphere near $L = 20$ linked to the main aurora (Clarke et al. 2004) are known to be intense. Is it possible that compressional waves radiate into the magnetotail as the dipole rotates and the $L = 20$ field lines rock up and down? This interpretation of phase delays in the flapping of the Jovian plasma sheet that are far too small to be explained by waves propagating down the plasma sheet seems worthy of further study.

5 What Force Dominates in Producing the Magnetodiscs of the Giant Planets?

We have noted that the stretched magnetic configuration of a magnetodisc requires an azimuthal current sheet and have introduced the equations that govern such a current (Eq. (1)). Figure 4 shows a meridian plane cut of a data-based model of the field (Khurana 1997). It is evident that the field begins to depart from a dipolar configuration somewhat inside of 20 R_J. In this section we consider how the plasma drives an azimuthal current. Contributions come from centrifugal acceleration of the equatorial plasma, from the gradient of plasma pressure, and from pressure anisotropy. As noted above, evidence supporting this conclusion was provided some decades ago both from force balance analysis (e.g., Mauk and Krimigis 1987) and from direct evidence of anisotropy of the flux of 50 keV and higher energy ions (Paranicas et al. 1991). Paranicas et al. found that the parallel pressure exceeded the perpendicular pressure in energetic particle distributions measured by Voyager 1 on three crossings of Jupiter's nightside plasma sheet at distances of 18.0, 23.1, and 35.45 R_J (Fig. 5). Energies measured in different channels require knowledge of the ion mass. In the distributions illustrated it is assumed that the ions are protons but it is likely that at least a portion of the particle population consists of heavy ions.

Fig. 4 A meridian plane cut through a data-based model of the field of Jupiter's magnetosphere (Khurana 1997) in the XZ plane. The field becomes non-dipolar well inside of 20 R_J (*dashed line*)

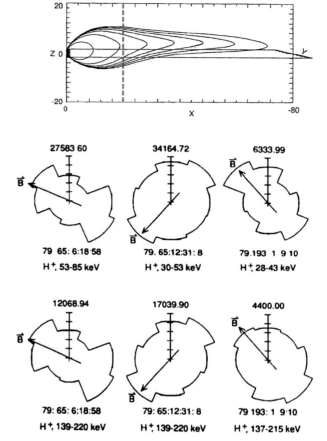

Fig. 5 From Paranicas et al. (1991), count rates of the Voyager 1 LECP detector in the scan plane and for different energy channels. The three columns show measurements made at 18.0, 23.1, and 35.45 R_J. For each distribution the counts per second are indicated and below are shown the times (in 1979) of the measurements and the energy range of the detector assuming that the ions are protons

Accepting the argument that the field distortion is produced dominantly by pressure anisotropy, one may seek to identify the mechanism that leads to that anisotropy. The centrifugal force has long been recognized as responsible for the confinement of low energy plasma to regions close to the equatorial part of plasma sheet flux tubes (see, for example, Bagenal et al. 1985; Moncuquet et al. 2002). Here I suggest a role for the centrifugal force not merely in equatorial confinement but also in creating pitch angle anisotropy.

In order to understand the stretched field configuration, one needs to consider all of the forces acting on the plasma. Equation (2) can be expressed as

$$-\nabla \left[p_\perp + \frac{B^2}{2\mu_0} \right] + \left(p_\parallel - p_\perp - \frac{B^2}{\mu_0} \right) \frac{\hat{\mathbf{n}}}{R_C} - \rho \Omega^2 r \sin(\vartheta)(\hat{\mathbf{z}} \times \hat{\boldsymbol{\varphi}}) = 0 \qquad (3)$$

where $(\mathbf{p}_\perp, p_\parallel)$ are the components of the pressure tensor perpendicular and parallel to $\mathbf{B}$, $\hat{\mathbf{z}}$ is a unit vector parallel to the spin axis, r is radial distance, ϑ is co-latitude, $\hat{\boldsymbol{\varphi}}$ is a unit vector in the azimuthal direction. Mauk and Krimigis (1987) found that on the day side of Jupiter inside of $\sim$22 R_J (where the field configuration differs little from that of a dipole field) the pressure gradient force, $-\nabla p_\perp$, is sufficient to balance the magnetic force but this force does not produce the stretched field configuration that develops beyond roughly 20 R_J.

At the equator, the radial component of Eq. (3) becomes

$$-\frac{\partial p_\perp}{\partial r} + \frac{(p_\parallel - p_\perp)}{R_C} + \rho\Omega^2 r = \frac{\partial}{\partial r}\left(\frac{B^2}{2\mu_0}\right) + \left(\frac{B^2}{\mu_0 R_C}\right) \tag{4}$$

showing the features of the plasma distribution that can control the equatorial curvature of the field line. I discuss them in the next section.

6 The Source of Pressure Anisotropy

In the magnetospheres of Jupiter and Saturn, the velocity space distributions are typically anisotropic. In the inner magnetosphere, the perpendicular pressure dominates. This is because the plasma is rich in pickup ions formed by ionization of neutrals introduced into the magnetosphere near Io's orbit at 6 R_J for Jupiter and near Enceladus's orbit at 4 R_S for Saturn. Pickup ions acquire a thermal speed equal to the speed of the flowing plasma in which they form, and, if the flow speed is comparable with or exceeds the thermal speed, particle distributions dominated by pickup ions satisfy $p_\perp > p_\parallel$. Thus at Jupiter, for example, distributions inside of 10 R_J are consistent with anisotropy $A = p_\parallel/p_\perp$ between 0.2 and 1 (Bagenal 1994). Something happens to the plasma between ~10 and ~20 R_J that greatly modifies the anisotropy. The change with L-shell for the heavy ion component of the plasma (mass per unit charge of 16 AMU) is particularly clear in Fig. 11 of Frank and Paterson (2004), reproduced here as Fig. 6. Plotted are the ion number density, the perpendicular temperature ($T_\perp$), and the anisotropy ($T_\parallel/T_\perp$) of the heavy ion component of the thermal plasma from data acquired near the rotational equator between 6.8 and 48.6 R_J on an outbound pass of Galileo near noon local time. The rocking magnetic equator passes over the spacecraft twice each rotation period at times marked with dotted vertical lines. Inside of ~25 R_J, at the equator crossings A is less than 1 (i.e., $p_\perp$ dominates), but its value increases markedly (to almost 10) between crossings. The strong variation arises because in the ~2.5 hours that follow each magnetic-equator crossing, the spacecraft moves onto field lines that cross the equator at increasingly large distances. With L taken as the equatorial crossing distance (in R_J) of a flux tube, the measurements imply that the thermal anisotropy increases with L. From Fig. 6 it becomes clear that beyond $L \approx 25$, the anisotropy is often >1 even at the equator. Beyond the inner magnetosphere, not only the suprathermal plasma considered by Paranicas et al. (1991) but the full plasma distribution is anisotropic with higher thermal speeds along field lines than across them, and the anisotropy increases with L. This suggests that plasma anisotropy is linked to radial transport.

Given a fairly steady plasma source near Io, Jovian plasma must move out through the magnetosphere. Let us consider the kinetic effects of rotation on outward-moving plasma. Most estimates suggest a displacement from near 6 R_J to ~20 R_J in ~30 days (e.g., Bagenal and Delamere 2011). Currents coupling the outward-moving plasma to the ionosphere maintain it in rotation at an angular velocity reduced from strict corotation, but in the middle magnetosphere typically greater than 60 % of strict corotation (Krupp et al. 2001). The effect of the outward displacement on the energy of ions of initial energy W_o bouncing on rapidly rotating flux tubes depends on their instantaneous position on the flux tube, their pitch angle, and their thermal energy. Outward displacement can be described by introducing an azimuthal electric field, E, derived from a scalar potential, $\Phi(\varphi)$, and assuming local axisymmetry. Consider ions of mass m, pitch angle α, and thermal energy, W ($W = W_\perp + W_\parallel$ with $W_\perp = W \sin^2 \alpha$) starting at $(r_o, \theta_o, \varphi_o)$. The outward displacement is assumed slow enough

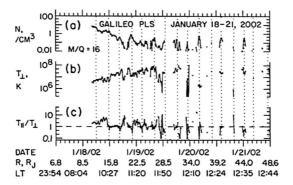

Fig. 6 A portion of Fig. 11 of Frank and Paterson (2004) showing properties (number density (cm^{-3}), perpendicular temperature (K) and anisotropy) of the heavy ion constituent (nominally 16 AMU per unit charge) of the Jovian plasma measured on an outbound equatorial pass of Galileo near noon. *Dotted lines* indicate current sheet crossings identified from magnetometer measurements. In the range between roughly 15 R$_J$ and $\sim$25 R$_J$ the anisotropy is <1 near the current sheet but >1 between equator crossings

to conserve the first adiabatic invariant, μ, so, except for extremely energetic particles, the perpendicular energy is expected to satisfy $W_\perp = \mu B$, implying that $T_\perp$ decreases with radial distance. (Strictly speaking, $W_\perp$ should be measured in the frame of the cold plasma, implying that $W_\perp = \frac{m}{2}(v_\perp - v_E)^2$, but the convective velocity, v_E, corresponding to outward displacement of 14 R$_J$ in $\sim$30 days is <0.4 km/s and can be ignored.) Figure 6 shows that, contrary to expectation, the perpendicular temperature increases with radial distance. Some of the increase of $T_\perp$ with r can be attributed to pickup ions, which are important in the range between 6 and 7.5 R$_J$, but Bagenal and Delamere (2011) note that much of the energy of the pickup population is radiated away in UV emissions and the issue of net energy in the region inside of 20 R$_J$ is complex. However, the energy of rotation of the plasma is at least half an order of magnitude larger than the warm plasma thermal energy density in the region between 6 and 20 R$_J$ (see Bagenal and Delamere 2011, Fig. 11), indicating that it is plausible to contemplate rotational acceleration as a source that can be tapped in order to maintain an elevated plasma temperature in the outer parts of the torus.

For an ion convecting outward in a time-stationary magnetic field in a rotating plasma, Northrop and Birmingham (1982) show that the change of parallel energy (from $W_{\|o}$ to $W_\|$) with displacement from (r_o, θ_o) to (r, θ) satisfies

$$W_\| - W_{\|o} = -\mu(B - B_o) - q\Delta\Phi + \frac{1}{2}m\Omega^2\left(r^2\sin^2\theta - r_o^2\sin^2\theta_o\right) \tag{5}$$

where we have set $W_\perp = \mu B$.

The first term on the right hand side of Eq. (5) is positive. The term proportional to $m\Omega^2$ is also positive for outward displacement and changes $W_\|$ by 270 (m_i/m_p) eV between 6 and 20 R$_J$. Small pitch angle particles gain parallel energy with outward displacement. The effect is mass dependent, so anisotropy increases more for the heavy ions than for the protons.

How would the centrifugal acceleration affect plasma on a stretching flux tube? Consider the bounce time for particles of energy W (keV) and mass 16 m$_p$ (m$_p$ is the proton mass). Because the particle bounce period decreases with increasing thermal energy, ions in the $\gtrsim$28 keV range studied by Paranicas et al. (1991) will respond adiabatically to centrifugal acceleration in the inner magnetosphere. Protons convecting out from 6 to 20 R$_J$ could gain about 0.6 keV from centrifugal acceleration, whereas field-aligned O$^+$ ions could gain

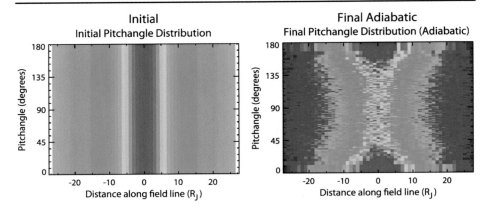

Fig. 7 From Fig. 14 of Vogt et al. (2014). The effect of adiabatic radial stretching of a flux tube on the pitch angle distribution of plasma. Color represents the pitch angle distribution as a function of distance along the field line (0 at the equator, positive northward), density of particles inside pitch angle bins of size $d(\cos \vartheta)$. The initial isotropic distribution appears as vertical color bands. As the flux tube stretches outward, the distribution near becomes more field-aligned, especially near the equator

$\sim$10 keV. It seems probable that the proton distributions are near isotropic but the heavy ion distributions are strongly field-aligned as a result of centrifugal acceleration in their bounce motion. The distributions obtained by Paranicas et al. (1991) (Fig. 5) probably register counts from both species, with the heavy ions dominantly responsible for the elevated count rate for pitch angles near 0° and 180°.

A full analysis of the effect of centrifugal acceleration on the thermal plasma would require use of a good field model and a particle code to follow bouncing particles in the expanding field. What such an analysis would be likely to reveal can be inferred by examining how the distribution of plasma on flux tubes in the middle magnetosphere changes as the plasma rotates from noon to dusk. Vogt et al. (2014) have analyzed the effect of rotation on Jovian plasma trapped on a single flux tube expanding from an equatorial crossing at 40 R_J to a final equatorial crossing at 55 R_J while rotating both at half the rate of corotation from noon to dusk and at 1 % of that rate (to represent slow rotation that produces an adiabatic particle response). Gradient-curvature drifts are ignored in this analysis, so the effect of centrifugal acceleration on the most energetic ions is underestimated. Figure 7 shows a figure from the Vogt et al. (2014) large scale kinetic simulation (Ashour-Abdalla et al. 1993) of the changes of a pitch angle distribution as result of the (slow) adiabatic outward expansion (from $L = 40$ to $L = 55$) in a rotating magnetosphere. The calculation confirms that the effect of stretching a flux tube in a rotating plasma is to increase the ratio of the parallel to the perpendicular energy.

The discussion of this section suggests a hypothetical scenario that could account for the structure of Jupiter's (and possibly Saturn's) magnetodisc. Near the Io (Enceladus) source, $p_\perp$ is greater than $p_\parallel$ because the plasma is rich in pickup ions. Outward stress is dominated by the pressure gradient force. In the 15–20 R_J equatorial region, both centrifugal and anisotropic pressure gradients increase field stretching. As plasma convects out, $p_\parallel$ becomes larger than $p_\perp$ through the preferential action of centrifugal acceleration on the parallel motion of heavy ions. By 20 R_J (some equivalent at Saturn, yet not identified), the ratio $p_\parallel/p_\perp$ has become large enough to account for the disk-like geometry of the plasma sheet. Beyond 20 R_J, $p_\parallel/p_\perp$ continues to increase and the plasma sheet becomes increasingly disc-like. The model predicts a dependence of plasma anisotropy on energy and ion species that has

Fig. 8 Measured values of the component of the magnetic field normal to the current sheet in the Jovian magnetosphere from Fig. 2 of Vogt et al. (2011). The *black curves* are probably magnetopause locations from Joy et al. (2002)

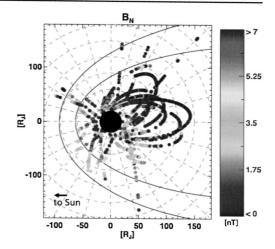

not yet been tested. It also suggests that the anisotropy will depend on the L-shell of the observation rather than the radial distance, and on the position along a flux tube, with the variations being especially significant where the fields undergo transitions from dipole-like to tail-like. Additional data analysis would help us understand the development of the pressure anisotropy that produces the magnetic structure of a magnetodisc.

7 Dawn-Dusk Asymmetries of the Plasma Sheet

A striking feature of the Jovian magnetodisc is its local time structure. Beyond of order 20 R_J, its north south thickness is far greater on the afternoon side of the magnetosphere (post noon to pre-midnight) than on the morning side (midnight to noon) (Kivelson and Khurana 2002; Vogt et al. 2011). This asymmetry is unlikely to arise purely through interaction with the solar wind because Earth's magnetosphere does not manifest such asymmetry. However, at Jupiter the solar wind confines rapidly rotating plasma within a boundary, and although that boundary is basically symmetrical about the noon midnight meridian, its effect on rotating plasma is not. As they rotate through the morning side, flux tubes lag their ionospheric roots. Near the equator, they move a high speed in a largely azimuthal sense as they rotate towards noon (Krupp et al. 2001). On the afternoon side, between noon and dusk the flux tubes must stretch radially to fill the volume within an expanding magnetopause. However, their flow speeds are far lower post noon than pre noon (Krupp et al. 2001). Inward and outward motion in a rotating system leads to acceleration and acceleration followed by pitch angle scattering implies heating and associated reduction of plasma confinement. It seems that such heating is taking place in the post noon sector of the magnetodisc. The component of **B** normal to the current sheet serves as a good proxy for the thickness of the current sheet at a given radial distance and its asymmetry about the noon midnight meridian is evident in Fig. 8. On the morning side, parallel pressure and centrifugal acceleration contribute to stretching flux tubes and thinning the plasma sheet, which reduces the normal component of **B**. On the evening side it seems that the flow slows and there is time for particles accelerated by rotation to scatter and increase the thermal pressure of the plasma. Correspondingly B_N increases and plasma can fill flux tubes over a greater north-south distance. It has been suggested that the mechanism

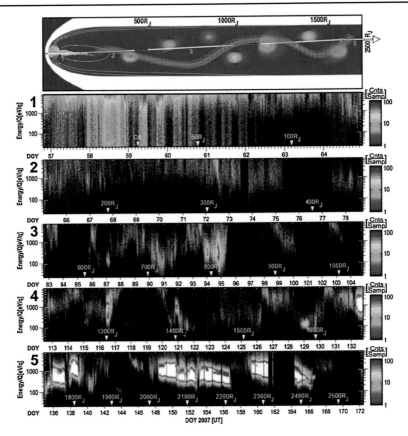

Fig. 9 Modified from McComas et al. (2007), Dynamic spectra (counts/sample in different energy per unit charge bands) from data of the New Horizons pass through the magnetotail of Jupiter. The five bottom panels display consecutive intervals labeled by day of year in 2007. The plasma energy density increases intermittently. The *top panel* shows the spacecraft trajectory through a schematic magnetotail. The *yellow* to *red* colored regions represent disjoint intervals of relatively high plasma content within which the count rate increases. On this schematic, we have superimposed an imagined fire-hose unstable flux tube, whose meandering could account for the intermittent appearance of elevated count rates

for heating the plasma is related to non-adiabatic effects of rotational acceleration as flux tubes rotate while moving outward between noon and dusk (Kivelson and Southwood 2005; Vogt et al. 2014) but observations are needed to confirm this interpretation.

8 Outflow in the Tail

Noted earlier in this chapter, is the need for outward transport of internally generated plasma in the magnetospheres of Jupiter and Saturn. Losses must ultimately balance sources and at the giant planets, transport down the nightside magnetodisc is the dominant loss mechanism. Other chapters discuss how loaded flux tubes rotating through the night side of the magnetosphere form plasmoids and carry off at least part of the plasma, as first discussed by Vasyliūnas (1983). This mechanism is easily accepted because it has some analogy to loss processes observed in the terrestrial magnetosphere. However, at Jupiter and Saturn, additional processes can enable the plasma to leave the system through the magnetotail.

In the giant planet magnetospheres, the outer portions of flux tubes rotating into the night side and stretching to large radial distances may not be able to maintain communication with the ionosphere. This is a matter that was initially discussed by Hill (1979) and later by Vasyliūnas (1994). If the flow in the magnetosphere changes, signals develop to transmit information to the ionosphere and the ionosphere must then communicate the requirement for an increased $\mathbf{j} \times \mathbf{B}$ force to restore a self-consistent flow to the equatorial magnetosphere. The communication requires finite time, and during that time, the plasma is moving outward. It is possible to imagine that the plasma outflow occurs too rapidly for the signals from the ionosphere to impose the required changes. In regions close to the planet, the delays are short and the plasma flow is controlled to a considerable degree by the ionosphere. However, when the ionosphere can no longer control the plasma, the situation can become exceptionally complex. This can happen in several ways that are worth thinking about.

Kivelson and Southwood (2005) have suggested that centrifugal force could stretch the field down the dusk flank of the magnetotail so that ultimately plasma might escape in very small scale bubbles or might no longer be tied to the field and diffuse off. Bagenal (2007) refers to the latter phenomenon as a "drizzle" of particles moving down tail. The process is particularly important when the pressure anisotropy becomes sufficiently large.

If the plasma is rotating, $p_\parallel$ continues to increase with r. The increase of anisotropy can, in principle, lead to two important changes in the dynamics of the magnetodisc. When the anisotropy becomes significant, the Alfvén speed is no longer defined by $v_{Ao} = B/(\mu_o \rho)^{1/2}$ but takes the form

$$v_A^2 = v_{Ao}^2 \left[1 + \frac{1}{2}(\beta_\perp - \beta_\parallel) \right] \quad \text{where } \beta_{\perp,\parallel} = p_{\perp,\parallel}/(B^2/2\mu_o) \tag{6}$$

For sufficiently large anisotropy, the familiar MHD wave modes that communicate between the ionosphere and the magnetosphere cease to propagate.[1] Again the implication is that the ionosphere no longer can exert an influence on the equatorial magnetosphere and the flow structure is controlled locally. Plasma loss down the tail would still be possible. How the lack of connection to the ionosphere would affect solar wind-driven reconnection is a question that has not been investigated.

Even before the anisotropy becomes large enough to block wave propagation, it may exceed the critical value $1 + B^2/\mu_o \rho$. In this case, the flux tube becomes fire-hose unstable, causing it to whip back and forth across the distant tail. It is possible that the effects of such unstable flux tubes have already been observed. The New Horizons spacecraft made a pass through Jupiter's magnetotail, exiting tailward of $\sim$1600 R_J (McComas et al. 2007). It encountered plasma intermittently on its down tail journey. The count rate in energy channels between 10s of eV and 10 keV increased aperiodically on its downtail journey as shown in Fig. 8. Various interpretations of the intermittent observation of increased counts have been proposed. The sporadic increases could arise from plasma sheet flapping or from bubbles of plasma moving down tail. Here we note the possibility that counts would intermittently increase if a fully developed firehose instability caused a flux tube of elevated plasma density to flick back and forth across the tail, occasionally passing over the New Horizons spacecraft, as illustrated schematically in the top panel of Fig. 9. Distinguishing the different interpretations of the pattern of plasma measurements from the data available may not be

---

[1] Thanks to David Southwood for inspiring this set of comments.

possible, but the possibility that it links to a fire hose instability seems worth consideration. The same behavior may appear in Saturn's magnetotail at large distances, but to this time the relevant measurements have not been made.

9 Summary

Although the plasma and field structure of the magnetodiscs of Jupiter and Saturn share elements of structure with the plasma sheet of Earth's magnetotail, their dynamics differ markedly, partly as a consequence of the important role of centrifugal acceleration. For these rapid rotators, the effects of rotation are seen in numerous features, including the development of anisotropy in the plasma distribution. Scales are so long that plasma may decouple from the ionosphere in the outer portions of the magnetodiscs (even without parallel electric fields). There are still unanswered questions related to what causes many of the dynamical changes observed in and around the magnetodiscs including plasma sheet flapping and the unorganized encounters with plasma in the very distant magnetotail of Jupiter. Further data analysis and additional measurements will be needed to enable us to understand more fully some of the special aspects of the dynamical processes discussed in this chapter.

Acknowledgements This work was supported in part by NNX12AK34G and 1416974 at the University of Michigan and NNX10AF16G at UCLA. Useful conversations with David Southwood, Xianzhe Jia, Krishan Khurana are gratefully acknowledged. Fran Bagenal provided insightful and helpful comments on an early draft.

References

M. Ashour-Abdalla, J.P. Berchem, J. Büchner, L.M. Zelenyi, Shaping of the magnetotail from the mantle: Global and local structuring. J. Geophys. Res. **98**, 5651–5686 (1993)

F. Bagenal, Empirical model of the Io plasma torus: Voyager measurements. J. Geophys. Res. **99**, 11,043 (1994)

F. Bagenal, The magnetosphere of Jupiter: Coupling the equator to the poles. J. Atmos. Terr. Phys. **69**, 387–402 (2007)

F. Bagenal, S. Bartlett, http://lasp.colorado.edu/mop/resources/graphics/ (2013)

F. Bagenal, P.A. Delamere, Flow of mass and energy in the magnetospheres of Jupiter and Saturn. J. Geophys. Res. **116**, A05209 (2011). doi:10.1029/2010JA016294

F. Bagenal, R.L. McNutt Jr., J.W. Belcher, H.S. Bridge, J.D. Sullivan, Revised ion temperatures for Voyager plasma measurements in the Io plasma torus. J. Geophys. Res. **90**(A2), 1755 (1985)

J.T. Clarke, D. Grodent, S.W.H. Cowley, E.J. Bunce, P. Zarka, J.E.P. Connerney, T. Satoh, Jupiter's aurora in Jupiter, in *The Planet, Satellites and Magnetosphere*, ed. by F. Bagenal, T.E. Dowling, W.B. McKinnon (Cambridge University Press, Cambridge, 2004), pp. 639–670. ISBN 0-521-81808-7

M.E. Davies et al., Report of the IAU/IAG/COSPAR Working Group on Cartographic Coordinates (1996)

I. de Pater, J.J. Lissauer, *Planetary Sciences*, 2nd edn. (Cambridge Univ. Press, New York, 2010)

L.A. Frank, W.R. Paterson, Plasmas observed near local noon in Jupiter's magnetosphere with the Galileo spacecraft. J. Geophys. Res. **109**, A11217 (2004). doi:10.1029/2002JA009795

L.A. Frank, W.R. Paterson, K.K. Khurana, Observations of thermal plasmas in Jupiter's magnetotail. J. Geophys. Res. **107**, A11003 (2002). doi:10.1029/2001JA000077

C.M. Hammond, M.G. Kivelson, R.J. Walker, Imaging the effect of dipole tilt on magnetotail boundaries. J. Geophys. Res. **99**, 6079 (1994). (UCLA IGPP Pub. No. 3667), 1993

T.W. Hill, Inertial limit on corotation. J. Geophys. Res. **84**, 6554 (1979)

X. Jia, M.G. Kivelson, Driving Saturn's magnetospheric periodicities from the upper atmosphere/ionosphere: Magnetotail response to dual sources. J. Geophys. Res. **117**, A11219 (2012). doi:10.1029/2012JA018183

X. Jia, M.G. Kivelson, T.I. Gombosi, Driving Saturn's magnetospheric periodicities from the upper atmosphere/ionosphere. J. Geophys. Res., Atmos. **117**, A04215 (2012). doi:10.1029/2011JA017367

S.P. Joy, M.G. Kivelson, R.J. Walker, K.K. Khurana, C.T. Russell, T. Ogino, Probabilistic models of the Jovian magnetopause and bow shock locations. J. Geophys. Res. **107**, A101309 (2002). doi:10.1029/2001JA009146

K.K. Khurana, A generalized hinged-magnetodisc model of Jupiter's nightside current sheet. J. Geophys. Res. **97**, 6269 (1992)

K.K. Khurana, Eular potential models of Jupiter's magnetospheric field. J. Geophys. Res. **102**, 11,195 (1997)

K.K. Khurana, H.K. Schwarzl, Global structure of Jupiter's magnetospheric current sheet. J. Geophys. Res. **110**, A07227 (2005). doi:10.1029/2004JA010757

K.K. Khurana, D.G. Mitchell, C.S. Arridge, M.K. Dougherty, C.T. Russell, C. Paranicas, N. Krupp, A.J. Coates, Sources of rotational signals in Saturn's magnetosphere. J. Geophys. Res. **114**, A02211 (2009). doi:10.1029/2008JA013312

M.G. Kivelson, K.K. Khurana, Properties of the magnetic field in the Jovian magnetotail. J. Geophys. Res. **107**(A8), 1196 (2002). doi:10.1029/2001JA000249

M.G. Kivelson, D.J. Southwood, Dynamical consequences of two modes of centrifugal instability in Jupiter's outer magnetosphere. J. Geophys. Res. **110**, A12209 (2005). doi:10.1029/2005JA011176

M.G. Kivelson, P.J. Coleman Jr., L. Froidevaux, R.L. Rosenberg, A time dependent model of the Jovian current sheet. J. Geophys. Res. **83**, 4823 (1978)

N. Krupp, A. Lagg, S. Livi, B. Wilken, J. Woch, E.C. Roelof, D.J. Williams, Global flows of energetic ions in Jupiter's equatorial plane: First-order approximation. J. Geophys. Res. **106**, 26,017 (2001)

B.H. Mauk, S.M. Krimigis, Radial force balance within Jupiter's dayside magnetosphere. J. Geophys. Res. **92**, 9931 (1987)

D.J. McComas et al., Diverse plasma populations and structures in Jupiter's magnetotail. Science **318**, 217 (2007)

M. Moncuquet, F. Bagenal, N. Meyer-Vernet, Latitudinal structure of outer Io plasma torus. J. Geophys. Res. **107**(A9), 1260 (2002). doi:10.1029/2001JA900124

T.G. Northrop, T.J. Birmingham, Adiabatic charged particle motion in rapidly rotating magnetospheres. J. Geophys. Res. **87**(A2), 661–669 (1982)

T.G. Northrop, C.K. Goertz, M.F. Thomsen, The magnetosphere of Jupiter as observed with Pioneer 10, 2, Nonrigid rotation of the magnetodisc. J. Geophys. Res. **79**, 3579 (1974)

C.P. Paranicas, B.H. Mauk, S.M. Krimigis, Pressure anisotropy and radial stress balance in the jovian neutral sheet. J. Geophys. Res. **96**, 21,135 (1991)

K. Szego, Z. Nemeth, G. Erdos, L. Foldy, M. Thomsen, D. Delapp, The plasma environment of Titan: The magnetodisc of Saturn near the encounters as derived from ion densities measured by the Cassini/CAPS plasma spectrometer. J. Geophys. Res. **116**, A10219 (2011). doi:10.1029/2011JA016629

V.M. Vasyliūnas, Plasma distribution and flow, in *Physics of the Jovian Magnetosphere*, ed. by A.J. Dessler (Cambridge Univ. Press, New York, 1983), p. 395

V.M. Vasyliūnas, Role of the plasma acceleration time in the dynamics of the Jovian magnetosphere. Geophys. Res. Lett. **21**(6), 401 (1994)

M.F. Vogt, M.G. Kivelson, K.K. Khurana, R.J. Walker, B. Bonfond, D. Grodent, A. Radioti, Improved mapping of Jupiter's auroral features to magnetospheric sources. J. Geophys. Res. **116**, A03220 (2011). doi:10.1029/2010JA016148

M. Vogt, M.G. Kivelson, K.K. Khurana, R.J. Walker, M. Ashour-Abdalla, Simulating the effect of centrifugal forces in Jupiter's magnetosphere. J. Geophys. Res. **119**, 1925 (2014). doi:10.1002/2013JA019381

DOI 10.1007/978-1-4939-3395-2_3
Reprinted from *Space Science Reviews* Journal, DOI 10.1007/s11214-014-0052-8

A Brief Review of Ultraviolet Auroral Emissions on Giant Planets

Denis Grodent

Received: 25 November 2013 / Accepted: 13 May 2014 / Published online: 24 May 2014
© Springer Science+Business Media Dordrecht 2014

Abstract The morphologies of the ultraviolet auroral emissions on the giant gas planets, Jupiter and Saturn, have conveniently been described with combinations of a restricted number of basic components. Although this simplified view is very handy for a gross depiction of the giant planets' aurorae, it fails to scrutinize the diversity and the dynamics of the actual features that are regularly observed with the available ultraviolet imagers and spectrographs. In the present review, the typical morphologies of Jupiter and Saturn's aurorae are represented with an updated and more accurate set of components. The use of sketches, rather than images, makes it possible to compile all these components in a single view and to put aside ultraviolet imaging technical issues that are blurring the emission sources, thus preventing one from disentangling the different auroral signatures. The ionospheric and magnetospheric processes to which these auroral features allude can then be more easily accounted. In addition, the use of components of the same kind for both planets may help to put forward similarities and differences between Jupiter and Saturn. The case of the ice giants Uranus and Neptune is much less compelling since their weak auroral emissions are very poorly documented and one can only speculate about their origin. This review presents a current perspective that will inevitably evolve in the future, especially with upcoming observing campaigns and forthcoming missions like Juno.

Keywords Jupiter · Saturn · Uranus · Neptune · Giant planets · Aurora · Ultraviolet · Magnetosphere

1 Introduction

1.1 General Characteristics of the Giant Gas Planets Aurorae

The giant gas planets, Jupiter and Saturn, are sharing several remarkable characteristics (Table 1). They are big, fast rotators and produce large internal magnetic fields. They are

D. Grodent (✉)
Université de Liège, Liège, Belgium
e-mail: d.grodent@ulg.ac.be

Table 1 Principal characteristics of the giant planets relevant to the auroral emissions compared to the Earth. Rotation periods of the gas giants correspond to the deep interior

	Mean solar irradiance (W m^{-2})	Equatorial radius (km)	Rotation period (h)	Surface magnetic field (nT)	Emitted UV auroral power (TW)
Earth	1366.1	6378	23.9345	30,600	0.01
Jupiter	50.5	71,492	9.9248	430,000	1
Saturn	15.04	60,268	10.6567	21,400	0.1
Uranus	3.7	25,559	17.24	22,800	0.001
Neptune	1.5	24,764	16.11	14,200	0.0001

much farther from the Sun than the Earth and thus receive relatively little energy from it. Their magnetosphere is tapped with internal plasma sources, and, an issue that is of direct concern to the present review, their hydrogen rich atmospheres display auroral emissions. This latter statement is particularly true for the gas giants Jupiter and Saturn, which emit auroral powers of 10^{12} W and 10^{11} W, respectively, compared to 10^{10} W for the Earth. The cases of icy giants Uranus and Neptune are less compelling because their much weaker auroral emissions (10^9 W and 10^8 W, respectively) have not been sufficiently observed. As a result, they will be concisely addressed in Sect. 4.

Another outstanding characteristic of Jupiter and Saturn comes from the fact that plasma flowing inside their magnetospheres is largely controlled by the corotation electric field. Contrary to the Earth, the contribution from the solar wind convection field may be disregarded, as it is much less important for Jupiter and for Saturn (see Table 1).

One immediate corollary to this domination by the corotation field is the rotating or subcorotating nature of the auroral features around the poles. Contrary to the Earth, where the bulk aurora is fixed with respect to the Sun, the aurora on Jupiter and Saturn is rotating at a significant fraction of planetary rigid rotation. It is said to corotate with the planetary magnetic field. Deviation from corotation may then be interpreted as the signature of an "unusual" magnetospheric process, where "unusual" means a process capable of disrupting the plasma's rotation around the planet with the magnetic field and/or generate beams of electrons at energies sufficiently large to excite atmospheric H_2 molecules or H atoms through collisions. As will be seen in the following sections, numerous such processes exist in Jupiter and Saturn's magnetospheres and may account for the majority of ultraviolet (UV) auroral emissions.

1.2 First Detections of Aurora on Giant Gas Planets

The UV aurorae on giant planets were first spotted more than three decades ago by the UVS spectrograph on board the two Voyager spacecraft. During the two flybys of Jupiter in 1979, auroral emissions from H_2 and H were unambiguously revealed (Broadfoot et al. 1979). A couple of years later, Voyager 2 flew by Saturn and obtained similar evidences of auroral activity (Broadfoot et al. 1981). The very limited spatial resolution of the UVS spectrograph did not make it possible to determine the precise spatial distributions of these aurorae. It is only thanks to the advent of the UV cameras on board the Hubble Space Telescope (FOC, WFPC1, WFPC2, STIS, ACS) that the accurate determination of the auroral morphology of Jupiter and Saturn became possible (see reviews of Clarke et al. (2004) for Jupiter and Kurth et al. (2009) for Saturn).

1.3 Atmospheric Origins of the Giant Planets UV Aurora

Comparisons between observed auroral spectra and models suggest that, on giant planets, auroral emissions observed in the UV range are principally the result of inelastic collisions between atmospheric H_2 molecules and energetic magnetospheric electrons precipitating in the atmosphere along magnetic field lines. These primary electrons gradually lose their energy to the ionosphere through ionization, dissociation and excitation of H_2 molecules. Ionization, the most efficient process, produces numerous secondary electrons that in turn impact H_2 molecules. The primary and secondary electrons interactions with H_2 mainly depend on their energy. They are governed by various cross-sections (ionizations, electronic excitations, vibrational and rotational excitations). In theory, any electron with energy above the excitation threshold of the B state of the H_2 molecule (~ 10 eV) has a chance to produce a UV photon of interest. The excitation cross section of the various excited states of H_2 maximize in the 20–160 eV range, meaning that primary electrons and mostly secondary electrons colliding with H_2 contribute to the production of auroral UV photons. Above ~ 50 eV, ionization and electronic excitation cross-sections follow similar electron energy dependences, however, ionization is one to two orders of magnitude more probable than excitation (Gustin et al. 2013).

The bulk of the UV auroral emission in the far ultraviolet (FUV) 70–180 nm range mainly results from electron impact excitation of H_2 to various excited rotational–vibrational–electronic states (see Gustin et al. (2009, 2012, 2013) for a thorough discussion on the origins of the auroral H_2 UV emissions). This process initiates de-excitation of excited H_2 through the emission of UV photons forming the Lyman and Werner bands, Lyman-alpha and continuum emissions. The energy degradation of the primary and secondary electrons in their course towards deeper atmospheric levels continues until they are thermalized in the ambient atmosphere. Direct excitation of atomic H produces Lyman-α emission. However, in the giant planets where H dominates H_2 at very high altitudes, the dominant source of H-Lyman-α emission is related to dissociative excitation of H_2 giving rise to fast-excited H fragments loosing their excess energy through emission of Ly-α as well as Lyman and Balmer series photons. In any case, dissociative excitation dominates the production of the auroral Lyman line and contributes ~ 99 % of the auroral UV spectrum of Jupiter (Grodent et al. 2001; Gustin et al. 2012) and Saturn.

1.4 Other Wavelength Ranges

In the following, we focus on ultraviolet emissions that are directly accessible to instruments like the UV cameras onboard the Hubble Space Telescope (the Advanced Camera for Surveys, ACS; the Space Telescope Imaging Spectrograph, STIS) for Jupiter and Saturn, or the Cassini Spacecraft (UltraViolet Imaging Spectrograph, UVIS) for Saturn. Nevertheless, it should be noted that aurorae on the giant planets are also glowing in other wavelengths including the radio (e.g. Lamy et al. 2009), infrared (e.g. Radioti et al. 2013a), visible (e.g. Vasavada et al. 1999) and X-Ray (e.g. Branduardi-Raymont et al. 2008) ranges. These radiations are more or less directly related to the UV emissions but they deserve specific treatments, beyond the scope of the present review paper, and provide complementary information. The interested reader will find useful information about auroral emissions in other wavelengths in the review paper by Badman et al. (2014).

2 Jupiter

2.1 Current Understanding

It is commonly admitted that the Jovian ultraviolet aurora consists of three main components; the main oval, the polar emissions (poleward of the main emission), and the satellites footprints (equatorward of the main emission).

The simplicity of this first order picture makes it very handy. However, it presents a skewed view of reality and does not reflect the variety of auroral structures appearing at Jupiter's poles. Furthermore, it leaves no room to the important dynamics of these features. Each auroral feature relates to one or more processes occurring in the magnetosphere. By oversimplifying the description of these auroral signatures, one might miss an important mechanism in the Jovian magnetosphere, or worse, improperly interpret it. The growing performances of the ultraviolet cameras successively installed on board the Hubble Space Telescope made it possible to go into the details of Jupiter's aurora. The STIS camera, for example, is able to reach a spatial resolution on the order of 100 km at the distance of Jupiter, roughly corresponding to a fraction of a percent of the characteristic size of the auroral region. STIS is also sensitive enough to the Jovian UV auroral emissions to permit temporal resolution of a few seconds, giving insight to the fastest processes taking place inside the magnetosphere and close to its boundary, or in the ionosphere. Its unprecedented spectral resolution of a fraction of an Angstrom (0.1 nm) makes it possible to probe the temperature and composition of the polar atmosphere. It also provides information on the precipitating particles giving rise to the aurora.

The present bottleneck for the interpretation of the auroral emissions resides in the remaining uncertainties on the magnetic field models. Although the magnetic field in the inner magnetosphere (the internal field) is relatively well constrained up to the orbit of Io, beyond this limit, the external field mainly originating from the current sheet and magnetopause currents becomes gradually dominant and unpredictable. Beyond the orbit of Ganymede, the last Galilean moon providing auroral constrains on the field, magnetic mapping of the auroral emissions becomes increasingly uncertain. At larger distances, near the magnetopause, magnetic models become rather speculative and complicate the deciphering of the poleward most auroral emissions. In the following sections, we give an overall description of the typical components of the Jovian UV aurora and try to relate them to most likely magnetospheric processes. It is probable that this overall picture will progress in the future, especially with the expected great harvest of the Juno and JUICE missions.

2.2 Main Components of Jupiter's Aurora

2.2.1 Northern and Southern Polar Regions

Gérard et al. (2013) analyzed quasi-simultaneous HST images of both hemispheres of Jupiter and found that most morphological auroral features identified in one hemisphere have a conjugate counterpart in the other hemisphere, with some significant differences in the power associated with conjugate regions (Fig. 1). Nevertheless, in the following we focus on the northern hemisphere of Jupiter. The main reason for this approach stems from the large hemispheric asymmetries in the internal magnetic field of Jupiter. The northern aurora is more tilted towards the equator than the south and, from Earth orbit, it appears with a better viewing geometry when the field is inclined towards the Earth. Another reason

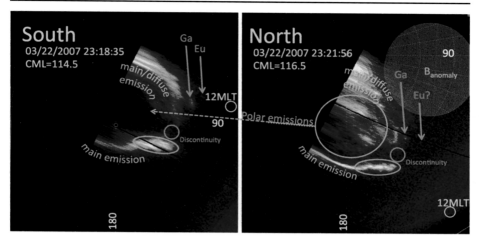

Fig. 1 Polar projections of typical ACS FUV images of Jupiter aurora obtained quasi-simultaneously (~3 min apart) in both hemispheres on 22 March 2007 (see text for detailed description)

may be linked with the presence of a possible magnetic anomaly in the northern hemisphere (Grodent et al. 2008). The presence of this anomaly locally perturbs the surface magnetic field and has the effect of a "magnifying glass" on the auroral emissions. In this ionospheric region, magnetic field lines are threading a smaller area in the magnetosphere and the auroral features appear more detached from each other, allowing easier disentangling. Figure 1 shows typical polar projections of ACS FUV images of Jupiter aurora obtained quasi-simultaneously (~3 min apart) in both hemispheres since the field of view (FOV) of ACS and STIS are too small to accommodate both hemispheres in the same image. The southern hemisphere is displayed for an observer looking through the planet from above the north pole. The images were obtained on 22 March 2007 during orbit J8-K8 (visit I8) of program GO-10862. Blue arrows point to the auroral footprints of Ganymede and Europa (Europa's footprint is barely visible). The blue circle points to a systematic discontinuity in the main emission. The green ellipse encircles a possible auroral signature of plasma injection. A small portion of the main emission (main emission) appears near 180°, leftward of the injection. Polar emissions appear in the northern hemisphere near 70° latitude (purple ellipse), while in this case no conjugate emission is visible in the south. The influence zone of a likely local magnetic anomaly in the north appears as a white-transparent disc (B_{anomaly}). Yellow circles marked 12MLT indicate the direction of magnetic noon at 15 R_J (1 Jovian radius = 1 R_J = 71,492 km), which corresponds to the longitude of the footprint of Ganymede when the moon is at 12LT. All longitudes are in System III (S3). Planetocentric parallels and meridians are drawn every 10°. The black line is the region occulted by the repeller wire. In this set of images, the Central Meridian Longitude (CML) is optimized for viewing both hemispheres quasi simultaneously.

The next figure (Fig. 2) sketches the typical FUV auroral components of the northern hemisphere of Jupiter that are commonly observed from Earth orbit. The various features are conveniently projected onto an orthographic polar map.

2.2.2 The Main Emission (Oval)

Overall Shape and Origin The main auroral oval, more correctly named main emission since it does not form an oval, is structured into a relative stable strip of emission around

Fig. 2 Sketch of the typical FUV auroral components of the northern hemisphere of Jupiter. Planetocentric parallels and System III meridians are drawn every 10°: 1. Main emission (oval); 2. Kink region; 3. Discontinuity; 4. Secondary emission; 5. Signatures of injections; 6. Io footprint (multiple) and tail; 7. Europa footprint and tail; 8. Ganymede footprint (multiple); 9. Polar active region; 10. Polar dark region; 11. Polar swirl region; 12. Polar auroral filament (PAF); 13. Dawn spots and arcs; 14. Midnight spot. The *upper gray shaded region* is not accessible to Earth orbit instruments

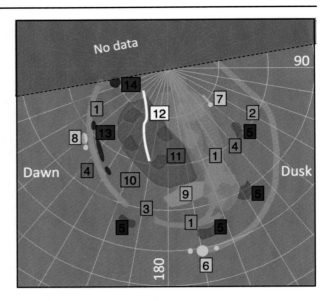

the magnetic poles. Grodent et al. (2003a) performed a long-term comparison of HST images and showed that the bulk of the auroral morphology is fixed in system-III longitude, meaning that it follows Jupiter's fast rotation. It forms a complex structure mixing multiple narrow arc-like structures, discontinuities, and diffuse patches of emission (orange structures marked "1" in Fig. 2). The global statistical shape of the northern aurora has been shown to be influenced by a magnetic anomaly, giving rise to a kink in the main emission (region marked "2" in Fig. 2). Such an anomaly has not been reported for the southern hemisphere. In general, the main auroral emissions vary in width between ∼100–500 km, though it is even broader at dusk. Its global emission is estimated to contribute almost 75 % of the Jovian auroral brightness integrated over the poles. The typical brightness exceeds ∼100 kR at UV wavelengths, peaking at up to several MR intensities (Gustin et al. 2006). In general, the dawnside portion (left side of Fig. 2) forms a relatively narrow arc, appearing almost continuous in UV images, the post-noon portion (upper left part of Fig. 2) consists of auroral patches and the dusk portion (right side of Fig. 2) appears to broaden and break from the main emission. The nightside sector of Jupiter's polar regions is not accessible to Earth orbit observatories like HST. Therefore, current knowledge of the nightside auroral morphology mainly relies on speculations.

Theoretical modeling suggests that the main Jovian auroral emission results from the magnetosphere-ionosphere coupling current system associated with the breakdown of rigid corotation in the middle magnetosphere and maps to the equatorial plane between 15 R_J (the orbit of Ganymede) and 40 R_J (Cowley and Bunce 2001; Hill 2001; Southwood and Kivelson 2001; Nichols and Cowley 2004). As the plasma diffuses outwards in the equatorial plane its angular velocity decreases due to conservation of the angular momentum. At a certain distance, breakdown of corotation occurs and a strong current system develops. When the plasma angular velocity becomes lower than that of the neutral atmosphere, ion-neutral collisions occur in the Pedersen layer of the ionosphere and produce a frictional torque that strives to spin up the plasma back to corotation. The current circuit is closed by a system of field-aligned currents which flow from the ionosphere to the equator (upward) in the inner part of the system, and return (downward) in the outer part. The upward field aligned currents are mainly carried by downward moving electrons. When these electrons fill in the loss cone, their velocity component parallel to the magnetic field is sufficient for the downward electrons to reach the jovian atmosphere where they lose their energy through collisions with the neutrals. Part of this precipitated energy is radiated away in the UV domain and produces the main auroral emission.

Recent work has also shown that the main auroral oval and the Ganymede auroral footprint move in latitude over periods of a few months (Bonfond et al. 2012a). This intriguing motion is not yet explained and could be a subtle combination of moon activity, mass loading rates, magnetodisc configuration, solar wind interaction and ionospheric conductivity.

Discontinuity   Radioti et al. (2008) reported the presence of a discontinuity in Jupiter's main emission, where the emission almost systematically drops abruptly to less than 10 % of the maximum value. HST UV images taken at different central meridian longitudes in both hemispheres show that the discontinuity (empty region marked "3" in Fig. 2) appears fixed in magnetic local time and map to a region of the equatorial plane between 08:00 and 13:00 LT. According to Galileo data, this sector threads downward field-aligned currents (Khurana 2001) presumably resulting from solar wind driven magnetospheric convection. Additionally, plasma flow measurements in the Jovian magnetosphere, inferred from Galileo (Krupp et al. 2001) show evidence of nearly corotating plasma in the dawn-to-dusk sector. According to corotation enforcement process described above, this would require weaker field-aligned currents (or reversed) and consequently fainter aurora emissions in the prenoon magnetic local time. The discontinuity may then originate from the reduced or/and downward field-aligned currents in that region.

2.2.3 The Secondary Emissions

Overall Shape and Origin   Secondary auroral emissions are appearing equatorward of the main emission and poleward of the Io footprint (Radioti et al. 2009a). They consist of emissions extending from the main emission towards lower latitudes, occasionally forming discrete arcs of emissions parallel to the main emission and/or patchy irregular structures (light blue diffuse arcs marked "4" in Fig. 2). They may also form isolated features that have been attributed to hot plasma injections in the middle magnetosphere. Together, they form the equatorward diffuse emissions (EDE).

Grodent et al. (2003a) suggested that theses emissions might account for the same corotation breakdown mechanism as for the main emission. The secondary emission may then represent a first step, at lower latitudes and lower intensities, of the process of corotation enforcement. It should be pointed out that the stepwise departure from rigid corotation is purely empirical, while the existing theoretical calculations, based on uniform magnetospheric plasma distribution, predict a smooth monotonic decline in plasma angular velocity with increasing latitude in the auroral ionosphere (e.g. Cowley et al. 2008a).

At Jupiter, Bhattacharya et al. (2001) suggested that wave particle interactions in a broad region in the magnetosphere (10 to 25 R_J) could lead to electron scattering and precipitation into the ionosphere contributing to the EDE. Tomás et al. (2004) related the transition of the electron pitch angle distribution (PAD) from pancake to bidirectional (observed within 10 to 17 R_J) to a discrete auroral emission equatorward of the main oval, under the assumption of electron scattering in the loss cone due to whistler mode waves. The brightness of the EDE usually ranges from 40 to 100 kR in the north and from 10 to 50 kR in the south. Based on the analysis of a large HST 1997–2007 dataset, it appeared that the EDE are almost always present, especially in the dusk side. The persistence of the EDE suggests that its origin is associated with a permanent magnetospheric feature such as the PAD boundary. Radioti et al.

(2009a) showed that the PAD boundary magnetically maps to the diffuse auroral emission region in the northern and southern hemisphere. Comparison of the derived precipitation energy flux with the observed brightness of the EDE further showed that the energy contained in the PAD boundary could account for the measured auroral emissions in both hemispheres.

Signatures of Injections As stated above, not all components of the EDE may be associated with a transition of the electron pitch angle distribution. Transient isolated auroral patches appearing equatorward of the main emission (purple patches marked "5" in Fig. 2) could be related to other mechanisms, such as electron scattering by whistler mode waves associated with anisotropic injection events (Xiao et al. 2003; Mauk et al. 2002) or with field aligned currents flowing along the boundary of a hot injected plasma cloud. These auroral signatures of injections may take the form of quasi-corotating shapeless features detaching from the main emission near the footpaths of Europa and Ganymede. At times, they overlap these satellites footprints, making their detection ambiguous. In the Jovian magnetosphere, the processes of plasma injection and interchange motion are thought to be associated with the radial inward transport of hot tenuous magnetotail plasma, compensating for the continuous opposite outward flow of cold iogenic plasma in such a manner as to conserve magnetic flux. To date, only one HST observation of an auroral injection signature could be unambiguously associated with an in situ Galileo detection of a cloud of injected energetic particles (Mauk et al. 2002). The fact that this case is unique does not stem from the rarity of the phenomenon; it is actually very frequent both in the HST and Galileo datasets, but from the lack of simultaneous Galileo—HST observations. Bonfond et al. (2012a) reported what appears to be an exceptional event. While the auroral injection signatures are usually confined between the main emission and the Io footpath, in June 2007 a large patch of UV emission was observed with HST in the northern hemisphere as far down as the expected location of the Io footprint. This feature appears to be the remnant of a large injection blob seen in the same sector in the southern hemisphere 34 hours before. Instead of simply overlapping the Io footprint (see below), this feature appears to have triggered a momentary substantial decrease of the Io footprint brightness. This behavior was suggested to result from the depleted nature of the flux tubes containing the sparse injected hot plasma that may have disrupted the Io-Jupiter interaction.

2.2.4 The Satellites Footprints

A comprehensive review of the different satellites footprints may be found in Bonfond (2012b).

The satellite's Ultraviolet auroral footprint appear in Jupiter's ionosphere close to the feet of the field lines passing through the satellites Io, Europa and Ganymede (Clarke et al. 2002) (yellow spots marked "6", "7" and "8" in Fig. 2, respectively). The observed morphology consists of either one, for Europa, or several distinct spots for Io and Ganymede (Bonfond et al. 2009, 2013) eventually followed by a trailing tail as is the case for Io and Europa (Clarke et al. 2002; Grodent et al. 2006) (yellow diffuse arc downstream of spots "6" and "7" in Fig. 2). It is actually anticipated that with a sufficiently sensitive instrument (which is not yet the case), one should observe the same features for all satellites footprints. Indeed, these small auroral features most probably stem from a common (universal) process in which the slow moving satellites pose obstacles to the fast corotating magnetospheric plasma flow. The magnetic perturbation associated with the continuous collisions of the plasma with the satellites' interaction regions propagates along the magnetic field lines as Alfvén waves. The locus of the perturbed points forms Alfvén wings directed towards both poles. On their way to the planet, the waves probably undergo filamentation (Chust et al. 2005; Hess et al. 2010a) and are partially reflected on plasma density gradients, especially at the plasma torus or the plasma sheet boundaries. The fraction of the waves escaping the torus or the sheet causes electron acceleration in both directions. These electrons ultimately precipitate into Jupiter's atmosphere where they produce the observed auroral signatures. Io's footprint brightness may reach up to 20 MR in the UV, representing a large local input of power to the upper atmosphere (Bonfond et al. 2013). The emissions from the Ganymede and Europa footprints are generally on the order of a few hundreds of kR in the UV. Bonfond et al. (2008, 2013), Jacobsen et al. (2007) and Hess et al. (2010a) proposed that the combination of Alfvén waves reflection and bidirectional electron acceleration may explain the relative motion of the different spots of the Io and Ganymede footprints, respectively, as well as the presence of electron beams affecting the ionization processes near Io (Dols et al. 2012; Saur et al. 2003).

The brightness of the Io and Ganymede footprint spots varies with the System III longitude of the satellite, i.e. with the location of the moon relative to the plasma torus/sheet center, with a ~10 h periodicity (Grodent et al. 2009; Bonfond et al. 2013). At Ganymede, a second time scale for brightness variations ranges from 10 to 40 min and was tentatively associated with interactions between Ganymede's mini-magnetosphere and local magnetospheric inhomogeneities, such as those produced by localized plasma injections. A similar process has been suggested to explain an exceptional drop of the Io footprint brightness (Bonfond et al. 2012a; Hess et al. 2013). The shortest time scale observed so far is on the order of 1 to 2 min. At Ganymede, these variations were suggested to be triggered by bursty reconnections at the satellite's magnetopause (Jia et al. 2010). Alternatively, they may be related to double layer regeneration as suggested for the Io footprint (Hess et al. 2010b; Bonfond et al. 2013).

The size of the ionospheric footprints appears to map to a region much wider than the moons. This implies that the satellite-magnetosphere interactions are not restricted to the satellites themselves, but more likely include either parts of the neutral cloud that surrounds and follows them in the case of Io and Europa or its mini-magnetosphere as is the case for Ganymede (Grodent et al. 2006, 2009; Bonfond 2010).

Several authors (Hill and Vasyliūnas 2002; Delamere and Bagenal 2003; Ergun et al. 2009) proposed that, contrary to the spots, Io's tail emission results from a steady state process owing to the progressive reacceleration of the plasma downstream of Io. On the other hand, MHD simulations indicate that it might actually be the result of multiple reflections of the Alfvén waves (Jacobsen et al. 2007). Grodent et al. (2006) observed a faint ~7500 km long tail following the spot when Europa is close to the center of the plasma sheet, suggesting that this auroral feature is the signature of an extended plasma plume downstream of Europa (Kivelson et al. 1999).

Although it is very much likely that there is an electrodynamic interaction between Callisto and Jupiter's magnetospheric environment that is similar to those at Io, Europa, and Ganymede (Menietti et al. 2001), so far, there is no strong evidence for a permanent Callisto footprint (Clarke et al. 2011). One possible reason for this lack of observation stems from Callisto orbiting at the distance mapping to the main auroral emission. As a result, the footprint cannot be disentangled from the much brighter main emission.

2.2.5 The Polar Emissions

Jupiter's polar auroral emissions, which include all auroral emission lying poleward of the main auroral emission, are directly linked to outer magnetosphere dynamics. Their origin continues to be debated. The polar emissions vary independently of the satellite and main auroral emission, and they appear to be ordered by magnetic local time (Grodent et al. 2003b), indicating potential external control by the solar wind. They are suggested to be magnetically connected to the outer magnetosphere and possibly related to a sector of the Dungey and/or Vasyliūnas cycle flows (Cowley et al. 2003; Grodent et al. 2003b; Stallard et al. 2003).

Based on their average brightness and temporal variability, the northern hemisphere UV polar emissions can be organized into three regions: the active, dark, and swirl regions (Grodent et al. 2003b) (regions marked "9", "10" and "11" in Fig. 2, respectively). Their shapes and locations vary with time and as Jupiter rotates.

Active Region   The active region is very dynamic and is characterized by the presence of flares, bright spots, and arc-like features. It is located just poleward of the main emission and maps roughly to the noon local time sector (green patch marked "9" in Fig. 2). There have been several interpretations of this region. Pallier and Prangé (2001) suggested that the bright spots of the active region are the signature of Jupiter's polar cusp, or possibly dayside aurora driven by an increase of the solar wind ram pressure. Waite et al. (2001) used the MHD model of Ogino et al. (1998) to map an observed polar flare to near the cushion region, ~40–60 R_J in the morning sector, and postulated that the flare could be produced by a magnetospheric disturbance due to a sharp increase in the solar wind dynamic pressure. Alternately, Grodent et al. (2003b) interpreted the polar flares as the signature of "explosive" magnetopause reconnection on the day side, based on their ~minutes-long characteristic time scale. They also suggest that the arc-like structures could be the signature of a Dungey cycle dayside x-line, following the arguments of Cowley et al. (2003). Recent observations showed that the flares could re-occur quasi-periodically every 3–2 minutes and this behavior has been tentatively associated with pulsed reconnections on the dayside magnetopause (Bonfond et al. 2011).

Dark Region   The dark region is located just poleward of the main oval in the dawn to pre-noon local time sector. As its name suggests, the dark region is an area that appears dark in the UV, displaying only a slight amount of emission (0–10 kR—Rayleighs) above the background level. The dark region displays a crescent shape that contracts and expands as Jupiter rotates, but appears fixed in local time (empty region marked "10" in Fig. 2).

Grodent et al. (2003b) associated the UV dark region with the Stallard et al. (2003) rotating Dark Polar Region (r-DPR), an area of subcorotating ionospheric flows, as measured by the Doppler shifts of infrared emission spectra. The dawn side r-DPR, and thus the dark region, is thought to be linked to the Vasyliūnas-cycle (Vasyliūnas 1983) sunward return flow of depleted flux tubes (Cowley et al. 2003). In the Vasyliūnas-cycle, mass-loaded flux tubes are stretched as they rotated into the night side; they eventually pinch off, and reconnection occurs in the midnight-predawn local time sector, releasing a plasmoid that can escape down the tail, while empty flux tubes rotate back around to the day side at a velocity close to that of corotation. Similarly, Southwood and Kivelson (2001) argued that the main oval emissions map to the plasma disk, which would mean that the dark region, just poleward of the main oval, maps to the cushion region. The cushion region is an area of southward-oriented and strongly fluctuating field in the outer magnetosphere in the post-dawn to noon local time sector where the field becomes more dipole-like than in the inner magnetosphere. It has been associated with empty flux tubes that were emptied by Vasyliūnas-type reconnection as they rotated through the night side (Kivelson and Southwood 2005).

Swirl Region   The swirl region is an area of patchy, transient emissions that exhibit turbulent, swirling motions. The swirl region is located poleward of the active and dark regions, and is roughly the center of the polar auroral emissions (red features marked "11" in Fig. 2). It is generally interpreted as mapping to open field lines. In comparing the UV and IR observations, Grodent et al. (2003b) associated the UV swirl region with the fixed Dark Polar region (f-DPR), an area in which the ionospheric flows are nearly stagnant in the magnetic pole reference frame (Stallard et al. 2003). The stagnant flows in the f-DPR (swirl region) then suggest that the area maps to open field lines associated with Dungey cycle return flows (Cowley et al. 2003), which are expected to flow across the ionosphere slowly because the Jovian magnetotail is ~hundreds or thousands of R_J in length. A long-lived quasi-Sun-aligned polar auroral filament (PAF) was observed on several occasions in the swirl region in images sequences obtained with HST (Nichols et al. 2009a) (white arc marked "12" in Fig. 2). This feature consists of two components: a sunward portion remaining approximately Sun-aligned and an anti-Sunward portion sub-rotating at a few tens of percent. This ~100 kR auroral feature appears to be independent of the local solar wind conditions. It was postulated that PAFs might be associated with large plasmoids slowly drifting down the magnetotail.

The magnetospheric mapping of these polar auroral regions was initially inferred from model magnetic fields (principally VIP4) that are known to be increasingly inaccurate beyond the orbit of Io. Instead of following these model magnetic field lines, Vogt et al. (2011) mapped equatorial regions to the ionosphere by requiring that the magnetic flux in some specified region at the equator equals the magnetic flux in the area to which it maps in the ionosphere. This mapping method directly takes into account the complexity of Jupiter's surface magnetic field, including the perturbation caused by a magnetic anomaly in the north and provides a more accurate mapping to the distant magnetosphere. Vogt et al. (2011) found that the polar auroral active region maps to field lines beyond the dayside magnetopause that can be interpreted as Jupiter's polar cusp; the swirl region maps to lobe field lines on the night side and can be interpreted as Jupiter's polar cap; the dark region spans both open and closed field lines and must be explained by multiple processes. Additionally, they concluded that the flux through most of the area inside the main oval matches the magnetic flux contained in the magnetotail lobes and is probably open to the solar wind.

Nightside and Polar Dawn Spots   Several detailed studies based on ultraviolet images taken with HST revealed transient auroral spots appearing in the dawn and midnight sectors along the poleward edge of the main emission (Grodent et al. 2004; Radioti et al. 2011a, 2011b). These polar auroral emissions take the form of multiple dawn arcs, polar dawn spots, or midnight spots (blue features marked "13" and "14" in Fig. 2, respectively). They were found to corotate with the planet, and their sizes, durations, locations, and 2 to 3 days re-occurrence period are consistent with auroral emissions triggered by internally driven tail reconnection. More specifically, based on a recent reanalysis of near simultaneous HST UV images and Galileo magnetic field observations, Radioti et al. (2011a, 2011b) proposed that the nightside spots, like the polar dawn spots, are triggered by the inward moving plasma flow released during magnetic reconnection at Jupiter's tail. These aurorae may then be related to the precipitation of plasma heated in the reconnection region and to the field-aligned currents that couple the changing angular momentum of the flux tubes between the magnetosphere and ionosphere. Radioti et al. (2011a, 2011b) and Kasahara et al. (2013) showed that the energy released by this process is sufficient to account for the observed spots emitted power (a fraction to several GW). Results from Ge et al. (2010), assuming an updated magnetosphere model, provide further direct evidence of a link between Jovian tail reconnection and polar auroral emissions. More precisely, they confirm that the ionospheric footpoints of tail dipolarization events are close to the polar dawn auroras. Vogt et al. (2014) performed an analysis of the magnetic signature of 43 tailward moving plasmoids and showed that their properties are consistent with a typical mass loss rate of $\sim$0.7–120 kg/s, much lower than the mass input rate from Io (suggesting that additional mass loss mechanisms may be significant), and a flux closure rate of $\sim$7–70 GWb/day, confirming that tail reconnection and plasmoids play an important role in flux transport at Jupiter.

3 Saturn

3.1 Current Understanding

Given the similarities between Jupiter and Saturn (Table 1), it is not surprising that the overall morphology of Saturn's UV auroral emissions resembles that of Jupiter. Saturn's aurora forms a variable ring of emission quasi-rotating around both magnetic poles and displays isolated intermittent structures, spots and arcs, on both sides of this principal component. However, closer inspection of these auroral features reveals significant differences with Jupiter, which will be discussed below. Following the marked response of Saturn's aurora to the changing solar wind conditions, it is often assumed that Saturn's auroral morphology is halfway between that of Jupiter and the Earth's, combining the usual Earth auroral components with Jupiter's corotating nature. This simplified view is certainly a useful starting point, but it can also be very misleading. Therefore, it is preferable to assume that Saturn's UV aurora is not an intermediate case but a case on its own, sharing some similarities with the Earth and Jupiter.

Like Jupiter, Saturn's UV aurora has been studied with the STIS and ACS cameras on board HST. Since, at opposition, the distance from Earth orbit to Saturn is about twice the distance to Jupiter, the spatial resolution is degraded by a factor of $\sim$2, roughly corresponding to 300 km per pixel. The auroral brightness on Saturn is usually one order of magnitude smaller than on Jupiter, with typical values ranging from 10 to 100 kR. The combination of fainter emissions and lower spatial resolution gives rise to images of lesser quality than for Jupiter. However, they are still detailed enough to reveal the complex and changing morphology of the aurora (e.g. Clarke et al. 2005). One of the most important lessons that we have learned from the HST–Cassini campaign that took place preceding the Saturn orbit insertion of Cassini in Jan. 2004 is the clear influence of the solar wind, most specifically its ram pressure directly measured by Cassini, on the global auroral morphology. During this campaign, the brightness of the aurora significantly increased in response to the arrival of large solar wind pressure pulses (e.g. Crary et al. 2005). On one occasion, the global morphology itself dramatically changed during the compression event. Within a few hours, the main emission ring, that was initially surrounding the pole, collapsed to a bright feature filling in a small region on the dawn side of the polar cap (e.g. Grodent et al. 2005; Badman et al. 2005).

The orbital insertion of Cassini marked the beginning of a new era in the exploration of Saturn. In particular, the UV aurora revealed itself to the ultraviolet imaging spectrograph (UVIS; Esposito et al. 2004). The high latitude orbits provided stunning views of the auroral emissions from almost above the poles. The data captured near Cassini's periapsis showed unexpected fine details of both poles, inaccessible to HST, of the auroral emissions that are greatly helping their interpretation. The UVIS instrument is observing the auroral emissions more frequently than the HST cameras do. Therefore, it is able to sample the auroral dynamics at various timescales. The most important advantage of UVIS stems from

Fig. 3 Typical pseudo-image of Saturn's northern FUV aurora obtained with the UVIS camera onboard the Cassini spacecraft on 6 Jan 2013 from 08:38 (SCT). Planetocentric parallels (from 60° to 90°) and local time meridians are drawn every 10°. The local time polar map is showing some of the typical components of Saturn's aurora: a bright dawn section of the main emission, noon feature detaching poleward from the main emission, diffuse dusk side main emission, and nightside outer emission. Dawn (06LT) is to the left; noon (12LT) is to the *bottom* of the figure. See text for a detailed description of the image

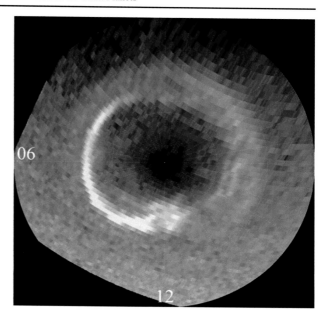

the combined use of the different instruments onboard the spacecraft. For example, it makes it possible to observe the auroral emissions from a vantage point threading the same magnetic flux tube (Bunce et al. 2014). As a result, it is now possible to simultaneously measure the characteristics of the energetic particles giving rise to the aurora and the aurora itself. UVIS is primarily a spectrograph. Its FUV and EUV channels are designed to obtain high-resolution spectra in the 56–191 nm range from which one may derive, for example, color ratios indicating the penetration depth of the impinging particles and thus estimate their energy (e.g. Gustin et al. 2012). Since the FUV and EUV channels both consist of a spectral slit, UVIS is not providing true images of the auroral region. The second spatial dimension is obtained by slewing the spacecraft in the direction perpendicular to the slit length. This motion allows one to spatially scan the auroral regions from which pseudo-2D images may be reconstructed (Grodent et al. 2011). Figure 3 displays a typical pseudo image of Saturn's FUV aurora obtained with the UVIS camera onboard the Cassini spacecraft. This composite image was reconstructed from a ~1 h observing sequence during which the UVIS spectral long-slit was scanned twice across the auroral region (the aurora was uninterruptedly accumulated in the spectral slit during 8 sec bins). It was obtained on 6 Jan 2013 from 08:38 (SCT). The sub-spacecraft latitude was close to 48° and the altitude was ~8.6 R_S, which offered an optimum view point of Saturn's north pole.

The increasing number of observations with HST and Cassini UVIS is at the basis of the growing complexity of possible auroral morphologies. These various auroral emission distributions may be ordered according to their spatial and dynamical characteristics. These in turn may be related to specific processes in the magnetosphere. Contrary to Jupiter, Saturn's UV aurora has been imaged at all local times, primarily thanks to the occasional high latitude vantage point of the Cassini spacecraft providing an optimum view of Saturn's poles. It should be mentioned that during April and May 2013, a new campaign took place during which coordinated observations of Saturn's aurora were made by the Cassini spacecraft and several Earth-based telescopes. The results of this campaign will be published in a journal special issue (2014 Icarus Special Issue: Saturn Auroral Campaign).

Fig. 4 Sketch of the typical UV auroral components observed at Saturn's poles with both HST STIS and ACS and the Cassini-UVIS spectrograph. The local time polar map displays parallels and meridians every 10°: 1. Main (ring of) emission; 2. Cusp emission; 3. Small scale spots and arcs; 4. Poleward auroral arcs; 5. Bifurcations; 6. Poleward auroral spots; 7. Signatures of injections; 8. Outer emission; 9. Enceladus footprint

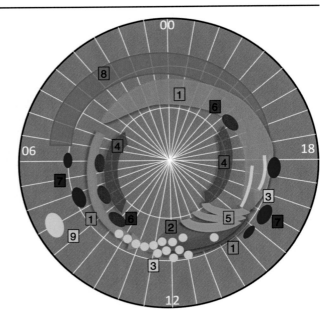

3.2 Main Components of Saturn's Aurora

Figure 4 shows the typical UV auroral components observed at Saturn's poles. They are sketched on a polar map independent from the observatory and apply to the northern and southern hemispheres. They may be organized as a function of their latitudinal location, local time and dynamical behavior. Other classifications are possible, but on the simplest level, one may divide the emissions in 4 categories; (1) the main ring of emission, (2) emissions poleward of the main emission, (3) emissions equatorward of the main emission, and (4) the Enceladus footprint.

3.2.1 The Main (Ring of) Emission

Like Jupiter, Saturn's aurora is harboring one principal component often referred to as the main ring of emission or main oval. It should be noted that, like Jupiter this main component, roughly contributing 2/3 of the total emission, is not forming a circle or an oval, not even a closed structure. Instead, it appears to consist of multiple structures of various sizes, often organized in a broken spiral. In the rest of the text we will refer to it as the main emission. It usually spreads around the poles at northern and southern latitudes larger than 70°, roughly corresponding to equatorial distances mapping to the ring current between ~10 and ~20 R_S (1 Saturnian radius = 1 R_S = 60,268 km; Badman et al. 2006). As stated above, its precise location was shown to respond to the solar wind activity. During quiet periods, the main emission is expanding to lower latitudes and during (or just after) active solar wind episodes, it is significantly contracting to larger latitudes. The expansion/contraction motion is not symmetrically affecting all longitudes simultaneously. As a matter of fact, this dynamical behavior and the overall spiral shape may be seen as signatures of the processes giving rise to the main emission. It is generally accepted that the main emission is associated with the flow shear between open and closed outer magnetosphere magnetic field lines rather than being directly due to the breakdown of plasma corotation due to mass-loading (Bunce et al. 2008;

Talboys et al. 2011). The distinction between closed and open magnetic field lines points to the role of solar-wind, as suggested in the modeling work proposed by Cowley et al. (2004a, 2004b, 2008b). The imbalance between dayside magnetopause reconnection with the solar wind, opening magnetic field lines, and magnetotail reconnection, closing field lines, would then explain Saturn's changing auroral morphology and its relation with solar wind activity. In addition to these large morphological modifications, HST-ACS observations obtained near Saturn's 2009 equinox (Nichols et al. 2010a, 2010b), provided images indicating that the location of the overall northern auroral region oscillates, with an amplitude of 1–2° consistent with that of the southern oval observed by Nichols et al. (2008). It was postulated that the cause of this oscillation is an external magnetospheric current system (Southwood and Kivelson 2007; Andrews et al. 2010).

The auroral brightness is usually varying from a few kR to several tens of kR and may occasionally reach values in excess of 100 kR. This is roughly one order of magnitude less than Jupiter's aurora. Badman et al. (2005) and Crary et al. (2005) showed that the emitted auroral power is directly correlated with the solar wind ram pressure and therefore anti-correlated with the size of the auroral region. During quiet solar wind episodes, the overall brightness may be so small that the auroral region is almost completely fading away (Gérard et al. 2006).

The characteristics that makes Saturn's aurora look like Jupiter's and different from the Earth's is its corotating nature. The bulk of the emission is found to rotate with the planet at approximately 70 % of rigid rotation (in the S3 longitude system). This velocity may not be representative of the whole emission, since isolated features are also observed to be quasi-fixed in local time (i.e. 0 % corotation). Some auroral features were shown to slow down from 70 % to ~20 % as they were approaching the sub-solar meridian (Grodent et al. 2005). This deceleration, accompanied with a significant poleward shift of several degrees of latitude, is still unexplained but might be associated with the process of dayside magnetopause reconnection.

As stated above, the main emission consists of several substructures that may be associated with different mechanisms. They are highlighted in Fig. 4 (orange features marked "1"). On the dawn side (left side), the emission is usually forming one or more relatively narrow arcs. These arcs can be very bright, they are actually the brightest observed features, and are corotating at 70 %. Most of the time, before these arcs rotated past 12LT, new ones replace them, giving the illusion of a permanent dawn side emission. However, on some occasions, the dawn side was not immediately replenished.

Owing to the lack of very long observing sequences, it is difficult to ascertain the origin of these arcs. It is possible that they are already present near midnight and light up as they are approaching 06LT. According to Cowley et al. (2005), these features may be explained by magnetotail reconnection near midnight with subsequent corotation of the planetward side of the reconnection accelerated plasma and field reorganization through dawn, then noon and dusk. An alternative mechanism was proposed to explain the direct causal link between ring current enhancements, taking the form of energetic neutral atoms emission (ENA), and auroral UV emissions (Mitchell et al. 2009). According to this, ring pressure asymmetry may generate sufficiently high currents that field aligned acceleration is required to supply the current carriers, resulting in recurrent bright auroral arcs that would favor the dawn sector. Later on, Nichols et al. (2010b) showed that the northern and southern dawnside auroral power exhibits a statistically significant variation, by factors of ~3, with maximum output occurring during peak Saturn kilometric radiation (SKR) power (Kurth et al. 2007, 2008), while there is evidence for weaker, opposite behavior in the duskside power. Such behavior may be indicative of modulation by the same external rotating current system as that postulated to explain the ~2° oscillation in the auroral oval location observed by Nichols et al. (2010a).

Once these rotating arcs leave the dawn sector and approach 12LT, they either continue their revolution around the pole, or give birth to a new kind of sub-structure. In the former case (orange features "1" at the bottom of Fig. 4), the arcs will preserve their overall shape but their brightness will continuously decrease. This may explain why the dawnside is usually found to be brighter than the duskside, although on some occasions the dawnside may get much dimmer than dusk, especially when features leaving the dawn side are not replaced with new ones. For the latter case, rotating arcs approaching the noon meridian may evolve into new types of structures, poleward of the main emission (dark green and light green features marked "2" and "5" in Fig. 4), and will be described in the next section. It should be pointed out that this evolution may only be caught during observing sequences spanning several hours. Since these are not very frequent, the association between dawn arcs and noon structures should be considered with caution. UVIS observations obtained on August 2008, when Cassini approached Saturn at an extremely small altitude of ~5 R$_S$, revealed details of the main emission in the noon to dusk sector at a spatial resolution close to 200 km (Grodent et al. 2011). These views are showing isolated features as small as 500 km across. They are taking the form of individual spots arranged in a "bunch of grapes" configuration near noon, and small narrow arcs near dusk (pale green spots and arcs marked "3" in Fig. 4). The latter arcs were tentatively associated with patterns of upward field aligned currents resulting from non uniform plasma flow in the equatorial plane while the spots were suggested to be the result of field aligned currents associated with vortices triggered by magnetopause Kelvin-Helmholtz waves. These close up views are very rare; therefore, it is currently difficult to ascertain whether the main emission is always formed of small-scale features or if UVIS captured an uncommon event. Alternatively, Meredith et al. (2013) suggested that such isolated patches appearing simultaneously in both hemispheres, as observed with HST, are consistent with field aligned currents associated with a second harmonic ULF FLR wave propagating eastward through the equatorial plasma.

3.2.2 Emissions Poleward of the Main Emission

Auroral features appearing poleward of the main emission may fit in three arbitrary categories. The first category comprises auroral features completely detached from the main emission and therefore presumably attached to open field lines. They are usually forming sporadic and faint arcs or branches at latitudes close to 80° (red arcs marked "4" in Fig. 4). Their origin is currently unknown, although analogies with similar features in Earth aurora suggest a possible association with Earth's theta aurora. The second category groups auroral structures that are still attached to the main emission (dark green region "2", light green arcs marked "5" in Fig. 4) and appear almost fixed in local time. The majority of these features are located near noon, suggesting that they are related to the process of magnetic reconnection near the nose of Saturn's magnetopause and were often termed "cusp aurora". At Saturn, auroral brightenings are frequently observed near noon (e.g. Gérard et al. 2004, 2005) (dark green feature "2" in Fig. 4). Following theoretical considerations, they were possibly attributed to reconnection with the solar wind magnetic field on the dayside magnetopause, similarly to the case of lobe cusp spots at Earth (Milan et al. 2000). Bunce et al. (2005) proposed that pulsed reconnection at the low-latitude dayside magnetopause for northward directed IMF (corresponding to the southward IMF case at Earth) is giving rise to pulsed twin-vortical flows in the magnetosphere and ionosphere in the vicinity of the OCFLB. These vortices build up field-aligned currents sufficient to produce the observed auroral enhancements near noon. During southward IMF conditions, reconnection cannot take place at low latitude, however, high-latitude lobe reconnection pulsed twin-vortical flows, bi-polar field aligned currents are expected and associated with auroral intensifications poleward of the OCFLB. During intermediate conditions, with a small northward and dominating east-west (B_y) component of the IMF, a mixed situation with reconnection at the high and the low latitude region may occur simultaneously. In addition, a non negligible B_y component may also favor reconnection on the flanks of the magnetopause giving rise to auroral brightenings appearing near the pre-noon or post-noon sectors.

An extended sequence of observations obtained with UVIS in 2008 revealed the evolution of an auroral feature starting as an intensification of the main emission near noon, similar to the cusp aurora described above, gradually detaching from the main emission (the OCFLB) in the poleward direction (Radioti et al. 2011a, 2011b, 2013b; Badman et al. 2013), and finally evolving into two arcs (other shorter UVIS sequences show up to 3 arcs) with one end attached to the main emission and the other end intruding the empty polar region at a smaller local time (light green arcs marked "5" in Fig. 4). These bifurcations of Saturn's main emission were tentatively attributed to consecutive reconnection events at the dayside magnetopause and were associated with open magnetic flux. Thorough inspection of the sequence of UV observations showed a concurrent equatorward motion, or expansion of the main ring of emission, such that the increase of the area poleward of the main emission (i.e. the polar cap size) is balanced by the area occupied by the bifurcations, therefore supporting the consecutive reconnections scenario and the possibility that dayside reconnection at Saturn can occur consecutively or simultaneously at several locations on the magnetopause with the reconnection lines following each other as they sweep along the flank of the magnetopause.

The same study from Radioti et al. (2011a, 2011b) pointed out transient spot-like structures appearing at the dawn and dusk poleward boundary of the main emission ring, establishing a third category of poleward features (dark blue spots marked "6" in Fig. 4). These small, isolated features are somewhat detached from the main emission and are therefore possibly connected to open magnetic field lines. Jackman et al. (2013) demonstrated that dipolarizations in the magnetotail following reconnection events might result in distinct, observable auroral signatures. They estimated that reconnection in the magnetotail can lead to rapid motion of newly closed field lines planetward and the diversion of the cross-tail current through the ionosphere, resulting in discrete auroral emission through hot plasma injection into and around the inner magnetosphere. The expected brightness of associated auroral signatures is on the order of 10 kR, somewhat smaller but still in reasonable agreement with the observed auroral spots of a few tens of kR. Jackman et al. (2013) further pointed out that the observed auroral spot that they considered in their study is a precursor to a larger intensification which followed about an hour later in the Cassini UVIS sequence, and which had previously been reported to be linked with recurrent energization from the tail (Mitchell et al. 2009).

3.2.3 Emissions Equatorward of the Main Emission

Two types of auroral structures appear equatorward of the main emission; spots and nightside extended arcs (outer emission). Spot features include the Enceladus footprint that will be addressed in the next section.

Spots Isolated transient UV auroral spots are occasionally observed in Saturn's ionosphere along the equatorward boundary of the main emission (Radioti et al. 2009b, 2013c) (purple spots marked "7" in Fig. 4). Their typical lifetime ranges from several minutes to a few tens of minutes. These relatively faint features—therefore difficult to detect—display typical brightness <10 kR, corresponding to emitted power on the order of 0.1 GW.

Quasi-simultaneous HST and Cassini observations suggested that these auroral spots are associated with the dynamics taking place in Saturn's magnetosphere. Most specifically, Cassini's in situ instruments detected signatures of energetic particle injections on magnetic field lines mapping close to the ionospheric region where, on the same day, HST observed the transient auroral spots. Radioti et al. (2009a, 2009b) proposed that the injection region may be directly coupled to Saturn's ionosphere by pitch angle diffusion and electron scattering by whistler waves, or by the electric current flowing along the boundary of the injected hot cloud. A more recent Cassini UVIS dataset made it possible to model the changing brightness distribution of such UV spot structures (Radioti et al. 2013b). Comparison of the brightness and size evolution of the simulated ionospheric signature, based on typical injected particles drift and plasma energy dispersion, with observed values demonstrated that these auroral spots behave as auroral signatures of an injection. Simultaneous Cassini observations of energetic neutral atoms (ENA) enhancements, indicative of a rotating heated plasma region, suggest that pitch angle diffusion and electron scattering may not be the only mechanism responsible for the observed auroral spots. Field aligned currents driven by pressure gradients along the boundaries of the injected hot plasma may also give rise to such auroral emissions.

Outer Emission   Recent observations of Saturn's aurora with the UVIS spectrograph onboard Cassini not only confirm the presence of a quasi-permanent partial ring of emission equatorward of Saturn's main auroral emission (Grodent et al., 2005, 2010) (light blue arc marked "8" in Fig. 4), but they also increase the number of positive cases and allow for a statistical analysis of the characteristics of this outer emission. This faint but distinct auroral feature appears at both hemispheres in the nightside sector. It magnetically maps to relatively large distances in the nightside magnetosphere, on the order of 9 R_S.

This auroral feature consists of one or more narrow arcs ($\sim 3°$ of latitude) of emission usually extending equatorward of the main emission from 18LT to 06LT through midnight, although some images show the emission extending down to 09LT. The emission is not uniform in longitude, the presence of patches allows one to estimate the level of corotation of the outer emission to $\sim 70\%$, similar to the main emission and compatible with a magnetospheric plasma source rotating at 7 to 10 R_S from Saturn.

It was initially thought that pitch angle scattering of electrons into the loss cone by whistler waves would be responsible for the outer auroral emission. Rough estimates suggested that a suprathermal electron population observed with Cassini (Schippers et al. 2008; Lewis et al. 2008) in the nightside sector between 7 and 10 R_S might power this process. However, a new analysis of 7 years of Cassini electron plasma data (Schippers et al. 2012) indicates the presence of layers of upward and downward field aligned currents. They appear to be part of a large-scale current system involving dayside-nightside asymmetries as well as trans-hemispheric variations. This system comprises a net upward current layer, carried by warm electrons, limited to the nightside sector which may as well generate the outer UV auroral emission.

3.2.4 The Enceladus Footprint

The detection threshold of the HST UV cameras and the amount of reflected sunlight leaking in these detectors are too large for a possible detection of Enceladus' footprint with STIS or ACS. On the other hand, the UVIS spectrograph is able to detect fainter emissions. On 26 August 2008, UVIS obtained three successive observations of Saturn's north pole unambiguously revealing the auroral footprint of Enceladus at a location consistent with the expected one (Pryor et al. 2011) (yellow spot marked "9" in Fig. 4). The spot brightness is on the order of 1 kR, just above UVIS detection threshold. The predicted southern footprint has not yet been detected for it is probably fainter than its northern counterpart, as is the case for the main aurora (Nichols et al. 2009b). The auroral spot size suggests that the emission is connected to an Enceladus interaction region at the equator extending as far as 20 Enceladus radii (R_E) downstream with a radial extent between 0 and 20 R_E, consistent with the extent of the plume resulting from Enceladus' cryo-volcanic activity. The Enceladus auroral footprint was shown to vary in brightness by a factor of about 3. The most likely cause for this observed large-scale variability is related to the time-variable cryo-volcanism from Enceladus' south polar vents, suggesting that plume activity was particularly high at the time of the UVIS observations. Two weeks prior this detection by UVIS, the in situ instruments of Cassini detected signatures of magnetic-field-aligned ion and electron beams with sufficient power to produce the observed auroral footprint of Enceladus. Observed changes in the characteristic energy of the field-aligned electron flux were tentatively associated with changes in the magnetic field perturbation suggesting an actual change in the total field-aligned current density. At Jupiter, the multiple components of the ultraviolet footprint of Io have been interpreted as being due to multiple reflections of a standing Alfvén wave current system driven by Io. It is possible that the flickering in energy of the beams observed downstream of Enceladus is the equatorial signature of a standing wave pattern like that observed at the Io footprint, suggesting a possible universal mechanism magnetically coupling a conducting moon to its parent planet's ionosphere.

4 The Ice Giants

4.1 Uranus

Like Jupiter and Saturn, Uranus' fast rotation provides the planet with a strong dynamo-magnetic field. Modeling of the interior of Uranus (Stanley and Bloxham 2006) suggests that the dynamo source region consists in a convecting thin shell surrounding a stably stratified fluid interior. This configuration is compatible with formation of a highly tilted (58.6°) and shifted (0.3 uranian radius) magnetic dipole as well as important multipolar components. This intricate asymmetric magnetic topology combines with the oddly inclined spin axis (98°) of Uranus to form a highly distorted magnetosphere interacting with the solar wind in a way that is changing during the course of the uranian day (Arridge et al. 2012). This complexity complicates the detection of auroral emissions, especially for a distant observer near Earth orbit.

The first unambiguous detection of aurora on Uranus was made possible by the unique flyby of the planet by Voyager 2 (V2) Spacecraft. On January 24, 1986, at the time of Uranus northern summer solstice, V2 was only 81,500 km from the cloud tops. Among the numerous observations performed with V2 instruments during this several-hour encounter, the extreme ultraviolet spectrometer (UVS) measured emissions in the 95–110 nm range near the magnetic poles (Broadfoot et al. 1986; Herbert and Sandel 1994). Those were attributed to auroral H_2 Lyman and Werner bands because at these wavelengths sunlight reflected by the uranian atmosphere is relatively weak. On the contrary, the auroral H Lyman α emission could not be discriminated from the much brighter dayside reflected solar H Lyman α light, nor from the nightside reflected interstellar medium H Lyman α.

Reconstruction from individual spatially resolved spectra taken at different times (Herbert 2009) provided an average map of the auroral emission, assumed to be time invariant. The bulk of the emission forms discrete spots around the north and south AH_5 model magnetic poles and map to closed field line regions near $L = 5$ (see Herbert (2009) for precise significance of L which slightly differs from the conventional McIlwain parameter). In addition, a pair of bright spots in the north auroral polar cap appears to map to $L \geq 20$, presumably on open field lines. In both cases, the magnetic longitude is such that the majority of the observed aurora connects to the magnetotail, suggesting substorm-like injection processes, possibly associated with the arrival of an interplanetary shock (Sittler et al. 1987), and an Earth-like partial ring current system.

The brightest UV auroral emissions thread field lines along which strong uranian kilometric radio emissions (UKR; e.g. Herbert and Sandel 1994) and whistler mode waves (Gurnett et al. 1986; Kurth and Gurnett 1991) were also concurrently observed with V2. This coincidence suggests that the auroral precipitation associated with the UV emission might stem from whistler mode plasma waves scattering magnetospheric electrons of several keV into the loss cone.

Since the rotation period of Uranus is not accurately known, the magnetic configuration inferred from V2 in 1986 is not sufficient to derive the present location of the magnetic poles. As a consequence, one does not know exactly where to search for the auroral signal, which is challenging new observation planning. Nevertheless, there have been several attempts to observe the uranian UV aurora from Earth orbit with HST, in 1998, 2005 and 2011, around Uranus equinox. Only the 2011 and 1998 HST datasets reveal unambiguous auroral emissions (Lamy et al. 2012). The 2011 aurora was shown to be potentially associated with a series of powerful CMEs emitted by the Sun two months earlier. These features appeared to form an extremely localized patch of weak emission, only visible in a few HST images. They were described as variable signatures with brightness comparable to that observed with V2. They are taking the form of spots or roughly continuous ring-like structures in the dayside. Lamy et al. (2012) suggested that the spots result from dayside reconnection with the IMF, while the ring-like structures would involve electron precipitation over a wide range of longitudes that might be related to a short-lived twisted configuration of the magnetotail.

Similar efforts were made to detect auroral signatures in the near-infrared wavelength range as part of a long term ground-based monitoring of Uranus' H_3^+ emissions (Melin et al. 2011). Whilst the aurora remains spatially unresolved in the infrared, probably owing to its small contrast with thermal emissions, observations conducted since 1992 show significant short-term variability. This variability is presumably caused by changes in particle precipitation flux and energy rather than by the slower variation of solar input energy. More recently, Melin et al. (2013) presented observations obtained in late 2011, simultaneously with HST UV observations, showing that Uranus' upper atmosphere had continued its long-term cooling trend beyond the 2007 equinox. This further suggests that Uranus thermospheric temperature is closely linked to the changing geometry of the solar wind and planetary magnetic field.

4.2 Neptune

The dynamo source region of Neptune is likely of the same nature as that of Uranus (Stanley and Bloxham 2006). Therefore, it is not surprising that its magnetic field is also highly asymmetric, tilted ($-47°$) and shifted (~ 0.5 Neptune Radius). Neptune's obliquity is significantly smaller than Uranus' ($29.6°$), yet it is still large enough to contribute to the complexity and variability of Neptune's magnetosphere.

Voyager 2 UVS detected marginal ultraviolet atmospheric emission from the nightside of Neptune (Broadfoot et al. 1989). This emission, consistent with H_2 band spectrum, forms two distinct features: a broad diffuse region extending from 55°S to 50°N near 60°W and a brighter, narrower region confined to high southern latitudes near 240°W (Sandel et al. 1990). This latter feature was tentatively attributed to auroral processes involving precipitation of energetic electrons trapped at L values ≥ 8 R_N (Mauk et al. 1994). It was also associated with a partial plasma torus formed by ionization of gas escaping from Triton's atmosphere (Hill and Dessler 1990); or, alternatively, with a magnetic anomaly effect (Paranicas and Cheng 1994). The latitudinally distributed emission near 60°W was suggested to result from precipitation of photoelectrons originating in the conjugate sunlit hemisphere. In any case, these emissions are so faint, a couple of Rayleighs in the H_2 band region shortward of Lyman α, that they were never detected from Earth orbit and are not expected to have any measurable infrared counterpart.

5 Conclusion

5.1 Jupiter and Saturn

Jupiter and Saturn are gas giant planets with strong magnetic fields and fast rotating H_2 dominated atmospheres. They both harbor conducting moons, one of which (Io and Enceladus, respectively) is a major internal plasma source taping their giant magnetospheres. All the ingredients are present on the two planets, in different proportions, to produce strong UV auroral emissions with, one might think, comparable morphologies. However, Jupiter and Saturn appear to respond differently to the interplanetary magnetic field and to changing solar wind conditions. These different responses are thought to impart noticeable dissimilarities on the UV auroral morphology and on the origin of some auroral features that, at first glance, are looking the same. The most striking case is the main emission. It is present on both planets and is forming a strip of emission around the pole that is partially corotating with the magnetic field. However, closer inspection of this main auroral feature reveals strong dissimilarities, such as the bifurcation of a fraction of the Saturnian main emission, which finds no equivalent in the Jovian aurora. This particular behavior points to the control of Saturn's main emission by processes related to the interaction of Saturn's magnetosphere with the solar wind. For Jupiter, this interaction appears to be much less important and is eclipsed by the corotation electric field. Despite these major differences, some auroral features appear to be common to gas giants. Among them, satellite magnetic footprints are easily recognizable since they detach from the rest of the emission. Injection of hot plasma in the middle magnetosphere is also a process common to both planets. Therefore, it is not surprising to find similar auroral signatures. The case of dayside and night side reconnection spot like signatures is less clear-cut since it involves magnetospheric mechanisms that are driven internally (Vasyliūnas-cycle) or externally (Dungey-cycle). The relative importance of these two cycles depends, again, on the significance of corotation enforcement compared to solar wind convection, which is different for Jupiter and Saturn.

5.2 Uranus and Neptune

Compared to Jupiter and Saturn, the UV aurorae on Uranus and Neptune are all poorly documented. The main reason stems from the weakness of the emission, making it extremely difficult to observe from Earth orbit, and from the complexity of the magnetic field and their rapidly changing distorted magnetosphere. Most of the observations were obtained with the Voyager 2 spacecraft that flew by Uranus in 1986 and by Neptune in 1989. On Uranus, the aurora forms discrete spots around both magnetic poles and map along closed and open magnetic field lines possibly connected to the magnetotail, suggesting substorm-like injection processes. On Neptune, the UV signal was marginally detected on the nightside of the planet. It consists of a broad diffuse region extending between northern and southern mid-latitudes and a narrower region confined to high southern latitudes. Only the latter was plausibly associated with auroral processes. Some recent observations with HST also revealed UV auroral emission on Uranus. This weak emission forms spots, tentatively attributed to reconnection with the IMF, or ring-like structures in the dayside possibly related to a short-lived twisted configuration of the magnetotail.

5.3 The Juno mission

Although our understanding of the auroral mechanisms prevailing at the giant planets is improving, thanks to the continuing observing, theoretical and modeling efforts, there are still numerous fundamental questions which need consideration. For example, the auroral particle acceleration processes above the atmosphere need to be confirmed; the actual role of solar wind in driving the magnetospheres is also a crucial point of debate. The NASA New Frontiers Juno mission will directly address some of these questions (Bolton et al. 2010). Juno will be the first spacecraft placed into an elliptical polar orbit around Jupiter following an insertion-orbit manoeuvre in July 2016. Juno's scientific payload consists of nine instruments, five of which are designed to determine the physical processes occurring in the high latitude magnetosphere of Jupiter, making it possible to directly relate them to auroral activity and to processes taking place in the low-latitude magnetosphere (Bagenal et al. 2014). Specifically, the magnetometer (MAG) will provide an accurate mapping of the magnetic field from the top of the ionosphere to the deep magnetosphere; the Jupiter Energetic particle Detector Instrument (JEDI) will measure the high energy and pitch angle of plasma sheet ions and electrons while the Jovian Auroral Distributions Experiment (JADE) will make the first characterization of the particles giving rise to aurora and will complement JEDI by observing the lower part of the energy spectrum; the plasma Waves (Waves) instrument will identify the regions of auroral currents and the auroral particles acceleration processes; at the same time, the Ultraviolet Spectrograph (UVS) will obtain spectral images of the UV aurora generated by the particles measured by JADE. In addition, the Jupiter InfraRed Auroral Mapper (JIRAM) will provide key information on the conditions prevailing in the auroral atmosphere and the visible camera (JunoCAM) will also observe the auroral emissions in the nightside sector. Thanks to its polar orbit and instruments suite Juno will be capable of simultaneously measuring key signatures of the efficient magnetosphere-ionosphere coupling at Jupiter. This knowledge will benefit to the case of Saturn and to some extent to Uranus and Neptune, as well as any giant magnetized planet surrounded by plasma.

Acknowledgements The author acknowledges the support of EUROPLANET RI project (Grant agreement No. 228319) funded by EU; and also the support of the International Space Science Institute (Bern). D. Grodent is partially supported by the PRODEX program managed by ESA in collaboration with the Belgian Federal Science Policy Office. This review is based on observations made with the Hubble Space Telescope obtained at the Space Telescope Science Institute, which is operated by AURA Inc., and on data obtained in the frame of the NASA/ESA Cassini Project. The Editor thanks the work of two anonymous referees.

References

D.J. Andrews, A.J. Coates, S.W.H. Cowley, M.K. Dougherty, L. Lamy, G. Provan, P. Zarka, Magnetospheric period oscillations at Saturn: comparison of equatorial and high-latitude magnetic field periods with north and south Saturn kilometric radiation periods. J. Geophys. Res. **115**, A12252 (2010). doi:10.1029/2010JA015666

C.S. Arridge et al., Uranus Pathfinder: exploring the origins and evolution of ice giant planets. Exp. Astron. **33**, 753–791 (2012). doi:10.1007/s10686-011-9251-4

S.V. Badman, E.J. Bunce, J.T. Clarke, S.W.H. Cowley, J.-C. Gérard, D. Grodent, S.E. Milan, Open flux estimates in Saturn's magnetosphere during the January 2004 Cassini-HST campaign, and implications for reconnection rates. J. Geophys. Res. **110**, A11216 (2005). doi:10.1029/2005JA011240

S.V. Badman, S.W.H. Cowley, J.-C. Gérard, D. Grodent, A statistical analysis of the location and width of Saturn's southern auroras. Ann. Geophys. **24**, 3533–3545 (2006). doi:10.5194/angeo-24-3533-2006

S.V. Badman, A. Masters, H. Hasegawa, M. Fujimoto, A. Radioti, D. Grodent, N. Sergis, M.K. Dougherty, A.J. Coates, Bursty magnetic reconnection at Saturn's magnetopause. Geophys. Res. Lett. **40**, 1027–1031 (2013). doi:10.1002/grl.50199

S.V. Badman, G. Branduardi-Raymont, M. Galand, S.L.G. Hess, N. Krupp, L. Lamy, H. Melin, C. Tao, Auroral processes at the giant planets: energy deposition, emission mechanisms, morphology and spectra. Space Sci. Rev. (2014). doi:10.1007/s11214-014-0042-x (this issue)

F. Bagenal, A. Adriani, F. Allegrini, S.J. Bolton, B. Bonfond, E.J. Bunce, J.E.P. Connerney, S.W.H. Cowley, R.W. Ebert, G.R. Gladstone, C.J. Hansen, W.S. Kurth, S.M. Levin, B.H. Mauk, D.J. McComas, C.P. Paranicas, D. Santos-Costas, R.M. Thorne, P. Valek, J.H. Waite, P. Zarka, Magnetospheric science objectives of the Juno mission. Space Sci. Rev. (2014). doi:10.1007/s11214-014-0036-8

B. Bhattacharya, R.M. Thorne, D.J. Williams, On the energy source for diffuse jovian auroral emissivity. Geophys. Res. Lett. **14**, 2751–2754 (2001)

S.J. Bolton (The Juno Science Team), The Juno mission, in *International Astronomical Union*, ed. by C. Barbieri et al.. Proceedings IAU Symposium, vol. 269 (Cambridge University Press, Cambridge, 2010). doi:10.1017/S1743921310007313

B. Bonfond, D. Grodent, J.-C. Gérard, A. Radioti, J. Saur, S. Jacobsen, UV Io footprint leading spot: a key feature for understanding the UV Io footprint multiplicity? Geophys. Res. Lett. **35**, L05107 (2008). doi:10.1029/2007GL032418

B. Bonfond, D. Grodent, J.-C. Gérard, A. Radioti, V. Dols, P.A. Delamere, J.T. Clarke, The Io UV footprint: location, inter-spot distances and tail vertical extent. J. Geophys. Res. **114**, A07224 (2009). doi:10.1029/2009JA014312

B. Bonfond, The 3-D extent of the Io UV footprint on Jupiter. J. Geophys. Res. **115**, A09217 (2010). doi:10.1029/2010JA015475

B. Bonfond, M.F. Vogt, J.-C. Gérard, D. Grodent, A. Radioti, V. Coumans, Quasi-periodic polar flares at Jupiter: a signature of pulsed dayside reconnections? Geophys. Res. Lett. **38**, L02104 (2011). doi:10.1029/2010GL045981

B. Bonfond, D. Grodent, J.-C. Gérard, T. Stallard, J.T. Clarke, M. Yoneda, A. Radioti, J. Gustin, Auroral evidence of Io's control over the magnetosphere of Jupiter. Geophys. Res. Lett. **39**, L01105 (2012a). doi:10.1029/2011GL050253

B. Bonfond, When moons create aurora: the satellite footprints on giant planets in auroral phenomenology and magnetospheric studies: earth and other planets, in Geophys. Monogr. Ser., vol. 197, ed. by A. Keiling, et al. (AGU, Washington, 2012b, to appear). pp. 133–140. doi:10.1029/2011GM001169

B. Bonfond, S. Hess, F. Bagenal, J.-C. Gérard, D. Grodent, A. Radioti, J. Gustin, J.T. Clarke, The multiple spots of the Ganymede auroral footprint. Geophys. Res. Lett. **40**, 4977–4981 (2013). doi:10.1002/grl.50989

G. Branduardi-Raymont, R.F. Elsner, M. Galand, D. Grodent, T.E. Cravens, P. Ford, G.R. Gladstone, J.H. Waite Jr., Spectral morphology of the X-ray emission from Jupiter's aurorae. J. Geophys. Res. **113**, A02202 (2008). doi:10.1029/2007JA012600

A.L. Broadfoot et al., Extreme ultraviolet observations from Voyager 1 encounter with Jupiter. Science **204**, 979–982 (1979). doi:10.1126/science.204.4396.979

A.L. Broadfoot et al., Extreme ultraviolet observations from Voyager 1 encounter with Saturn. Science **212**, 206–211 (1981). doi:10.1126/science.212.4491.206

A.L. Broadfoot et al., Ultraviolet spectrometer observations of Uranus. Science **233**(4759), 74–79 (1986). doi:10.1126/science.233.4759.74

A.L. Broadfoot et al., Ultraviolet spectrometer observations of Neptune and Triton. Science **246**(4936), 1459–1466 (1989). doi:10.1126/science.246.4936.1459

E.J. Bunce, S.W.H. Cowley, S.E. Milan, Interplanetary magnetic field control of Saturn's polar cusp aurora. Ann. Geophys. **23**, 1405–1431 (2005)

E.J. Bunce et al., Origin of Saturn's aurora: simultaneous observations by Cassini and the Hubble space telescope. J. Geophys. Res. **113**, A09209 (2008). doi:10.1029/2008JA013257

E.J. Bunce, D. Grodent, S.L. Jinks, C.S. Arridge, D.J. Andrews, S.V. Badman, S.W.H. Cowley, M.K. Dougherty, W.S. Kurth, D.G. Mitchell, G. Provan, Cassini nightside observations of the oscillatory motion of Saturn's northern auroral oval. J. Geophys. Res. (2014) (in press). doi:10.1029/2013JA019527

T. Chust, A. Roux, W.S. Kurth, D.A. Gurnett, M.G. Kivelson, K.K. Khurana, Are Io's Alfvén wings filamented? Galileo observations. Planet. Space Sci. **53**, 395–412 (2005). doi:10.1016/j.pss.2004.09.021

J.T. Clarke et al., Ultraviolet emissions from the magnetic footprints of Io, Ganymede and Europa on Jupiter. Nature **415**, 997–1000 (2002)

J.T. Clarke, D. Grodent, S.W.H. Cowley, E.J. Bunce, P. Zarka, J.E.P. Connerney, T. Satoh, Jupiter's aurora, in *Jupiter: The Planet, Satellites and Magnetosphere* (Cambridge Univ. Press, Cambridge, 2004), pp. 639–670

J.T. Clarke et al., Morphological differences between Saturn's ultraviolet aurorae and those of earth and Jupiter. Nature **433**, 717–719 (2005). doi:10.1038/nature03331

J.T. Clarke, S. Wannawichian, N. Hernandez, B. Bonfond, J.-C. Gérard, D. Grodent, Detection of auroral emissions from Callisto's magnetic footprint at Jupiter. Poster presented at the EPSC-DPS Joint Meeting 2011, EPSC abstracts, vol. 6, EPSC-DPS2011-1468 (2011)

S.W.H. Cowley, E.J. Bunce, Origin of the main auroral oval in Jupiter's coupled magnetosphere-ionosphere system. Planet. Space Sci. **49**, 1067–1088 (2001)

S.W.H. Cowley, E.J. Bunce, T.S. Stallard, S. Miller, Jupiter's polar ionospheric flows: theoretical interpretation. Geophys. Res. Lett. **30**(5), 1220 (2003). doi:10.1029/2002GL016030

S.W.H. Cowley, E.J. Bunce, J.M. O'Rourke, A simple quantitative model of plasma flows and currents in Saturn's polar ionosphere. J. Geophys. Res. **109**, A05212 (2004a). doi:10.1029/2003JA010375

S.W.H. Cowley, E.J. Bunce, R. Prangé, Saturn's polar ionospheric flows and their relation to the main auroral oval. Ann. Geophys. **22**, 1379–1394 (2004b). doi:10.5194/angeo-22-1379-2004

S.W.H. Cowley, S.V. Badman, E.J. Bunce, J.T. Clarke, J.-C. Gérard, D. Grodent, C.M. Jackman, S.E. Milan, T.K. Yeoman, Reconnection in a rotation-dominated magnetosphere and its relation to Saturn's auroral dynamics. J. Geophys. Res. **110**, A02201 (2005). doi:10.1029/2004JA010796

S.W.H. Cowley, A.J. Deason, E.J. Bunce, Axi-symmetric models of auroral current systems in Jupiter's magnetosphere with predictions for the Juno mission. Ann. Geophys. **26**, 4051–4074 (2008a). doi:10.5194/angeo-26-4051-2008

S.W.H. Cowley, C.S. Arridge, E.J. Bunce, J.T. Clarke, A.J. Coates, M.K. Dougherty, J.-C. Gérard, D. Grodent, J.D. Nichols, D.L. Talboys, Auroral current systems in Saturn's magnetosphere: comparison of theoretical models with Cassini and HST observations. Ann. Geophys. **26**, 2613–2630 (2008b). doi:10.5194/angeo-26-2613-2008

F.J. Crary et al., Solar wind dynamic pressure and electric field as the main factors controlling Saturn's aurorae. Nature **433**, 720–722 (2005). doi:10.1038/nature03333

P.A. Delamere, F. Bagenal, Modeling variability of plasma conditions in the Io torus. J. Geophys. Res. **108**(A7), 1276 (2003). doi:10.1029/2002JA009706

V. Dols, P.A. Delamere, F. Bagenal, W.S. Kurth, W.R. Paterson, Asymmetry of Io's outer atmosphere: constraints from five Galileo flybys. J. Geophys. Res. **117**, E10010 (2012). doi:10.1029/2012JE004076

R.E. Ergun, L. Ray, P.A. Delamere, F. Bagenal, V. Dols, Y.-J. Su, Generation of parallel electric fields in the Jupiter–Io torus wake region. J. Geophys. Res. **114**, A05201 (2009). doi:10.1029/2008JA013968

L.W. Esposito et al., The Cassini ultraviolet imaging spectrograph investigation. Space Sci. Rev. **115**, 299–361 (2004). doi:10.1007/s11214-004-1455-8

Y.S. Ge, C.T. Russell, K.K. Khurana, Reconnection sites in Jupiter's magnetotail and relation to jovian auroras. Planet. Space Sci. **58**, 1455–1469 (2010). doi:10.1016/j.pss.2010.06.013

J.-C. Gérard, D. Grodent, J. Gustin, A. Saglam, J.T. Clarke, J.T. Trauger, Characteristics of Saturn's FUV aurora observed with the space telescope imaging spectrograph. J. Geophys. Res. **109**, A09207 (2004). doi:10.1029/2004JA010513

J.-C. Gérard, E.J. Bunce, D. Grodent, S.W.H. Cowley, J.T. Clarke, S.V. Badman, Signature of Saturn's auroral cusp: simultaneous Hubble space telescope FUV observations and upstream solar wind monitoring. J. Geophys. Res. **110**, A11201 (2005). doi:10.1029/2005JA011094

J.-C. Gérard, D. Grodent, A. Radioti, B. Bonfond, J.T. Clarke, Hubble observations of Jupiter's north–south conjugate ultraviolet aurora. Icarus **226**, 1559–1567 (2013). doi:10.1016/j.icarus.2013.08.017

J.-C. Gérard et al., Saturn's auroral morphology and activity during quiet magnetospheric conditions. J. Geophys. Res. **111**, A12210 (2006). doi:10.1029/2006JA011965

D. Grodent, J.H. Waite Jr., J.-C. Gérard, A self-consistent model of the jovian auroral thermal structure. J. Geophys. Res. **106**(A7), 12933–12952 (2001). doi:10.1029/2000JA900129

D. Grodent, J.T. Clarke, J. Kim, J.H. Waite Jr., S.W.H. Cowley, Jupiter's main auroral oval observed with HST-STIS. J. Geophys. Res. **108**(A11), 1389 (2003a). doi:10.1029/2003JA009921

D. Grodent, J.T. Clarke, J.H. Waite Jr., S.W.H. Cowley, J.-C. Gerard, J. Kim, Jupiter's polar auroral emissions. J. Geophys. Res. **108**(A10), 1366 (2003b). doi:10.1029/2003JA010017

D. Grodent, J.-C. Gérard, J.T. Clarke, G.R. Gladstone, J.H. Waite Jr., A possible auroral signature of a magnetotail reconnection process on Jupiter. J. Geophys. Res. **109**, A05201 (2004). doi:10.1029/2003JA010341

46

D. Grodent, J.-C. Gérard, S.W.H. Cowley, E.J. Bunce, J.T. Clarke, Variable morphology of Saturn's southern ultraviolet aurora. J. Geophys. Res. **110**, A07215 (2005). doi:10.1029/2004JA010983

D. Grodent, J.-C. Gérard, J. Gustin, B.H. Mauk, J.E.P. Connerney, J.T. Clarke, Europa's FUV auroral tail on Jupiter. Geophys. Res. Lett. **33**, L06201 (2006). doi:10.1029/2005GL025487

D. Grodent, B. Bonfond, J.-C. Gérard, A. Radioti, J. Gustin, J.T. Clarke, J. Nichols, J.E.P. Connerney, Auroral evidence of a localized magnetic anomaly in Jupiter's northern hemisphere. J. Geophys. Res. **113**, A09201 (2008). doi:10.1029/2008JA013185

D. Grodent, B. Bonfond, A. Radioti, J.-C. Gérard, X. Jia, J.D. Nichols, J.T. Clarke, Auroral footprint of Ganymede. J. Geophys. Res. **114**, A07212 (2009). doi:10.1029/2009JA014289

D. Grodent, A. Radioti, B. Bonfond, J.-C. Gérard, On the origin of Saturn's outer auroral emission. J. Geophys. Res. **115**, A08219 (2010). doi:10.1029/2009JA014901

D. Grodent, J. Gustin, J.-C. Gérard, A. Radioti, B. Bonfond, W.R. Pryor, Small-scale structures in Saturn's ultraviolet aurora. J. Geophys. Res. **116**, A09225 (2011). doi:10.1029/2011JA016818

D.A. Gurnett, W.S. Kurth, F.L. Scarf, R.L. Poynter, First plasma wave observations at Uranus. Science **233**(4759), 106–109 (1986). doi:10.1126/science.233.4759.106

J. Gustin, S.W.H. Cowley, J.-C. Gérard, G.R. Gladstone, D. Grodent, J.T. Clarke, Characteristics of jovian morning bright FUV aurora from Hubble space telescope/space telescope imaging spectrograph imaging and spectral observations. J. Geophys. Res. **111**, A09220 (2006). doi:10.1029/2006JA011730

J. Gustin, J.-C. Gérard, W. Pryor, P.D. Feldman, D. Grodent, G. Holsclaw, Characteristics of Saturn's polar atmosphere and auroral electrons derived from HST/STIS, FUSE and Cassini/UVIS spectra. Icarus **200**, 176–187 (2009). doi:10.1016/j.icarus.2008.11.013

J. Gustin, B. Bonfond, D. Grodent, J.-C. Gérard, Conversion from HST ACS and STIS auroral counts into brightness, precipitated power, and radiated power for H_2 giant planets. J. Geophys. Res. **117**, A07316 (2012). doi:10.1029/2012JA017607

J. Gustin, J.-C. Gérard, D. Grodent, G.R. Gladstone, J.T. Clarke, W.R. Pryor, V. Dols, B. Bonfond, A. Radioti, L. Lamy, J.M. Ajello, Effects of methane on giant planet's UV emissions and implications for the auroral characteristics. J. Mol. Spectrosc. **291**, 108–117 (2013). doi:10.1016/j.jms.2013.03.010

S.L.G. Hess, P. Delamere, V. Dols, B. Bonfond, D. Swift, Power transmission and particle acceleration along the Io flux tube. J. Geophys. Res. **115**, A06205 (2010a). doi:10.1029/2009JA014928

S.L.G. Hess, A. Pétin, P. Zarka, B. Bonfond, B. Cecconi, Lead angles and emitting electron energies of Io-controlled decameter radio arcs. Planet. Space Sci. **58**(10), 1188–1198 (2010b). doi:10.1016/j.pss.2010.04.011

S.L.G. Hess, B. Bonfond, P.A. Delamere, How could the Io footprint disappear? Planet. Space Sc. (2013, in press). doi:10.1016/j.pss.2013.08.014

F. Herbert, B.R. Sandel, The uranian aurora and its relationship to the magnetosphere. J. Geophys. Res. **99**(A3), 4143–4160 (1994). doi:10.1029/93JA02673

F. Herbert, Aurora and magnetic field of Uranus. J. Geophys. Res. **114**, A11206 (2009). doi:10.1029/2009JA014394

T.W. Hill, A.J. Dessler, Convection in Neptune's magnetosphere. Geophys. Res. Lett. **17**, 1677–1680 (1990). doi:10.1029/GL017i010p01677

T.W. Hill, The jovian auroral oval. J. Geophys. Res. **106**, 8101–8107 (2001)

T.W. Hill, V.M. Vasyliūnas, Jovian auroral signature of Io's corotational wake. J. Geophys. Res. **107**(A12), 1464 (2002). doi:10.1029/2002JA009514

C.M. Jackman, N. Achilleo, S.W.H. Cowley, E.J. Bunce, A. Radioti, D. Grodent, S.V. Badman, M.K. Dougherty, W. Pryor, Auroral counterpart of magnetic field dipolarizations in Saturn's tail. Planet. Space Sci. **82–83**, 34–42 (2013). 2013. doi:10.1016/j.pss.2013.03.010

S. Jacobsen, F.M. Neubauer, J. Saur, N. Schilling, Io's nonlinear MHD-wave field in the heterogeneous jovian magnetosphere. Geophys. Res. Lett. **34**, L10202 (2007). doi:10.1029/2006GL029187

X. Jia, R.J. Walker, M.G. Kivelson, K.K. Khurana, J.A. Linker, Dynamics of Ganymede's magnetopause: intermittent reconnection under steady external conditions. J. Geophys. Res. **115**, A12202 (2010). doi:10.1029/2010JA015771

S. Kasahara, E.A. Kronberg, T. Kimura, C. Tao, S.V. Badman, A. Masters, A. Retinò, N. Krupp, M. Fujimoto, Asymmetric distribution of reconnection jet fronts in the jovian nightside magnetosphere. J. Geophys. Res. **118**, 375–384 (2013). doi:10.1029/2012JA018130

K.K. Khurana, Influence of solar wind on Jupiter's magnetosphere deduced from currents in the equatorial plane. J. Geophys. Res. **106**(A11), 25999–26016 (2001). doi:10.1029/2000JA000352

M.G. Kivelson, K.K. Khurana, D.J. Stevenson, L. Bennett, S. Joy, C.T. Russell, R.J. Walker, C. Zimmer, C. Polanskey, Europa and Callisto: induced or intrinsic fields in a periodically varying plasma environment. J. Geophys. Res. **104**(A3), 4609–4625 (1999). doi:10.1029/1998JA900095

M.G. Kivelson, D.J. Southwood, Dynamical consequences of two modes of centrifugal instability in Jupiter's outer magnetosphere. J. Geophys. Res. **110**, A12209 (2005). doi:10.1029/2005JA011176

N. Krupp, A. Lagg, S. Livi, B. Wilken, J. Woch, E.C. Roelof, D.J. Williams, Global flows of energetic ions in Jupiter's equatorial plane: first-order approximation. J. Geophys. Res. **106**(A11), 26017–26032 (2001). doi:10.1029/2000JA900138

W.S. Kurth, D.A. Gurnett, Plasma waves in planetary magnetospheres. J. Geophys. Res. **96**(S01), 18977–18991 (1991). doi:10.1029/91JA01819

W.S. Kurth, A. Lecacheux, T.F. Averkamp, J.B. Groene, D.A. Gurnett, A saturnian longitude system based on a variable kilometric radiation period. Geophys. Res. Lett. **34**, L02201 (2007). doi:10.1029/2006GL028336

W.S. Kurth, T.F. Averkamp, D.A. Gurnett, J.B. Groene, A. Lecacheux, An update to a saturnian longitude system based on kilometric radio emissions. J. Geophys. Res. **113**, A05222 (2008). doi:10.1029/2007JA012861

W.S. Kurth, E.J. Bunce, J.T. Clarke, F.J. Crary, D.C. Grodent, A.P. Ingersoll, U.A. Dyudina, L. Lamy, D.G. Mitchell, A.M. Persoon, W.R. Pryor, J. Saur, T. Stallard, Auroral processes, in *Saturn from Cassini-Huygens*, vol. 12, ed. by M. Dougherty et al. (Springer, Dordrecht, 2009), pp. 333–374

L. Lamy, B. Cecconi, R. Prangé, P. Zarka, J.D. Nichols, J.T. Clarke, An auroral oval at the footprint of Saturn's kilometric radio sources, colocated with the UV aurorae. J. Geophys. Res. **114**, A10212 (2009). doi:10.1029/2009JA014401

L. Lamy et al., Earth-based detection of Uranus' aurorae. Geophys. Res. Lett. **39**, L07105 (2012). doi:10.1029/2012GL051312

G.R. Lewis, N. André, C.S. Arridge, A.J. Coates, L.K. Gilbert, D.R. Linder, A.M. Rymer, Derivation of density and temperature from the Cassini-Huygens CAPS electron spectrometer. Planet. Space Sci. **56**, 901–912 (2008). doi:10.1016/j.pss.2007.12.017

B.H. Mauk, S.M. Krimigis, M.H. Acuña, Neptune's inner magnetosphere and aurora: energetic particle constraints. J. Geophys. Res. **99**(A8), 14781–14788 (1994). doi:10.1029/94JA00735

B.H. Mauk, J.T. Clarke, D. Grodent, J.H. Waite Jr., C.P. Paranicas, D.J. Williams, Transient aurora on Jupiter from injections of magnetospheric electrons. Nature **415**, 1003–1005 (2002)

H. Melin, T. Stallard, S. Miller, L.M. Trafton, T. Encrenaz, T.R. Geballe, Seasonal variability in the ionosphere of Uranus. Astrophys. J. **729**, 134 (2011). doi:10.1088/0004-637X/729/2/134

H. Melin, T. Stallard, S. Miller, T.R. Geballe, L.R. Trafton, J. O'Donoghue, Post-equinoctial observations of the ionosphere of Uranus. Icarus **223**(2), 741–748 (2013). doi:10.1016/j.icarus.2013.01.012

J.D. Menietti, D.A. Gurnett, I. Christopher, Control of jovian radio emission by Callisto. Geophys. Res. Lett. **28**, 3047–3050 (2001). doi:10.1029/2001GL012965

C.J. Meredith, S.W.H. Cowley, K.C. Hansen, J.D. Nichols, T.K. Yeoman, Simultaneous conjugate observations of small-scale structures in Saturn's dayside ultraviolet auroras: implications for physical origins. J. Geophys. Res. **118**, 2244–2266 (2013). doi:10.1002/jgra.50270

S.E. Milan, M. Lester, S.W.H. Cowley, M. Brittnacher, Dayside convection and auroral morphology during an interval of northward interplanetary magnetic field. Ann. Geophys. **18**, 436–444 (2000)

D.G. Mitchell et al., Recurrent energization of plasma in the midnight-to-dawn quadrant of Saturn's magnetosphere, and its relationship to auroral UV and radio emissions. Planet. Space Sci. **57**, 1732–1742 (2009). doi:10.1016/j.pss.2009.04.002

J.D. Nichols, S.W.H. Cowley, Magnetosphere-ionosphere coupling currents in Jupiter's middle magnetosphere: effect of precipitation-induced enhancement of the ionospheric Pedersen conductivity. Ann. Geophys. **22**, 1799–1827 (2004). doi:10.5194/angeo-22-1799-2004

J.D. Nichols, J.T. Clarke, S.W.H. Cowley, J. Duval, A.J. Farmer, J.-C. Gérard, D. Grodent, S. Wannawichian, Oscillation of Saturn's southern auroral oval. J. Geophys. Res. **113**, A11205 (2008). doi:10.1029/2008JA013444

J.D. Nichols, J.T. Clarke, J.C. Gérard, D. Grodent, Observations of jovian polar auroral filaments. Geophys. Res. Lett. **36**, L08101 (2009a). doi:10.1029/2009GL037578

J.D. Nichols, S.V. Badman, E.J. Bunce, J.T. Clarke, S.W.H. Cowley, F.J. Crary, M.K. Dougherty, J.-C. Gérard, D. Grodent, K.C. Hansen, W.S. Kurth, D.G. Mitchell, W.R. Pryor, T.S. Stallard, D.L. Talboys, S. Wannawichian, Saturn's equinoctial auroras. Geophys. Res. Lett. **36**, L24102 (2009b). doi:10.1029/2009GL041491

J.D. Nichols, S.W.H. Cowley, L. Lamy, Dawn-dusk oscillation of Saturn's conjugate auroral ovals. Geophys. Res. Lett. **37**, L24102 (2010a). doi:10.1029/2010GL045818

J.D. Nichols, B. Cecconi, J.T. Clarke, S.W.H. Cowley, J.-C. Gérard, A. Grocott, D. Grodent, L. Lamy, P. Zarka, Variation of Saturn's UV aurora with SKR phase. Geophys. Res. Lett. **37**, L15102 (2010b). doi:10.1029/2010GL044057

T. Ogino, R.J. Walker, M.G. Kivelson, A global magnetohydrodynamic simulation of the jovian magnetosphere. J. Geophys. Res. **103**(A1), 225–235 (1998). doi:10.1029/97JA02247

L. Pallier, R. Prangé, More about the structure of the high latitude jovian aurorae. Planet. Space Sci. **49**, 1159–1173 (2001)

C. Paranicas, A.F. Cheng, Drift shells and aurora computed using the O8 magnetic field model for Neptune. J. Geophys. Res. **99**(A10), 19433–19440 (1994). doi:10.1029/94JA01573

W.R. Pryor et al., The auroral footprint of Enceladus on Saturn. Nature **472**, 331–333 (2011). doi:10.1038/nature09928

A. Radioti, J.-C. Gérard, D. Grodent, B. Bonfond, N. Krupp, J. Woch, Discontinuity in Jupiter's main auroral oval. J. Geophys. Res. **113**, A01215 (2008). doi:10.1029/2007JA012610

A. Radioti, A.T. Tomás, D. Grodent, J.-C. Gérard, J. Gustin, B. Bonfond, N. Krupp, J. Woch, J.D. Menietti, Equatorward diffuse auroral emissions at Jupiter: simultaneous HST and Galileo observations. Geophys. Res. Lett. **36**, L07101 (2009a). doi:10.1029/2009GL037857

A. Radioti, D. Grodent, J.-C. Gérard, E. Roussos, C. Paranicas, B. Bonfond, D.G. Mitchell, N. Krupp, S. Krimigis, J.T. Clarke, Transient auroral features at Saturn: signatures of energetic particle injections in the magnetosphere. J. Geophys. Res. **114**, A03210 (2009b). doi:10.1029/2008JA013632

A. Radioti, D. Grodent, J.-C. Gérard, M.F. Vogt, M. Lystrup, B. Bonfond, Nightside reconnection at Jupiter: auroral and magnetic field observations from 26 July 1998. J. Geophys. Res. **116**, A03221 (2011a). doi:10.1029/2010JA016200

A. Radioti, D. Grodent, J.-C. Gérard, S.E. Milan, B. Bonfond, J. Gustin, W. Pryor, Bifurcations of the main auroral ring at Saturn: ionospheric signatures of consecutive reconnection events at the magnetopause. J. Geophys. Res. **116**, A11209 (2011b). doi:10.1029/2011JA016661

A. Radioti, M. Lystrup, B. Bonfond, D. Grodent, J.-C. Gérard, Jupiter's aurora in ultraviolet and infrared: simultaneous observations with the Hubble space telescope and the NASA infrared telescope facility. J. Geophys. Res. **118**, 2286–2295 (2013a). doi:10.1002/jgra.50245

A. Radioti, E. Roussos, D. Grodent, J.-C. Gérard, N. Krupp, D.G. Mitchell, J. Gustin, B. Bonfond, W. Pryor, Signatures of magnetospheric injections in Saturn's aurora. J. Geophys. Res. **118**, 1922–1933 (2013b). doi:10.1002/jgra.50161

A. Radioti, D. Grodent, J.-C. Gérard, B. Bonfond, J. Gustin, W. Pryor, J.M. Jasinski, C.S. Arridge, Auroral signatures of multiple magnetopause reconnection at Saturn. Geophys. Res. Lett. **40**, 4498–4502 (2013c). doi:10.1002/grl.50889

B.R. Sandel, F. Herbert, A.J. Dessler, T.W. Hill, Aurora and airglow on the night side of Neptune. Geophys. Res. Lett. **17**, 1693 (1990). doi:10.1029/GL017i010p01693

J. Saur, D.F. Strobel, F.M. Neubauer, M.E. Summers, The ion mass loading rate at Io. Icarus **163**, 456–468 (2003). doi:10.1016/S0019-1035(03)00085-X

P. Schippers et al., Multi-instrument analysis of electron populations in Saturn's magnetosphere. J. Geophys. Res. **113**, A07208 (2008). doi:10.1029/2008JA013098

P. Schippers, N. André, D.A. Gurnett, G.R. Lewis, A.M. Persoon, A.J. Coates, Identification of electron field-aligned current systems in Saturn's magnetosphere. J. Geophys. Res. **117**, A05204 (2012). doi:10.1029/2011JA017352

E.C. Sittler Jr., K.W. Ogilvie, R. Selesnick, Survey of electrons in the uranian magnetosphere: Voyager 2 observations. J. Geophys. Res. **92**, 15263–15281 (1987). doi:10.1029/JA092iA13p15263

D.J. Southwood, M.G. Kivelson, A new perspective concerning the influence of the solar wind on the jovian magnetosphere. J. Geophys. Res. **106**(A4), 6123–6130 (2001). doi:10.1029/2000JA000236

D.J. Southwood, M.G. Kivelson, Saturnian magnetospheric dynamics: elucidation of a camshaft model. J. Geophys. Res. **112**, A12222 (2007). doi:10.1029/2007JA012254

T.S. Stallard, S. Miller, S.W.H. Cowley, E.J. Bunce, Jupiter's polar ionospheric flows: measured intensity and velocity variations poleward of the main auroral oval. Geophys. Res. Lett. **30**(5), 1221 (2003). doi:10.1029/2002GL016031

S. Stanley, J. Bloxham, Numerical dynamo models of Uranus and Neptune's magnetic fields. Icarus **184**, 556–572 (2006). doi:10.1016/j.icarus.2006.05.005

D.L. Talboys, E.J. Bunce, S.W.H. Cowley, C.S. Arridge, A.J. Coates, M.K. Dougherty, Statistical characteristics of field-aligned currents in Saturn's nightside magnetosphere. J. Geophys. Res. **116**, A04213 (2011). doi:10.1029/2010JA016102

A.T. Tomás, J. Woch, N. Krupp, A. Lagg, K.-H. Glassmeier, W.S. Kurth, Energetic electrons in the inner part of the jovian magnetosphere and their relation to auroral emissions. J. Geophys. Res. **109**, A06203 (2004). doi:10.1029/2004JA010405

A.R. Vasavada, A.H. Bouchez, A.P. Ingersoll, B. Little, C.D. Anger (the Galileo SSI Team), Jupiter's visible aurora and Io footprint. J. Geophys. Res. **104**, 27133–27142 (1999)

V.M. Vasyliūnas, Plasma distribution and flow, in *Physics of the Jovian Magnetosphere*, ed. by A.J. Dessler (Cambridge Univ. Press, New York, 1983), pp. 395–453. doi:10.1017/CBO9780511564574.013

M.F. Vogt, M.G. Kivelson, K.K. Khurana, R.J. Walker, B. Bonfond, D. Grodent, A. Radioti, Improved mapping of Jupiter's auroral features to magnetospheric sources. J. Geophys. Res. **116**, A03220 (2011). doi:10.1029/2010JA016148

M.F. Vogt, C.M. Jackman, J.A. Slavin, E.J. Bunce, S.W.H. Cowley, M.G. Kivelson, K.K. Khurana, Structure and statistical properties of plasmoids in Jupiter's magnetotail. J. Geophys. Res. **119**, 821–843 (2014). doi:10.1002/2013JA019393

H. Waite et al., An auroral flare at Jupiter. Nature **410**(6830), 787–789 (2001)

F. Xiao, R.M. Thorne, D.A. Gurnett, D.J. Williams, Whistler-mode excitation and electron scattering during an interchange event near Io. Geophys. Res. Lett. **30**(14), 1749 (2003). doi:10.1029/2003GL017123

🖄 Springer

DOI 10.1007/978-1-4939-3395-2_4
Reprinted from *Space Science Reviews* Journal, DOI 10.1007/s11214-014-0075-1

Solar Wind and Internally Driven Dynamics: Influences on Magnetodiscs and Auroral Responses

**P.A. Delamere · F. Bagenal · C. Paranicas · A. Masters ·
A. Radioti · B. Bonfond · L. Ray · X. Jia · J. Nichols ·
C. Arridge**

Received: 10 February 2014 / Accepted: 6 July 2014 / Published online: 26 August 2014
© Springer Science+Business Media Dordrecht 2014

Abstract The dynamics of the giant planet magnetodiscs are strongly influenced by planetary rotation. Yet the solar wind must ultimately remove plasma from these rapidly rotating magnetodiscs at the same rate that plasma is transported radially outward from the source regions: the Io and Enceladus plasma tori. It is not clear how the solar wind influences magnetospheric dynamics when the dynamics are dominated by rotation. However, auroral observations provide important clues. We review magnetodisc sources and radial transport and the solar wind interaction with the giant magnetospheres of Jupiter and Saturn in an attempt to connect auroral features with specific drivers. We provide a discussion of auroral signatures that are related to the solar wind interaction and summarize with a discussion of global magnetospheric dynamics as illustrated by global MHD simulations. Many questions remain and it is the intent of this review to highlight several of the most compelling questions for future research.

Keywords Neutral clouds · Plasma torus · Momentum loading · Dynamics · Aurora ·
Magnetodiscs · Reconnection · Solar wind

1 Introduction

The dynamics of giant planet magnetospheres are largely dominated by planetary rotation. The centrifugal confinement of plasma originating from Io and Enceladus, located deep in the inner magnetosphere, is fundamental to the formation of these structures. The equatorially-confined plasma carries azimuthal currents that distort the magnetic field into a thin disc-like structure called the magnetodisc (see discussion by M. Kivelson, this issue). The outward plasma transport mechanism in these rapidly-rotating magnetospheres is

P.A. Delamere (✉) · F. Bagenal · C. Paranicas · A. Masters · A. Radioti · B. Bonfond · L. Ray · X. Jia ·
J. Nichols · C. Arridge
Geophysical Institute & Physics Department, University of Alaska Fairbanks, 903 Koyukuk Drive,
PO Box 757320, Fairbanks, AK 99775, USA
e-mail: Peter.Delamere@gi.alaska.edu

thought to be the centrifugally-driven flux tube interchange instability (akin to the gravitational Rayleigh-Taylor instability). Eventually, the plasma is lost to the magnetotail, modulated by the variable upstream solar wind environment (Arridge et al. 2011; Jackman and Arridge 2011). Reconnection likely plays a key role in generating disconnected plasma blobs (plasmoids) that can be carried away by the solar wind. The Vasyliunas cycle is based on centrifugal stretching of the magnetodisc in the tail region, allowing the current sheet to thin and reconnection to operate (Vasyliunas 1983). Dayside reconnection can drive a Dungey cycle of reconnection that on long time scales requires tail reconnection to conserve flux (Dungey 1961). This solar wind-driven reconnection could be important for magnetodisc dynamics. Finally, the tangential stresses generated at the magnetopause boundary by the solar wind's viscous interaction can also influence dynamics (Axford and Hines 1961; Delamere and Bagenal 2010). All of these processes lead to mass loss from the magnetodisc.

Mass transport is a two-way process in planetary magnetospheres and cannot be considered independent of magnetic flux or internal energy transport. While net mass flux can point into or out of the magnetosphere, magnetic flux must be balanced on long time scales. At Jupiter and Saturn, roughly 500 kg/s and 50 kg/s are lost to the solar wind, respectively (Delamere and Bagenal 2013). During radial transport, plasma is heating non-adiabatically, requiring an input energy of 3–16 TW at Jupiter and 75–630 GW at Saturn (Bagenal and Delamere 2011). The detailed physical transport mechanisms and associated heating are poorly understood. However, ultimately the transport physics must take into consideration the interaction of the giant magnetodiscs with the solar wind.

The purpose of this article is to review the internally-driven dynamics and the solar wind interaction. We review: mass and energy transport in planetary magnetodiscs (Sect. 2), magnetopause boundary processes that facilitate the solar wind interaction (Sect. 3), and observational clues found in the auroral emissions (Sect. 4). Finally, we conclude with a discussion of what global simulations reveal about the solar wind interaction (Sect. 5) and how magnetodisc plasma is lost to the solar wind.

2 Mass and Energy Flow in Planetary Magnetodiscs

2.1 Sources

Quantifying the transport of mass and energy through planetary magnetodiscs is crucial for untangling internal and external (i.e. solar wind) drivers of magnetospheric dynamics. For example, the radial mass transport rate, $\dot{M}$, together with ionospheric conductivity characterizes the magnetosphere-ionosphere coupling that can generate internally-driven aurora (e.g. Hill 1979; Nichols and Cowley 2004; Ray et al. 2010). Non-adiabatic heating of the plasma during radial transport affects the plasma-β at the magnetopause boundary that, in part, determines the nature of the solar wind interaction (e.g. Masters et al. 2012a). The equilibrium scale (i.e. subsolar distance of the magnetopause boundary) of the magnetosphere is dependent on the momentum transfer rate from the solar wind, requiring that plasma mass loss to the solar wind balances the internal plasma source rate (Delamere and Bagenal 2013). In this section we discuss the flow of mass and energy in the context of magnetosphere-ionosphere coupling and the superthermal plasma populations of planetary magnetodiscs.

Located deep in the inner magnetospheres of Jupiter and Saturn, Io and Enceladus are the respective sources of neutral gas that supply plasma to these planetary magnetodiscs. The fate of these neutral gases is very different. At Jupiter nearly all of the iogenic neutral gas is ionized and either removed from the system as fast escaping neutrals as a result of charge exchange or transported radially outward as plasma. Of the initial $\sim$ tonne per second (1000 kg/s) of neutral gas generated by Io, only 500 kg/s survives as plasma for the magnetodisc. At Saturn, the low thermal electron temperatures near Enceladus result in a neutral dominated gas in Saturn's inner magnetospheres. The expansion and redistribution of this neutral torus eventually supplies Saturn's magnetodisc at larger radial distances, where the electrons are hot enough to ionize. Ultimately, only $\sim$50 kg/s of the initial $\sim$200 kg/s of water vapor spewing from the Encledus geysers is transport radially as plasma through the magnetodisc. A complete discussion of the satellite neutral clouds and evolution of the plasma tori is given by Achilleos et al. (2014, this issue).

2.2 Plasma Transport and Magnetosphere-Ionosphere Coupling

At both Jupiter and Saturn, radial transport of plasma is driven primarily by centrifugal forces. In this process, which is similar to the gravitationally driven Rayleigh-Taylor instability, rapidly rotating cold, dense plasma experiences an outward centrifugal force. As the flux tubes containing cold, dense plasma move radially outward, flux tubes with hot, tenuous plasma are transported towards the planet to conserve magnetic flux, while reducing the centrifugal potential energy of the system (see review by Thomas et al. 2004). Radial transport out of the Io torus occurs very slowly, with plasma residing inside of 10 R_J for $\sim$23–50 days (Delamere and Bagenal 2003).

As plasma is radially transported outwards, conservation of angular momentum dictates that it slows in its angular motion. However, the planetary magnetic field is frozen-in to the magnetospheric plasma. Therefore, the slowing magnetospheric plasma exerts a stress on the magnetic field, bending it backwards. Simultaneously, field-aligned currents develop, supporting the magnetic field geometry and transferring angular momentum from the planetary atmosphere to the magnetospheric plasma. These currents flow radially through the magnetospheric equator, and the ensuing $\mathbf{J} \times \mathbf{B}$ forces accelerates the magnetospheric plasma towards corotation. The current circuit closes in the planetary ionosphere, with the latitudinal current directed equatorwards, thus imposing a $\mathbf{J} \times \mathbf{B}$ force in the ionosphere opposite to that in the magnetosphere. At Jupiter, the main auroral emission, observed in both IR and UV wavelengths, is a signature of this current circuit, as the upward field-aligned currents that feed the magnetospheric plasma are associated with planetward moving electrons that excite the planetary atmosphere. Low latitude aurora have also been identified on Saturn in the infrared and ultraviolet domain, which could correspond to the main emissions on Jupiter (Stallard et al. 2008; Grodent et al. 2010). While omnipresent, Jupiter's main auroral is more dynamic than originally thought; However, the constant production—and subsequent transport—of Iogenic plasma lends itself to an exploration of the steady state system, which can then be expanded to understand how the solar wind influences the main auroral emission.

Hill (1979) first explored the above scenario, predicting the radial profile of the magnetospheric plasma's angular velocity through balancing the ionospheric and magnetosphere torques. His analytic description assumed a dipole planetary magnetic field and constant ionospheric Pedersen conductance, Σ_P. He found that the departure from corotation of the magnetospheric plasma angular velocity could be characterized by a value,

$$L = \left(\frac{\pi \Sigma_P R_p^2 B_P^2}{\dot{M}} \right)^{1/4} \tag{1}$$

where R_P is the planetary radius, B_P is the magnetic field strength at the ionosphere, and $\dot{M}$ is the radial mass transport rate in kg/s. For radial distance larger the L, the lag from corotation of the plasma would be significant. It is clear to see from Eq. (1) that the location of the departure from corotation depends on the ratio of the ionospheric Pedersen conductance to the radial mass transport rate.

Since the seminal analysis of Hill (1979), there have been many studies expanding the simplifying assumptions of the original analysis (e.g. Huang and Hill 1989; Pontius 1997; Cowley and Bunce 2001; Nichols and Cowley 2004; Smith and Aylward 2009; Ray et al. 2010; Yates et al. 2012). The giant planet systems are more complicated than a dipole magnetic field and a planetary atmosphere with constant Pedersen conductance. In steady state the main factors that influence the ability of the planetary atmosphere to communicate angular momentum to its surrounding magnetospheric plasma, and hence field-aligned current strengths and subsequent auroral emission location and brightness, are (1) the rotation rate of the planetary thermosphere; (2) the modification of the ionospheric Pedersen conductance with auroral electron precipitation; (3) the magnitude of the north-south component of the planetary magnetic field; and (4) the latitudinal distribution of current carriers along the magnetic field and possible accompanying potential drops.

Ultimately, in the case of Jupiter, the angular momentum that is transferred out to the magnetospheric plasma comes from the planet's deep interior. Angular momentum is transported from the deep interior to the thermosphere via advection and eddy diffusion (Smith and Aylward 2008; Huang and Hill 1989). Thermospheric flows are then coupled to the ionosphere through ion-neutral collisions. If the neutral thermosphere is significantly subcorotating, then that will affect the maximum planetary angular velocity that is communicated to the magnetosphere. The Pedersen conductance, in part, controls the magnitude of the ionospheric currents, and therefore the field-aligned currents that transport angular momentum outwards to the magnetospheric plasma. Electron precipitation into the planetary atmosphere will enhance the Pedersen conductance, acting as a positive feedback mechanism, which encourages the transfer of angular momentum (Nichols and Cowley 2004; Ray et al. 2010). In both the jovian and saturnian magnetospheres, the planetary magnetic field is distended by a current sheet that results in a decreased field strength from that of a dipole. The departure of the field from a dipole configuration depends on the plasma pressure, and will therefore change with local plasma parameters (Caudal 1986; Achilleos et al. 2010b, 2010a; Nichols 2011). Diminishing magnitude of the north-south component of the planetary magnetic field will hinder the delivery of angular momentum to the magnetospheric plasma, affecting the $\mathbf{J} \times \mathbf{B}$ force on the plasma (Pontius 1997).

Finally, both Jupiter and Saturn are rapid rotators, therefore their magnetospheric plasma experiences strong centrifugal forces. Cold, heavy ions will be confined to the equatorial plane whilst lighter ions, such as protons, and electrons will have more mobility along the magnetic field. The ensuing charge separation results in an ambipolar electric field that acts to restrict electrons in their planetward motion and pull positively charge particles up off the equatorial plane. Ionospheric plasma is restricted to high latitudes because of gravitational forces. Therefore, at high latitudes, there is a minimum in the plasma density along the magnetic flux tube (Su et al. 2003; Ray et al. 2009). Should the magnetospheric plasma's demand for angular momentum, and hence current, from the planet exceed the thermal electron current density, field-aligned potentials will develop at the location of the plasma density minimum. Changes in the magnitude of the field-aligned potentials with latitude, or equivalently equatorial radius, will modify the electric field mapping between the ionosphere and the magnetosphere, allowing for rotational decoupling between the two regions. Another effect of the field-aligned potentials is to spread the transfer of angular momentum over a broader region in the magnetospheric equator (Ray et al. 2010). Note that this process does not necessarily broaden the width of the auroral emission at the planet because of the distension of the planetary magnetic field due to the current sheet. Jupiter's sulfur dioxide-based chemistry, and hence lower abundance of protons relative to the Saturn's water group-based system, means that this effect is stronger in the jovian system.

At higher energies (i.e. superthermal particles) the nature of the sources of particles and their transport can be quite different from the picture we just put forward. For instance, energetic protons above a certain energy are believed to be created via the Cosmic ray albedo neutron decay (Crand) mechanism (e.g., Cooper and Simpson 1980). These particles are the decay products of neutrons created when cosmic rays have access to Saturn and its rings. Recently, Roussos et al. (2011) linked energetic proton fluxes to the phase of the solar cycle to provide more evidence for the Crand mechanism.

At higher energies, there is evidence of both the local interchange, described above for thermal plasma, and the injection of global distributions of hot ions. Mitchell (2015) described the latter type as signatures of current sheet collapse. These processes could in fact be larger scale manifestations of interchange. But at least from the point of view of in situ data, there are qualitative differences between larger scale injections, that can be imaged in ENAs, and flux tube interchange, that occurs closer to the planet.

It is also not well understood how flux tube interchange and/or current sheet collapse relate to transport on smaller spatial scale, e.g., micro-diffusion. Microdiffusion is believed to be important, for instance, in explaining the rate of fill in of satellite micro signatures. These depletions created by satellite absorptions fill in even when injections are not present. It is important to keep in mind that there are multiple modes of transport in these systems.

2.3 Suprathermal Particles in Planetary Magnetodiscs

The solar wind flows approximately radially outward from the Sun. The ion plasma contains light and heavy ions in various charge states, including both pick-up Hydrogen (H^+) and Helium (i.e., He^+) and solar Hydrogen (H^+) and He^{++} (alpha particles). The pickup ions are derived from interstellar neutrals. Helium ions found in planetary magnetospheres provide a good tracer of the access of solar wind. For instance, Haggerty et al. (2009) have examined the relative proportion of the main energetic ions as a function of distance down Jupiter's magnetotail. They show that closer to the planet, the charged particles become more dominated by Iogenic composition, but also reveal the deep penetration of solar wind Helium.

The velocity distribution of the quiescent solar wind electrons can be modeled using a Maxwellian for the cold plasma core, a kappa distribution for the halo (Vasyliunas 1968), and a power law for the energetic tail (see Fig. 1). The velocity distribution of solar wind H^+ measured in the solar wind at 5.2 AU (appropriate for Jupiter) also includes an interstellar pickup proton population (Fig. 2).

The kappa distribution is used throughout the planetary magnetospheres and the heliosphere to describe suprathermal particles (e.g., Dialynas et al. 2009; Decker et al. 2005). It is like a Maxwellian, but at small kappa parameters, it deviates in important ways (see Fig. 3). The suprathermals can often play a leading role in the total particle pressure in planetary magnetospheres (Sergis et al. 2010), and not surprisingly because it is a pressure-driven current, the planetary ring current (Krimigis et al. 2007). Particles that generate auroral signatures are typically in the suprathermal energy range.

Kappa distributions are observed regularly in magnetospheric contexts but the source of these distributions is not well understood. In their very useful paper, Pierrard and Lazar (2010) outline some of the theory behind the distribution and its applications in the solar system. They point out, for instance, that, "an isotropic kappa distribution (instead of a Maxwellian) in a planetary (context) leads to a number density n decreasing as a power law

Fig. 1 Electron distribution of the quiet time solar wind from Wang et al. (2012)

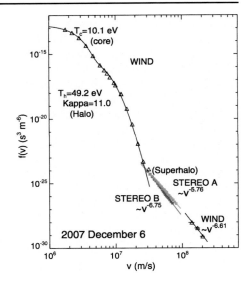

Fig. 2 Proton velocity distribution (*plot symbols*) measured in the quiet solar wind by Ulysses at an average helioradius of 5.2 AU. In addition to the core, halo and power-law tail components, the proton distribution includes pickup protons whose distribution is flat from low velocities up to about twice the solar wind speed (the ordinate has dimensions $s^3 \, km^{-6}$) (Mason and Gloeckler 2012)

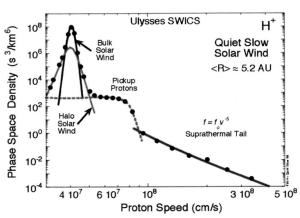

(instead of exponentially) with the radial distance r and a temperature T increasing with radial distance (instead of being constant)...."

In addition to how the distribution may evolve as a function of radial distance from the planet, it is also worth mentioning the important role of neutrals on suprathermal populations. Neutrals from Enceladus populate a wide region of the inner and middle magnetosphere of Saturn. These neutrals can cause suprathermal singly-charged ions to be lost from the magnetosphere via charge exchange. In this process, the more energetic ion leaves the system as an energetic neutral atom, leaving behind a cold ion. In fact, data reveal that in Saturn's inner magnetosphere, very low fluxes of trapped ions are present, except right at the corotation speed (Young et al. 2005, their Fig. 1). Corotation speed ions also undergo charge-exchange with ambient neutrals but this only replaces one thermal ion with another. Paranicas et al. (2008) also found that energetic ions can be largely absent in Saturn's inner magnetosphere due to charge exchange. They show that often the energetic ions that are detected are due to fresh injections.

Springer

Fig. 3 From the work of Pierrard and Lazar (2010) showing examples of the kappa distribution for different kappa parameters. This figure illustrates that as kappa goes to infinity, the distribution is a Maxwellian

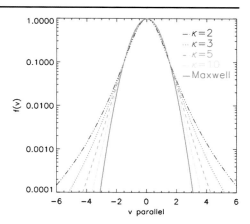

The situation for suprathermal electrons at Saturn is somewhat different. These particles can be energized by heating from the cold plasma range (for instance through Coulomb collisions with the corotating ions) and by invariant conservation as part of the inward motion that takes place during injections/interchange. Suprathermal electrons can lose energy in interactions with the neutral gas. They scatter off of all the charges in the gas/dust medium and lose energy by, for instance, providing excitations to the atoms and molecules in the medium. They are also efficient at ionizing and lose energy this way.

To summarize, the role of suprathermals in the inner magnetospheres of Jupiter and Saturn is probably vastly different. At Jupiter, the distribution function of charged particles is probably very robust between the cold plasma and the energetic particle range outward of the confined neutral gas distributions. However, at Saturn, there are important losses of ions in the suprathermal energy range and above to charge exchange with the expanded neutral gas distribution (Cassidy and Johnson 2010; Fleshman et al. 2012). Furthermore, even suprathermal and higher energy electrons at Saturn can be cooled by the presence of the neutral gas. Sergis et al. (2010), for instance, showed that the corotating cold plasma is very critical in the inner magnetosphere of Saturn and more energetic particles start to play a role in the pressure at larger distances.

3 Magnetopause Boundary Processes

The question of exactly how the solar wind interacts with the giant planet magnetospheres has been a major subject of debate. Although we have a significant amount still to learn about this topic, recent studies have caused our understanding to evolve.

At the magnetopause boundary of a planetary magnetosphere there is a direct interaction between the shock-processed solar wind plasma and the planetary magnetized plasma environment. Understanding the solar wind-magnetosphere interaction at any planet must be based on a clear understanding of the physical processes operating at the magnetopause. These processes lead to mass, momentum, and energy transport into the magnetosphere.

In this section we discuss the processes operating at the magnetopause boundary of both Jupiter's and Saturn's magnetosphere. As we will see, current understanding suggests that the nature of these processes varies between planets, making the assumption of an Earth-like solar wind interaction very likely invalid.

3.1 Magnetic Reconnection

One of the major magnetopause processes responsible for energy transport is magnetic reconnection (Dungey 1961; Vasyliunas 1975; Russell 1975) (see the review by Paschmann 2008 and references therein). This is a fundamental process where plasma and coupled magnetic field lines flow toward a reconnection 'x-line', where they become decoupled and are essentially 'cut' and 'reconnected', before flowing away from the x-line and recoupling with the plasma. This process changes the topology of the magnetic field and releases magnetic energy. In the case of Earth's magnetosphere, reconnection at the magnetopause is the principal driver of energy flow through the magnetosphere.

The physics of the reconnection process remains the subject of much research. It is well-established that the reconnection rate decreases as the ratio of plasma to magnetic pressure (the plasma β) adjacent to the current sheet increases (e.g. Sonnerup and Ledley 1974; Anderson et al. 1997); however, recent work suggests that certain plasma conditions can suppress the onset of reconnection itself. An increasing difference in plasma β across a current sheet has been shown to suppress reconnection onset for an increasing range of magnetic shears across the boundary (Phan et al. 2010, 2013). This is due to the potential for particle drifts within the current sheet to prevent the flow pattern required for reconnection from being established (Swisdak et al. 2003, 2010). This effect is hereafter referred to as diamagnetic suppression. In addition, it has been suggested that a super-Alfvénic flow shear in the direction of the reconnecting field can also suppress reconnection, similarly related to disruption of reconnection-imposed flows (e.g. Cassak and Otto 2011). This effect is hereafter referred to as flow shear suppression.

We know the Mach numbers of planetary bow shocks increase with distance from the Sun (e.g. Russell et al. 1982), which produce different solar wind plasma conditions adjacent to each planetary magnetopause. We also know that there is non-negligible plasma pressure in the outer magnetospheres of Jupiter and Saturn, combined with significant flows in the sense of planetary rotation. When considering the nature of magnetic reconnection at both Jupiter's and Saturn's magnetopause, an assessment of the effect of these differing conditions is essential.

Spacecraft observations at Jupiter's magnetopause have revealed evidence for the occurrence of reconnection (Walker and Russell 1985; Huddleston et al. 1997). Identification of encounters with the reconnection-related phenomenon of Flux Transfer Events (FTEs) (Russell and Elphic 1978; Walker and Russell 1985) is a particularly clear indication that the process does operate. However, spacecraft observations at Jupiter's magnetopause remain limited, meaning that we cannot yet build a statistical picture of where, and under what conditions, reconnection occurs.

McComas and Bagenal (2007) argued that the long time scale associated with an open field line being dragged to the magnetotail by the solar wind compared to the time scale of Interplanetary Magnetic Field (IMF) sector changes means that such open field lines can be closed through further magnetopause reconnection. The proposed ability of this scenario to keep the amount of open flux in the system small was challenged by Cowley et al. (2008), primarily on the basis of a comparison of expected rates of magnetic flux opening and closure due to magnetopause reconnection at different locations. This concept remains the subject of debate (McComas and Bagenal 2008).

Most recently, Desroche et al. (2012) used idealized models of conditions at Jupiter's magnetopause to draw conclusions about the solar wind-magnetosphere interaction. Included in their investigation was an assessment of how favorable conditions at Jupiter's magnetopause are for reconnection onset (see Fig. 4). They found that the large flow shears

Fig. 4 Assessment of how favorable conditions at Jupiter's magnetopause are for magnetic reconnection onset. In all panels the magnetopause surface is viewed from along the solar wind flow direction. The panel surrounded by a *blue rectangle* shows an assessment of flow shear suppression, and the panels surrounded by a *red rectangle* show assessments of diamagnetic suppression (for different values of the plasma β in the magnetosphere, β_{MSP}). Regions of each surface *shaded in red* are regions where reconnection onset is possible. Adapted from Desroche et al. (2012)

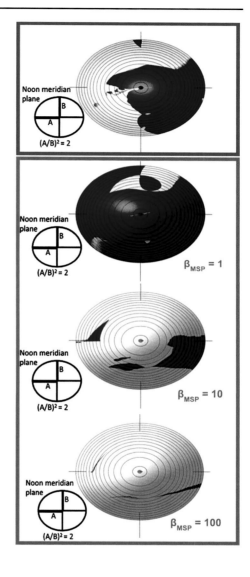

across the dawn flank magnetopause generally prohibit reconnection due to flow shear suppression. Furthermore, by considering different values of the (poorly constrained) plasma β in Jupiter's near-magnetopause magnetosphere they showed that diamagnetic suppression may also be severe.

In the case of Saturn's magnetosphere, in situ evidence for magnetopause reconnection has also been reported (Huddleston et al. 1997; McAndrews et al. 2008; Lai et al. 2012; Fuselier et al. 2014), and it has been suggested that some dayside auroral features are caused by bursts of magnetopause reconnection (Radioti et al. 2011a; Badman et al. 2012a, 2013). However, unlike Jupiter, no examples of FTEs have been identified to date (Lai et al. 2012), and neither Saturn's auroral power nor the thickness on the magnetospheric boundary layer adjacent to the magnetopause show a clear response to the orientation of the IMF (unlike the case of Earth's magnetopause reconnection-driven magnetosphere) (Crary et al. 2005; Clarke et al. 2009; Masters et al. 2011).

Indirect evidence for dayside magnetopause reconnection at Saturn has been provided by in situ observations by Cassini at high latitudes in the magnetosphere. The in situ data shows the presence of magnetosheath plasma at the expected location of the cusp deep within the magnetosphere and with dispersions characteristic of magnetopause reconnection. Jasinski et al. (2014) have used ion energy-pitch angle dispersions to argue for a magnetopause reconnection interpretation using data from the northern hemisphere. Furthermore, they observed stepped time-energy ion dispersions that argue for pulsed reconnection, as suggested by the rapid variations in clock angle inside CIRs (Jackman et al. 2005).

The southern polar cusp was studied using in situ data from Cassini in January and February 2007 (Arridge et al. 2014). In these events energy-pitch angle dispersions were also suggested for magnetopause reconnection. These case studies demonstrated injections in the cusp under a range of solar wind dynamic pressure conditions thus showing that the solar wind interaction has an important component under more rarefield solar wind conditions. The in situ data also contains evidence for a boundary layer between the open field lines in the cusp and the field-aligned currents driving the main auroral emission thus suggesting that the main auroral emission and the OCB are not exactly co-located. Evidence was also presented that the cusp oscillated in position in phase with Saturn's global magnetospheric periodicities.

Note that although spacecraft observations in the vicinity of Saturn's magnetopause are currently more extensive than those made at Jupiter's magnetopause, the Saturn data sets also suffer from limited instrument fields-of-view. This almost always prohibits the assessment of whether or not the clearest evidence for magnetopause reconnection in the form of reconnection outflows (jets) were present at a spacecraft magnetopause encounter. As a result, a comprehensive statistical picture of where and why magnetopause reconnection occurs is also not available for Saturn's magnetopause.

However, Masters et al. (2012a) assessed the diamagnetic suppression condition at Saturn's magnetopause using spacecraft observations of local plasma β conditions at the boundary. They found that diamagnetic suppression is likely to be severe, primarily due to high-β conditions in the solar wind adjacent to Saturn's magnetopause (see Fig. 5). This high-β environment is consistent with the high-Mach number of Saturn's bow shock. It is thus expected that reconnection at Saturn's magnetopause is generally limited to regions of the magnetopause surface where the fields on either side of the boundary become almost antiparallel (note that this is presently consistent with observed reconnection signatures, e.g. Fuselier et al. 2014). More recently, Masters et al. (2014) reported the results of a study that was similarly based on measured near-magnetopause conditions. These authors inferred reconnection electric field strengths (assuming the satisfaction of onset conditions), and suggested that the resulting reconnection voltages applied to Saturn's magnetosphere are generally not high enough to represent a major driver of system dynamics.

As we have discussed, near-magnetopause plasma conditions at both Jupiter and Saturn differ from those at Earth's magnetopause. Faster flows in the magnetosphere can suppress reconnection, and higher plasma β on either side of the boundary can also limit the onset of the reconnection process. Overall, current understanding suggests that the coupling between the solar wind and both Jupiter's and Saturn's magnetospheres via magnetopause reconnection is weaker than that at Earth. How Jupiter and Saturn compare with each other remains unclear.

🕮 Springer

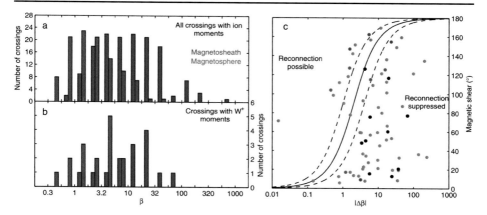

Fig. 5 Measured plasma β conditions at Saturn's magnetopause and an assessment of the importance of diamagnetic suppression of magnetopause reconnection. Taken from Masters et al. (2012a)

3.2 Shear-Flow Driven Instabilities

Energy transport across the magnetopause can also be achieved through the growth of instabilities that are driven by the flow shears often present at the boundary. The most well studied of these is the Kelvin-Helmholtz (K-H) instability (e.g. Dungey 1955; Chandrasekhar 1961). This is a fundamental instability that can grow in a wide range of fluid environments. For sufficiently high flow shear across a boundary between two fluids, like a planetary magnetopause, the boundary can become K-H unstable. A seed perturbation of a K-H unstable interface will grow with time, rather than be suppressed, leading to surface waves in the linear phase of the instability, and complex boundary vortices in the subsequent nonlinear phase. There is substantial evidence for the development of K-H-driven waves and vortices on Earth's magnetopause (e.g. Hasegawa et al. 2004).

In this planetary magnetopause context, the role of K-H instability in promoting the growth of other instabilities has received much research attention, in particular magnetic reconnection (Nykyri and Otto 2001; Nakamura and Fujimoto 2005; Nakamura et al. 2006). Reconnection within rolled-up K-H vortices has been observed at Earth's magnetopause (Nykyri et al. 2006; Hasegawa et al. 2009), and conditions in these complex structures is expected to enhance plasma diffusion across the interface (e.g. Cowee et al. 2010). In case of the terrestrial magnetosphere, solar wind energy transport into the magnetosphere under northward IMF conditions is thought to occur primarily through the flanks of the magnetopause as a result of the growth of the K-H instability (e.g. Hasegawa et al. 2004).

The different plasma conditions at the magnetopause of both Jupiter and Saturn compared to at Earth's magnetopause are expected to also affect shear flow-driven instabilities like the K-H instability. Although the magnetosheath flow is similar, magnetospheric flows just inside each giant planet magnetopause are appreciable, and directed from dawn to dusk, having a strong effect on the flow shear. In addition, different plasma densities and magnetic field strengths and orientations can affect the K-H stability of these boundaries (e.g. Chandrasekhar 1961; Southwood 1968). It is reasonable to expect that the dawn flank of both Jupiter's and Saturn's magnetopause is K-H unstable, due to the larger flow shear, and thus that more K-H-related perturbations should be present in this region (Galopeau et al. 1995).

Wave activity on Jupiter's magnetopause has been identified based on multiple spacecraft crossings of the boundary (Huddleston et al. 1997); however, as for investigation of other

Fig. 6 Assessment of the stability of Jupiter's magnetopause to the growth of the Kelvin-Helmholtz (K-H) instability. In all panels the magnetopause surface is viewed from along the solar wind flow direction, and *shaded regions* of each surface are regions predicted to be K-H unstable. Different panels consider different levels of magnetospheric polar flattening. Adapted from Desroche et al. (2012)

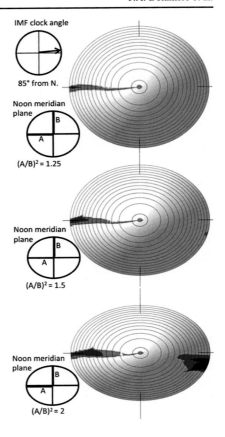

processes operating at Jupiter's magnetopause, limited spacecraft data sets prevent large statistical analyses. Note that separating wave-driving mechanisms is difficult, i.e. solar wind pressure fluctuations can cause waves as well as K-H instability growth.

In their assessment of what conditions at Jupiter's magnetopause mean for how the solar wind interacts with the magnetosphere, Desroche et al. (2012) consider the K-H stability of the boundary. They found that polar flattening of the magnetosphere (caused by centrifugal confinement of magnetospheric plasma in roughly the plane of the planetary equator) can have a significant effect on the flow in the magnetosheath, and that, as expected, the dawn flank of the magnetopause should be far more K-H unstable than the dusk flank, due to the difference in flow shears (see Fig. 6).

Delamere and Bagenal (2010) suggested that solar wind driving of Jupiter's magnetosphere is predominantly due to viscous processes, like growth of the K-H instability, operating at the magnetopause. This model is akin to that of Axford and Hines (1961). Delamere and Bagenal (2010) suggested that rather than a global cycle of reconnection where flux is opened at the dayside magnetopause and closed in the magnetotail, flux is predominantly opened and closed intermittently in small-scale structures in turbulent interaction regions on the flanks of the magnetosphere. K-H vortices and associated reconnection is a key element of this understanding of Jupiter's magnetospheric dynamics.

Statistical studies of perturbations of Saturn's magnetopause have provided a clearer picture of K-H instability growth at this magnetospheric boundary. Case studies initially re-

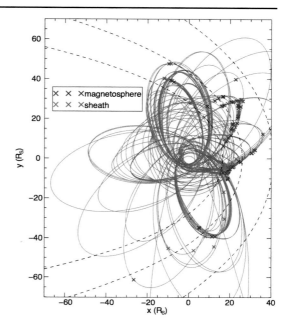

Fig. 7 Location of potential signatures of Kelvin-Helmholtz (K-H) vortices on Saturn's magnetopause. Approximately the planetary equatorial plane is shown, with the center of Saturn at the origin, and the x-axis pointing towards the Sun. *Dashed curves* indicate the range of possible locations of Saturn's magnetopause. The Cassini spacecraft trajectory is shown as a *solid gray line*. The locations of identified signatures are indicated, and their color indicates whether they correspond to occasions when the spacecraft was in the magnetosheath or the magnetosphere. Taken from Delamere et al. (2013)

vealed evidence for waves on both the dawn and dusk sides of the magnetopause (Huddleston et al. 1997; Masters et al. 2009; Cutler et al. 2011), before a recent statistical study by Masters et al. (2012b) showed that waves are a common feature of Saturn's magnetopause dynamics. Contrary to our expectation of a dawn-dusk asymmetry, roughly equal wave occurrence was found at dawn and dusk. The tailward propagation of the identified waves, generally perpendicular to the local magnetospheric magnetic field, suggests that the K-H instability is a dominant wave driving mechanism. Most recently, Mistry et al. (2014) used simple modeling to propose that the amplitude of K-H waves is generally comparable to the amplitude of global boundary oscillations at a period close to that of planetary rotation. A case study of a spacecraft encounter with a K-H vortex near the boundary of Saturn's magnetosphere was presented by Masters et al. (2010), and analysed in further detail by Delamere et al. (2011) and Wilson et al. (2012). Masters et al. (2010) suggested that such structures can generate spots of auroral emission, which are similar to observed auroral features (Grodent et al. 2011). A more comprehensive study of K-H vortices was recently reported by Delamere et al. (2013). These authors identified the distinctive magnetic field signatures of K-H vortices to identify many possible vortex events, which mostly occurred on the dusk flank, also contrary to expectations (see Fig. 7). Delamere et al. (2013) suggested that this is because K-H perturbations grow in the prenoon to dusk region, before moving to dusk with the local centre of mass where they are encountered as vortices.

Numerical models have also been used to address the Saturn magnetopause K-H instability problem. Magnetohydrodynamic simulations of Saturn's magnetosphere suggest K-H vortex formation at both dawn and dusk (Fukazawa et al. 2007a, 2007b; Walker et al. 2011), although the reason for this may differ from the reason for the observed absence of the expected local time asymmetry. In addition, hybrid simulations suggest appreciable vortex-induced diffusion and energy transport across Saturn's magnetopause (Delamere et al. 2011), which could drive magnetospheric dynamics.

These studies of shear-flow-driven instabilities at the magnetopauses of Jupiter and Saturn suggest some differences with the terrestrial case, which remain to be comprehensively explained. However, it is very likely that the growth of these instabilities is widespread, particularly in the case of Saturn where statistical observation based on spacecraft observations have been carried out.

4 Auroral Signatures

4.1 Dynamics of Internally-Driven Aurora

Auroral emissions at Jupiter and Saturn originate from the impact of charged particles in their hydrogen-dominated atmosphere (see review of the atmospheric response to these charged particles in Badman et al. 2014a). The aurora at Jupiter is generally described as formed of three components, the satellite footprints, the main oval and the polar emissions. A fourth component should however be added to this trio: the outer emissions, i.e. the emissions located equatorward from the main oval, but not directly related with the satellite footprints. All these components vary with time, often independently. But sometimes the combination of the observed changes proves to be a valuable tool to unveil the large scale magnetospheric processes at play. The auroral processes that dominate at Jupiter are also present at Saturn. For example, Enceladus also creates an auroral footprint on Saturn (Pryor et al. 2011). Additionally, low latitude aurora have also been identified on Saturn in the infrared and ultraviolet domain, which could correspond to the main emissions and to the outer emission on Jupiter, respectively (Stallard et al. 2008; Grodent et al. 2010).

4.1.1 The Satellite Footprints

At Jupiter, satellite footprints are the auroral signature of the electro-magnetic interaction between the Galilean moons Io, Europa and Ganymede on the one hand and the Jovian magnetosphere on the other hand. They appear as individual or a series of auroral spots located close to the field lines connected to moons, sometimes accompanied with arcs mapping to the moons' orbit (see review by Bonfond 2012). While the location of the Io footprint appears to remain constant with time for a given System III longitude, the Ganymede footprint may shift from one epoch to another (see Fig. 8). This shift is the consequence of changes in the magnetic field mapping and an equatorward motion corresponds to variations of either the current sheet density or thickness (Grodent et al. 2008). Ganymede and its footprint thus constitute precious landmarks to disentangle motions of auroral features owing to the magnetic field topology from those owing to motion of the source region in the magnetosphere.

Based on ultraviolet observations from the UVIS instrument on board Cassini, Pryor et al. (2011) identified the presence of the Enceladus footprint on three pseudo-images of the northern polar region. However, this footprint is particularly weak and cases of non-detection far outnumber the positive detections.

4.1.2 Outer Emissions

The outer emissions generally designate the auroral emissions located equator-ward from the main oval, but excluding the satellite footprints. They appear in three form: arcs, patches and diffuse emissions. It is thought that these outer emissions are caused by two different mechanisms: pitch angle diffusion and inward injections of hot plasma.

Based on simultaneous in-situ and auroral observations by Galileo and HST, Radioti et al. (2009a) attributed the faint arcs and elongated diffuse emissions to electrons being diffused in the loss cone at the pitch angle diffusion boundary.

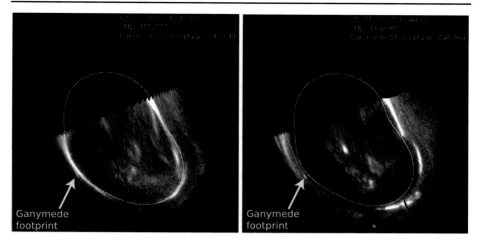

Fig. 8 Polar projection of the northern Jovian hemisphere aurora on February 27th (*left panel*) and on May 21st 2007 (*right panel*). The observing geometry was very similar, with CMLs of 155.3° and 159.7° respectively and a Ganymede S3 longitude of 246.6° and 247.0° respectively. Nevertheless, the Ganymede footprint is outside the main emission in the *first image* and inside it in the *second case*, suggesting that the source region has moved inside the Ganymede orbit (15 R_J). Additionally, the Ganymede footprint location moved 0.5° equatorwards implying an increased stretching of the magnetic field lines. The *white line* is the reference oval from February 2007

The patchy blobs, appearing either individually or in groups, have been attributed by Mauk et al. (2002) to injections of hot plasma coming from the outer magnetosphere and moving rapidly radially inward as a counterpart to the cold plasma moving out of the inner magnetosphere. Cassini energetic electron measurements (Radioti et al. 2009b) and numerical simulations together with simultaneous UV and ENA emissions (Radioti et al. 2013b) also indicate that injected plasma populations can create auroral emissions at Saturn. These studies suggest that, in the Kronian system, pitch angle diffusion associated with electron scattering by whistler-mode waves is the main driver of the UV auroral emissions associated with injections, while field aligned currents driven by the pressure gradient along the boundaries of the cloud might have a smaller contribution.

The morphological differences between these features is not always clear and some studies have mixed these two types of emissions under the generic terms of low-latitude or outer emissions. The brightness of the outer emissions does not seem to be correlated with the solar wind input, but appears to have considerably increased as the main oval expanded in spring 2007 (Nichols et al. 2009a; Bonfond et al. 2012). At Saturn, Grodent et al. (2010) reported the presence of a diffuse outer auroral oval, based on ultraviolet images from HST. Such an oval would map between 4 and 11 R_S and could be related to the precipitation of hot electrons diffusing in the loss cone.

4.1.3 Main Emissions/Main Oval

The brightest feature of the jovian aurora is a ring of emission centered on each magnetic pole, which is usually called the main oval or main emissions. Indeed, the oval usually appears broken and sometimes shows forks and parallel arcs. Moreover, the part of the main emissions corresponding to the pre-noon sector is generally 5 to 10 times weaker than the remaining of the aurora in the UV domain (Radioti et al. 2008a). While the UV and infrared main emissions are colocated, the brightness variations along the oval do not appear correlated (Radioti et al. 2013a). In the X-rays, the signature of the main emission differs from those of the polar emissions. The photons originating from the main oval are more energetic (>2 keV) and caused by electron bremsstrahlung while the softer photons (<2 keV) from the polar regions are related to the precipitation of highly stripped heavy ions (Branduardi-Raymont et al. 2008). The origin of the main emissions is related to the corotation breakdown, the large scale magnetosphere-ionosphere coupling and the associated field-aligned currents (e.g. Nichols 2011; Ray et al. 2012 and reference therein).

Gustin et al. (2004) analysed far UV spectra from the STIS instrument on-board HST and studied the relationship between the brightness of the emissions and the methane absorption (through color ratios). They showed that the energy flux of the precipitating electrons was usually (but not always) correlated to the energy of these electrons, as expected from an acceleration caused by field aligned currents. They also showed that this correlation was much less clear for the polar emissions. They did not show any systematic dependence of the electron energy with local time, but the orientation of the observations was always similar (i.e. Central Meridian Longitude around 160° for the north and around 80° for the south).

Observed in the early FUV images of the jovian aurora (Ballester et al. 1996; Clarke et al. 1998) and then regularly afterwards, these dramatic enhancements of the dawn arc of the main emissions are still poorly understood. Gustin et al. (2006) found that while exclusively found at dawn, the brightness enhancements appeared to rotate with the magnetic field. They also showed that the brightness and the color ratio of these storms are correlated on timescales of tens of minutes. Moreover, during the large HST campaign in spring 2007, Nichols et al. (2009a) did not notice any relationship between the occurrence of dawn storms and the solar wind, contrary to other features (see below).

A correlation of the auroral emitted power with the solar wind pressure has been reported by Nichols et al. (2007), Clarke et al. (2009), Nichols et al. (2009a) (see Fig. 9), even if it is lower than at Saturn. This difference between the two planets was expected since the main emissions at Jupiter result from currents related to the corotation lag of outward drifting plasma. Actually, first order models predicted that solar wind compression regions would induce an increase in the angular velocity of the equatorial plasma and decrease the currents related to the lag from corotation, thus resulting in a dimmer aurora (e.g. Southwood and Kivelson 2001). Contrary to these expectations, Nichols et al. (2007) reported a brightening of the main emission corresponding to a period when the magnetosphere first modestly shrunk and then expanded, based on images acquired in 2000 while Cassini was upstream of Jupiter. Clarke et al. (2009) compared the brightness of the whole jovian aurora with solar wind conditions during the large 2007 HST campaign and came to a similar conclusion, i.e. a correlation of the auroral brightness with the solar wind pressure. Using the same 2007 data set, Nichols et al. (2009a) separated the aurora into distinct regions: the low latitude emissions, the main emissions, and the high latitude emissions, in order to identify the component of the aurora which responded the most to solar wind input. The outer region does not appear to be correlated with the solar wind, but enhancements of the main emissions and, to a lesser extent, of parts of the polar emissions are associated with solar wind compression regions, as in Nichols et al. (2007).

The solution of this issue may lie in the detailed timing of the response to the arrival of a compression region, as more detailed models suggest that it may result in successive phases of auroral dimming and brightening (Cowley et al. 2007). Indeed, estimates of the solar wind condition at Jupiter based on Earth-based measurements has always been challenging; the most tricky part being the prediction of the arrival time of the shocks. The delay between

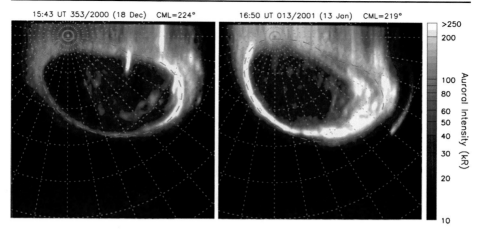

Fig. 9 HST images obtained during the Cassini Jupiter fly-by epoch and corresponding to a solar wind rarefaction region (*left*) and compression region (*right*) (adapted from Nichols et al. 2007)

the arrival of a compression region at the dayside front of the magnetopause and the auroral response is thus unclear with the current dataset.

The radial distance where the main emissions are mapping appears to vary with both local time and time. Grodent et al. (2003a) studied the location of the main oval on five sets of images from the northern hemisphere and reported a shift in its location as a function of CML. Since System III-fixed features appear to migrate poleward as the planet rotates, these images suggest that the L-shells related to the main emissions are closer at dusk then at dawn. The authors explained this difference with a different mass outflow rate as a function of the local time. Vogt et al. (2011) built a local time dependent magnetic field mapping model to map and interpret the auroral emissions located poleward of the Ganymede footprint. One of their main results was that the main emissions mapped closer to Jupiter at dawn rather than at dusk, contrary to the previous conclusion.

Additionally, Grodent et al. (2008) has shown that the location of the main emissions, even when observed in very similar configurations, could significantly (equivalent of 3° of latitude) change from one observation to another. On the same pair of images, the location of the Ganymede footprint had also shifted in the same direction. Based on images from the large HST campaign from spring 2007, Bonfond et al. (2012) also studied the location of the main emissions and showed that the main oval location continuously expended from February to June. On top of this long-term trend, the location of the main emissions showed day-by-day variations. Both studies showed that the Ganymede footprint had also moved in the same direction as the main oval, indicating that the apparent motion of the main oval was at least partially due to a stretching of the magnetic field lines. At the same time, these authors noticed that the occurrence of very bright outer emissions also increased from February to June. They interpreted these events as the outcome of an increased Io torus and plasma sheet density caused by an increased volcanism on Io, supporting the conclusions of Nichols (2011). This enhanced loading of the system would lead to both an increase of the plasma sheet density, explaining the shift of the Ganymede footprint, and an increase of the plasma outflow rate, explaining the expansion of the main oval. Moreover, an increased injection occurrence rate, in order to conserve the magnetic flux in the inner magnetosphere, would explain the increased occurrence of large injection blobs.

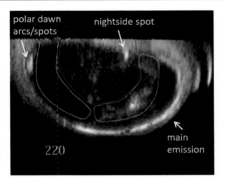

Fig. 10 Polar projection of the northern auroral region obtained from the Hubble Space Telescope on December 16, 2000, at CML = 220 deg. The main emission is indicated by the *arrow*. The shape and position of the three main polar regions are shown: the dark region (*yellow contour*), the swirl region (*red contour*), and the active region (*green contour*). The polar dawn arcs/spots and the nightside spot are indicated. Parallels and meridians are drawn every 10 deg. The CML is marked with a *vertical green dashed line* and longitude 180 deg is highlighted with a *red dashed line*. The *red dot* locates the magnetic footprint of Ganymede (VIP4 model) as the orbital longitude of the satellite matches the CML and therefore indicates the direction of magnetic noon at 15 R_J. Adapted from Grodent et al. (2003b)

At Saturn, observations of H_3^+ emissions in the infrared domain demonstrated the presence of an auroral oval mapping to a radial distance of 4 R_S (Stallard et al. 2008). This infrared secondary oval is thus located at lower latitude than the ultraviolet oval described above. The location and the faint emissions are consistant with models of a current system associated with the corotation breakdown of the magnetospheric plasma, similar to the Jupiter case (see also Ray et al. 2013).

4.2 Dynamics of the Polar Aurora

4.2.1 Jupiter

The emissions located poleward of the main emission at Jupiter, the polar emissions, are suggested to be magnetically connected to the middle and outer magnetosphere and possibly related to a sector of Dungey and/or Vasyliunas cycle flows (Cowley et al. 2003; Grodent et al. 2003b; Stallard et al. 2003). They are classified into three main regions: the swirl, the dark and the active region (Fig. 10).

The swirl region is the polar-most region of the polar emissions. It usually displays weak and very dynamic patches of emission that sometimes give the impression of a rotating crown on animations made of successive images. Its IR counterpart is the fixed- Dark Polar region (Stallard et al. 2003). According to the magnetic field mapping model by Vogt et al. (2011), this area is a region of open flux, which is consistent with the fact that the ionospheric flow are fixed relative to the sun, though Delamere and Bagenal (2010) argued that this region may be tied to intermittently open flux stemming from a viscous interaction along the dawn flank.

Polar auroral filaments have been observed in the same region. Nichols et al. (2009b) reported the finding of thin rectilinear and quasi-sun aligned features in the swirl region. The sunward head of the filaments remains fixed in local time while the anti-sun-ward part sub-corotates. These filaments were seen in 7 % of the 2007 HST observations and their occurrence appears to be independent from the solar wind conditions. Nichols et al. (2009b) suggested that these emissions could be related to magnetotail dynamics.

The dark region is usually devoid of significant UV emissions. It appears to correspond to a region that is also relatively dark in the IR: the rotating Polar Dark Region (r-DPR). This region has been associated with downward going field aligned currents, which would explain why the region is dark (Grodent et al. 2003b). However, the brightness of the IR emissions in the r-DPR still represent 30 to 40 % of those of the main emissions relative brighter in IR compared to the low emission in UV (Stallard et al. 2003). Recent studies suggested that this difference is explained in terms of soft precipitation and most probable Joule heating (Radioti et al. 2013a). Moreover, the mapping model from Vogt et al. (2011) suggests that half of the dark region corresponds to closed field lines while the other half corresponds to opened field lines, in contradiction with the apparent uniformity of this region. Delamere and Bagenal (2013) suggested that this region contains closed flux with much of the missing flux from Vogt et al. (2011) stored in a wing along the dawn flank. The FUV-dark region also appears dark in the X-rays (Branduardi-Raymont et al. 2008).

At the equatorward edge of the dark region, at the dawn and midnight flank of the main emission auroral observations have shown the occasional appearance of spotty transient emissions (Fig. 10). In particular, parallel arc structures are observed to be located in the dawn sector (Grodent et al. 2003a) and isolated spots to appear in the dusk-midnight sector, poleward of the main emission (Grodent et al. 2004). The dawn arcs and the nightside spots were proposed to be triggered by reconnection processes in the jovian magnetotail, given their observed location and properties. An analysis based on daily UV auroral observations (Radioti et al. 2008b) revealed the presence of periodic auroral features ("polar dawn spots", similar to the dawn arcs (Grodent et al. 2003a). They consist of transient auroral emissions in the polar dawn region, with a characteristic recurrence period of 2–3 days. Because of their periodic recurrence and observed location, the polar dawn spots were interpreted as auroral signatures of internally driven magnetic reconnection in the jovian magnetotail (Vasyliunas cycle). Particularly, they were associated with the inward moving flow bursts released during magnetotail reconnection in Jupiter's tail (Radioti et al. 2010). The association of the polar dawn auroral spots with tail reconnection was further confirmed by Ge et al. (2010). The authors magnetically mapped tail reconnection events into Jupiter's ionosphere, by tracing field lines using a jovian magnetosphere model (Khurana 1997).

More recently, Radioti et al. (2011b) reported observations of a dusk side spot occurring at nearly the same time as a reconnection signature was observed in the Galileo magnetometer data (Vogt et al. 2010). This spot was mapped using an updated mapping model to an equatorial position close to the Galileo spacecraft and inside of a statistical X-line, further confirming the association of the auroral spots with inward flow from tail reconnection. Not only UV but also IR emissions bear the signature of tail reconnection. Comparison of near-simultaneous UV and IR observations on 26 July 1998 has revealed a bright IR polar spot, which could be a possible signature of tail reconnection (Radioti et al. 2011b). The IR spot appears within an interval of 30 minutes from the ultraviolet, poleward of the main emission in the ionosphere and in the post-dusk sector planetward of the tail reconnection X-line in the equatorial plane. Finally, auroral observations can provide a hint as to the extent of the tail X-line. Near-simultaneous HST auroral and Galileo observations demonstrated that ionospheric signatures of inward moving flows released during tail reconnection are instantaneously detected over a wide local time range (Radioti et al. 2011b). However, whether reconnection at Jupiter's tail can result in a simultaneous release of flow bursts over a large local time sector is a question still to be resolved by future missions to Jupiter and/or remote observations.

The brightest polar region is called the active region and is located inside the dusk flank of the main emission. It usually displays a mix of quiet and dynamic auroral spots, blobs

Fig. 11 Polar projection of Saturn's northern aurora obtained with the FUV channel of UVIS onboard Cassini on January 21, 2009. Noon is to the *bottom* and dusk to the *right*. The *grid* shows latitudes at intervals of 10° and meridians of 40°. *Arrows* indicate the main emission and bifurcations of the main emission. Adapted from Radioti et al. (2011a)

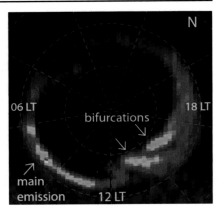

and arcs (Grodent et al. 2003b). It's IR counterpart appears to be the Bright Polar Region (Stallard et al. 2003).

In the active region, extremely bright flares are observed. These are localized transient features, which exhibit intense brightenings. Waite et al. (2001) reported an extreme case where the brightness reached 40 MR. Elsner et al. (2005) compared the timing and the location of the UV and the Xray flares and found that flare brightenings in both wavelengths were quasi-simultaneous, but not exactly co-located. Simultaneous UV and X-Ray observations demonstrated that the <2 keV photons appear co-located in the "active region", while there is absence of high energy X-ray photons in this area (Branduardi-Raymont et al. 2008). Finally, Bonfond et al. (2011) reported two cases of quasiperiodic variations of these flare emissions, with a re-occurrence time of 2–3 minutes. They also noted a rapid dawnward propagation of the flares in one of the cases. They tentatively attributed these periodic flares to a signature of pulsed component reconnection on the dayside magnetosphere. However, the recent finding of similar periodicities in high-energy electron data in the outer and middle magnetosphere put this interpretation into question. Moreover, Pallier and Prange (2001) suggested that some of the emissions in the active region could be associated with the cusp.

4.2.2 Saturn

Hubble Space Telescope (HST) observations demonstrated that Saturn's aurora responds to solar wind changes (Grodent et al. 2005; Clarke et al. 2009) and its brightness and shape varies with time. The main auroral emission at Saturn (Fig. 11) is suggested to be produced by the magnetosphere-solar wind interaction, through the shear in rotational flow across the open closed field line boundary (OCFLB) (e.g. Cowley and Bunce 2003; Bunce et al. 2008). Saturn's auroral morphology is, to a large extent, controlled by the balance between the magnetic field reconnection rate at the dayside magnetopause and the reconnection rate in the nightside tail (Cowley et al. 2004; Badman et al. 2005, 2014b). Recently, Ultraviolet Imaging Spectrograph (UVIS) observations revealed the presence of small-scale structures in the dayside main auroral emissions indicative of magnetopause Kelvin-Helmholtz instabilities (Grodent et al. 2011).

Observations (Gérard et al. 2004, 2005) and theoretical studies (Bunce et al. 2005a) showed that bright FUV emissions at Saturn observed occasionally near noon are probably associated with reconnection occurring at the dayside magnetopause, similar to the "lobe cusp spot" at Earth (i.e. emissions located at the cusp magnetic foot point, Fuselier et al. 2002). Specifically, it was proposed by Bunce et al. (2005a) that pulsed reconnection at the low-latitude dayside magnetopause for northward directed Interplanetary Magnetic Field (IMF) is giving rise to pulsed twin-vortical flows in the magnetosphere and ionosphere in the vicinity of the OCFLB. For the case of southward IMF and high-latitude lobe reconnection pulsed twin-vortical flows, bi-polar field-aligned currents are expected and associated with auroral intensifications poleward of the OCFLB. Recently, Cassini UVIS revealed the presence of bifurcations of the main dayside auroral emission, which are interpreted as signatures of consecutive reconnection events at Saturn's magnetopause (Radioti et al. 2011a). The authors suggested that magnetopause reconnection can lead to a significant increase of the open flux within a couple of days. In particular, it was estimated that each reconnection event opens ~10 % of the flux contained within the polar cap. Further studies based on Cassini multi-instrument observations, including auroral UV and IR data, confirmed that the auroral arcs are related to bursty reconnection at Saturn involving upward field aligned currents (Badman et al. 2012b) and suggested that bursty reconnection at Saturn is efficient at transporting flux (Badman et al. 2013). Additionally, auroral observations have shown evidence of magnetopause reconnection at multiple sites along the same magnetic flux tube similar to the terrestrial case (Fasel et al. 1993), which give rise to successive rebrightenings of auroral structures (Radioti et al. 2013c).

In addition to dayside reconnection, tail reconnection leaves its signature in the aurora at Saturn. Changes in open flux obtained from the auroral images and comparison with open flux estimated from the upstream interplanetary data allowed the estimation of the average tail reconnection rates at Saturn (Badman et al. 2005). Also, small spots of auroral emission lying near the main emission, observed by the UVIS instrument onboard Cassini, are suggested to be associated with dipolarizations in the tail (Jackman et al. 2013). These auroral features are suggested to be the precursor to a more intense activity associated with recurrent energization via particle injections from the tail following reconnection and plasmoid formation (Mitchell et al. 2009).

Auroral dawn enhancements expanding in the polar auroral region have been recently reported in different studies and related to tail reconnection. Nichols et al. (2014) presented HST auroral intensifications in the dawn auroral sector, propagating at 330 % rigid corotation from near 01 h LT toward 08 h LT. They suggested that these emissions are indicative of ongoing, bursty reconnection of lobe flux in the magnetotail, with flux closure rates of 280 kV. Similar events of intense dawn auroral activity were recently observed by the UVIS instrument on board Cassini (Radioti et al. 2014a) and was characterized by significant flux closure with a rate ranging from 200 to 1000 kV. Additionally, Radioti et al. (2014b) revealed multiple intensifications within an enhanced auroral dawn region suggesting an x-line in the tail, which extends from 02 to 05 LT. The localized enhancements evolved in arc and spot-like small scale features, which were suggested to be related to plasma flows enhanced from reconnection which diverge into multiple narrow channels then spread azimuthally and radially. They proposed that the evolution of tail reconnection at Saturn may be pictured by an ensemble of numerous narrow current wedges or that inward transport initiated in the reconnection region could be explained by multiple localized flow burst events. Badman (2014) reported on Saturn's auroral morphology during a solar wind compression event. The authors suggested that their observations were evidence of tail reconnection events, initially involving Vasyliunas-type reconnection of closed mass loaded magnetotail field lines and then proceeding onto open field lines, causing contraction of the polar cap region.

Enhancements in energetic neutral atom (ENA) emission and Saturn kilometric radiation (SKR) data, together with auroral observations from HST and UVIS reported the initiation of several acceleration events in the midnight to dawn quadrant at radial distances of 15 to 20 R_S, related to tail reconnection (Mitchell et al. 2009).

Fig. 12 Polar projection of Saturn's aurora as captured from UVIS onboard Cassini on DOY 224, 2008. The *arrow* indicates the first observation of Saturn's transpolar arc. Its formation is possibly related to tail reconnection (adapted from Radioti et al. 2014a)

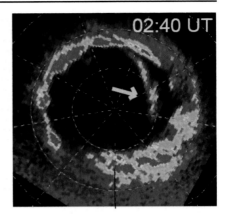

Finally, UVIS auroral observations revealed the first and only observation of an Earth-like transpolar arc at Saturn (Fig. 12) (Radioti et al. 2014a). Transpolar arcs are features which extend from the nightside auroral oval into the open magnetic field line region (polar cap) and they represent the optical signatures of magnetotail dynamics (e.g. Frank et al. 1982; Kullen 2000; Milan et al. 2005). The authors suggested that the formation of the transpolar arc at Saturn is possibly related to tail reconnection similarly to the terrestrial case (Milan et al. 2005). However, the rarity of the occurrence of the transpolar arc at Saturn indicates that the conditions for its formation are rarely met at the giant planet, contrary to the Earth.

4.3 Solar Wind Influence

4.3.1 Jupiter

It has long been known that Jupiter's radio emissions associated with auroral processes are controlled by conditions in the interplanetary medium incident on Jupiter's magnetosphere (Terasawa et al. 1978; Barrow et al. 1986; Zarka and Genova 1983; Genova et al. 1987; Bose and Bhattacharya 2003; Echer et al. 2010; Hess et al. 2012). The first evidence that Jupiter's auroras themselves are modulated by conditions in the interplanetary medium was reported by Baron et al. (1996), who analysed NASA IRTF observations of the planet's H_3^+ emissions in conjunction with Ulysses solar wind data over a $\sim$100 day interval near the Jupiter encounter in 1992. They showed that the change in solar wind dynamic pressure between their auroral observations was reasonably well correlated with the observed total intensity of the H_3^+ auroral emission. More detailed understanding was obtained during the Cassini Jupiter flyby in late 2000–early 2001. A combination of Cassini and Galileo spacecraft *in situ* measurements and remote sensing, along with a program of Hubble Space Telescope (HST) Space Telescope Imaging Spectrograph (STIS) observations near to closest approach revealed in detail the response of the auroral emissions to changing conditions in the interplanetary medium (Gurnett et al. 2002; Pryor et al. 2005; Nichols et al. 2007). The solar wind during the interval of the encounter, which occurred at solar maximum, was dominated by coronal mass ejections (CMEs), corotating interaction regions (CIRs) and large amplitude stream interactions resulting in deep several-day solar wind rarefaction regions punctuated by strong compressions as highlighted by the interplanetary magnetic field (IMF) magnitude shown in Fig. 13(a), taken from Gurnett et al. (2002). These authors showed that, when

Fig. 13 Plot showing
interplanetary medium, radio and
EUV data from (**a**) the Cassini
MAG instrument, (**b**) the solar
wind ion densities from the
Cassini CAPS instrument, (**c**) the
integrated (0.5 ± 5.6 MHz)
hectometric radiation intensities
(1-h averages) from the Galileo
PWS instrument, and (**d**) the
disk-integrated extreme
ultraviolet auroral H_2 band
(110.8 ± 113.1 nm) intensities
for the Cassini ultraviolet
imaging spectrograph (UVIS)
instrument for event A. The
arrows show the times at which
the interplanetary shock was
detected by Cassini and Galileo.
Reproduced from Gurnett et al.
(2002)

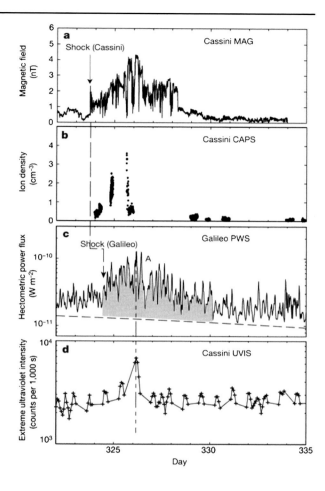

corrected for solar wind travel time, the jovian extreme ultraviolet (EUV) emission and hectometric radiation exhibited peaks in intensity of factors of 2–4 above the background near to the time of maximum solar wind density as shown in Figs. 13(b–c).

While the global response of the auroral intensity to the interplanetary conditions was thus revealed by the Cassini EUV data, the detailed morphological variation was elucidated by the contemporaneous HST Space Telescope Imaging Spectrometer (STIS) images of the far ultraviolet (FUV) aurora discussed by Grodent et al. (2003a,b) and Nichols et al. (2007). The conditions in the interplanetary medium observed at the time of HST observations, suitably propagated from the Cassini spacecraft to the planet's ionosphere, are shown in Fig. 14, adapted from Nichols et al. (2007). The plot is colour-coded such that red data points indicate solar wind, blue represents magnetosheath, and green signifies magnetosphere. The vertical grey lines indicate the times of the HST images. It is apparent that most of the HST images were obtained during solar wind rarefaction regions with dynamic pressure of order $\sim$0.01 nPa, corresponding to an expanded magnetosphere with a sub-solar distance of $\sim$70–100 R_J. Only one visit on 13 January corresponds to a compression event with dynamic pressure >0.1 nPa and a correspondingly compressed magnetosphere of sub-solar extent $\sim$50–60 R_J. In Fig. 14 we then also show two representative images of the north-

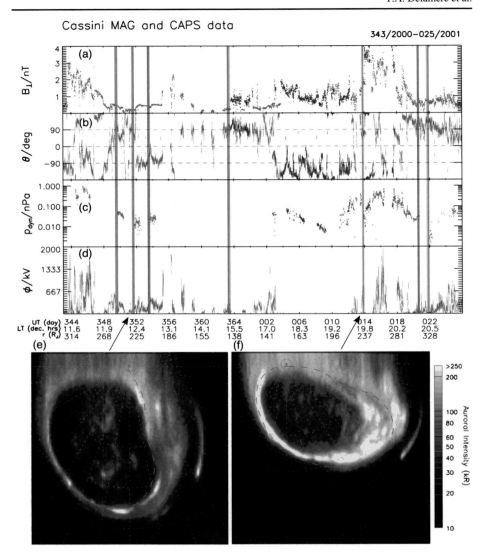

Fig. 14 Plot showing the conditions in the interplanetary medium as measured by Cassini over days 343/2000 to 025/2001, along with representative HST images of Jupiter's auroras obtained during the interval. From top to bottom the panels show: (**a**) the perpendicular IMF magnitude $B_\perp$ in nT, (**b**) the IMF clock angle relative to Jupiterôs spin axis θ in degrees, (**c**) the dynamic pressure of the solar wind p_{dyn} in nPa on a log scale, (**d**) and the estimated dayside reconnection voltage ϕ in kV calculated using the algorithm of Nichols et al. (2006). Data are plotted versus estimated time of impact on Jupiterôs ionosphere, along with the local time and range of Cassini relative to Jupiter. In the plots for θ and ϕ each 'data point' is stretched into a *vertical line* representing the effect produced by the $\sim \pm 9.5$ deg diurnal variation of Jupiterôs dipole axis offset. The *vertical grey lines* indicate the times of emission of the auroral photons observed by HST. Panels (**e**) and (**f**) show representative images of Jupiterôs northern UV aurora with CML obtained by the HST during the millennium campaign, projected onto a latitude-longitude grid viewed from above the north pole. The CML of each image, i.e., approximately the sunward direction, is aligned toward the bottom of each image, such that dawn is to the left, dusk to the right, and midnight to the top. The intensity scale is logarithmic and saturated at 250 kR. The Grodent et al. (2003a) reference main oval is shown by the *dashed red line*. A $10° \times 10°$ jovigraphic grid is overlaid. The images were obtained on 351/2000 and 013/2001. Adapted from Nichols et al. (2007)

ern auroras (the hemisphere most visible from Earth due to the non-dipolar nature of the jovian internal magnetic field), corresponding to rarefaction and compression solar wind conditions.

The basic morphology of Jupiter's auroras has been discussed in Sects. 4.1 and 4.2, such that here we concentrate only on the differences between the two solar wind states. The auroras associated with the rarefaction region shown in Fig. 14(e) exhibit the 'classic' morphology, i.e. the main oval is narrow and well-defined between System III (SIII) longitudes ~160–240°, typically lying in the dawn to noon sector at the central meridian longitudes for which the northern auroras are observed, while it is broad and diffuse in the 'kink' sector at smaller SIII longitudes. Poleward of the main oval lies a relatively dark region, more typically existing on the dawn side but occasionally extending around to the dusk side, and poleward still lies a region of patchy, variable emission which is particularly bright in an active region near noon. The images obtained during the compression region on 13 January, shown in Fig. 14(f), were in comparison obtained at high CML value, such that the viewing angle is less optimal than for the rarefaction regions, allowing good views only of the narrow main oval region. However, it is obvious that the main oval is significantly brighter and expanded poleward along its entire length, merging with bright auroral forms in the active region, and the polar patchy emission is also extended equatorward down to the main oval. The enhanced auroral power associated with the compression was thus a result of brightened and/or expanded main oval and polar emissions. It should be noted that, although these enhanced auroras were overall associated with a compression region, Nichols et al. (2007) noted that there was significant variation in the interplanetary conditions during this interval, such that it remains ambiguous as to whether the brightening was associated with a transient compression or expansion of the magnetosphere.

Further understanding of the response of Jupiter's auroras was obtained in 2007, when Jupiter's auroras were observed once per day using the Advanced Camera for Surveys (ACS) onboard HST for two month-long programs, the first in February/March during the New Horizons flyby and the second in May/June near opposition (Clarke et al. 2009; Nichols et al. 2009a). Using these data Clarke et al. (2009) considered the variation in total emitted FUV power in comparison to changes in the solar wind conditions as propagated from Earth using the 1-D MHD model of Zieger and Hansen (2008). In total, the two intervals included six solar wind forward shocks and three reverse shocks, the former associated with sharp increases in dynamic pressure occurring at compression region onset but which may also occur within larger merged compression regions, while the latter corresponds to sharp decreases in dynamic pressure at the tail end of compressions. With some uncertainty in the timing of the modelled propagation, Clarke et al. (2009) showed that forward shocks were typically associated with increases in total auroral power, while reverse shocks did not induce any significant change in auroral output. The variation of the individual components of auroral emission over these intervals was considered by Nichols et al. (2009a), from whose study some representative results are displayed in Figs. 15 and 16. Specifically, Figs. 15(a–c) show the power variation from the 3 different regions of the auroras delimited by the solid yellow lines in Fig. 16, i.e. the high latitude region, the main oval, and the low latitude region, and panels (d) and (e) show the solar wind dynamic pressure and IMF magnitude estimated using the model of Zieger and Hansen (2008). In the bottom two panels the solid lines show the model result with the original timings, while the dotted lines show the results shifted by +2.1 days, calibrated to an interplanetary forward shock observed by the New Horizons spacecraft on day 53 (Clarke et al. 2009). The dark grey regions show the estimated arrival time of the forward shocks within $1 - \sigma$ uncertainty of the MHD model timings, and the light grey regions are similar but for the shifted timings.

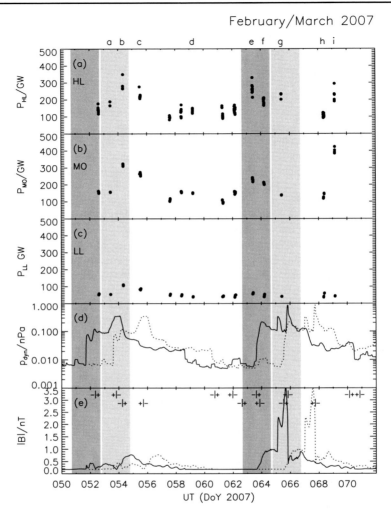

Fig. 15 Plots showing the power emitted from the different auroral regions, along with the modelled solar wind conditions for the first HST campaign in February/March 2007. Specifically, we show (**a**) the power emitted from the high latitude region P_{HL} in GW, (**b**) the power emitted from the main oval region P_{MO} in GW, (**c**) the power emitted from the low latitude region P_{LL} in GW, (**d**) the solar wind dynamic pressure in nPa, and (**e**) the IMF magnitude $|B|$ in nT. The individual points in panels (**a**)–(**c**) represent the powers obtained for each image. The *solid lines* in the MHD model panels show the original model timings, while the *dotted line* show the timings shifted by +2.1 days. The *dark grey regions* shows the estimated arrival time of the forward shocks within 1 standard deviation uncertainty of the MHD model timings, and the *light grey regions* are similar but for the shifted timings. Also shown in panel (**e**) are the estimated locations of the heliospheric sector boundaries, along with the sign of B_T either side. The original timing is on top, while the shifted timing is below. Reproduced from Nichols et al. (2009a)

As discussed by Clarke et al. (2009), three overall brightness enhancements were observed, two corresponding forward shocks (the first using the shifted model timings, the second using the original timing), while the third, a dawn storm, does not correspond obviously to any solar wind event. The auroral emissions corresponding to the data points labelled at the

⚛ Springer

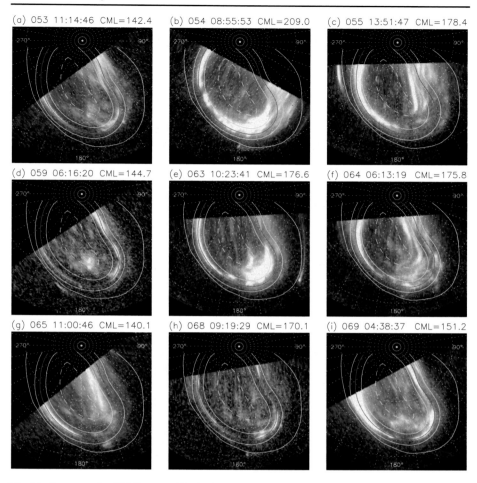

Fig. 16 Representative HST images of Jupiter's northern auroras corresponding to the visits labeled at the top of Fig. 15. The projection view is from above the north pole, and the image is displayed with a log color scale saturated at 500 kR. The *red line* shows the reference main oval as given by Table 1 in Nichols et al. (2009a). The *solid yellow lines* show the boundaries between the high latitude region, the main oval and the low latitude emission. The *dashed yellow line* indicates the boundary between the polar inner and polar outer regions. The *yellow points* indicate a 10° × 10° planetocentric latitude—SIII longitude grid. The image is oriented such that SIII longitude 180° is directed toward the bottom. Reproduced from Nichols et al. (2009a)

top of Fig. 15(a) are shown in Fig. 16. Using these data, Nichols et al. (2009a) showed that Jupiter's auroras respond to solar wind compression region onset in a broadly repeatable manner, which can be summarised as follows:

- The total emitted power from the main oval increases by factors of ∼2–3.
- The main oval is brightened along longitudes >165°, and is shifted poleward by ∼ 1° and expanded, as evidenced in Figs. 16(b) and (c), and (e–g).
- In contrast, there is little emission at longitudes < 165°, and any auroras are patchy and disordered.
- Under-sampling notwithstanding, the main oval apparently persists in this disturbed state for 2–3 days following compression region onset.

- The poleward emission varies broadly with the main oval, but with significant variation superposed thereon.
- The noon active region is brightened for ~1 day and merges with the poleward-shifted main oval, as shown in Figs. 16(b) and (e)
- Bright poleward dusk arcs, sometimes multiple in nature, are apparent for ~2 days following compression region onset, as evidenced in Figs. 16(c) and (f).
- The high latitude patchy auroras vary independently from the lower latitude polar auroras and the main oval, causing the superposed variation mentioned above.
- Dawn storms, for which the main characteristic is exceptionally bright dawn-side main oval auroras, but in which all auroral components brighten simultaneously as shown in Fig. 16(i), occur sporadically with no apparent solar wind trigger.

As well as these changes in the morphology of Jupiter's auroras in response to varying conditions in the interplanetary medium, there are a number of auroral forms which exhibit local time alignment or are thought to be associated with the solar wind interaction. The active region located near noon has been observed to 'flare' from a few kR to ~10 MR over a few minutes (Waite et al. 2001), and transient 'inner ovals' a few degrees poleward of the main oval have also been reported (Ballester et al. 1996; Pallier and Prange 2001; Nichols et al. 2007), similar to those thought to exist at the open-closed field line boundary (Cowley et al. 2005a, 2007). It has been suggested that at least some of the polar auroras may be related to the solar wind interaction at the dayside (Clarke et al. 1998; Pallier and Prange 2001; Waite et al. 2001; Grodent et al. 2003b). This idea was explored theoretically by Bunce et al. (2004), who showed that pulsed dayside reconnection with a ~45 min period during intervals of strong solar wind interaction could excite adjacent high latitude regions of UV and X-ray emission, as observed by Gladstone et al. (2002) and Elsner et al. (2005). Pulsed reconnection is also postulated to be responsible for periodic polar flares observed in the southern auroras, which exhibit a 2–3 minute periodicity and propagate swiftly from dusk to dawn (Bonfond et al. 2011). Radioti et al. (2008b) presented observations of periodic polar dawn spots of auroral emission just poleward of the main oval, which exhibit a 2–3 day periodicity, and were attributed to nightside reconnection, while Radioti et al. (2008a) showed that the main oval exhibits a persistent dim region ('discontinuity') in the pre-noon sector thought to represent a decrease in field-aligned current intensity owing to increased equatorial plasma angular velocity due to confinement by the magnetopause. Quasi-sun-aligned polar auroral filaments were also reported by Nichols et al. (2009b), which are superficially similar to terrestrial trans-polar arcs but probably not generated in the same manner. They were observed over an interval of 6 days in 2007 and appear independent of incident solar wind conditions, and were thus postulated to map significantly down the tail. The origin of this feature is unknown and it is hoped that the Juno and JUICE spacecraft will shed light on Jupiter's enigmatic polar auroras.

4.3.2 Saturn

As for Jupiter, the modulation of Saturn's auroral current system by the incident solar wind was first recognized through observations of the associated radio emissions; Voyager observations showed that the intensity of the Saturn Kilometric Radiation (SKR) was correlated with solar wind dynamic pressure (Desch and Kaiser 1981). The UV spectrometer onboard Voyager also revealed auroral emissions at a latitude consistent with the open-closed field-line boundary (Sandel and Broadfoot 1981). The detailed morphology of Saturn's auroras was revealed by later HST images, with Badman et al. (2006) and Bunce et al. (2008) confirming the co-location with the open-closed field-line boundary and Gérard et al. (2005)

Fig. 17 Comparison between HST images and solar wind conditions propagated to Saturn for the period 2530 January 2004. Reproduced from Crary et al. (2005)

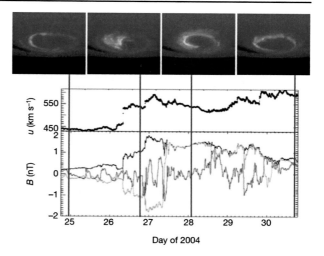

Day of 2004

identifying a transient noon auroral feature on the poleward edge of the main oval with the cusp. Transient duskside poleward forms observed in the south were associated by Radioti et al. (2009b) with energetic particle injections, although later analysis of similar features by Meredith et al. (2013), who examined equinoctial HST images of Saturn's simultaneous conjugate auroras, showed them to be hemispherically asymmetric and thus ascribed them to dayside reconnection of the rotating planetary field with B_y-dominated IMF.

The breakthrough in understanding the major effect of the solar wind on Saturn's auroras occurred as the Cassini spacecraft was approaching the planet in January 2004, and a large CIR passed over the spacecraft, corresponding to significantly increased solar wind velocity, dynamic pressure and IMF magnitude as shown in Fig. 17. As also shown in the figure, contemporaneous HST/STIS images of Saturn's southern auroras revealed that in response to these enhancements the dawnside auroras brightened significantly and expanded poleward, filling the dawnside polar cap with emission, while the radius of the oval shrank in proportion to the brightness (Crary et al. 2005; Clarke et al. 2005). Using the large HST/ACS program in 2007/2008, this was shown by Clarke et al. (2009) to be a repeatable morphological response to solar wind compression region onset, with the total power typically increasing by factors of ∼2–3, as shown in Figs. 18 and 19. Further analysis of the SKR intensity over an extended interval has also revealed a positive correlation with solar wind pressure (Rucker et al. 2008). This morphology has been interpreted theoretically as a manifestation of compression-induced tail reconnection (Cowley et al. 2005b), while the effect of IMF direction has also been shown to be a significant factor in controlling the radius of the auroral oval (Belenkaya et al. 2010, 2011).

Since orbit insertion, the Cassini spacecraft has yielded significant information regarding the solar wind effect on Saturn's auroras. The UV Imaging Spectrometer (UVIS) and Visible and Infrared Mapping Spectrometer (VIMS) onboard Cassini have also revealed details of small-scale features in Saturn's auroras, some of which are associated with the solar wind interaction. Grodent et al. (2011) showed that Saturn's main oval comprises individual auroral patches, which they associated with the Kelvin-Helmholtz instability, although Meredith et al. (2013) later showed that, rather than being conjugate, patches in the two hemispheres are magnetically interlaced, and therefore identified them instead with the field-aligned currents of eastward-propagating ULF waves. Radioti et al. (2011b) analysed bifurcations in the main oval, and associated these with pulsed dayside reconnection,

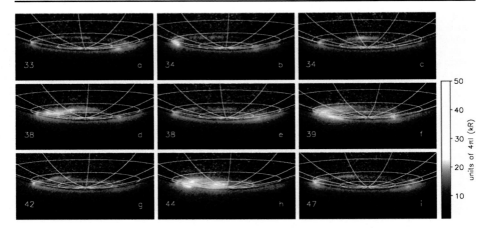

Fig. 18 Sample UV images of Saturnôs south pole in February 2008 with quiet and disturbed conditions. The *left-hand number* is day of year in 2008, and the part label letters correspond to the lettering at the top of Fig. 19. All frames were obtained with ACS, with a limiting sensitivity of 1 to 2 kR after modelling and subtraction of reflected solar emissions. Reproduced from Clarke et al. (2009)

while Badman et al. (2012b) identified small-scale equatorward moving H_3^+ auroral forms, as shown in Fig. 20, with dayside reconnection although it was unclear as to whether the reconnection was at high- or low-latitudes. Recently, Badman et al. (2013) have presented observations of poleward features near noon, in conjunction with in situ evidence of recent dayside reconnection with the B_y-dominated IMF. Overall, therefore a considerable body of evidence has now been built up suggesting the solar wind plays important roles, both directly and indirectly, in modulating the auroral morphology at the solar system's two giant planets.

5 Global Modeling of the Giant Planet Magnetospheres

5.1 Global Modeling Techniques and Limitations

The most commonly used models in simulating planetary magnetospheres are magnetohydrodynamic (MHD) models in which the plasma is treated as a magnetized fluid. While an MHD model does not treat plasma kinetic effects (e.g. gradient/curvature drift and microscale wave-particle interactions) and the small-scale processes (e.g. reconnection) are facilitated by numerical resistivity, it has been shown to be capable of providing a reasonably good description of the large-scale structure and dynamics of planetary magnetospheres. Moreover, because MHD models can usually cover a large-size simulation domain and, at the same time, achieve reasonably high spatial resolution at relatively low computational costs (compared to kinetic/particle codes), it remains, at present, the most feasible tool for global simulations of planetary magnetospheres. This is especially the case for the gas giants, Jupiter and Saturn, because of the vast spatial and long temporal scales involved in the two magnetospheric systems.

There have been a number of global MHD models applied to modeling the magnetospheres of Jupiter (e.g. Ogino et al. 1998; Walker and Ogino 2003; Moriguchi et al. 2008; Fukazawa et al. 2005, 2006, 2010) and Saturn (e.g., Hansen et al. 2005; Fukazawa et al.

Fig. 19 Total auroral power from Saturn's south polar region, best fit auroral oval radius, and SKR emission spectrum compared with propagated solar wind velocity and dynamic pressure in February 2008. Solar wind values obtained by propagation from Earth-based measurements have arrival times shifted 2.6 days later to match the time when Cassini measured a strong compression of the magnetosphere on DOY 38. Oval radius values were obtained by fitting a circle to the low-latitude edge of the observed auroral emissions. SKR emission measurements are from the Cassini RPWS instrument. Solar wind values were obtained by propagation from Earth-based measurements, with forward shock times $\pm1\sigma$ uncertainties shaded. Reproduced from Clarke et al. (2009)

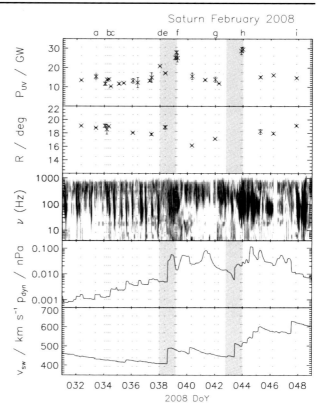

2007a,b; Kidder et al. 2012; Zieger et al. 2010; Walker et al. 2011; Jia and Kivelson 2012; Jia et al. 2012a,b; Winglee et al. 2013). In developing a global magnetosphere model for the giant planets, two aspects that are of particular importance are the rapid planetary rotation and strong internal plasma sources associated with the moons (in particular, Io in the case of Jupiter, and, Enceladus in the case of Saturn). All the global MHD models applied to the gas giants are designed to include these important factors to some extent, although they differ in many aspects, such as the assumption about the internal plasma sources, simulation boundary conditions, and the way of modeling the coupling between the magnetosphere and ionosphere. In the following, we provide an overview of the various global MHD simulations of the giant planet magnetospheres and highlight some important findings from global MHD models regarding the magnetospheric configuration and responses to solar wind driving.

5.2 Global Modeling Results

5.2.1 Global Configuration

Because of the rapid planetary rotation and strong internal plasma sources, the configurations of Jupiter's and Saturn's magnetospheres differ from that of a solar wind driven magnetosphere, such as the Earth's magnetosphere, where such effects do not play a significant role in shaping the magnetosphere under normal circumstances. There have been considerable efforts devoted to constructing empirical models of the magnetospheric boundaries

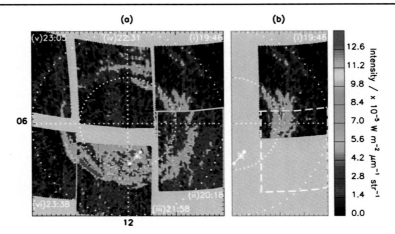

Fig. 20 (**a**) A mosaic of six Cassini VIMS images of Saturn's infrared aurora taken on 2008-320. The start time of each image is marked at its edge: (i) 19:46 UT, (ii) 20:18 UT, (iii) 21:58 UT, (iv) 22:31 UT, (v) 23:05 UT and (vi) 23:38 UT. The *white grid* marks latitudes at intervals of 10° and the noon-midnight and dawn-dusk meridians. The *white line delimited by dots* shows Cassiniôs ionospheric footprint during 12:0024:00 UT on 2008-320. The *white asterisk* marks Cassiniôs footprint at 22:00 UT on DOY 320. (**b**) Image i taken by Cassini VIMS showing the area overlapped by the following image ii whose outline is marked by the *white dashed line*. Cassiniôs ionospheric footprint is shown here for clarity. Reproduced from Badman et al. (2012b)

based on in-situ observations. Complementary to those data-based empirical models, global MHD models of the giant planets that incorporate the aforementioned important effects may provide useful, quantitative information about the global shape of the magnetosphere as well as how it varies with both the external and internal conditions.

In a series of global simulation studies of Jupiter's magnetosphere (Ogino et al. 1998; Walker et al. 2005; Fukazawa et al. 2006), the authors have examined the modeled locations of Jupiter's magnetopause and bow shock for cases with and without the planetary rotation included, and also compared model results for low and high solar wind dynamic pressure conditions. By comparing the magnetopause locations from simulation runs with and without rotation included, their model results clearly demonstrate the effect of centrifugal inflation of Jupiter's magnetosphere (e.g., Hill et al. 1974). They also find that for the range of dynamic pressures used in their simulations, the modeled magnetopause locations, in general, fall within the range inferred from spacecraft observations.

Similar modeling studies on magnetospheric boundaries have been carried out for Saturn. For example, Jia et al. (2012b) have compared the standoff distances of Saturn's bow shock and magnetopause between their global MHD model and data-based empirical model. The non-steady solar wind input used in the Jia et al. MHD simulation allows for comparison for a wide range of solar wind dynamic pressures as well as for a variety of IMF conditions. As shown in Fig. 21, the sub-solar magnetopause location predicted by the MHD model agrees well with the prediction by the empirical model of Kanani et al. (2010) developed using Cassini data. For various external conditions considered, not only the mean location but also the variation of the magnetopause location show good agreement between the two models. The MHD model result also confirms the finding obtained from earlier analysis using Cassini data (e.g., Arridge et al. 2006; Kanani et al. 2010) that Saturn's magnetopause is neither as rigid as the Earth's nor as compressible as Jupiter's. An interesting feature predicted by both the MHD and empirical models is that as the dynamic pressure becomes weaker, Saturn's

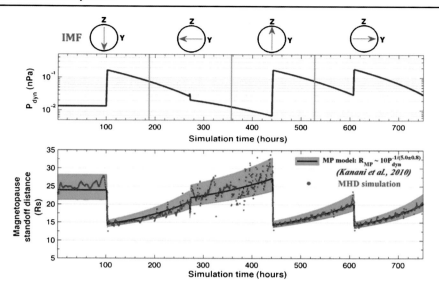

Fig. 21 Comparison of the magnetopause standoff distance between the global MHD simulation of Jia et al. (2012a) and the data-based empirical model of Kanani et al. (2010) constructed based on Cassini observations. (*top*) The solar wind conditions used as model input, including the IMF orientation (indicated by the *red arrows*) and the dynamic pressure. The *orange lines* mark the times when the discontinuities of IMF rotation arrive at the nose of the magnetopause. (*bottom*) The *red dots* show the standoff distance extracted from the MHD simulation and the *blue trace* along with the *green shaded area* indicate the standoff distance predicted by the magnetopause model of Kanani et al. (2010)

magnetopause boundary location tends to exhibit larger variations. The global MHD simulation also suggests that the location of Saturn's magnetopause is less sensitive to changes in the IMF orientation than to the dynamic pressure, a result also obtained from another global simulation study by Fukazawa et al. (2007b).

In addition to the sub-solar locations of the magnetopause and bow shock, another important aspect concerning the shape of the magnetosphere is its dimensions in the dawn-dusk direction and in the north-south direction. Because of the presence of a disk-like current sheet at Jupiter, Jupiter's magnetospheric boundaries are expected to exhibit strong polar flattening (e.g., Slavin et al. 1985; Huddleston et al. 1998; Pilkington et al. 2014), meaning that the magnetosphere extends further from the planet in the dawn-dusk direction than in the north-south direction. Walker et al. (2005) have specifically discussed the issue of polar flattening by use of their Jupiter simulations considering a variety of external conditions. They find that, while polar flattening is present in their simulations, the level of flattening appears to depend on various factors including both the strength and orientation of the IMF and the upstream solar wind dynamic pressure. Their model results suggest that polar flattening is more prominent for lower solar wind dynamic pressure. While more observational data along with modeling efforts are clearly needed to better characterize the polar flattening effect at Jupiter, comparatively little is known about this effect at Saturn. Future studies combining analysis of Cassini measurements with global magnetosphere models certainly need to be undertaken to assess the effect of polar flattening at Saturn.

5.2.2 Global Dynamics (Magnetospheric Responses to Solar Wind Driving)

Global MHD models have also been used extensively to study the dynamics of the giant planet magnetospheres. For rapidly rotating magnetospheres like those of Jupiter and Saturn, the interplay between rotationally driven processes and solar wind driven processes is an important aspect in considering global magnetospheric dynamics. Towards this end, global MHD models have been used to investigate the relative importance of various internal and external processes in driving global magnetospheric dynamics, though the detailed internal magnetodisc force balance involving both thermal and superthermal plasma populations described by Achilleos et al. (2010a,b) cannot be self-consistently modeled in MHD.

For Jupiter, while it is generally thought that the magnetosphere is driven largely by internal processes, the extent to which the solar wind affects the global dynamics of the magnetosphere remains uncertain. Fukazawa et al. (2006) performed a set of global MHD simulations to systematically examine the effect of the IMF and solar wind dynamic pressure on Jupiter's magnetospheric structure and dynamics. They find that the dynamic pressure controls the overall size of the magnetosphere and can sometimes influence the properties of Jupiter's plasma sheet through altering the size of the cushion region, an intermediate region of enhanced magnetic field strength between the dayside magnetopause and the dense plasma sheet (see discussion by Went et al. 2011 and Kivelson and Southwood 2005). As the dynamic pressure becomes larger, the cushion region becomes narrower such that changes in the external solar wind can have more direct and stronger effects on the plasma sheet. The external solar wind conditions are also found in their global simulation to be able to impose strong influences on the global plasma convection and the magnetospheric dynamics.

Among other things, a manifestation of the external influence is the control on the way magnetic reconnection occurs in Jupiter's magnetotail. Their model results show that both the IMF and dynamic pressure play an important role in determining the location of the tail neutral line, with the neutral line being closer to the planet for higher dynamic pressure and stronger IMF cases. The external conditions also affect how frequently tail reconnection takes place. Under low solar wind pressure and weak IMF conditions, tail reconnection tends to occur in a periodic manner with periods of the order of tens of hours. The Galileo Energetic Particle Detectors occasionally observed periodic flow bursts (likely associated with reconnection) in the jovian magnetotail with a periodicity of 2 to 3 days (e.g., Krupp et al. 1998; Woch et al. 2002; Kronberg et al. 2005), however, the origin of these periodic flow bursts is not well understood. The New Horizons spacecraft traversed Jupiter's magnetotail and the Solar Wind Around Pluto (SWAP) instrument showed diverse plasma populations and structures. The quasi-periodic fluctuations seen by SWAP at a 3- to 4-day period were thought to be caused by plasmoids moving down the tail (McComas et al. 2007). Given that the repetitive tail reconnection events seen in the global simulations have periods comparable to that associated with the observed flow bursts, Fukazawa et al. (2006) have argued that it is possible that the observed periodic flow bursts were produced by periodic formation of tail X-line driven by the solar wind under certain external conditions.

While it is still debated whether or not the solar wind significantly impacts Jupiter's magnetosphere, it is clear that Saturn's magnetosphere responds strongly to solar wind forcing (Sect. 4.3.2). The effect of the solar wind forcing on Saturn's magnetosphere has also been studied using global MHD simulations including both single-fluid (Fukazawa et al. 2007a, 2007b; Walker et al. 2011; Zieger et al. 2010; Jia et al. 2012b) and multi-fluid MHD models (Kidder et al. 2012). Some of the modeling studies have focused on the large-scale behavior of the magnetosphere under steady solar wind conditions, while others have considered relatively realistic external conditions by using time-varying solar wind input. For instance,

Kidder et al. (2012) examined the response of Saturn's magnetosphere to IMF rotation and solar wind pressure enhancement while including the affect of a warped magnetodisc generated by mechanical stresses delivered from the solar wind due to the angle between the solar wind flow and Saturn's spin axis (Arridge et al. 2008; Carbary et al. 2010). They found that both types of external forcing can trigger tail reconnection forming plasmoids in their simulation. Motivated by various observational studies (e.g., Clarke et al. 2005; Crary et al. 2005; Cowley et al. 2005b; Bunce et al. 2005b), Jia et al. (2012b) conducted a global simulation study using a solar wind input that contains features, such as compression and rarefaction, typical of the Corotating Interaction Regions (CIRs) formed in the solar wind near Saturn (Jackman et al. 2008) in order to characterize the dynamical response of the magnetosphere to different types of solar wind disturbances. Among other things, the Jia et al. study specifically investigated the role of tail reconnection in driving dynamics in Saturn's magnetosphere and how tail reconnection takes place under different external conditions.

There are two types of magnetic reconnection identified in the Jia et al. global simulation. The first corresponds to the so-called "Vasyliunas-cycle" reconnection, which is an internal process intrinsic to a rotationally driven magnetosphere (Vasyliunas 1983) where the centrifugal acceleration of mass-loaded flux tubes forced by the planetary rotation gives rise to reconnection on closed magnetic field lines. An important product of this process is the formation of plasmoids, which provide a means for removing plasma from the magnetosphere. The second type identified in the simulation refers to the so-called "Dungey-cycle" reconnection that involves open field lines previously stored in the tail lobes. The two types of reconnection appear to have different field and plasma characteristics in the reconnection products. In particular, the case involving Dungey-cycle reconnection, where reconnection proceeds to the lobe field lines above the plasma sheet, typically results in hotter and more depleted flux tubes with faster bulk flows in the outflows from the reconnection site compared to those produced directly by the Vasyliunas-cycle reconnection. On their return from the tail reconnection site to the dayside, the hot, tenuous, and rapidly moving flux tubes associated with lobe field reconnection may generate significant disturbances in the magnetosphere and the polar ionosphere, particularly on the dawn side, such as generating strong field-aligned currents that would be expected to cause auroral brightening (Badman and Cowley 2007; Mitchell 2015).

The interplay between the Vasyliunas cycle and the Dungey cycle leads to a complex picture of global magnetospheric convection at Saturn. Figure 22 shows an equatorial view of the plasma convection pattern seen in the global simulation of Jia et al. (2012b) under conditions of strong solar wind driving. Although this figure shows only a snapshot from the simulation that should not be viewed as representing a steady state (especially in the case of varying solar wind conditions), it does reveal notable features typically seen when both the Vasyliunas and the Dungey cycles are at work. The pure Vasyliunas-cycle X-line (see the closed magnetic loops shown by the orange lines) is confined to a limited region in the pre-midnight sector while the Dungey-cycle X-line (across which tailward and planetward fast flows are present), albeit variable both in space and time, is seen mainly in the post-midnight region, adjacent to the Vasyliunas-cycle X-line. In addition to the X-line geometries, another important feature of the resulting convection pattern is that the magnetotail on both the dawnside and duskside flanks contains regions of mass-loaded flux tubes streaming down the tail. These regions appear to be rather important for the loss of magnetospheric plasma, which will be further discussed below.

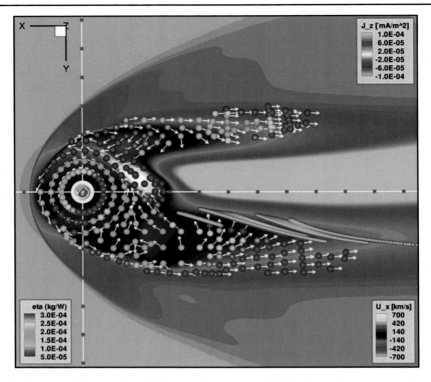

Fig. 22 An equatorial view of the global convection pattern and distribution of plasma content in Saturn's magnetosphere under strong solar driving conditions (from Jia et al. 2012b). The background colors represent contours of the horizontal flow velocity (V_x) and the *color contours* on a circular disk surrounding Saturn show field-aligned currents intensity in the northern ionosphere (mapped to radial distance of 4 R_S for illustration purposes). The intersections of sampled closed field lines with the equatorial plane are plotted as *balls* color-coded with their corresponding flux tube content. Also plotted are unit flow vectors of the closed field lines depicting the direction of their motion. The *orange traces* show some representative field lines that form closed loops. *Grey squares* mark off every 10 R_S along the axes

5.2.3 Plasma Loss from the Magnetosphere

For both Jupiter and Saturn, where their moons are continuously adding a significant amount of plasma into the magnetosphere, a fundamental problem is how the magnetospheric plasma is lost from the system. Quantifying such a process through analysis of in-situ measurements turns out to be rather difficult given the highly limited data coverage. Simple order-of-magnitude estimates based on the characteristics of observed tail reconnection and plasmoid events (Bagenal 2007; Bagenal and Delamere 2011) imply that at both planets, plasma loss through large-scale plasmoid release appears insufficient to account for the total loss of plasma required to balance the mass input by internal plasma sources. Global magnetosphere simulations may provide useful insight into how the plasma loss process(es) occur in a rapidly rotating magnetosphere. Kidder et al. (2012) showed that plasmoid formation can be triggered by external forcing (e.g. flipping the IMF direction or a pulse in dynamic pressure). Zieger et al. (2010) and Jia et al. (2012b) have attempted to quantify the mass release processes identified in their global MHD model of Saturn's magnetosphere. Their calculation indicates that on average large-scale plasmoid releases only account for a small

🖄 Springer

fraction ($\sim$10 %) of the total loss of plasma added to the magnetosphere by the moon Enceladus and its extended neutral cloud. However, they find that a significant fraction of the planetary plasma appears to be lost through processes (e.g., small-scale plasmoids) near the flanks of the magnetotail that contain flux tubes filled with magnetospheric plasma streaming down the tail (see Fig. 22). The breaking-off of those flux tubes (likely through small-scale plasmoid release) at large distances suggest these regions are crucially important for releasing plasma from the magnetosphere, a situation similar to that proposed by Kivelson and Southwood (2005) in a jovian context. Selective escape of hot plasma due to its kinetic properties was discussed by Sergis et al. (2013). Islands of heavy water groups ions were found in the magnetosheath and upstream on the bow shock, indicating that some plasma can escape at the dayside magnetopause. In addition, Krupp et al. (2002) discusses leakage of energetic particles on Jupiter's dusk flank and Thomsen et al. (2007) assesses superthermal ions in Saturn's foreshock region.

6 Summary

The interaction of giant planet magnetodiscs with the solar wind is tied to the momentum transfer rate from the solar wind to the magnetodisc. In steady state, the scale of the interaction must be sufficient to remove plasma supplied by Io ($\sim$100 s kg/s) and by Enceladus ($\sim$10 s kg/s) for Jupiter and Saturn respectively (Delamere and Bagenal 2013). Yet, the details of how plasma is transported from the inner to the outer magnetosphere is poorly understood and remains a fundamental and unsolved problem in magnetodisc physics. Ultimately, magnetic reconnection on both large-scale (Dungey/Vasyliunas cycle) and small-scale (along the flanks of the magnetosphere) together with diffusive processes at the magnetopause boundary facilitate the loss of plasma from the magnetodisc to the solar wind. Auroral signatures provide key diagnostics of this interaction and auroral evidence suggests that large-scale plasmoid formation in the tail is not the primary loss mechanism. Instead, small-scale drizzle along the dusk flank akin to a planetary wind may be the primary pathway (Bagenal 2007; Thomsen et al. 2014). Global-scale MHD simulations produce drizzle along the dusk flank, but we note that this process is a numerical artifact on the grid scale and cannot provide insight into scale-scale physics of the problem. We summarize with the following questions for future research efforts:

– What are the properties of radial plasma transport and magnetic flux circulation in the giant planet magnetodiscs?
– How are thin current sheets generated in the magnetodiscs, allowing reconnection to operate (Vasyliunas 1983; Zimbardo 1991)? What is the relative importance of centrifugal stresses (internal) vs. solar wind-induced stresses (external)?
– How is plasma heated during radial transport?
– What is the prevailing mechanism for mass loss on the dusk flank (i.e. diffusive or small-scale reconnection)?

These are some of the issues that will hopefully be addressed during the end-of-mission polar orbit of the Cassini spacecraft and the arrival of the polar-orbiting Juno spacecraft at Jupiter.

Acknowledgements   Peter Delamere acknowledges support by NASA grant NNX13AH309G. Xianzhe Jia is supported by the NASA Cassini Data Analysis Program through grant NNX12AK34G and the NASA Outer Planets Research Program through grant NNX12AM74G, and by the NASA Cassini mission under contract 1409449 with JPL. Licia Ray acknowledges supported by NSF Grant 1064635. Chris Arridge is supported by the "Royal Society University Research Fellowship".

References

N. Achilleos, P. Guio, C.S. Arridge, A model of force balance in Saturn's magnetodisc. Mon. Not. R. Astron. Soc. **401**, 2349–2371 (2010a). doi:10.1111/j.1365-2966.2009.15865.x

N. Achilleos, P. Guio, C.S. Arridge, N. Sergis, R.J. Wilson, M.F. Thomsen, A.J. Coates, Influence of hot plasma pressure on the global structure of Saturn's magnetodisk. Geophys. Res. Lett. **37**, L20201 (2010b). doi:10.1029/2010GL045159

N. Achilleos et al., Space Sci. Rev. (2014, this issue)

B.J. Anderson, T.-D. Phan, S.A. Fuselier, Relationships between plasma depletion and subsolar reconnection. J. Geophys. Res. **102**, 9531–9542 (1997). doi:10.1029/97JA00173

C.S. Arridge, N. Achilleos, M.K. Dougherty, K.K. Khurana, C.T. Russell, Modeling the size and shape of Saturn's magnetopause with variable dynamic pressure. J. Geophys. Res. (Space Phys.) **111**, A11227 (2006). doi:10.1029/2005JA011574

C.S. Arridge, K.K. Khurana, C.T. Russell, D.J. Southwood, N. Achilleos, M.K. Dougherty, A.J. Coates, H.K. Leinweber, Warping of Saturn's magnetospheric and magnetotail current sheets. J. Geophys. Res. (Space Phys.) **113**, A08217 (2008). doi:10.1029/2007JA012963

C.S. Arridge, N. André, C.L. Bertucci, P. Garnier, C.M. Jackman, Z. Németh, A.M. Rymer, N. Sergis, K. Szego, A.J. Coates, F.J. Crary, Upstream of Saturn and Titan. Space Sci. Rev. **162**, 25–83 (2011). doi:10.1007/s11214-011-9849-x

C. Arridge, N. Achilleos, Y. Bogdanova, E.J. Bunce, S. Cowley, M. Dougherty, A. Fazakerley, G. Jones, J. Jasinski, K. Khuana, S. Krimigis, N. Krupp, L. Lamy, J. Leisner, E. Roussos, C. Russell, P. Zarka, Cassini observations of Saturn's southern polar cusp. J. Geophys. Res. (2014, submitted)

W.I. Axford, C.O. Hines, A unifying theory of high latitude geophysical phenomena and geomagnetic storms. Can. J. Phys. **39**, 1433 (1961)

S.V. Badman, E.J. Bunce, J.T. Clarke, S.W.H. Cowley, J.-C. Gérard, D. Grodent, S.E. Milan, Open flux estimates in Saturn's magnetosphere during the January 2004 Cassini-HST campaign, and implications for reconnection rates. J. Geophys. Res. (Space Phys.) **110**, A11216 (2005). doi:10.1029/2005JA011240

S.V. Badman, S.W.H. Cowley, J.-C. Gérard, D. Grodent, A statistical analysis of the location and width of Saturn's southern auroras. Ann. Geophys. **24**(12), 3533–3545 (2006)

S.V. Badman, S.W.H. Cowley, Significance of Dungey-cycle flows in Jupiter's and Saturn's magnetospheres, and their identification on closed equatorial field lines. Ann. Geophys. **25**, 941–951 (2007). doi:10.5194/angeo-25-941-2007

S.V. Badman, D.J. Andrews, S.W.H. Cowley, L. Lamy, G. Provan, C. Tao, S. Kasahara, T. Kimura, M. Fujimoto, H. Melin, T. Stallard, R.H. Brown, K.H. Baines, Rotational modulation and local time dependence of Saturn's infrared H_3^+ auroral intensity. J. Geophys. Res. (Space Phys.) **117**, A09228 (2012a). doi:10.1029/2012JA017990

S.V. Badman, N. Achilleos, C.S. Arridge, K.H. Baines, R.H. Brown, E.J. Bunce, A.J. Coates, S.W.H. Cowley, M.K. Dougherty, M. Fujimoto, G. Hospodarsky, S. Kasahara, T. Kimura, H. Melin, D.G. Mitchell, T. Stallard, C. Tao, Cassini observations of ion and electron beams at Saturn and their relationship to infrared auroral arcs. J. Geophys. Res. **117**(A1), A01211 (2012b). doi:10.1029/2011JA017222

S.V. Badman, A. Masters, H. Hasegawa, M. Fujimoto, A. Radioti, D. Grodent, N. Sergis, M.K. Dougherty, A.J. Coates, Bursty magnetic reconnection at Saturn's magnetopause. Geophys. Res. Lett. **40**, 1027–1031 (2013)

S.V. Badman, Saturn's auroral morphology and field-aligned currents during a solar wind compression. Icarus (2014, submitted)

S.V. Badman, G. Branduardi-Raymont, M. Galand, S.L.G. Hess, N. Krupp, L. Lamy, H. Melin, C. Tao, Auroral processes at the giant planets: energy deposition, emission mechanisms, morphology and spectra. Space Sci. Rev. (2014a). doi:10.1007/s11214-014-0042-x

S.V. Badman, C.M. Jackman, J.D. Nichols, J.T. Clarke, J.-C. Gérard, Open flux in Saturn's magnetosphere. Icarus **231**, 137–145 (2014b). doi:10.1016/j.icarus.2013.12.004

F. Bagenal, The magnetosphere of Jupiter: coupling the equator to the poles. J. Atmos. Sol.-Terr. Phys. **69**, 387–402 (2007). doi:10.1016/j.jastp.2006.08.012

F. Bagenal, P.A. Delamere, Flow of mass and energy in the magnetospheres of Jupiter and Saturn. J. Geophys. Res. (Space Phys.) **116**, A05209 (2011). doi:10.1029/2010JA016294

G.E. Ballester, J.T. Clarke, J.T. Trauger, W.M. Harris, K.R. Stapelfeldt, D. Crisp, R.W. Evans, E.B. Burgh, C.J. Burrows, S. Casertano, J.S. Gallagher III, R.E. Griffiths, J.J. Hester, J.G. Hoessel, J.A. Holtzman, J.E. Krist, V. Meadows, J.R. Mould, R. Sahai, P.A. Scowen, A.M. Watson, J.A. Westphal, Time-resolved observations of Jupiter's far-ultraviolet aurora. Science **274**, 409–412 (1996)

R.L. Baron, T. Owen, J.E.P. Connerney, T. Satoh, J. Harrington, Solar wind control of Jupiter's H_3^+ auroras. Icarus **120**, 437–442 (1996). doi:10.1006/icar.1996.0063

C. Barrow, M. Desch, F. Genova, Solar-wind control of Jupiter decametric radio-emission. Astron. Astrophys. **165**(1–2), 244–250 (1986)

E.S. Belenkaya, I.I. Alexeev, M.S. Blokhina, E.J. Bunce, S.W.H. Cowley, J.D. Nichols, V.V. Kalegaev, V.G. Petrov, G. Provan, IMF dependence of Saturn's auroras: modelling study of HST and Cassini data from 12–15 February 2008. Ann. Geophys. **28**, 1559–1570 (2010). doi:10.5194/angeo-28-1559-2010

E.S. Belenkaya, S.W.H. Cowley, J.D. Nichols, M.S. Blokhina, V.V. Kalegaev, Magnetospheric mapping of the dayside UV auroral oval at Saturn using simultaneous HST images, Cassini IMF data, and a global magnetic field model. Ann. Geophys. **29**, 1233–1246 (2011). doi:10.5194/angeo-29-1233-2011

B. Bonfond, When moons create aurora: the satellite footprints on giant planets, in *Auroral Phenomenology and Magnetospheric Processes: Earth and Other Planets*. Geophysical Monograph Series, vol. 197 (Am. Geophys. Union, Washington, 2012), pp. 133–140

B. Bonfond, M.F. Vogt, J.-C. Gérard, D. Grodent, A. Radioti, V. Coumans, Quasi-periodic polar flares at Jupiter: a signature of pulsed dayside reconnections? Geophys. Res. Lett. **38**(2), L02104 (2011). doi:10.1029/2010GL045981

B. Bonfond, D. Grodent, J.-C. Gérard, T. Stallard, J.T. Clarke, M. Yoneda, A. Radioti, J. Gustin, Auroral evidence of Io's control over the magnetosphere of Jupiter. Geophys. Res. Lett. **39**, L01105 (2012). doi:10.1029/2011GL050253

S. Bose, A. Bhattacharya, Solar plasma activated decametric radio emission of non-Io origin from Jupiter magnetosphere. Indian J. Radio Space Phys. **32**(1), 43–51 (2003)

G. Branduardi-Raymont, R.F. Elsner, M. Galand, D. Grodent, T.E. Cravens, P. Ford, G.R. Gladstone, J.H. Waite, Spectral morphology of the X-ray emission from Jupiter's aurorae. J. Geophys. Res. (Space Phys.) **113**, A02202 (2008). doi:10.1029/2007JA012600

E.J. Bunce, S.W.H. Cowley, T.K. Yeoman, Jovian cusp processes: implications for the polar aurora. J. Geophys. Res. **109**(A18), A09S13 (2004). doi:10.1029/2003JA010280

E.J. Bunce, S.W.H. Cowley, S.E. Milan, Interplanetary magnetic field control of Saturn's polar cusp aurora. Ann. Geophys. **23**, 1405–1431 (2005a). doi:10.5194/angeo-23-1405-2005

E.J. Bunce, S.W.H. Cowley, D.M. Wright, A.J. Coates, M.K. Dougherty, N. Krupp, W.S. Kurth, A.M. Rymer, In situ observations of a solar wind compression-induced hot plasma injection in Saturn's tail. Geophys. Res. Lett. **32**, L20S04 (2005b). doi:10.1029/2005GL022888

E.J. Bunce, C.S. Arridge, J.T. Clarke, A.J. Coates, S.W.H. Cowley, M.K. Dougherty, J.-C. Gérard, D. Grodent, K.C. Hansen, J.D. Nichols, D.J. Southwood, D.L. Talboys, Origin of Saturn's aurora: simultaneous observations by Cassini and the Hubble Space Telescope. J. Geophys. Res. **113**(A12), A09209 (2008). doi:10.1029/2008JA013257

J.F. Carbary, N. Achilleos, C.S. Arridge, K.K. Khurana, M.K. Dougherty, Global configuration of Saturn's magnetic field derived from observations. Geophys. Res. Lett. **37**, L21806 (2010). doi:10.1029/2010GL044622

P.A. Cassak, A. Otto, Scaling of the magnetic reconnection rate with symmetric shear flow. Phys. Plasmas **18**(7), 074501 (2011). doi:10.1063/1.3609771

T.A. Cassidy, R.E. Johnson, Collisional spreading of Enceladus's neutral cloud. Icarus **209**, 696–703 (2010). doi:10.1016/j.icarus.2010.04.010

G. Caudal, A self-consistent model of Jupiter's magnetodisc including the effects of centrifugal force and pressure. J. Geophys. Res. **91**, 4201–4221 (1986). doi:10.1029/JA091iA04p04201

S. Chandrasekhar, *Hydrodynamic and Hydromagnetic Stability* (1961)

J.T. Clarke, L. Ben Jaffel, J.-C. Gérard, Hubble Space Telescope imaging of Jupiter's UV aurora during the Galileo orbiter mission. J. Geophys. Res. **103**, 20,217–20,236 (1998). doi:10.1029/98JE01130

J.T. Clarke, J.-C. Gérard, D. Grodent, S. Wannawichian, J. Gustin, J. Connerney, F.J. Crary, M.K. Dougherty, W.S. Kurth, S.W.H. Cowley, E.J. Bunce, T.W. Hill, J. Kim, Morphological differences between Saturn's ultraviolet aurorae and those of Earth and Jupiter. Nature **433**(7027), 717–719 (2005). doi:10.1038/nature03331

J.T. Clarke, J.D. Nichols, J.-C. Gérard, D. Grodent, K.C. Hansen, W.S. Kurth, G.R. Gladstone, J. Duval, S. Wannawichian, E.J. Bunce, S.W.H. Cowley, F.J. Crary, M.K. Dougherty, L. Lamy, D. Mitchell, W.R. Pryor, K. Retherford, T.S. Stallard, B. Zieger, P. Zarka, B. Cecconi, The response of Jupiter's and Saturn's auroral activity to the solar wind. J. Geophys. Res. **114**, A05210 (2009). doi:10.1029/2008JA013694

J.F. Cooper, J.A. Simpson, Sources of high-energy protons in Saturn's magnetosphere. J. Geophys. Res. **85**, 5793–5802 (1980). doi:10.1029/JA085iA11p05793

M.M. Cowee, D. Winske, S.P. Gary, Hybrid simulations of plasma transport by Kelvin-Helmholtz instability at the magnetopause: density variations and magnetic shear. J. Geophys. Res. (Space Phys.) **115**, A06214 (2010). doi:10.1029/2009JA015011

S. Cowley, E. Bunce, R. Prangé, Saturn's polar ionospheric flows and their relation to the main auroral oval. Ann. Geophys. **22**, 1379–1394 (2004). doi:10.5194/angeo-22-1379-2004

S.W.H. Cowley, E.J. Bunce, Origin of the main auroral oval in Jupiter's coupled magnetosphere-ionosphere system. Planet. Space Sci. **49**, 1067–1088 (2001). doi:10.1016/S0032-0633(00)00167-7

S.W.H. Cowley, E.J. Bunce, Corotation-driven magnetosphere-ionosphere coupling currents in Saturn's magnetosphere and their relation to the auroras. Ann. Geophys. **21**, 1691–1707 (2003). doi:10.5194/angeo-21-1691-2003

S.W.H. Cowley, E.J. Bunce, T.S. Stallard, S. Miller, Jupiter's polar ionospheric flows: theoretical interpretation. Geophys. Res. Lett. **30**, 1220 (2003). doi:10.1029/2002GL016030

S.W.H. Cowley, I.I. Alexeev, E.S. Belenkaya, E.J. Bunce, C.E. Cottis, V.V. Kalegaev, J.D. Nichols, R. Prangé, F.J. Wilson, A simple axisymmetric model of magnetosphere-ionosphere coupling currents in Jupiter's polar ionosphere. J. Geophys. Res. **110**(A9), A11209 (2005a). doi:10.1029/2005JA011237

S.W.H. Cowley, S.V. Badman, E.J. Bunce, J.T. Clarke, J.-C. Gérard, D. Grodent, C.M. Jackman, S.E. Milan, T.K. Yeoman, Reconnection in a rotation-dominated magnetosphere and its relation to Saturn's auroral dynamics. J. Geophys. Res. **110**(A9), A02201 (2005b). doi:10.1029/2004JA010796

S.W.H. Cowley, J.D. Nichols, D.J. Andrews, Modulation of Jupiter's plasma flow, polar currents, and auroral precipitation by solar wind-induced compressions and expansions of the magnetosphere: a simple theoretical model. Ann. Geophys. **25**, 1433–1463 (2007)

S.W.H. Cowley, S.V. Badman, S.M. Imber, S.E. Milan, Comment on "Jupiter: a fundamentally different magnetospheric interaction with the solar wind" by D.J. McComas and F. Bagenal. Geophys. Res. Lett. **35**, L10101 (2008). doi:10.1029/2007GL032645

F.J. Crary, J.T. Clarke, M.K. Dougherty, P.G. Hanlon, K.C. Hansen, J.T. Steinberg, B.L. Barraclough, A.J. Coates, J.-C. Gérard, D. Grodent, W.S. Kurth, D.G. Mitchell, A.M. Rymer, D.T. Young, Solar wind dynamic pressure and electric field as the main factors controlling Saturn's aurorae. Nature **433**, 720–722 (2005). doi:10.1038/nature03333

J.C. Cutler, M.K. Dougherty, E. Lucek, A. Masters, Evidence of surface wave on the dusk flank of Saturn's magnetopause possibly caused by the Kelvin-Helmholtz instability. J. Geophys. Res. (Space Phys.) **116**, A10220 (2011). doi:10.1029/2011JA016643

R.B. Decker, S.M. Krimigis, E.C. Roelof, M.E. Hill, T.P. Armstrong, G. Gloeckler, D.C. Hamilton, L.J. Lanzerotti, Voyager 1 in the foreshock, termination shock, and heliosheath. Science **309**, 2020–2024 (2005). doi:10.1126/science.1117569

P.A. Delamere, F. Bagenal, Modeling variability of plasma conditions in the Io torus. J. Geophys. Res. **108**, 5-1 (2003) doi:10.1029/2002JA009706

P.A. Delamere, F. Bagenal, Solar wind interaction with Jupiter's magnetosphere. J. Geophys. Res. **115**, A10201 (2010). doi:10.1029/2010JA015347

P.A. Delamere, F. Bagenal, Magnetotail structure of the giant magnetospheres: implications of the viscous interaction with the solar wind. J. Geophys. Res. (Space Phys.) **118**, 7045–7053 (2013). doi:10.1002/2013JA019179

P.A. Delamere, R.J. Wilson, A. Masters, Kelvin-Helmholtz instability at Saturn's magnetopause: hybrid simulations. J. Geophys. Res. (Space Phys.) **116**, A10222 (2011). doi:10.1029/2011JA016724

P.A. Delamere, R.J. Wilson, S. Eriksson, F. Bagenal, Magnetic signatures of Kelvin-Helmholtz vortices on Saturn's magnetopause: global survey. J. Geophys. Res. (Space Phys.) **118**, 393–404 (2013). doi:10.1029/2012JA018197

M.D. Desch, M.L. Kaiser, Voyager measurements of the rotation period of Saturn's magnetic field. Geophys. Res. Lett. **8**(3), 253–256 (1981)

M. Desroche, F. Bagenal, P.A. Delamere, N. Erkaev, Conditions at the expanded Jovian magnetopause and implications for the solar wind interaction. J. Geophys. Res. (Space Phys.) **117**, A07202 (2012). doi:10.1029/2012JA017621

K. Dialynas, S.M. Krimigis, D.G. Mitchell, D.C. Hamilton, N. Krupp, P.C. Brandt, Energetic ion spectral characteristics in the Saturnian magnetosphere using Cassini/MIMI measurements. J. Geophys. Res. (Space Phys.) **114**, A01212 (2009). doi:10.1029/2008JA013761

J.W. Dungey, Electrodynamics of the outer atmosphere, in *Proceedings of the Ionosphere Conference* (The Physical Society of London, 1955), p. 225

J.W. Dungey, Interplanetary magnetic field and the auroral zones. Phys. Rev. Lett. **6**(2), 47–48 (1961). doi:10.1103/PhysRevLett.6.47

E. Echer, P. Zarka, W.D. Gonzalez, A. Morioka, L. Denis, Solar wind effects on Jupiter non-Io DAM emissions during Ulysses distant encounter (2003–2004). Astron. Astrophys. **519**, A84 (2010). doi:10.1051/0004-6361/200913305

R.F. Elsner, N. Lugaz, J.H. Waite, T.E. Cravens, G.R. Gladstone, P. Ford, D. Grodent, A. Bhardwaj, R.J. MacDowall, M.D. Desch, T. Majeed, Simultaneous Chandra X ray, Hubble Space Telescope ultraviolet, and Ulysses radio observations of Jupiter's aurora. J. Geophys. Res. **110**(A9), A01207 (2005). doi:10.1029/2004JA010717

G.J. Fasel, L.C. Lee, R.W. Smith, A mechanism for the multiple brightenings of dayside poleward-moving auroral forms. Geophys. Res. Lett. **20**, 2247–2250 (1993). doi:10.1029/93GL02487

B.L. Fleshman, P.A. Delamere, F. Bagenal, T. Cassidy, The roles of charge exchange and dissociation in spreading Saturn's neutral clouds. J. Geophys. Res., Planets **117**, E05007 (2012). doi:10.1029/2011JE003996

L.A. Frank, J.D. Craven, J.L. Burch, J.D. Winningham, Polar views of the earth's aurora with Dynamics Explorer. Geophys. Res. Lett. **9**, 1001–1004 (1982). doi:10.1029/GL009i009p01001

K. Fukazawa, T. Ogino, R.J. Walker, Dynamics of the Jovian magnetosphere for northward interplanetary magnetic field (IMF). Geophys. Res. Lett. **32**, L03202 (2005). doi:10.1029/2004GL021392

K. Fukazawa, T. Ogino, R.J. Walker, Configuration and dynamics of the Jovian magnetosphere. J. Geophys. Res. (Space Phys.) **111**, A10207 (2006). doi:10.1029/2006JA011874

K. Fukazawa, T. Ogino, R.J. Walker, Vortex-associated reconnection for northward IMF in the Kronian magnetosphere. Geophys. Res. Lett. **34**, L23201 (2007a). doi:10.1029/2007GL031784

K. Fukazawa, S.-i. Ogi, T. Ogino, R.J. Walker, Magnetospheric convection at Saturn as a function of IMF BZ. Geophys. Res. Lett. **34**, 1105 (2007b). doi:10.1029/2006GL028373

K. Fukazawa, T. Ogino, R.J. Walker, A simulation study of dynamics in the distant Jovian magnetotail. J. Geophys. Res. (Space Phys.) **115**, A09219 (2010). doi:10.1029/2009JA015228

S.A. Fuselier, H.U. Frey, K.J. Trattner, S.B. Mende, J.L. Burch, Cusp aurora dependence on interplanetary magnetic field B_z. J. Geophys. Res. (Space Phys.) **107**, 1111 (2002). doi:10.1029/2001JA900165

S.A. Fuselier, R. Frahm, W.S. Lewis, A. Masters, J. Mukherjee, S.M. Petrinec, I.J. Sillanpaa, The location of magnetic reconnection at Saturn's magnetopause: a comparison with Earth. J. Geophys. Res. (Space Phys.) **119**, 2563–2578 (2014). doi:10.1002/2013JA019684

P.H.M. Galopeau, P. Zarka, D.L. Quéau, Source location of Saturn's kilometric radiation: the Kelvin-Helmholtz instability hypothesis. J. Geophys. Res. **100**, 26,397–26,410 (1995). doi:10.1029/95JE02132

Y.S. Ge, C.T. Russell, K.K. Khurana, Reconnection sites in Jupiter's magnetotail and relation to Jovian auroras. Planet. Space Sci. **58**, 1455–1469 (2010). doi:10.1016/j.pss.2010.06.013

F. Genova, P. Zarka, C. Barrow, Voyager and Nancay observations of the Jovian radio-emission at different frequencies—solar-wind effect and source extent. Astron. Astrophys. **182**, 159–162 (1987)

J.-C. Gérard, D. Grodent, J. Gustin, A. Saglam, J.T. Clarke, J.T. Trauger, Characteristics of Saturn's FUV aurora observed with the Space Telescope Imaging Spectrograph. J. Geophys. Res. (Space Phys.) **109**, A09207 (2004). doi:10.1029/2004JA010513

J.-C. Gérard, E.J. Bunce, D. Grodent, S.W.H. Cowley, J.T. Clarke, S.V. Badman, Signature of Saturn's auroral cusp: simultaneous Hubble Space Telescope FUV observations and upstream solar wind monitoring. J. Geophys. Res. **110**(A9), A11201 (2005). doi:10.1029/2005JA011094

G.R. Gladstone, J.H. Waite, D. Grodent, W. Lewis, F.J. Crary, R.F. Elsner, M. Weisskopf, T. Majeed, J. Jahn, A. Bhardwaj, J.T. Clarke, D.T. Young, M.K. Dougherty, S.A. Espinosa, T.E. Cravens, A pulsating auroral X-ray hot spot on Jupiter. Nature **415**(6875), 1000–1003 (2002)

D. Grodent, J.T. Clarke, J. Kim, J.H. Waite, S.W.H. Cowley, Jupiter's main auroral oval observed with HST-STIS. J. Geophys. Res. **108**, 1389 (2003a). doi:10.1029/2003JA009921

D. Grodent, J.T. Clarke, J.H. Waite, S.W.H. Cowley, J.-C. Gérard, J. Kim, Jupiter's polar auroral emissions. J. Geophys. Res. **108**, 1366 (2003b). doi:10.1029/2003JA010017

D. Grodent, J.-C. Gérard, J.T. Clarke, G.R. Gladstone, J.H. Waite, A possible auroral signature of a magnetotail reconnection process on Jupiter. J. Geophys. Res. (Space Phys.) **109**, A05201 (2004). doi:10.1029/2003JA010341

D. Grodent, J.-C. Gérard, S.W.H. Cowley, E.J. Bunce, J.T. Clarke, Variable morphology of Saturn's southern ultraviolet aurora. J. Geophys. Res. (Space Phys.) **110**, A07215 (2005). doi:10.1029/2004JA010983

D. Grodent, J.-C. Gérard, A. Radioti, B. Bonfond, A. Saglam, Jupiter's changing auroral location. J. Geophys. Res. **113**, 1206 (2008). doi:10.1029/2007JA012601

D. Grodent, A. Radioti, B. Bonfond, J.-C. Gérard, On the origin of Saturn's outer auroral emission. J. Geophys. Res. (Space Phys.) **115**, A08219 (2010). doi:10.1029/2009JA014901

D. Grodent, J. Gustin, J.C. Gérard, A. Radioti, B. Bonfond, W.R. Pryor, Small-scale structures in Saturn's ultraviolet aurora. J. Geophys. Res. **116**(A9), A09225 (2011). doi:10.1029/2011JA016818

D.A. Gurnett, W.S. Kurth, G.B. Hospodarsky, A.M. Persoon, P. Zarka, A. Lecacheux, S.J. Bolton, M.D. Desch, W.M. Farrell, M.L. Kaiser, H.-P. Ladreiter, H.O. Rucker, P. Galopeau, P. Louarn, D.T. Young, W.R. Pryor, M.K. Dougherty, Control of Jupiter's radio emission and aurorae by the solar wind. Nature **415**, 985–987 (2002)

J. Gustin, J.-C. Gérard, D. Grodent, S.W.H. Cowley, J.T. Clarke, A. Grard, Energy-flux relationship in the FUV Jovian aurora deduced from HST-STIS spectral observations. J. Geophys. Res. **109**, 10,205 (2004). doi:10.1029/2003JA010365

J. Gustin, S.W.H. Cowley, J.-C. Gérard, G.R. Gladstone, D. Grodent, J.T. Clarke, Characteristics of Jovian morning bright FUV aurora from Hubble Space Telescope/Space Telescope Imaging Spectrograph imaging and spectral observations. J. Geophys. Res. (Space Phys.) **111**, A09220 (2006). doi:10.1029/2006JA011730

D.K. Haggerty, M.E. Hill, R.L. McNutt, C. Paranicas, Composition of energetic particles in the Jovian magnetotail. J. Geophys. Res. (Space Phys.) **114**, A02208 (2009). doi:10.1029/2008JA013659

K.C. Hansen, A.J. Ridley, G.B. Hospodarsky, N. Achilleos, M.K. Dougherty, T.I. Gombosi, G. Tóth, Global MHD simulations of Saturn's magnetosphere at the time of Cassini approach. Geophys. Res. Lett. **32**, L20S06 (2005). doi:10.1029/2005GL022835

H. Hasegawa, M. Fujimoto, T.-D. Phan, H. Rème, A. Balogh, M.W. Dunlop, C. Hashimoto, R. TanDokoro, Transport of solar wind into Earth's magnetosphere through rolled-up Kelvin-Helmholtz vortices. Nature **430**, 755–758 (2004). doi:10.1038/nature02799

H. Hasegawa, A. Retinò, A. Vaivads, Y. Khotyaintsev, M. André, T.K.M. Nakamura, W.-L. Teh, B.U.Ö. Sonnerup, S.J. Schwartz, Y. Seki, M. Fujimoto, Y. Saito, H. Rème, P. Canu, Kelvin-Helmholtz waves at the Earth's magnetopause: multiscale development and associated reconnection. J. Geophys. Res. (Space Phys.) **114**, A12207 (2009). doi:10.1029/2009JA014042

S.L.G. Hess, E. Echer, P. Zarka, Solar wind pressure effects on Jupiter decametric radio emissions independent of Io. Planet. Space Sci. **70**, 114–125 (2012). doi:10.1016/j.pss.2012.05.011

T.W. Hill, Inertial limit on corotation. J. Geophys. Res. **84**, 6554–6558 (1979). doi:10.1029/JA084iA11p06554

T.W. Hill, A.J. Dessler, F.C. Michel, Configuration of the Jovian magnetosphere. Geophys. Res. Lett. **1**, 3–6 (1974). doi:10.1029/GL001i001p00003

T.S. Huang, T.W. Hill, Corotation lag of the Jovian atmosphere, ionosphere, and magnetosphere. J. Geophys. Res. **94**, 3761–3765 (1989). doi:10.1029/JA094iA04p03761

D.E. Huddleston, C.T. Russell, G. Le, A. Szabo, Magnetopause structure and the role of reconnection at the outer planets. J. Geophys. Res. **102**, 24,289–24,004 (1997). doi:10.1029/97JA02416

D.E. Huddleston, C.T. Russell, M.G. Kivelson, K.K. Khurana, L. Bennett, Location and shape of the Jovian magnetopause and bow shock. J. Geophys. Res. **103**, 20,075–20,082 (1998). doi:10.1029/98JE00394

C.M. Jackman, N. Achilleos, E.J. Bunce, B. Cecconi, J.T. Clarke, S.W.H. Cowley, W.S. Kurth, P. Zarka, Interplanetary conditions and magnetospheric dynamics during the Cassini orbit insertion fly-through of Saturn's magnetosphere. J. Geophys. Res. (Space Phys.) **110**, A10212 (2005). doi:10.1029/2005JA011054

C.M. Jackman, R.J. Forsyth, M.K. Dougherty, The overall configuration of the interplanetary magnetic field upstream of Saturn as revealed by Cassini observations. J. Geophys. Res. (Space Phys.) **113**, A08114 (2008). doi:10.1029/2008JA013083

C.M. Jackman, C.S. Arridge, Solar cycle effects on the dynamics of Jupiter's and Saturn's magnetospheres. Solar Phys. **274**(1–2), 481–502 (2011). doi:10.1007/s11207-011-9748-z

C.M. Jackman, N. Achilleos, S.W.H. Cowley, E.J. Bunce, A. Radioti, D. Grodent, S.V. Badman, M.K. Dougherty, W. Pryor, Auroral counterpart of magnetic field dipolarizations in Saturn's tail. Planet. Space Sci. **82**, 34–42 (2013). doi:10.1016/j.pss.2013.03.010

J. Jasinski, C. Arridge, L. Lamy, J. Leisner, M. Thomsen, D. Mitchell, A. Coates, A. Radioti, G. Jones, E. Roussos, N. Krupp, D. Grodent, M. Dougherty, J. Waite, Cusp observation at Saturn's high latitude magnetosphere by the Cassini spacecraft. Geophys. Res. Lett. (2014). doi:10.1002/2014GL059319

X. Jia, M.G. Kivelson, Driving Saturn's magnetospheric periodicities from the upper atmosphere/ionosphere: magnetotail response to dual sources. J. Geophys. Res. (Space Phys.) **117**, A11219 (2012). doi:10.1029/2012JA018183

X. Jia, M.G. Kivelson, T.I. Gombosi, Driving Saturn's magnetospheric periodicities from the upper atmosphere/ionosphere. J. Geophys. Res. (Space Phys.) **117**, A04215 (2012a). doi:10.1029/2011JA017367

X. Jia, K.C. Hansen, T.I. Gombosi, M.G. Kivelson, G. Tóth, D.L. DeZeeuw, A.J. Ridley, Magnetospheric configuration and dynamics of Saturn's magnetosphere: a global MHD simulation. J. Geophys. Res. (Space Phys.) **117**, A05225 (2012b). doi:10.1029/2012JA017575

S.J. Kanani, C.S. Arridge, G.H. Jones, A.N. Fazakerley, H.J. McAndrews, N. Sergis, S.M. Krimigis, M.K. Dougherty, A.J. Coates, D.T. Young, K.C. Hansen, N. Krupp, A new form of Saturn's magnetopause using a dynamic pressure balance model, based on in situ, multi-instrument Cassini measurements. J. Geophys. Res. (Space Phys.) **115**, 6207 (2010). doi:10.1029/2009JA014262

K.K. Khurana, Euler potential models of Jupiter's magnetospheric field. J. Geophys. Res. **102**, 11,295–11,306 (1997). doi:10.1029/97JA00563

A. Kidder, C.S. Paty, R.M. Winglee, E.M. Harnett, External triggering of plasmoid development at Saturn. J. Geophys. Res. (Space Phys.) **117**, A07206 (2012). doi:10.1029/2012JA017625

M.G. Kivelson, D.J. Southwood, Dynamical consequences of two modes of centrifugal instability in Jupiter's outer magnetosphere. J. Geophys. Res. (Space Phys.) **110**, A12209 (2005). doi:10.1029/2005JA011176

S.M. Krimigis, N. Sergis, D.G. Mitchell, D.C. Hamilton, N. Krupp, A dynamic, rotating ring current around Saturn. Nature **450**, 1050–1053 (2007). doi:10.1038/nature06425

E.A. Kronberg, J. Woch, N. Krupp, A. Lagg, K.K. Khurana, K.-H. Glassmeier, Mass release at Jupiter: substorm-like processes in the Jovian magnetotail. J. Geophys. Res. (Space Phys.) **110**, A03211 (2005). doi:10.1029/2004JA010777

N. Krupp, J. Woch, A. Lagg, B. Wilken, S. Livi, D.J. Williams, Energetic particle bursts in the predawn Jovian magnetotail. Geophys. Res. Lett. **25**, 1249–1252 (1998). doi:10.1029/98GL00863

N. Krupp, J. Woch, A. Lagg, S.A. Espinosa, S. Livi, S.M. Krimigis, D.G. Mitchell, D.J. Williams, A.F. Cheng, B.H. Mauk, R.W. McEntire, T.P. Armstrong, D.C. Hamilton, G. Gloeckler, J. Dandouras, L.J. Lanzerotti, Leakage of energetic particles from Jupiter's dusk magnetosphere: dual spacecraft observations. Geophys. Res. Lett. **29**, 1736 (2002). doi:10.1029/2001GL014290

A. Kullen, The connection between transpolar arcs and magnetotail rotation. Geophys. Res. Lett. **27**, 73–76 (2000). doi:10.1029/1999GL010675

H.R. Lai, H.Y.W.C.T. Russell, C.S. Arridge, M.K.M.K. Dougherty, Reconnection at the magnetopause of Saturn: perspective from fte occurrence and magnetosphere size. J. Geophys. Res. **117**(A5), A05222 (2012)

G.M. Mason, G. Gloeckler, Power law distributions of suprathermal ions in the quiet solar wind. Space Sci. Rev. **172**, 241–251 (2012). doi:10.1007/s11214-010-9741-0. http://adsabs.harvard.edu/abs/2012SSRv..172..241M

A. Masters, N. Achilleos, C. Bertucci, M.K. Dougherty, S.J. Kanani, C.S. Arridge, H.J. McAndrews, A.J. Coates, Surface waves on Saturn's dawn flank magnetopause driven by the Kelvin-Helmholtz instability. Planet. Space Sci. **57**, 1769–1778 (2009). doi:10.1016/j.pss.2009.02.010

A. Masters, N.A. Achilleos, N. Sergis, M.K. Dougherty, M.G. Kivelson, C.S. Arridge, S.M. Krimigis, H.J. McAndrews, M.F. Thomsen, S.J. Kanani, N. Krupp, A.J. Coates, Cassini observations of a Kelvin-Helmholtz vortex in Saturn's outer magnetosphere. J. Geophys. Res. (2010). doi:10.1029/2010JA015351

A. Masters, D.G. Mitchell, A.J. Coates, M.K. Dougherty, Saturn's low-latitude boundary layer: 1. Properties and variability. J. Geophys. Res. (Space Phys.) **116**, A06210 (2011). doi:10.1029/2010JA016421

A. Masters, J.P. Eastwood, M. Swisdak, M.F. Thomsen, C.T. Russell, N. Sergis, F.J. Crary, M.K. Dougherty, A.J. Coates, S.M. Krimigis, The importance of plasma β conditions for magnetic reconnection at Saturn's magnetopause. Geophys. Res. Lett. **39**, L08103 (2012a). doi:10.1029/2012GL051372

A. Masters, N. Achilleos, J.C. Cutler, A.J. Coates, M.K. Dougherty, G.H. Jones, Surface waves on Saturn's magnetopause. Planet. Space Sci. **65**, 109–121 (2012b). doi:10.1016/j.pss.2012.02.007

A. Masters, M. Fujimoto, H. Hasegawa, C.T. Russell, A.J. Coates, M.K. Dougherty, Can magnetopause reconnection drive Saturn's magnetosphere? Geophys. Res. Lett. **41**, 1862–1868 (2014). doi:10.1002/2014GL059288

B.H. Mauk, J.T. Clarke, D. Grodent, J.H. Waite, C.P. Paranicas, D.J. Williams, Transient aurora on Jupiter from injections of magnetospheric electrons. Nature **415**, 1003–1005 (2002)

H.J. McAndrews, C.J. Owen, M.F. Thomsen, B. Lavraud, A.J. Coates, M.K. Dougherty, D.T. Young, Evidence for reconnection at Saturn's magnetopause. J. Geophys. Res. (Space Phys.) **113**, A04210 (2008). doi:10.1029/2007JA012581

D.J. McComas, F. Bagenal, Jupiter: a fundamentally different magnetospheric interaction with the solar wind. Geophys. Res. Lett. **34**, 20,106 (2007). doi:10.1029/2007GL031078

D.J. McComas, F. Allegrini, F. Bagenal, F. Crary, R.W. Ebert, H. Elliott, A. Stern, P. Valek, Diverse plasma populations and structures in Jupiter's magnetotail. Science **318**, 217 (2007). doi:10.1126/science.1147393

D.J. McComas, F. Bagenal, Reply to comment by S.W.H. Cowley et al. on "Jupiter: a fundamentally different magnetospheric interaction with the solar wind". Geophys. Res. Lett. **35**, L10103 (2008). doi:10.1029/2008GL034351

C.J. Meredith, S.W.H. Cowley, K.C. Hansen, J.D. Nichols, T.K. Yeoman, Simultaneous conjugate observations of small-scale structures in Saturn's dayside ultraviolet auroras—implications for physical origins. J. Geophys. Res. (2013). doi:10.1002/jgra.50270

S.E. Milan, B. Hubert, A. Grocott, Formation and motion of a transpolar arc in response to dayside and nightside reconnection. J. Geophys. Res. (Space Phys.) **110**, A01212 (2005). doi:10.1029/2004JA010835

R. Mistry, M.K. Dougherty, A. Masters, A.H. Sulaiman, E.J. Allen, Separating drivers of Saturnian magnetopause motion. J. Geophys. Res. (Space Phys.) **119**, 1514–1522 (2014). doi:10.1002/2013JA019489

D.G. Mitchell, Injection, interchange and reconnection: energetic particle observations in Saturn's magnetotail, in *Magnetotails in the Solar System*, ed. by A. Keiling, C. Jackman, P.A. Delamere. AGU Monographs (American Geophysical Union, Washington, 2015). ISBN 978-1-118-84234-8

D.G. Mitchell, S.M. Krimigis, C. Paranicas, P.C. Brandt, J.F. Carbary, E.C. Roelof, W.S. Kurth, D.A. Gurnett, J.T. Clarke, J.D. Nichols, J.-C. Gérard, D.C. Grodent, M.K. Dougherty, W.R. Pryor, Recurrent energization of plasma in the midnight-to-dawn quadrant of Saturn's magnetosphere, and its relationship to auroral UV and radio emissions. Planet. Space Sci. **57**, 1732–1742 (2009). doi:10.1016/j.pss.2009.04.002

T. Moriguchi, A. Nakamizo, T. Tanaka, T. Obara, H. Shimazu, Current systems in the Jovian magnetosphere. J. Geophys. Res. (Space Phys.) **113**, A05204 (2008). doi:10.1029/2007JA012751

T.K.M. Nakamura, M. Fujimoto, Magnetic reconnection within rolled-up MHD-scale Kelvin-Helmholtz vortices: two-fluid simulations including finite electron inertial effects. Geophys. Res. Lett. **32**, L21102 (2005). doi:10.1029/2005GL023362

T.K.M. Nakamura, M. Fujimoto, A. Otto, Magnetic reconnection induced by weak Kelvin-Helmholtz instability and the formation of the low-latitude boundary layer. Geophys. Res. Lett. **33**, L14106 (2006). doi:10.1029/2006GL026318

J. Nichols, S. Cowley, Magnetosphere-ionosphere coupling currents in Jupiter's middle magnetosphere: effect of precipitation-induced enhancement of the ionospheric Pedersen conductivity. Ann. Geophys. **22**, 1799–1827 (2004). doi:10.5194/angeo-22-1799-2004

J.D. Nichols, Magnetosphere-ionosphere coupling in Jupiter's middle magnetosphere: computations including a self-consistent current sheet magnetic field model. J. Geophys. Res. (Space Phys.) **116**, A10232 (2011). doi:10.1029/2011JA016922

J.D. Nichols, S.W.H. Cowley, D.J. McComas, Magnetopause reconnection rate estimates for Jupiter's magnetosphere based on interplanetary measurements at ~5AU. Ann. Geophys. **24**, 393–406 (2006)

J.D. Nichols, E.J. Bunce, J.T. Clarke, S.W.H. Cowley, J.-C. Gérard, D. Grodent, W.R. Pryor, Response of Jupiter's UV auroras to interplanetary conditions as observed by the Hubble Space Telescope during the Cassini flyby campaign. J. Geophys. Res. **112**(A11), A02203 (2007). doi:10.1029/2006JA012005

J.D. Nichols, J.T. Clarke, J.-C. Gérard, D. Grodent, K.C. Hansen, Variation of different components of Jupiter's auroral emission. J. Geophys. Res. **114**, A06210 (2009a). doi:10.1029/2009JA014051

J.D. Nichols, J.T. Clarke, J.C. Gérard, D. Grodent, Observations of Jovian polar auroral filaments. Geophys. Res. Lett. **36**, L08101 (2009b). doi:10.1029/2009GL037578

J.D. Nichols, S. Badman, K. Baines, R. Brown, E. Bunce, J. Clarke, S. Cowley, F. Crary, M. Dougherty, J. Grard, A. Grocott, D. Grodent, W. Kurth, H. Melin, D. Mitchell, W. Pryor, T. Stallard, Dynamic auroral storms on Saturn as observed by the Hubble Space Telescope. Geophys. Res. Lett. (2014). doi:10.1002/2014GL060186

K. Nykyri, A. Otto, Plasma transport at the magnetospheric boundary due to reconnection in Kelvin-Helmholtz vortices. Geophys. Res. Lett. **28**, 3565–3568 (2001). doi:10.1029/2001GL013239

K. Nykyri, A. Otto, B. Lavraud, C. Mouikis, L.M. Kistler, A. Balogh, H. Rème, Cluster observations of reconnection due to the Kelvin-Helmholtz instability at the dawnside magnetospheric flank. Ann. Geophys. **24**, 2619–2643 (2006)

T. Ogino, R.J. Walker, M.G. Kivelson, A global magnetohydrodynamic simulation of the Jovian magnetosphere. J. Geophys. Res. **103**, 225 (1998). doi:10.1029/97JA02247

L. Pallier, R. Prange, More about the structure of the high latitude Jovian aurorae. Planet. Space Sci. **49**(10–11), 1159–1173 (2001)

C. Paranicas, D.G. Mitchell, S.M. Krimigis, D.C. Hamilton, E. Roussos, N. Krupp, G.H. Jones, R.E. Johnson, J.F. Cooper, T.P. Armstrong, Sources and losses of energetic protons in Saturn's magnetosphere. Icarus **197**, 519–525 (2008). doi:10.1016/j.icarus.2008.05.011

G. Paschmann, Recent in-situ observations of magnetic reconnection in near-Earth space. Geophys. Res. Lett. **35**, L19109 (2008). doi:10.1029/2008GL035297

T.D. Phan, J.T. Gosling, G. Paschmann, C. Pasma, J.F. Drake, M. Øieroset, D. Larson, R.P. Lin, M.S. Davis, The dependence of magnetic reconnection on plasma β and magnetic shear: evidence from solar wind observations. Astrophys. J. Lett. **719**, L199–L203 (2010). doi:10.1088/2041-8205/719/2/L199

T.D. Phan, G. Paschmann, J.T. Gosling, M. Oieroset, M. Fujimoto, J.F. Drake, V. Angelopoulos, The dependence of magnetic reconnection on plasma β and magnetic shear: evidence from magnetopause observations. Geophys. Res. Lett. **40**, 11–16 (2013). doi:10.1029/2012GL054528

V. Pierrard, M. Lazar, Kappa distributions: theory and applications in space plasmas. Sol. Phys. **267**, 153–174 (2010). doi:10.1007/s11207-010-9640-2

N.M. Pilkington, N. Achilleos, C.S. Arridge, A. Masters, N. Sergis, A.J. Coates, M.K. Dougherty, Polar confinement of Saturn's magnetosphere revealed by in situ Cassini observations. J. Geophys. Res. (Space Phys.) **119**, 2858–2875 (2014). doi:10.1002/2014JA019774

D.H. Pontius, Radial mass transport and rotational dynamics. J. Geophys. Res. **102**, 7137–7150 (1997). doi:10.1029/97JA00289

W.R. Pryor, A.I.F. Stewart, L.W. Esposito, W.E. McClintock, J.E. Colwell, A.J. Jouchoux, A.J. Steffl, D.E. Shemansky, J.M. Ajello, R.A. West, C.J. Hansen, B.T. Tsurutani, W.S. Kurth, G.B. Hospodarsky, D.A. Gurnett, K.C. Hansen, J.H. Waite, F.J. Crary, D.T. Young, N. Krupp, J.T. Clarke, D. Grodent, M.K. Dougherty, Cassini UVIS observations of Jupiter's auroral variability. Icarus **178**, 312–326 (2005). doi:10.1016/j.icarus.2005.05.021

W.R. Pryor, A.M. Rymer, D.G. Mitchell, T.W. Hill, D.T. Young, J. Saur, G.H. Jones, S. Jacobsen, S.W.H. Cowley, B.H. Mauk, A.J. Coates, J. Gustin, D. Grodent, J.-C. Gérard, L. Lamy, J.D. Nichols, S.M. Krimigis, L.W. Esposito, M.K. Dougherty, A.J. Jouchoux, A.I.F. Stewart, W.E. McClintock, G.M. Holsclaw, J.M. Ajello, J.E. Colwell, A.R. Hendrix, F.J. Crary, J.T. Clarke, X. Zhou, The auroral footprint of Enceladus on Saturn. Nature **472**, 331–333 (2011). doi:10.1038/nature09928

A. Radioti, J.-C. Gérard, D. Grodent, B. Bonfond, N. Krupp, J. Woch, Discontinuity in Jupiter's main auroral oval. J. Geophys. Res. **113**, 1215 (2008a). doi:10.1029/2007JA012610

A. Radioti, D. Grodent, J.C. Gérard, B. Bonfond, J.T. Clarke, Auroral polar dawn spots: Signatures of internally driven reconnection processes at Jupiter's magnetotail. Geophys. Res. Lett. **35**(3) (2008b)

A. Radioti, A.T. Tomás, D. Grondent, J.C. Gérard, J. Gustin, B. Bonfond, N. Krupp, J. Woch, J.D. Menietti, Equatorward diffuse auroral emissions at Jupiter: simultaneous HST and Galileo observations. Geophys. Res. Lett. **36**, 7101 (2009a)

A. Radioti, J.-C. Gérard, E. Roussos, C.P. Paranicas, B. Bonfond, D.G. Mitchell, N. Krupp, S.M. Krimigis, J.T. Clarke, Transient auroral features at Saturn: signatures of energetic particle injections in the magnetosphere. J. Geophys. Res. **114**, A03210 (2009b). doi:10.1029/2008JA013632

A. Radioti, D. Grodent, J.-C. Gérard, B. Bonfond, Auroral signatures of flow bursts released during magnetotail reconnection at Jupiter. J. Geophys. Res. (Space Phys.) **115**, A07214 (2010). doi:10.1029/2009JA014844

A. Radioti, D. Grodent, J.-C. Gérard, S.E. Milan, B. Bonfond, J. Gustin, W. Pryor, Bifurcations of the main auroral ring at Saturn: ionospheric signatures of consecutive reconnection events at the magnetopause. J. Geophys. Res. (Space Phys.) **116**, A11209 (2011a). doi:10.1029/2011JA016661

A. Radioti, D. Grodent, J.C. Gerard, M.F. Vogt, M. Lystrup, B. Bonfond, Nightside reconnection at Jupiter: auroral and magnetic field observations from 26 July 1998. J. Geophys. Res. **116**, A03221 (2011b). doi:10.1029/2010JA016200

A. Radioti, M. Lystrup, B. Bonfond, D. Grodent, J.-C. Gérard, Jupiter's aurora in ultraviolet and infrared: simultaneous observations with the Hubble Space Telescope and the NASA infrared telescope facility. J. Geophys. Res. (Space Phys.) **118**, 2286–2295 (2013a). doi:10.1002/jgra.50245

A. Radioti, E. Roussos, D. Grodent, J.-C. Gérard, N. Krupp, D.G. Mitchell, J. Gustin, B. Bonfond, W. Pryor, Signatures of magnetospheric injections in Saturn's aurora. J. Geophys. Res. (Space Phys.) **118**, 1922–1933 (2013b). doi:10.1002/jgra.50161

A. Radioti, D. Grodent, J.-C. Gérard, B. Bonfond, J. Gustin, W. Pryor, J.M. Jasinski, C.S. Arridge, Auroral signatures of multiple magnetopause reconnection at Saturn. Geophys. Res. Lett. **40**, 4498–4502 (2013c). doi:10.1002/grl.50889

A. Radioti, D. Grodent, J. Gérard, S. Milan, R. Fear, C. Jackman, B. Bonfond, W. Pryor, Saturn's elusive transpolar arc. Geophys. Res. Lett. (2014a, submitted)

A.D. Radioti, Grodent, X. Jia, J.-C. Gérard, B. Bonfond, W. Pryor, J. Gustin, D. Mitchell, C. Jackman, A remarkable magnetotail reconnection event at Saturn, as observed by UVIS/Cassini. Icarus (2014b, submitted)

L.C. Ray, Y.-J. Su, R.E. Ergun, P.A. Delamere, F. Bagenal, Current-voltage relation of a centrifugally confined plasma. J. Geophys. Res. (Space Phys.) **114**, A04214 (2009). doi:10.1029/2008JA013969

L.C. Ray, R.E. Ergun, P.A. Delamere, F. Bagenal, Magnetosphere-ionosphere coupling at Jupiter: effect of field-aligned potentials on angular momentum transport. J. Geophys. Res. (Space Phys.) **115**, A09211 (2010). doi:10.1029/2010JA015423

L.C. Ray, M. Galand, L.E. Moore, B.L. Fleshman, Characterizing the limitations to the coupling between Saturn's ionosphere and middle magnetosphere. J. Geophys. Res. (Space Phys.) **117**, A07210 (2012). doi:10.1029/2012JA017735

L.C. Ray, M. Galand, P.A. Delamere, B.L. Fleshman, Current-voltage relation for the Saturnian system. J. Geophys. Res. (Space Phys.) **118**, 3214–3222 (2013). doi:10.1002/jgra.50330

E. Roussos, N. Krupp, C.P. Paranicas, P. Kollmann, D.G. Mitchell, S.M. Krimigis, T.P. Armstrong, D.R. Went, M.K. Dougherty, G.H. Jones, Long- and short-term variability of Saturn's ionic radiation belts. J. Geophys. Res. (Space Phys.) **116**, A02217 (2011). doi:10.1029/2010JA015954

H.O. Rucker, M. Panchenko, K.C. Hansen, U. Taubenschuss, M.Y. Boudjada, W.S. Kurth, M.K. Dougherty, J.T. Steinberg, P. Zarka, P.H.M. Galopeau, D.J. McComas, C.H. Barrow, Saturn kilometric radiation as a monitor for the solar wind? Adv. Space Res. **42**, 40–47 (2008). doi:10.1016/j.asr.2008.02.008

C.T. Russell, Reconnection, in *Physics of Solar Planetary Environments* (Am. Geophys. Union, Washington, 1975), pp. 526–540

C.T. Russell, R.C. Elphic, Initial ISEE magnetometer results—magnetopause observations. Space Sci. Rev. **22**, 681–715 (1978). doi:10.1007/BF00212619

C.T. Russell, M.M. Hoppe, W.A. Livesey, Overshoots in planetary bow shocks. Nature **296**, 45–48 (1982). doi:10.1038/296045a0

B. Sandel, A. Broadfoot, Morphology of Saturn's aurora. Nature **292**, 679–682 (1981)

N. Sergis, S.M. Krimigis, E.C. Roelof, C.S. Arridge, A.M. Rymer, D.G. Mitchell, D.C. Hamilton, N. Krupp, M.F. Thomsen, M.K. Dougherty, A.J. Coates, D.T. Young, Particle pressure, inertial force, and ring current density profiles in the magnetosphere of Saturn, based on Cassini measurements. Geophys. Res. Lett. **37**, L02102 (2010). doi:10.1029/2009GL041920

N. Sergis, C.M. Jackman, A. Masters, S.M. Krimigis, M.F. Thomsen, D.C. Hamilton, D.G. Mitchell, M.K. Dougherty, A.J. Coates, Particle and magnetic field properties of the Saturnian magnetosheath: presence and upstream escape of hot magnetospheric plasma. J. Geophys. Res. (Space Phys.) **118**, 1620–1634 (2013). doi:10.1002/jgra.50164

J.A. Slavin, E.J. Smith, J.R. Spreiter, S.S. Stahara, Solar wind flow about the outer planets—gas dynamic modeling of the Jupiter and Saturn bow shocks. J. Geophys. Res. **90**, 6275–6286 (1985). doi:10.1029/JA090iA07p06275

C.G.A. Smith, A.D. Aylward, Coupled rotational dynamics of Saturn's thermosphere and magnetosphere: a thermospheric modelling study. Ann. Geophys. **26**, 1007–1027 (2008). doi:10.5194/angeo-26-1007-2008

C.G.A. Smith, A.D. Aylward, Coupled rotational dynamics of Jupiter's thermosphere and magnetosphere. Ann. Geophys. **27**, 199–230 (2009). doi:10.5194/angeo-27-199-2009

B.U.O. Sonnerup, B.G. Ledley, Magnetopause rotational forms. J. Geophys. Res. **79**, 4309–4314 (1974). doi:10.1029/JA079i028p04309

D.J. Southwood, The hydromagnetic stability of the magnetospheric boundary. Planet. Space Sci. **16**, 587–605 (1968). doi:10.1016/0032-0633(68)90100-1

D.J. Southwood, M.G. Kivelson, A new perspective concerning the influence of the solar wind on the Jovian magnetosphere. J. Geophys. Res. **106**, 6123–6130 (2001). doi:10.1029/2000JA000236

T.S. Stallard, S. Miller, S.W.H. Cowley, E.J. Bunce, Jupiter's polar ionospheric flows: measured intensity and velocity variations poleward of the main auroral oval. Geophys. Res. Lett. **30**, 1221 (2003). doi:10.1029/2002GL016031

T. Stallard, S. Miller, H. Melin, M. Lystrup, S.W.H. Cowley, E.J. Bunce, N. Achilleos, M. Dougherty, Jovian-like aurorae on Saturn. Nature **453**, 1083–1085 (2008). doi:10.1038/nature07077

Y.-J. Su, R.E. Ergun, F. Bagenal, P.A. Delamere, Io-related Jovian auroral arcs: modeling parallel electric fields. J. Geophys. Res. (Space Phys.) **108**, 1094 (2003). doi:10.1029/2002JA009247

M. Swisdak, B.N. Rogers, J.F. Drake, M.A. Shay, Diamagnetic suppression of component magnetic reconnection at the magnetopause. J. Geophys. Res. (Space Phys.) **108**, 1218 (2003). doi:10.1029/2002JA009726

M. Swisdak, M. Opher, J.F. Drake, F. Alouani Bibi, The vector direction of the interstellar magnetic field outside the heliosphere. Astrophys. J. **710**, 1769–1775 (2010). doi:10.1088/0004-637X/710/2/1769

T. Terasawa, K. Maezawa, S. Machida, Solar-wind effect on Jupiter non-Io-related radio-emission. Nature **273**(5658), 131–132 (1978)

N. Thomas, F. Bagenal, T.W. Hill, J.K. Wilson, *The Io Neutral Clouds and Plasma Torus* (2004), pp. 561–591

M.F. Thomsen, J.P. Dilorenzo, D.J. McComas, D.T. Young, F.J. Crary, D. Delapp, D.B. Reisenfeld, N. Andre, Assessment of the magnetospheric contribution to the suprathermal ions in Saturn's foreshock region. J. Geophys. Res. (Space Phys.) **112**, A05220 (2007). doi:10.1029/2006JA012084

M.F. Thomsen, C.M. Jackman, R.L. Tokar, R.J. Wilson, Plasma flows in Saturn's nightside magnetosphere. J. Geophys. Res. (2014). doi:10.1002/2014JA019912

V.M. Vasyliunas, A survey of low-energy electrons in the evening sector of the magnetosphere with OGO 1 and OGO 3. J. Geophys. Res. **73**, 2839–2885 (1968)

V.M. Vasyliunas, Theoretical models of magnetic field line merging. I. Rev. Geophys. Space Phys. **13**, 303–336 (1975). doi:10.1029/RG013i001p00303

V.M. Vasyliunas, Physics of the Jovian magnetosphere, in *Plasma Distribution and Flow* (1983), pp. 395–453

M.F. Vogt, M.G. Kivelson, K.K. Khurana, S.P. Joy, R.J. Walker, Reconnection and flows in the Jovian magnetotail as inferred from magnetometer observations. J. Geophys. Res. (Space Phys.) **115**, A06219 (2010). doi:10.1029/2009JA015098

M.F. Vogt, M.G. Kivelson, K.K. Khurana, R.J. Walker, B. Bonfond, D. Grodent, A. Radioti, Improved mapping of Jupiter's auroral features to magnetospheric sources. J. Geophys. Res. (Space Phys.) **116**, A03220 (2011). doi:10.1029/2010JA016148

J.H. Waite, G.R. Gladstone, W.S. Lewis, R. Goldstein, D.J. McComas, P. Riley, R.J. Walker, P. Robertson, S. Desai, J.T. Clarke, D.T. Young, An auroral flare at Jupiter. Nature **410**, 787–789 (2001)

R.J. Walker, T. Ogino, A simulation study of currents in the Jovian magnetosphere. Planet. Space Sci. **51**, 295–307 (2003). doi:10.1016/S0032-0633(03)00018-7

R.J. Walker, C.T. Russell, Flux transfer events at the Jovian magnetopause. J. Geophys. Res. **90**, 7397–7404 (1985). doi:10.1029/JA090iA08p07397

R.J. Walker, S.P. Joy, M.G. Kivelson, K. Khurana, T. Ogino, K. Fukazawa, The locations and shapes of Jupiter's bow shock and magnetopause, in *AIP Conf. Proc.*, vol. 781 (2005)

R.J. Walker, K. Fukazawa, T. Ogino, D. Morozoff, A simulation study of Kelvin-Helmholtz waves at Saturn's magnetopause. J. Geophys. Res. (Space Phys.) **116**, A03203 (2011). doi:10.1029/2010JA015905

 ⚛ Springer

L. Wang, R.P. Lin, C. Salem, M. Pulupa, D.E. Larson, P.H. Yoon, J.G. Luhmann, Quiet-time interplanetary $\sim$2–20 keV superhalo electrons at solar minimum. Astrophys. J. Lett. **753**, L23 (2012). doi:10.1088/2041-8205/753/1/L23

D.R. Went, M.G. Kivelson, N. Achilleos, C.S. Arridge, M.K. Dougherty, Outer magnetospheric structure: Jupiter and Saturn compared. J. Geophys. Res. (Space Phys.) **116**, A04224 (2011). doi:10.1029/2010JA016045

R.M. Winglee, A. Kidder, E. Harnett, N. Ifland, C. Paty, D. Snowden, Generation of periodic signatures at Saturn through Titan's interaction with the centrifugal interchange instability. J. Geophys. Res. (Space Phys.) **118**, 4253–4269 (2013). doi:10.1002/jgra.50397

R.J. Wilson, P.A. Delamere, F. Bagenal, A. Masters, Kelvin-Helmholtz instability at Saturn's magnetopause: Cassini ion data analysis. J. Geophys. Res. (Space Phys.) **117**, A03212 (2012). doi:10.1029/2011JA016723

J. Woch, N. Krupp, A. Lagg, Particle bursts in the Jovian magnetosphere: evidence for a near-Jupiter neutral line. Geophys. Res. Lett. **29**, 1138 (2002). doi:10.1029/2001GL014080

J.N. Yates, N. Achilleos, P. Guio, Influence of upstream solar wind on thermospheric flows at Jupiter. Planet. Space Sci. **61**, 15–31 (2012). doi:10.1016/j.pss.2011.08.007

D.T. Young, J.-J. Berthelier, M. Blanc, J.L. Burch, S. Bolton, A.J. Coates, F.J. Crary, R. Goldstein, M. Grande, T.W. Hill, R.E. Johnson, R.A. Baragiola, V. Kelha, D.J. McComas, K. Mursula, E.C. Sittler, K.R. Svenes, K. Szegö, P. Tanskanen, M.F. Thomsen, S. Bakshi, B.L. Barraclough, Z. Bebesi, D. Delapp, M.W. Dunlop, J.T. Gosling, J.D. Furman, L.K. Gilbert, D. Glenn, C. Holmlund, J.-M. Illiano, G.R. Lewis, D.R. Linder, S. Maurice, H.J. McAndrews, B.T. Narheim, E. Pallier, D. Reisenfeld, A.M. Rymer, H.T. Smith, R.L. Tokar, J. Vilppola, C. Zinsmeyer, Composition and dynamics of plasma in Saturn's magnetosphere. Science **307**, 1262–1266 (2005). doi:10.1126/science.1106151

P. Zarka, F. Genova, Low-frequency jovian emission and solar-wind magnetic-sector structure. Nature **306**, 767–768 (1983)

B. Zieger, K.C. Hansen, Statistical validation of a solar wind propagation model from 1 to 10 AU. J. Geophys. Res. **113**(A12), A08107 (2008). doi:10.1029/2008JA013046

B. Zieger, K.C. Hansen, T.I. Gombosi, D.L. De Zeeuw, Periodic plasma escape from the mass-loaded Kronian magnetosphere. J. Geophys. Res. (Space Phys.) **115**, A08208 (2010). doi:10.1029/2009JA014951

G. Zimbardo, Collisionless reconnection in Jupiter's magnetotail. Geophys. Res. Lett. **18**, 741–744 (1991). doi:10.1029/91GL00472

DOI 10.1007/978-1-4939-3395-2_5
Reprinted from *Space Science Reviews* Journal, DOI 10.1007/s11214-014-0042-x

Auroral Processes at the Giant Planets: Energy Deposition, Emission Mechanisms, Morphology and Spectra

**Sarah V. Badman · Graziella Branduardi-Raymont ·
Marina Galand · Sébastien L.G. Hess · Norbert Krupp ·
Laurent Lamy · Henrik Melin · Chihiro Tao**

Received: 21 November 2013 / Accepted: 17 March 2014 / Published online: 2 April 2014
© Springer Science+Business Media Dordrecht 2014

Abstract The ionospheric response to auroral precipitation at the giant planets is reviewed, using models and observations. The emission processes for aurorae at radio, infrared, visible, ultraviolet, and X-ray wavelengths are described, and exemplified using ground- and space-based observations. Comparisons between the emissions at different wavelengths are

S.V. Badman (✉) · H. Melin
Department of Physics and Astronomy, University of Leicester, University Road,
Leicester, LE1 7RH, UK
e-mail: s.badman@lancaster.ac.uk

Present address:
S.V. Badman
Department of Physics, Lancaster University, Bailrigg, Lancaster, LA1 4YB, UK

G. Branduardi-Raymont
Mullard Space Science Laboratory, University College London, Holmbury St Mary, Dorking, Surrey,
RH5 6NT, UK

M. Galand
Department of Physics, Imperial College, London SW7 2AK, UK

S.L.G. Hess
LATMOS, Université Versailles-St Quentin, IPSL/CNRS, 11 Boulevard d'Alembert,
78280 Guyancourt, France

N. Krupp
Max-Planck-Institut für Sonnensystemforschung, 37077 Göttingen, Germany

L. Lamy
LESIA, Observatoire de Paris, CNRS, Université Pierre et Marie Curie, Université Paris Diderot,
Meudon, France

C. Tao
LPP, CNRS-Ecole Polytechnique-UPMC, Route de Saclay, 92120 Palaiseau, France

Present address:
C. Tao
IRAP, CNRS, Université de Toulouse, Toulouse, France

made, where possible, and interpreted in terms of precipitating particle characteristics or atmospheric conditions. Finally, the spatial distributions and dynamics of the various components of the aurorae (moon footprints, low-latitude, main oval, polar) are related to magnetospheric processes and boundaries, using theory, in situ, and remote observations, with the aim of distinguishing between those related to internally-driven dynamics, and those related to the solar wind interaction.

Keywords Giant planet · Aurora · Magnetodisk

1 Introduction: Key Magnetospheric Regions and Interactions

The magnetospheres of the outer planets are huge plasma laboratories in space. They are driven by the fast rotation of the planet with its strong internal magnetic field, combined with powerful internal plasma sources (the satellites Io and Europa in the case of Jupiter, and Enceladus at Saturn). Several comprehensive reviews of outer planet magnetospheres and their dynamics have been published (e.g. Dessler 1983; Bagenal et al. 2004; Dougherty et al. 2009) and in this introductory section we only briefly overview the key magnetospheric regions and their dynamics, before describing in detail in the subsequent sections the auroral emissions generated at different wavelengths, and how they are utilised to diagnose the magnetospheric dynamics.

1.1 Jupiter

Our knowledge of the global configuration and dynamics of the Jovian magnetosphere is based on measurements taken onboard spacecraft flying through the Jovian system (Pioneer 10 and 11, Voyager 1 and 2, Ulysses, Cassini, New Horizons) and especially from results of the orbiting spacecraft Galileo.

Figure 1 shows a sketch of the key regions and magnetospheric interactions of the Jovian magnetosphere. Traditionally the magnetosphere is subdivided into the inner, middle and outer magnetosphere. In the inner magnetosphere orbits the volcanic moon Io (at $6R_J$ radial distance), which is the main source of oxygen and sulphur neutrals in the magnetosphere, and the moon Europa (at $9R_J$ radial distance) where hydrogen and possibly oxygen originate. Both moons create a torus along their orbit around the planet in which neutrals are ionized to form plasma tori. While the mass added to the magnetosphere from the moons plays an important role in driving dynamics and auroral emissions throughout the magnetosphere (described below), the moons also have a local interaction with the Jovian magnetic field, resulting in auroral footprints at the ionospheric end of the connecting flux tubes. The interaction occurs because the satellites form obstacles to the corotating plasma flow, which is moving faster than their Keplerian orbital velocities. The perturbation of the plasma and field around the moon propagates along the magnetic field as Alfvén waves, interacting with electrons, which finally precipitate into the ionosphere and generate aurora (e.g. Kivelson 2004). At Jupiter the footprints of Io, Europa, and Ganymede have been identified, while the footprint of Callisto is mostly hidden underneath the main oval (Connerney et al. 1993; Clarke et al. 2002). The observed footprints take the form of spots (multiple spots in the cases of Io and Ganymede) and also have trails of enhanced emissions, or 'wakes', behind the footprint itself (e.g. Bonfond et al. 2008, 2013).

Due to the centrifugal force of the fast rotating planet, plasma moves radially outward from the tori in the inner magnetosphere. The magnetic field lines frozen in to the plasma

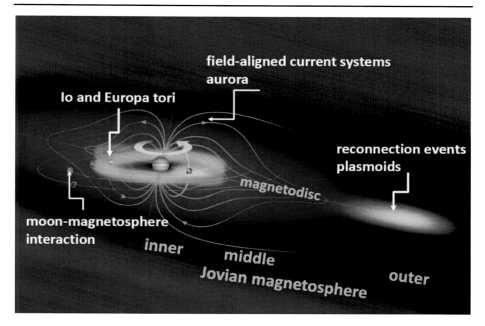

Fig. 1 Sketch of the key magnetospheric regions in the Jovian magnetosphere. Credit: max planck institute for solar system research

in the middle magnetosphere are therefore continuously stretched outward near the equator and deviate significantly from a dipole configuration. Oppositely-directed field lines come close together, and a stable configuration can only be reached through formation of a current sheet between the oppositely-directed fields, and an associated plasmasheet. An equatorially confined magnetodisc is formed, which wobbles up and down with respect to the equator due to the 9.6° tilt between Jupiter's magnetic dipole axis and the planetary rotation axis. The magnetodisc is relatively thin in the dawn sector ($2R_J$ half thickness) and thicker on the dusk side ($7.6R_J$ half thickness) (Khurana et al. 2004).

As the plasma moves outward through the magnetosphere, it also slows. This means that the magnetic field frozen in to the plasma in the magnetodisk is sub-corotating, yet these field lines have their ends fixed in the ionosphere, where collisions between atmospheric neutrals rotating with the planet and ions can occur. The planet therefore supplies angular momentum to the magnetosphere, attempting to spin the field and plasma back up to corotation. The angular momentum is transferred by a field aligned current system, which is directed upward from the ionosphere, radially outward in the equatorial middle magnetosphere (such that the $\mathbf{j} \times \mathbf{B}$ force acts in the direction of planetary rotation), returning downward to the ionosphere at higher latitudes, and closing through an equatorward ionospheric current. The portion of the current directed upward from the ionosphere, carried by downgoing electrons, is responsible for Jupiter's main auroral oval (Cowley and Bunce 2001; Hill 2001).

The radial distance where the plasma begins to depart from rigid corotation, i.e. where the ionosphere can no longer impart sufficient angular momentum, seems to be dependent on local time. It is further out in the pre-dawn sector, at $40R_J$, compared to $20\text{–}25R_J$ in the dusk sector, which may be related to the distribution of mass-loading and loss in the magnetosphere (Vasyliunas 1983; Krupp et al. 2001; Woch et al. 2004). Therefore, while Jupiter's main emission is relatively stable over time, its intensity and location can be affected by the location and magnitude of corotation breakdown in the magnetosphere, which in turn can be affected by, e.g. volcanic activity at Io or solar wind compression of the magnetosphere. These processes are discussed in more detail in Sect. 3.1.

In the outer magnetosphere the field lines are stretched and sub-corotating. When the current sheet becomes particularly thin, reconnection can occur between oppositely-directed field lines. This ultimately results in the release of a plasmoid downtail and the contraction of the newly-reconnected field line back toward the planet. Reconnection in the magneto-tail could occur only on closed, stretched field lines, or continue onto open, lobe field lines (Vasyliunas 1983; Cowley et al. 2003). In situ measurements show that reconnection preferentially occurs at radial distances of $60–80 R_J$ and its signatures are sometimes observed with a periodicity of 2–3 days (Krupp et al. 1998; Woch et al. 1998; Louarn et al. 1998; McComas and Bagenal 2007; Hill et al. 2009; Vogt et al. 2010). One possible scenario to explain the periodicity, involving a cycle of mass loading and unloading, was first pointed out by Krupp et al. (1998) and Woch et al. (1998). They suggested that, after reconnection, the emptied field lines take approximately a day to snap back radially inwards towards the planet, and azimuthally in the direction of planetary rotation, before the mass-loading cycle starts again. The field lines moving radially inward after reconnection can have auroral signatures in the ionosphere, poleward of the main oval, related to field-aligned currents linking the dipolarised field line to the ionosphere (Grodent et al. 2004; Kasahara et al. 2011).

Even though the solar wind interaction at Jupiter does not play the most important role in terms of dynamics, compared to rotationally-driven dynamics, evidence of solar wind driving and auroral signatures have been identified in the high latitude and outermost regions (see Sect. 6.1). Currently two basic scenarios are discussed: (i) an open magnetosphere where magnetic flux opened during reconnection at the dayside magnetopause is transported across the polar region into the magnetotail with a return planetward flow on the dawnside of the tail (Cowley et al. 2003; Badman and Cowley 2007), and (ii) a magnetosphere where magnetic flux is opened and closed intermittently in small-scale structures on the flanks of the magnetosphere, via a viscous interaction between heavy, dense plasma inside the magnetosphere and light, tenuous plasma in the solar wind, with a velocity shear between them (Delamere and Bagenal 2010). Support for a solar wind interaction at Jupiter is also reported in MHD simulations by Fukazawa et al. (2010), where periodic plasmoid releases are present in the simulation only occur if the solar wind dynamic pressure is low enough. Corresponding auroral signatures of magnetopause reconnection and an open field region have been identified at Jupiter (e.g. Pallier and Prangé 2001; Cowley et al. 2003), but some mysteries remain, including the origin of dynamic, transient emissions seen in both the UV and IR 'bright polar region', which is thought to map to open, and thus plasma-depleted, field lines.

1.2 Saturn

Saturn's magnetosphere has been visited by the flyby missions Pioneer 11, Voyager 1 and 2, and by Cassini as the first orbiting spacecraft around the ringed planet. Figure 2 shows a sketch of Saturn's magnetosphere, indicating the key magnetospheric regions and plasma populations.

Saturn also has major sources of neutrals inside the magnetosphere, primarily the moon Enceladus, which releases water ice and dust grains into the Kronian magnetosphere through active geysers in the southern polar region, at a rate of up to a few hundred $kg\,s^{-1}$ (e.g.

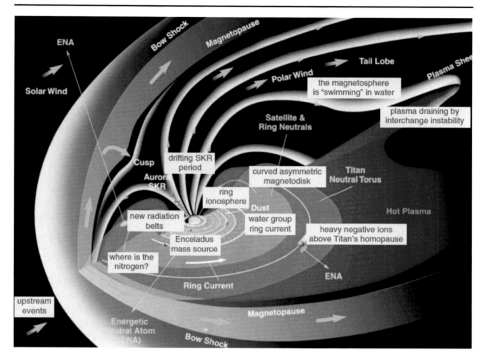

Fig. 2 Sketch of the key magnetospheric regions and plasma populations in the Kronian magnetosphere (from Gombosi et al. 2009)

Hansen et al. 2006). Like the Io-Jupiter interaction, Dougherty et al. (2006) showed that when some of this water is ionised near Enceladus, it perturbs the magnetic field, resulting in a field-aligned current linking the near-moon environment to Saturn's ionosphere. The associated auroral spot has been identified in a few UV images of Saturn's northern ionosphere (Pryor et al. 2011). Titan, orbiting at $\sim 20 R_S$, was previously thought to be a significant particle source, especially for nitrogen, but has been revealed to be only a minor plasma source for the magnetosphere by Cassini measurements (Smith et al. 2007b).

Saturn's magnetosphere is also rotation-dominated and forms an equatorially-stretched asymmetric magnetodisc. Plasma is observed to sub-corotate throughout Saturn's magnetosphere (Wilson et al. 2009; Thomsen et al. 2010), and a relatively faint auroral arc has been identified in the infrared observations of the conjugate latitudes in the ionosphere as the signature of corotation-enforcement currents (Stallard et al. 2010). The main auroral emission lies at higher latitudes, and is driven by field-aligned currents associated with the flow shear between anti-sunward, open and outer magnetospheric field lines, and sub-corotating middle magnetosphere field lines (Cowley et al. 2005; Bunce et al. 2008). Transport processes like interchange motion and injection events are continuously present in the Kronian magnetosphere, showing the highly dynamic nature of Saturn's magnetosphere (Mitchell et al. 2009a). Observational studies have linked diffuse auroral enhancements, equatorward of the main oval, with injection or particle scattering events in the magnetodisc (Radioti et al. 2009, 2013b; Grodent et al. 2010).

The Kronian magnetosphere is overall rotationally-dominated, however, solar wind parameters do play a role in its dynamics (Mauk et al. 2009). One example is the fact that solar wind compression regions can trigger injection events in the nightside of the Kronian magnetosphere, which are observed in particle and auroral data (Clarke et al. 2005; Bunce et al. 2005b; Mitchell et al. 2009b). Reconnection events, dipolarisations, and ejected plasmoids have also been identified in Saturn's magnetosphere (Jackman et al. 2011, 2013), but were not found to occur quasi-periodically as in the case of Jupiter. Again, the most poleward auroral emissions seem to reflect these events, indicating their occurrence in the outer magnetosphere (Clarke et al. 2005; Grodent et al. 2005; Jackman et al. 2013).

In the sections below we describe in detail how particles originating in the magnetosphere and solar wind impact on the atmosphere and cause auroral emissions at different wavelengths, and how these emissions can reveal the magnetospheric dynamics, including differences between magnetodisk- and solar wind-driven events.

2 Response of the Ionosphere to Auroral Forcing at the Giant Planets

Particles, momentum and energy are exchanged between the planetary upper atmosphere and magnetosphere via the ionosphere in the high latitude regions. There is a net momentum transferred from the atmosphere to the magnetosphere, while energy through, for instance particle precipitation, is deposited from the magnetosphere to the atmosphere (e.g. Hill 1979, 2001; Cowley and Bunce 2001). These particles primarily originate from moons (e.g., Io and Europa at Jupiter, and Enceladus at Saturn), and to a lesser extent from the planetary atmosphere and the solar wind (e.g., polar regions at Jupiter). Some of the ions resulting from ionization of the moon's gas torus are neutralized through charge exchange and leave the system; the others are picked up by the planetary magnetic field closely rotating at the planet's rotation rate and flow outward through the planetary magnetosphere (Bagenal and Delamere 2011). The resulting upward currents, flowing from the atmosphere to the magnetosphere, that supply the required angular momentum accelerate the particles, increasing their energy and energy flux (e.g. Ray et al. 2010, 2012a). Particles can also precipitate as a result of wave-particle interactions (e.g. Radioti et al. 2009).

When the energized particles reach the high latitude upper atmosphere, they collide with the atmospheric species, depositing energy through ionization, excitation and dissociation of the neutral gas. This yields the so-called 'auroral emissions' defined as the photomanifestation of the interaction of energetic, extra-atmospheric particles with an atmosphere (e.g. Bhardwaj and Gladstone 2000; Galand and Chakrabarti 2002; Fox et al. 2008; Slanger et al. 2008). Auroral particle degradation results in an increase in ionospheric densities and electrical conductances (e.g. Millward et al. 2002; Hiraki and Tao 2008; Galand et al. 2011). Ionospheric currents, which allow closure of the magnetospheric current system, are enhanced and induce strong Joule heating of the high-latitude thermosphere (e.g. Miller et al. 2005; Smith et al. 2005; Müller-Wodarg et al. 2012). This high-latitude atmospheric heating is a key player in the energy crisis at the giant planets (e.g. Yelle and Miller 2004). In other words, particle precipitation, which can be traced via auroral emissions, plays a critical role in the thermosphere-ionosphere system and its coupling to the magnetosphere.

2.1 Energy Deposition of Precipitating Auroral Particles

2.1.1 Energetic Electrons

The incident auroral electron characteristics derived from the spectroscopic analysis of the ultraviolet auroral emissions (see Sect. 3.1.1) are summarized in Table 1 for the main auroral ovals of Jupiter and Saturn.

104

Table 1 Characteristics of the mean energy and energy flux of the auroral electrons incident at the top of the atmosphere over the main auroral ovals of Jupiter and Saturn. These characteristics have been derived from recent analyses of ultraviolet auroral emissions

Mean electron energy E_{prec} (keV)	Electron energy flux Q_{prec} (mW m^{-2})	Reference
Jupiter		
30–200 Typically 75	2–30	Gustin et al. (2004b)
0.01–3 (soft) 15–22 (hard)	–	Ajello et al. (2005)
460 (dawn storm)	90 (dawn storm)	Gustin et al. (2006)
Saturn		
–	1.9–3.2 (dawn), 4.2–7.7 (pre-noon), 0.3–1.5 (afternoon), <0.4 (dusk), 0.3–0.8 (pre-midnight)	Cowley et al. (2004b)
12 ± 3	7.5 (pre-noon max), 5 (midnight)	Gérard et al. (2004)
1–5, 5–30[a]	–	Gérard et al. (2009)
13–18 (STIS); 10 (Cassini/UVIS/FUV); <15 (FUSE) Typically 10	0.3–1.4 (STIS)[b] Typically 1	Gustin et al. (2009)
–	0.9	Gustin et al. (2012)
≤ 21	–	Gérard et al. (2013)
10–20 (Cassini/UVIS)	≤ 1–17	Lamy et al. (2013)

[a]The two sets of values correspond to two different atmospheric models used for the analysis. The energy values quoted correspond to the characteristics energy, which is equal to half the mean energy if the energy distribution is assumed to be Maxwellian. An energy range of 0.3–2 is quoted in Table 1 of Gérard et al. (2013) for the analysis of Gérard et al. (2009), which most likely is a typo error.

[b]Applying a 10 kR–1 mW m^{-2} conversion factor to the total auroral brightness in the H_2 Lyman and Werner bands (e.g., Gustin et al. 2012).

Models of Suprathermal Electron Transport   Auroral, energetic electrons interact with the atmospheric neutrals through elastic scattering and inelastic collisions, the latter including ionization, excitation, dissociation or a combination of them. Ionization yields the production of secondary electrons, which can in their turn interact with the atmosphere. Furthermore, suprathermal electrons interact with the thermal, ionospheric electrons through Coulomb collisions. This yields an increase in the ionospheric electron temperature (e.g. Grodent et al. 2001; Galand et al. 2011).

As a result of the interaction with the atmospheric species, the suprathermal electrons undergo degradation in energy and redistribution in pitch angle, defined as the angle between the electron velocity and the local magnetic field. As the energy loss is a function of the electron energy, and secondary electrons are added towards lower energies, the initial electron energy distribution at the top of the atmosphere changes, as the electrons penetrate deeper in the atmosphere. The calculation of the distribution of electrons in both position and velocity space is required. Three approaches have been applied to auroral electrons at Jupiter and Saturn, all assuming steady-state conditions and the guiding center approximation (Rees 1989):

- The 'Continuous Slowing Down Approximation' (CSDA) method assumes that the energy loss is a continuous rather than a discrete process (Gérard and Singh 1982; Singhal et al. 1992; Rego et al. 1994; Prangé et al. 1995; Dalgarno et al. 1999). The variation dE in electron energy per path length ds in an atmosphere composed of species k with neutral density n_k and energy loss L_k is given by:

$$\frac{dE}{ds} = -\Sigma_k n_k(s) L_k(E) \tag{1}$$

The method, simple to implement, requires—in order to be able to integrate Eq. (1)—that either the atmospheric composition is independent of altitude (e.g., Dalgarno et al. 1999) or that atmospheric species have energy losses proportional to each other (e.g., Rego et al. 1994). The method is limited to high energies where the assumption of a continuous loss is justified and scattering is neglected. The CSDA method allows the calculation of the profiles in altitude of ionization and excitation rates.

- An alternative method is to utilise transport models based on the explicit, direct solution of the Boltzmann equation, which can use a two-stream approach (up/down) (Waite 1981; Waite et al. 1983; Achilleos et al. 1998; Grodent et al. 2001; Gustin et al. 2009) or multi-stream approach (more than two pitch angles considered) (Kim et al. 1992; Perry et al. 1999; Menager et al. 2010; Galand et al. 2011). The Boltzmann equation expresses the conservation of the number of particles in the phase space, as given by:

$$\frac{df}{dt} + f\nabla_{\underline{v}} \cdot \frac{\underline{F}}{m} = \left(\frac{\delta f}{\delta t}\right)_{coll} + S_{ext} \tag{2}$$

where $f(\underline{r}, \underline{v}, t)$ is the suprathermal electron distribution at position $\underline{r}$, velocity $\underline{v}$ and time t. The second term on the LHS takes into account the effect of any dissipative forces $\underline{F}$. The first term on the RHS represents variation due to collisions and the second term is associated with external sources (e.g., photoelectrons, secondary electrons from an ion beam).

The Boltzmann equation is solved in terms of the suprathermal electron intensity ($I_e = \frac{v^2}{m} f$), which is a measurable quantity. The phase space is usually reduced to three dimensions, path length s along the magnetic field line, kinetic energy E, and cosine μ of the pitch angle θ. Scattering is included. Beside ionization, excitation, and dissociation rates this method allows the calculation of thermal electron heating rates.

- Monte Carlo simulations refer to a stochastic method based on the collision-by-collision algorithm (Hiraki and Tao 2008; Gérard et al. 2009; Tao et al. 2011). A large number of particles is considered and followed in the simulated atmosphere. The Monte Carlo approach avoids the use of an energy grid, which can be of great interest for problems with electron energies ranging over five orders of magnitude. Its drawback is that it is computationally expensive, since it requires a large number of particles to reduce the statistical noise. At Jupiter and Saturn, only excitation, ionization and dissociation processes have been included; thermal electron heating, which is efficient at low energies (<1 eV), has not been considered.

Suprathermal electron transport models are driven by the electron intensity at the top of the atmosphere, which is a function of energy and pitch angle. The energy distribution is usually assumed to be Maxwellian (or a combination of several), Gaussian or monoenergetic, though any distribution can be considered. The initial pitch angle distribution is often assumed to be isotropic over the downward hemisphere or field-aligned. Prangé et al. (1995) showed that anisotropy affects the excitation rates and color ratios. Nevertheless, the effect is attenuated when elastic scattering is included (Hiraki and Tao 2008). Anisotropy does not affect significantly the H Ly α spectral profile (Menager et al. 2010). When defining the incident distribution, the energy flux Q_{prec} should be defined over the downward hemisphere, as follows:

$$Q_{prec} = 2\pi \int_{E_{min}}^{E_{max}} dE . E \int_0^1 d\mu . \mu . I_e^{prec}(E, \mu) \tag{3}$$

where $I_e^{prec}(E, \mu)$ is the intensity of the incident electrons. Depending on the magnetic field orientation, the integration over angle can be from 0 to 1, or from 0 to -1 (0 and (-1) wrongly switched in Eq. (1) of Galand et al. 2011). For an isotropic beam, Eq. (3) is reduced to: $Q_{prec} = \pi \int_{E_{min}}^{E_{max}} dE . E . I_e^{prec}(E)$, while for a field-aligned beam, Eq. (3) is two times larger.

Validation of these models at Jupiter and Saturn is performed by ensuring particle and energy conservation. Models have been compared in terms of atmospheric column above the maximum energy deposition altitude (Galand et al. 2011) and of electron production rate (see section 'Electron Production Rate'). The former shows a 20 % agreement between the results of Gustin et al. (2009) and those of Galand et al. (2011) for a pure H_2 atmosphere except around 20 keV. This anomaly has not yet been explained. There is no apparent reason for a sharp change around 20 keV, as seen in the work by Gustin et al. (2009).

Most of the auroral electron energy is lost through collisions with neutrals and about 50 % of the total energy input is used to heat the atmosphere (Grodent et al. 2001). The percentage of energy lost through collisions with neutrals increases with the electron energy (Menager et al. 2010; Galand et al. 2011). At Saturn, for 10 keV electrons, 89 % of the energy is lost that way with the remaining transferred to thermal electrons (7 %) or escaping as a result of collisional scattering (4 %). Among the energy lost with neutrals, more than 90 % is lost through collisions with H_2 including 50 % used for ionizing H_2 and producing H_2^+ (Galand et al. 2011). In addition to auroral emissions produced by the excitation of atmospheric species from the UV to the IR (see Sect. 3), suprathermal electrons produce Bremsstrahlung emissions themselves in the hard X-ray range, as detected in the auroral zones of Jupiter (see Sect. 4).

Electron Production Rate   The electron production rate $P_e(z)$ induced by auroral electrons is derived from the suprathermal electron intensity $I_e(z, E, \mu)$ calculated as a function of altitude z, energy E, and pitch angle μ (see section 'Models of Suprathermal Electron Transport'), as follows:

$$P_e(z) = 2\pi \sum_k n_k(z) \int_{-1}^1 d\mu \int_{E_{th}}^{E_{max}} dE \, \sigma_k^{ioni}(E) I_e(z, E, \mu) \tag{4}$$

where $n_k(z)$ is the number density of the neutral species k at altitude z and $\sigma_k^{ioni}(E)$ is the total ionization cross section of the neutral species k by electrons of energy E. E_{th} represents the ionization threshold of a single, non-dissociative ionization. Double ionization is not considered here. Volume excitation rates can be calculated in a way similar to Eq. (4) except that the ionization cross section is replaced by the excitation cross section. Under solar

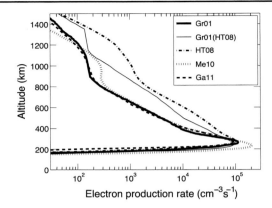

Fig. 3 Comparison of the profile in altitude of the electron production rate at Jupiter between the models of Grodent et al. (2001) [Gr01], Hiraki and Tao (2008) [HT08], Menager et al. (2010) [Me10] and Galand et al. (2011) [Ga11]. The incident electron energy distribution is a triple Maxwellian applied to discrete aurora as defined by Grodent et al. (2001). The thin line annotated "Gr01(HT08)" corresponds to the electron production rate profile for Grodent et al. (2001) shown in Hiraki and Tao (2008). The 100 km altitude level is assumed to correspond to a pressure of 1 mbar

illumination, photo-ionisation by EUV solar radiation (0.1–100 nm) and electron-impact ionization by photoelectrons and their secondaries (e.g. Kim and Fox 1991; Galand et al. 2009; Menager et al. 2010) occurs.

The electron production rate is proportional to the energy flux Q_{prec} of the incident electrons. The altitude of the peak production decreases with the initial energy of the energetic electrons. The more energetic an electron is, the more collisions are required to have it thermalized. A comparison between electron production rates derived from different models using a triple Maxwellian energy distribution for the incident electrons is shown in Fig. 3. There is a very good agreement between the profiles obtained by Grodent et al. (2001) (thick, solid line) and Galand et al. (2011) (dashed line) with less than 7 % difference at the peak. There are large differences above the peak altitude between these two profiles and the one by Hiraki and Tao (2008) (dash-dotted line). The reason is most likely due to different altitude profiles used for the thermospheric densities. In Grodent et al. (2001) the neutral density profiles are given as a function of pressure. A pressure-altitude conversion is required in order to calculate the auroral electron transport on an altitude grid. Galand et al. (2011) derived a conversion between these two quantities assuming hydrostatic equilibrium, ideal gas law, and using the pressure-density profiles given in Grodent et al. (2001). The 100 km altitude level is also assumed to correspond to a pressure of 1 mbar. Hiraki and Tao (2008) used a different altitude-pressure conversion and derived the profile in altitude for Grodent et al. (2001) shown as the thin, solid line in Fig. 3. The profile derived by Menager et al. (2010) (dotted line in Fig. 3) agrees overall with the profile by Grodent et al. (2001) (thick, solid line) except around 1000 km, and near the peak by a factor larger than 2. It is not clear what the source of discrepancy is.

Hiraki and Tao (2008) successfully compared their electron production profile with the one derived by Rego et al. (1994) for 10 keV electrons. Menager et al. (2010) compared their ion production rates at the peak against those presented by Perry et al. (1999). They found 10 % difference for electrons, but larger differences for H^+ (produced from H_2) and hydrocarbon ions.

2.1.2 Energetic Ions

Models of Suprathermal Ion Transport Beside protons of planetary or solar origin (e.g. Patterson et al. 2001), sulfur ions from Io's torus and/or oxygen ions from the icy moons (e.g., Europa, Enceladus) are also present in Jupiter's and Saturn's magnetospheres (Lanzerotti et al. 1992; Bagenal and Delamere 2011). Like for electrons, suprathermal ions collide with atmospheric neutrals yielding scattering, ionization, excitation, and dissociation (or a combination of them). The secondary electrons produced through particle-impact ionization can have enough energy to interact in their turn with atmospheric species. Each type of ion species interacts differently with the atmosphere. For instance, an incident proton beam loses most of its energy through ionization, while an incident oxygen beam does not lose more than 50 % in ionization (Ishimoto and Torr 1987).

The energy degradation of ions is complicated by charge-changing reactions. For example, an energetic proton can capture an electron and become an energetic H atom. In its turn, this H atom can interact with the atmospheric species and/or get stripped of its electron and become a proton again:

$$
\begin{aligned}
\mathrm{H}^+ + \mathrm{H}_2 &\rightarrow \mathrm{H} + \mathrm{H}_2^+ &\quad& \text{Capture} \\
\mathrm{H} + \mathrm{H}_2 &\rightarrow \mathrm{H}^+ + \mathrm{H}_2 + e^- &\quad& \text{Stripping}
\end{aligned}
\tag{5}
$$

Therefore, unlike the case of electrons, more than one charge state needs to be considered: 2 in the case of an incident proton beam (e.g., 0 for H, 1 for H^+); many more in the case of oxygen with stripping collisions potentially producing high charge state ions (e.g., O^{7+} and O^{8+}) (Cravens and Ozak 2012). Furthermore, another complication occurs when a significant part of the incident ion beam is neutralized. As neutral species are not affected by the magnetic field, the neutral beam spreads spatially (in particular latitudinally), which may result in an attenuation of the ion intensity at the centre of the beam (e.g. Lorentzen 2000).

Beside exciting atmospheric neutrals resulting in auroral emissions similar to those produced by electron-induced aurora, ion precipitations have unique signatures distinct from electron precipitations, when the excited species is the energetic ion (or neutral) species itself. For instance, soft X-ray, K-shell emission provides the main evidence that acceleration and precipitation of energetic heavy ions—with energies larger than MeVs—are taking place on Jupiter (Cravens and Ozak 2012); see also Sect. 4. Doppler-shifted H emissions produced by energetic H atoms are a signature of proton precipitation. While in the N_2-dominated terrestrial atmosphere, such a signature is easily detectable (e.g. Galand and Chakrabarti 2006), it is not the case in an H_2-dominated atmosphere. Suprathermal particles induce strong H emissions with photons undergoing frequency shift. This results in a wide spectral profile around H lines (Prangé et al. 1995; Rego et al. 1999). So far no unambiguous detection of a Doppler-shifted component emitted by the energetic H atoms has been made in the H Ly α spectral profile, though it has been speculated (Prangé et al. 1995). Model predictions have shown that the contribution of the Doppler-shifted wing is decreasing with increasing energies and is expected to be small for incident MeV protons. Therefore, its non-detection thus far does not mean that auroral protons do not contribute to the UV emissions at Jupiter and Saturn (Rego et al. 1999).

Two of the three types of approaches used for modeling suprathermal electron transport and energy degradation (see section 'Models of Suprathermal Electron Transport') have been applied to suprathermal ions at Jupiter and Saturn (e.g. Ozak et al. 2010):

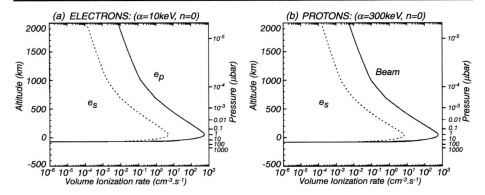

Fig. 4 Profiles in altitude (*left axis*) and pressure (*right axis*) of the volume ionization rate—that is, electron production rate profiles, as double ionization is negligible—due to primary particles (*solid line*) and secondary electrons (*dotted line*) in the case of 10 keV electron incident beam (*left*) and 300 keV proton incident beam. The energy flux of the incident particles is assumed to be the same (from Rego et al. 1994)

– CSDA (e.g. Rego et al. 1994; Cravens et al. 1995; Horanyi et al. 1988). The beam is assumed to be in charge equilibrium, which is not always valid (Rego et al. 1994). For a pure H_2 atmosphere, the equilibrium fractions of H and H^+ at energy E are given by:

$$F_H(E) = \frac{\sigma_{H_2}^{10}(E)}{\sigma_{H_2}^{10}(E) + \sigma_{H_2}^{01}(E)}$$

$$F_{H^+}(E) = \frac{\sigma_{H_2}^{01}(E)}{\sigma_{H_2}^{10}(E) + \sigma_{H_2}^{01}(E)}$$

(6)

where $\sigma_{H_2}^{10}(E)$ and $\sigma_{H_2}^{01}(E)$ are the electron capture cross section and the electron stripping cross section for H_2, respectively. As the former becomes increasingly dominant towards lower energies, the fraction F_H of H atoms—given by Eq. (6)—increases as well. In a gas mixture, the effective equilibrium fraction F is derived from the sum of the equilibrium fraction of each neutral species weighted by its volume mixing ratio.

– Monte Carlo simulations (e.g. Kharchenko et al. 1998, 2006, 2008; Hui et al. 2009, 2010b). Unlike for CSDA, no assumption is made on the charge state fraction, the particle charge state being recorded after each collision. In addition, the spreading of the beam is computed explicitly (when 3D simulations are carried out), although it is computationally demanding. In additions it requires as input the latitudinal width of the incident ion beam, which is poorly constrained at the Earth and is not known at the giant planets.

Comparison Between Electron and Ion Energy Deposition   Comparisons between auroral electrons and protons have been carried out in terms of electron production rates and excitation rates (Rego et al. 1994) and of color ratio and H Ly α spectral profiles (Rego et al. 1999). The probability of collisions with neutral species differs between electrons and protons. As a result, for a given mean energy auroral electrons penetrate deeper in the atmosphere compared with protons. Nevertheless electrons and protons are not expected to have similar energies at the top of the atmosphere. Protons are anticipated to be more energetic, which may compensate for this difference in collision probability.

For a given energy flux, incident 10 keV electrons and 300 keV protons produce very similar volume ionization rate profiles in altitude, as illustrated in Fig. 4. They deposit energy at a similar altitude and the ionization rate is the same at the peak. Only at high altitudes do auroral electrons have a higher ionization rate than protons, up to 40 % at 2000 km. In addition, for both populations the contribution to ionization from secondary electrons is negligible, representing only 1 % of the total ionization. By contrast, the excitation rates depend upon the nature of the precipitating particles. For instance, in a proton beam the secondary electrons are the main contributor to the total excitation of H Ly α, while in an electron beam the contributions from primary and secondary electrons are similar (Rego et al. 1994). The dependence in energy of the color ratio between two H_2 emission bands also varies with the nature of the particles. For a given mean energy, electrons penetrate deeper in the atmosphere, which means that the spectral band around 160 nm, which suffers from hydrocarbon absorption, undergoes stronger attenuation than in the case of protons (Rego et al. 1999). Color ratios inform on the altitude of deposition with similar values found for particles depositing their energy at the same altitude. Caution needs to be applied when deriving the initial energy of the particles (see Sect. 3.1.1). As soft electrons have similar color ratios as hard protons, the presence of protons, even modest, may yield a significant underestimation of the electron mean energy if the incident beam is assumed to be pure electrons. A similar effect has been observed in the auroral regions at Earth (Galand and Lummerzheim 2004).

2.2 Ionospheric Response to Auroral Forcing

2.2.1 Electron Densities

Observations of Electron Density Profiles in altitude of the electron density are obtained by radio occultations. In this technique the spacecraft is emitting a radio signal which traverses the planetary atmosphere before being received by large radio telescopes on the ground at Earth. As the spacecraft is passing behind the planet as seen from Earth, the signal is refracted by free electrons in the ionosphere. For the outer planets, only measurements at dawn and dusk are possible. The number density of ionospheric electrons is derived from the dimming of the signal. The latest update of the electron density profiles obtained at Jupiter, as measured by Voyager and Galileo and analyzed through a detailed, multi-path technique, is presented by Yelle and Miller (2004) (see Figs. 5(a) and 5(b)). Most profiles have a peak around $0.5–2 \times 10^{11}$ m^{-3} (or $0.5–2 \times 10^5$ cm^{-3}) at an altitude between 1500 and 2000 km. The Voyager (panel (b)) and Galileo/0 ingress (G0N) (panel (a)) profiles with large peak electron densities located at low altitudes occurred at dusk, while Galileo/0 egress (G0X) with a low peak electron density at higher altitudes occurred at dawn. The local-time dependence could be explained by the difference in magnitude and location of the peak (see section 'Diurnal Variation' below). However, the Galileo/3 and 4 do not exhibit such a behaviour (Yelle and Miller 2004). The characteristics of the electron density profiles at Jupiter measured mostly at mid-latitudes in the southern hemisphere do not seem to correlate with any obvious geophysical parameters (McConnell et al. 1982; Yelle and Miller 2004).

At Saturn, radio occultations of the ionosphere have provided electron density profiles from Pioneer 11 (Kliore et al. 1980) and the two Voyagers (Lindal et al. 1985), as well as from the Radio Science Sub-System (RSS) onboard Cassini in the near-equatorial regions and at mid- and high latitudes (Nagy et al. 2006; Kliore et al. 2009) (see also Matcheva and Barrow 2012). Figure 5(c) shows average profiles for each of these three regions. Most profiles have peaks between 3×10^2 and 3×10^4 cm^{-3}. These values are lower than those at Jupiter, which is located closer to the Sun. The peak electron density increases with latitude

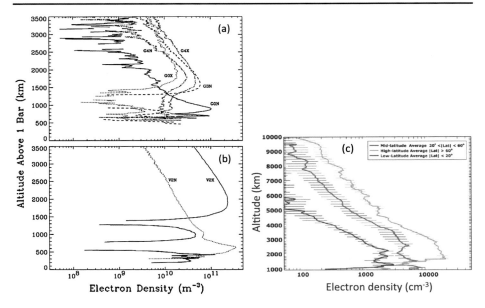

Fig. 5 *Left*: electron density profiles in the Jovian ionosphere as measured by Galileo (panel (**a**), *top*) over the 23–43°S range and Voyager (panel (**b**), *bottom*) over the 50–67°S range (Yelle and Miller 2004). *Right* (panel (**c**)): Electron density profiles in the Saturn ionosphere, averaged over low- (*red*), mid- (*green*) and high- (*orange*) latitudes from all Cassini occultations (Kliore et al. 2009)

(see section 'Latitudinal Distribution' below). At low latitudes, the peak is found to have higher electron density values at dusk than dawn (see section 'Diurnal Variation'). No clear dawn-dusk asymmetry is however seen at mid-latitudes (Nagy et al. 2009).

While the overall magnitude of the density profiles is captured by ionospheric models (e.g. Moore et al. 2004, 2006; Galand et al. 2009), gravity waves need to be invoked to explain the highly-structure vertical profiles at Jupiter and Saturn (Barrow and Matcheva 2011, 2013; Matcheva and Barrow 2012). At low latitudes above the homopause, a surge in water inflow may also contribute to the presence of 'bite-outs' in the profiles observed at Saturn (Moore and Mendillo 2007).

Impulsive radio bursts at Saturn, referred as Saturn Electrostatic Discharges (SEDs), have been detected by the two Voyagers (Warwick et al. 1981, 1982) and the Cassini/Radio and Plasma Wave Science (RPWS) instrument (Gurnett et al. 2005; Fischer et al. 2006, 2007). These discharges are produced by lightning occurring in convective-looking clouds at mid-latitudes (e.g. Dyudina et al. 2010). Peak electron densities are derived from the measurement of the low-frequency cutoff below which the radio waves, which traverse Saturn's ionosphere on their way to the spacecraft, are not detected (Fischer et al. 2011). The diurnal variation of the peak electron density derived from SEDs analysis is discussed in section 'Diurnal Variation'.

At Jupiter, no high frequency radio component—similar to SEDs—above the cutoff frequency of the ionosphere has been detected (e.g. Fischer et al. 2008). This non-detection has been explained as the result of the strong absorption of the radio waves in Jupiter's lower ionospheric layers (Zarka 1985) and of the decrease of the spectral power of Jovian spherics with increasing frequency (Farrell et al. 1999).

Ionospheric Models The continuity equation, which expresses the conservation of the number of particles, allows the calculation of the number density of species i:

$$\frac{\partial n_i}{\partial t} + \nabla \cdot (n_i \underline{u}_i) = P_i - L_i \tag{7}$$

where $\underline{u}_i$ is the bulk velocity, and P_i and L_i are the production and loss rates, respectively, of species i. XUV solar radiation is the main source of electrons at low and mid latitudes and particle precipitation dominates in the auroral regions. Ions are also produced through chemical reactions, such as charge exchange. Loss rates include ion-neutral reactions and, for molecular ions, ion-electron dissociative recombination (e.g. Kim and Fox 1991; Moses and Bass 2000). Ionospheric models have been developed for the outer planets, as reviewed by Waite et al. (1997) with more recent models proposed for Jupiter (Achilleos et al. 1998; Perry et al. 1999; Grodent et al. 2001; Millward et al. 2002; Tao et al. 2010; Barrow and Matcheva 2011) and Saturn (Moses and Bass 2000; Moore et al. 2004, 2006, 2008, 2010, 2012; Galand et al. 2009, 2011; Tao et al. 2011; Barrow and Matcheva 2013).

The second term in Eq. (7) represents transport processes, such as plasma diffusion. It becomes increasingly important with altitude, controlling the upper part of the ionosphere ($\geq$2300 km above the 1 bar level at Saturn (Moore et al. 2004). Neutral winds can also have a significant effect by redistributing the plasma from one region to another. Horizontal winds will move the plasma vertically along the magnetic field lines. Galand et al. (2011) showed that in the auroral regions where the dip angle is large, the main contributor is the vertical component of the thermospheric wind, which decreases the electron density peak magnitude by as much as 75 %.

In the lower ionosphere, transport timescales are significantly larger than chemical loss timescales and the photochemical regime dominates. Assuming also steady-state conditions, the continuity equation (7) is thus reduced to:

$$P_i = L_i \tag{8}$$

Under photochemical equilibrium, the H^+ to H_3^+ number density ratio is proportional to the electron density (e.g. Moore et al. 2004). This means that molecular ions will be more abundant at low latitudes where the electron density is reduced compared with high latitudes where it has significantly larger values (see section 'Latitudinal Distribution').

Introducing an effective recombination coefficient $\bar{\alpha}$ defined as the recombination coefficient of individual ion species weighted by their number density, Eq. (8) applied to ionospheric electrons becomes:

$$P_e = \bar{\alpha} n_e^2, \quad \text{that is, } n_e = \sqrt{\frac{P_e}{\bar{\alpha}}} \tag{9}$$

In the absence of (or under limited) solar illumination, P_e is proportional to the energy flux Q_{prec} of the precipitating particles (see Sect. 2.1.1). Therefore, Eq. (9) means that: $n_e \propto (Q_{prec})^{1/2}$ (for constant precipitation over time). This assumes that the photochemical regime dominates, which is fulfilled in the lower ionosphere.

As H and H_2 are the dominant neutral species in this region, H_2^+ and H^+ are the main ions produced through photo-ionization and electron-impact ionization. H_2^+ is very reactive and quickly interacts with H_2 to become H_3^+. H_3^+ is lost through electron dissociative recombination, which significantly depletes in density during the night (e.g. Kim and Fox 1994, see section 'Diurnal Variation').

Early ionospheric models at the giant planets predicted that the long-lived H^+ would be the dominant species and overestimated the peak electron density (McElroy 1973). Loss mechanisms have been introduced in order to match the observed peak electron densities: (1) charge-exchange of H^+ with vibrationally-excited H_2 (e.g. McConnell et al. 1982; Moses and Bass 2000; Moore et al. 2010); (2) forced vertical motion of the plasma (e.g. McConnell et al. 1982; Majeed and McConnell 1991); (3) water inflow especially at Saturn (Connerney and Waite 1984; Majeed and McConnell 1991; Moses and Bass 2000; Moore et al. 2006, see section 'Latitudinal Distribution').

H^+ is efficiently lost when H_2 is in vibrationally-excited levels as follows:

$$H^+ + H_2(\nu \geq 4) \rightarrow H_2^+ + H \tag{10}$$

The reaction rate, k1, for reaction (10) is now well established with a value of 10^{-9} cm^3 s^{-1} (Krstić 2002; Huestis 2008). In photochemical models usually an effective reaction rate k_1^* is used which is defined as:

$$k_1^* = k_1 \frac{n_{H_2}^{\nu \geq 4}}{n_{H_2}} \tag{11}$$

where n_{H_2} is the total H_2 number density and $n_{H_2}^{\nu \geq 4}$ is the H_2 number density in a vibrationally-excited level $\nu \geq 4$.

Assessing the relative amount of H_2 in vibrational levels of non-LTE origin carries out large uncertainties, which limit the estimate of the electron density (Nagy et al. 2009 and references therein). Estimations have been proposed by Moses and Bass (2000), and more recent updates have been presented by Moore et al. (2010, 2012) in order to best match the Cassini/RSS observations. k_1^* is expected to increase in the auroral regions (e.g. Cravens 1987).

Below the homopause, H^+ efficiently reacts with hydrocarbons (Kim and Fox 1994; Moses and Bass 2000). This loss is twice as fast as that of H_3^+ resulting in the dominance of molecular ions, in particular hydrocarbon and metallic ions, in this region (Moses and Bass 2000; Kim et al. 2001).

Diurnal Variation   Based on radio occultation observations, low-latitude profiles obtained at Saturn exhibit a strong dawn/dusk asymmetry with lower peak electron density observed at dawn, as illustrated in Fig. 6(a). The presence of water at low latitudes yields a depletion in H^+ and molecular ions become dominant, at least in the late morning up to early afternoon sector (Moore et al. 2006, see also section 'Latitudinal Distribution'). At sunrise, the molecular ion density builds up quickly, faster than H^+ density. The loss of H_2^+ producing H_3^+ is faster than that producing H^+ (e.g. Moses and Bass 2000; Galand et al. 2009). At sunset, the molecular ion density decays quickly as a result of electron dissociative recombination.

At mid-latitudes, on the one hand, electron density profiles obtained from radio occultations at Jupiter and at Saturn do not exhibit any clear local-time dependence (see section 'Observations of Electron Density'). This may be due to additional processes (Nagy et al. 2009), such as dynamic effects combined with the less efficient electron-ion recombination in the absence of water (Moore and Mendillo 2007). On the other hand, diurnal variations in the mid-latitude electron density have been obtained at Saturn from the analysis of SEDs (see section 'Observations of Electron Density'). Very large reductions in the peak electron density from mid-day to mid-night have been inferred, as illustrated in Fig. 6(b). They extend over more than 2 orders of magnitude for the Voyager era (dotted and dashed lines) and

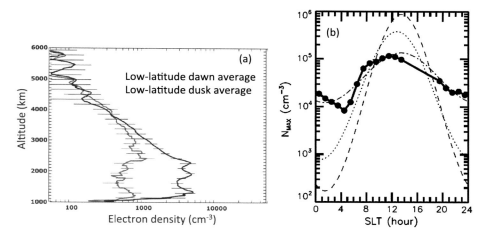

Fig. 6 *Left* (**a**): average electron density profiles obtained by Cassini/RSS at low latitudes at dawn (*pink*) and dusk (*blue*) (Kliore et al. 2009); *right* (**b**): diurnal trend in the peak electron density from Cassini/RPWS SED analysis (Fischer et al. 2011) (dots linked by a *thick, solid line*). The *dash-dotted line* represents a fit to the Cassini dataset. The *dotted* and *dashed lines* are fits for the Kaiser et al. (1984a) and Zarka (1985) diurnal maximum electron density trends from the Voyager era (Moore et al. 2012)

to one order of magnitude for Cassini (dots and dash-dotted line). The peak electron densities are found to exhibit a solar-zenith-angle dependency most pronounced at dawn (Fischer et al. 2011). Even when considering the more moderate Cassini results, photochemical models cannot reproduce the observed diurnal variations using the current best estimates for the production and loss sources (Moore et al. 2012). Most likely the large diurnal variations observed from SEDs are related to the sharp peaks frequently seen in the Cassini/RSS radio occultation measurements and located below the main ionospheric peak (Kliore et al. 2009).

Latitudinal Distribution   While at Jupiter most electron density profiles analyzed in detail have been measured at mid-latitudes (see section 'Observations of Electron Density'), the wealth of radio occultations obtained by Cassini/RSS allows the derivation of the latitudinal behavior in ionospheric properties. With the decrease in solar illumination with latitude, the peak electron density and total electron content (TEC) are expected to decrease with increasing latitude. Kliore et al. (2009) and Moore et al. (2010) showed that the reverse trend is observed, as illustrated in Fig. 7. The decrease in TEC from mid latitudes towards the equator is likely due to the inflow of water from the rings and icy moons (Connerney and Waite 1984; Moore et al. 2006, 2010). The solar-driven model (solid lines) reproduces well the Cassini/RSS values (symbols) at low-latitudes when this additional loss process is included (Moore et al. 2006). The addition of water converts H^+ to H_3O^+ via the very short-lived H_2O^+:

$$H^+ + H_2O \rightarrow H_2O^+ + H$$
$$H_2O^+ + H_2O/H_2 \rightarrow H_3O^+ + OH/H$$

(12)

As a result in the shift from atomic, long-lived H^+ to molecular, shorter-lived H_3O^+, the electron density is reduced (Connerney and Waite 1984; Moore et al. 2006; Moore and Mendillo 2007; Müller-Wodarg et al. 2012). Moore et al. (2010) obtained a best agreement

Fig. 7 Latitudinal variation of the total electron content (TEC) from the Cassini/RSS radio occultation observations (Kliore et al. 2009) and from solar-driven model simulations (Moore et al. 2010)

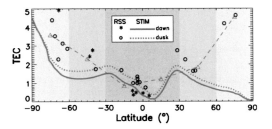

in latitudinal profiles of TEC between model and observations when imposing for the water flux a Gaussian profile centered on the equator with a peak value of 0.5×10^7 cm^{-2} s^{-1} and a full width half maximum (FWHM) of 23.5° latitude. The bulk of the gaseous water in the Saturnian system is expected to come as a neutral species from the icy moons, especially Enceladus—which replenishes the E ring (e.g. Jurac and Richardson 2005). Some of the water produced at Enceladus is lost to Saturn, mostly in the equatorial region (Fleshman et al. 2012). Though part of the water influx is neutral, there is recent evidence for water ion precipitation at low- and mid-latitudes (O'Donoghue et al. 2013).

At high latitudes, the increase in H_2 ($\nu \geq 4$) (Cravens 1987) results in a more efficient removal of H^+ through charge-exchange with H_2 (see section 'Ionospheric Models'). Nevertheless this increased loss does not compensate for the additional source in ionization induced by particle precipitation (e.g. Millward et al. 2002; Galand et al. 2011; Tao et al. 2011). As a result large values for the electron densities and TEC are observed (symbols in Fig. 7). The main peak in the electron density profile at high latitudes is associated with H^+, replacing H_3^+ lost by electron recombination due to the large electron density. H_3^+ drives the ionospheric peak—at least in the late morning and early afternoon sector—in the absence of particle precipitation (e.g. Galand et al. 2011) and at lower latitudes at solar minimum or in the presence of ring shadowing (Moore et al. 2004; Müller-Wodarg et al. 2012).

2.2.2 Ionospheric Electrical Conductances

Electrical, ionospheric conductivities are associated with particle mobility in the direction perpendicular to the planetary magnetic field and parallel (Pedersen) or perpendicular (Hall) to the ionospheric electric field. They have been calculated using ionospheric models applied to Jupiter (e.g. Millward et al. 2002; Hiraki and Tao 2008; Smith and Aylward 2009; Tao et al. 2010) and to Saturn (e.g. Moore et al. 2010; Galand et al. 2011; Ray et al. 2012b; Müller-Wodarg et al. 2012) or derived using Cassini/RSS electron density (Moore et al. 2010). The Pedersen conductivity profiles are strongly peaked in altitude, as illustrated in Fig. 8. The Pedersen conducting layer associated with a current carried by ions is located in the region where the ion gyrofrequency is similar to the ion-neutral collision frequency. It corresponds to a region in the lower ionosphere dominated by molecular ions, close to the homopause. The production and mobility of these ion species therefore control the conductances (Millward et al. 2002; Moore et al. 2010; Galand et al. 2011). The Hall conductivity layer, associated with a current carried by electrons, is broader than the Pedersen layer. It is located at lower altitudes below the homopause (e.g. Galand et al. 2011) where the chemistry with hydrocarbons becomes important and complex (e.g. Kim and Fox 1994; Moses and Bass 2000). Despite an auroral forcing at Jupiter stronger than at Saturn, the Jovian conductivities have peak magnitudes smaller than the Kronian ones (see Fig. 8). This is due to the differences in the magnetic field strength between both planets.

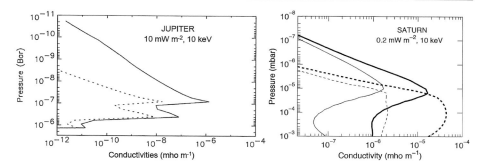

Fig. 8 Ionospheric Pedersen (*solid line*) and Hall (*dashed lines*) conductivities profiles as a function of pressure: (*left*) at Jupiter (Millward et al. 2002) for 10 keV electrons ($Q_{prec} = 10$ mW m^{-2}); (*right*) at Saturn for solar illumination only (78°S, equinox, solar minimum) (*thin lines*) and for solar and auroral 10 keV electrons ($Q_{prec} = 0.2$ mW m^{-2}) (*thick lines*) (adapted from Galand et al. 2011)

Ionospheric conductances—or height-integrated conductivities—are most intense at low and mid-latitudes on the dayside due to solar-induced ionization and in the auroral regions over all local times due to particle precipitation (Moore et al. 2010; Galand et al. 2011; Müller-Wodarg et al. 2012). Outside these regions, the lower ionosphere, where the conductivities peak, is strongly depleted due to the absence of ionisation, though not totally, providing a low-level background conductance.

Table 2 summarizes the different values published in the literature for ionospheric, Pedersen conductances calculated using ionospheric models applied to the auroral oval regions. For a given auroral forcing (10 keV electrons with an energy flux of 1 mW m^{-2}), Pedersen conductances at Jupiter are two orders of magnitude smaller than the values at Saturn. Ionospheric composition and integration altitude regions for the conductivities are similar at both planets. The difference in conductances comes primarily from the difference in magnetic field strength. Jupiter's magnetic field, the strongest planetary field encountered in the Solar System, is 20 times as strong as Saturn's. The conductances are dependent on the magnetic field strength through the angular gyrofrequency. If the magnetic field of Saturn is multiplied by a factor 20, the conductances are found to decrease by a factor 150 to 200 (Galand et al. 2011).

Nevertheless auroral forcing is not the same at Jupiter and at Saturn. The initial energy and energy flux of the auroral electrons is higher at Jupiter, as illustrated in Table 1. Typical values for the auroral electron characteristics are highlighted in bold in Table 2 and yield Pedersen conductance values in the 1.5–2 mho range at Jupiter and in the 10–15 mho range at Saturn (see Table 2). These Jovian values are supported by estimations from field-aligned potential models—using parameterized Pedersen conductance relations: they spread from 0.7 to 1.5 mho (Smith and Aylward 2009; Tao et al. 2009; Ray et al. 2010). Therefore, difference in auroral forcing reduces the difference in conductances between Jupiter and Saturn down to one order of magnitude. At Jupiter, for 10 keV electrons with an energy flux of 10 mW m^{-2}, Millward et al. (2002) found a value for the Pedersen conductance of 0.12 mho, while Hiraki and Tao (2008) derived a value of 0.5 mho (see Table 2). Differences may be associated with differences in the induced location of the Pedersen conductive layer, in the set of ion species considered ($[H_2^+, H_3^+]$ versus $[H_2^+, H_3^+, H^+]$) and assumptions made (photochemical equilibrium versus transport included). Bougher et al. (2005) derived conductance values more than an order magnitude larger than those obtained by Millward et al. (2002) for 10 keV and 100 keV (see Table 2).

Table 2 Pedersen conductance Σ_P calculated using energy deposition and ionospheric models and presented as a function of the ionization source (Sun, auroral electrons) over the main auroral oval. The characteristics of the auroral electrons are given in terms of the initial mean energy and energy flux. Typical values for the auroral characteristics (see Table 1) and the associated conductance are shown in *bold*

Energy source E_{prec} (keV), Q_{prec} (mW m^{-2})	Pedersen conductance Σ_P (mho)	Reference [atmospheric model]
Jupiter		
Electrons [10, 1]	0.04	Millward et al. (2002)
Electrons [10, 10]	0.12	[3D GCM]
Electrons [10, 100]	0.62	
Electrons [60, 10]	**1.75**	
Electrons [22, 100]+[3,10]+	9 (NH)[a]	Bougher et al. (2005)
Electrons [0.1, 0.5]	12.5 (SH)[a]	[3D GCM]
Electrons [1, 1]	0.008	Hiraki and Tao (2008)
Electrons [10, 10]	0.5	[1D ionospheric model]
Saturn		
Solar only (Main oval: noon, 78°, equinox, solar minimum)	0.7	Galand et al. (2011) [1D ionospheric model using 3D neutral output]
Solar + **Electrons [10, 1]**	**11.5**	
Solar + Electrons [10, 0.2]	5	
Solar + Electrons [2, 0.2]	10	

[a]NH and SH stands for northern hemisphere and southern hemisphere, respectively.

Though the mean energy of the particle is 22 keV (for the bulk population) this does not explain the large difference. Differences in magnetic field models may be the reason.

As for background, solar-driven Pedersen conductance values, Tao et al. (2009) computed values of the order of 0.01 mho in the auroral regions of Jupiter, reaching 0.11 mho on the dayside at low Jovian latitudes. At Saturn, they are of the order of 0.5 mho over the auroral main oval (Galand et al. 2011) increasing to a few mho at 60° latitude (Ray et al. 2012b). Over the whole range of latitudes (which does not include the main ovals) Moore et al. (2010) assessed Pedersen conductances from Cassini/RSS observations and obtained values as high as 8 mho when the full altitude profiles with sharp, narrow peaks in the low ionosphere are included. However, when considering only the topside ionosphere above 1200 km, solely driven by solar illumination, the Pedersen conductances are reduced to values below 1.5 mho.

Ionospheric conductances vary with the initial energy of the particles. Auroral electrons with low (high) energies penetrate above (below) the conductivity layer and are therefore not as effective to increase conductances. At Jupiter, Millward et al. (2002) found that 60 keV electrons, which deposit their energy near the homopause, are the most effective at enhancing electrical conductances. For larger energies, the induced electron density derived by Millward et al. (2002) decreases: as hydrocarbons are neglected in the model, the major ion is H_3^+, which is quickly destroyed through dissociative recombination. Hiraki and Tao (2008) also found that for a given energy flux Pedersen conductance increases with the initial electron energy. They derived a $(E_{prec})^{1.65}$ dependence with saturation around 300 keV, which corresponds to the upper limit of their model validity range. Furthermore, the Pedersen conductance depends on the pitch angle θ of the incident electrons. Hiraki and Tao

(2008) found that the conductance decreases by as much as 40 % with increasing θ, as electrons with larger pitch angles do not penetrate as deep, further away from the conductance layer. At Saturn, Galand et al. (2011) found that auroral electrons with mean energy of 2–3 keV are the most effective at maximizing the Pedersen conductance, while the electron mean energy needs to be increased by more than 20 keV to maximize the Hall conductance. At very low energies (less than a few 100 eV) the contribution by auroral particles was found to be so low—as they only reach very high altitudes—that conductances are driven by solar illumination (Galand et al. 2011).

In presence of intense, hard aurora (energy flux $Q_{prec} > 0.04$ mW m^{-2} (at Saturn) and mean energy $E_{prec} \geq 10$ keV), the electron density closely follows the energy flux Q_{prec} of the auroral electrons (Millward et al. 2002; Moore et al. 2010; Galand et al. 2011; Müller-Wodarg et al. 2012): $\Sigma \propto (Q_{prec})^{1/2}$. It is not surprising as the conductances are roughly proportional to the main ion density, that is, approximatively to n_e. In addition, $n_e \propto (Q_{prec})^{1/2}$, at least in the region where conductivities peak (see section 'Ionospheric Models'). Millward et al. (2002) found the following dependence for the Pedersen and Hall conductances (in mho) induced by 10 keV electrons at Jupiter:

$$
\begin{aligned}
\log_{10} \Sigma_{\mathrm{P}} &= \alpha_{\mathrm{P}} \log_{10} Q_{prec} + \beta_{\mathrm{P}} [\log_{10} Q_{prec}]^{1/2} + \gamma_{\mathrm{P}} \\
\log_{10} \Sigma_{\mathrm{H}} &= \alpha_{\mathrm{H}} \log_{10} Q_{prec} + \beta_{\mathrm{H}} [\log_{10} Q_{prec}]^{1/2} + \gamma_{\mathrm{H}}
\end{aligned}
\tag{13}
$$

where Q_{prec} is given in mW m^{-2}, $\alpha_{\mathrm{P}} = 0.437$, $\beta_{\mathrm{P}} = 0.089$, and $\gamma_{\mathrm{P}} = -1.438$ and $\alpha_{\mathrm{H}} = 0.244$, $\beta_{\mathrm{H}} = 0.121$, and $\gamma_{\mathrm{H}} = -3.118$.

When the energy Q_{prec} varies with local time (Lamy et al. 2009; Badman et al. 2012b) the response of the ionosphere needs to be taken into account through a time shift. Galand et al. (2011) found the following dependence for the Pedersen and Hall conductances (in mho) induced by 10 keV electrons at Saturn:

$$
\begin{aligned}
\Sigma_{\mathrm{P}}(t) &= 11.5 \big[Q_{prec}(t - \Delta t_{\mathrm{P}}) \big]^{1/2} \\
\Sigma_{\mathrm{H}}(t) &= 24.7 \big[Q_{prec}(t - \Delta t_{\mathrm{H}}) \big]^{1/2}
\end{aligned}
\tag{14}
$$

where Q_{prec} is given in mW m^{-2}, and Δt_{P} and Δt_{H} are 10 min 12 s and 4 min 26 s (Earth minutes), respectively. The shift is a function of the energy of the incident particles, increasing significantly for smaller energies, which correspond to auroral electrons reaching higher altitudes where the ionospheric response is slower (e.g. Millward et al. 2002). Finally, the dependence of the ionospheric conductances has also been proposed in terms of field-aligned current (FAC) by Nichols and Cowley (2004) and Ray et al. (2010) (based on the ionospheric modeling by Millward et al. 2002) and by Hiraki and Tao (2008), Tao et al. (2010). Such models allow feedback of the ionospheric conductance on the FAC.

Ionospheric conductances depend indirectly on the ionospheric electric field present in the auroral regions. When the latter is increased, Joule heating is enhanced resulting in an increase in temperatures in the upper atmosphere. The enhancement in thermospheric temperature with electric field depends on the ionospheric conductivities, that is, on the auroral electron energy flux. The heating of the upper atmosphere yields its expansion. This means that conductances are calculated over a larger vertical integral, resulting in an enhancement of their values (Müller-Wodarg et al. 2012). For instance, for 10 keV electrons with an energy flux Q_{prec} of 1.2 mW m^{-2}, when the electric field strength is increased from 80 mV m^{-1} to 100 mV m^{-1}, the thermospheric temperatures increase from 450 K to 850 K (by a factor of ~ 1.9) and the Pedersen conductance is enhanced by 50 %. This increase in conductance with the electric field depends on the electron energy flux, increasing with Q_{prec}.

Effective ionospheric conductances have been introduced in order to take into account, in current models (e.g. Nichols and Cowley 2004), the rotational slippage of the neutral atmosphere from rigid corotation due to ion-neutral frictional drag (Huang and Hill 1989; Bunce et al. 2003). The effective Pedersen conductance is defined as (Cowley et al. 2004a):

$$\Sigma_P^* = (1 - k)\Sigma_P \qquad (15)$$

with the parameter k defined as:

$$k = \frac{\Omega - \omega_n}{\Omega - \omega_i} \qquad (16)$$

where Ω is the planetary angular velocity, ω_n is the angular velocity of the neutral atmosphere, and ω_i is the angular velocity of the plasma. General Circulation Models (GCMs) calculate the 'true' ionospheric conductances (e.g. Millward et al. 2002; Galand et al. 2011; Müller-Wodarg et al. 2012). They include ion-neutral drag and associated neutral dynamics and can therefore be used for assessing the parameter k. At Jupiter, the derived values of k are around 0.5 (Cowley et al. 2004a) ranging from 0.3 to 0.8 throughout the whole outer regions (Millward et al. 2005; Smith et al. 2005). Millward et al. (2005) showed that the parameter k increases when the incident electron energy increases, while it decreases when the equatorward auroral voltage is enhanced. Tao et al. (2009) found regions where the slippage can yield a height-dependent k_z parameter larger than 1: this corresponds to a region where the neutral wind velocity is larger than the ion drift velocity in the planetary rotation frame caused by Coriolis forces and viscosity. For regions where $k_z < 1$, they derived values for k between 0.25 and 0.35 over the 63–73° latitude region. Finally, Smith and Aylward (2009) found negative values of k in the ionospheric regions mapping to magnetospheric radii inside of $20R_J$ as a result of super-rotation of the neutrals.

At Saturn, Smith and Aylward (2008) derived values between 0 and 0.6 at high latitudes with a mean value of 0.4 over the auroral oval, very close to the value of 0.5 derived by Galand et al. (2011). Smith and Aylward (2008) also found that the k-parameter becomes negative at 25° co-latitude at a result of super-rotation of the neutrals in this region.

2.3 Auroral Emission Processes

Figure 9 outlines the sequence of processes which occur after auroral particles precipitate into the H_2-dominant atmosphere leading to the radiation of UV, VIS and IR emissions.

2.3.1 UV Emission Processes: Production and Radiation Transfer

UV photons are emitted from electron-excited molecules and/or atoms when they de-excite to their ground states. Jupiter and Saturn's UV emissions mainly consist of H Lyman α and H_2 Lyman and Werner bands excited by precipitating electron impacts. The excitation rates to the B and C states are directly related to the strength of the H_2 Lyman and Werner bands, respectively. The effect of the quenching of B and C states is small (Gérard and Singh 1982). The emission intensity of the transition band from the upper v' to the lower v'' state, $I_{v',v''}^W$, in the Werner system is given by

$$I_{v',v''}^W = I_c q_{v',0}^{X \to C} A_{v'v''}^{C \to X} / \Sigma_{v''} A_{v'v''} \qquad (17)$$

Fig. 9 Flowchart of emission processes after auroral particle precipitation into the H_2-dominated atmosphere

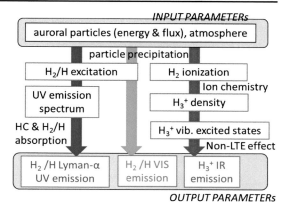

where I_c is the total intensity of the C state; $q_{v',0}^{X \to C}$ represents the Frank-Condon factors for the excitation rate of the C state into the v' level; $A_{v'v''}^{C \to X}$ is the Einstein coefficient for the transition from v' to v''. The fraction $A_{v'v''}^{C \to X} / \Sigma_{v''} A_{v'v''}$ corresponds to the branching ratio for the line. The emission intensity of a $v' \to v''$ transition band in the Lyman system is also given by the total intensity of the B state with a contribution from the E and F states of 25 % (Gérard and Singh 1982). Transitions from other excited states of H_2 (B', B", D, D') also contribute to EUV emission in the wavelength range 80–120 nm (Gustin et al. 2004a). Lyman α is also estimated to contribute <10 % of the total UV emissions (Perry et al. 1999).

UV auroral emissions at wavelengths <130 nm and <120 nm are absorbed and modified by hydrocarbon molecules and H_2, respectively. The spectrum after absorption depends on the optical depth and is a function of the absorption cross section and the column density of the absorber above the emission region as follows:

$$I_{\text{after},\lambda} = I_{\text{before},\lambda} \exp(-\tau_\lambda), \quad \tau_\lambda = \int \sigma_\lambda N_z ds, \tag{18}$$

where $I_{\text{before},\lambda}$ is the spectrum before absorption, τ_λ is the optical depth, σ_λ and N_z are the absorption cross section and density of absorber, respectively, and ds is taken along the path of the emitted photon. Synthetic H_2 spectra before and after H_2 self-absorption are shown in Figs. 10(a) and 10(b), respectively. The absorption cross section of hydrocarbons depends on wavelength as shown for methane by the black line in Fig. 10(c). The blue and green lines in Fig. 10(c) show combined H Lyman α and H_2 spectra before and after absorption by methane.

2.3.2 Infrared Emission Processes: Production and Non-LTE Effects

Auroral electron precipitation is the dominant ionization source in the high latitudes in addition to the solar EUV in the dayside. These ionization processes stimulate ion chemistry in the ionosphere. The major chemical reactions are depicted in Fig. 11.

H_3^+ is one of the important species for IR emission from Jupiter and Saturn. Emission from H_3^+ is not a direct result of particle precipitation, but is a chemical product formed via the ionisation of molecular hydrogen as shown in Fig. 11. It is formed via this very efficient and exothermic process:

$$H_2 + H_2^+ \to H_3^+ + H \tag{19}$$

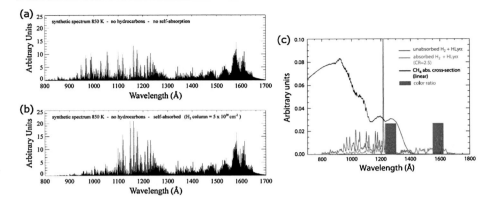

Fig. 10 Estimated spectrum (**a**) before and (**b**) after the H_2 self-absorption due to a H_2 column density of 5×10^{20} cm^{-2}, and (**c**) H Lyman-α and H_2 spectrum before (*blue*) and after (*green*) absorption by hydrocarbons for the case when the intensity ratio $I(1550–1620$ Å$)/I(1230–1300$ Å$) = 2.5$. The absorption cross section of the main absorber, methane, is shown by the *black line*. The *blue line* is a laboratory spectrum obtained from impact of 100 eV electrons on H_2 gas at 300 K, simulating an intrinsic non-absorbed auroral emission spectrum (from Gustin et al. 2004a, 2013.)

Fig. 11 Ion chemistry in the ionospheres of Jupiter and Saturn showing the main reactions described in the text

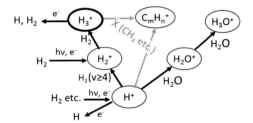

which means that in an environment rich in molecular hydrogen, such as the upper atmosphere of a gas giant, the production of H_3^+ is a tracer of energy injected into the system. Molecular hydrogen can also be produced via more novel paths:

$$H^+ + H_2(v \geq 4) \rightarrow H_2^+ + H \tag{20}$$

followed by reaction (19) that forms H_3^+. The life-time of H^+ is longer than that of H_3^+ by a factor of 10–100 (Kim and Fox 1994). Unfortunately, the mixing ratio of $H_2/H_2(v \geq 4)$ is unconstrained by three orders of magnitude (Majeed and McConnell 1991) so how effective reaction (20) is to shorten the life-time of H^+ is unclear.

Generally, H_3^+ is lost by these reactions:

$$H_3^+ + e^- \rightarrow H_2 + H \tag{21}$$

$$H_3^+ + e^- \rightarrow H + H + H \tag{22}$$

$$H_3^+ + X \rightarrow HX + +H_2 \tag{23}$$

where reaction (23) is extremely efficient if X is a species with more protons than H_2 (Flower 1990). Since the thermosphere of Jupiter is mostly H and H_2, reaction (23) is only important at very low altitudes, where hydrocarbons very efficiently quench any population of H_3^+.

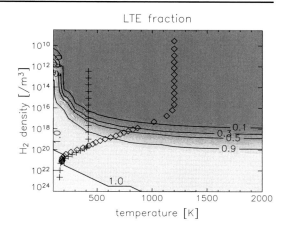

Fig. 12 Contour maps of LTE fraction as a function of temperature and H$_2$ density. Parameters for Jupiter's and Saturn's atmosphere are shown by diamonds and pluses, respectively

Above the homopause, the loss of H$_3^+$ is mainly governed by reactions (21) and (22), and the H$_3^+$ life-time becomes a function of electron density. At the auroral latitudes, the life-time of H$_3^+$ is about 10 s, whereas it is about 10^3 s at lower latitudes (Achilleos et al. 1998). At Saturn, auroral life-times are about 500 s (Melin et al. 2011).

H$_3^+$ is excited vibrationally following collisions with background H$_2$ under high thermospheric temperature. The population of these vibrationally excited states is determined by the balance between collisional excitation/de-excitation and radiation transitions, i.e., IR radiation. This IR radiation effect, combined with a decrease of the H$_2$ density, i.e., a decrease in collisional excitation, at high altitudes produces a deviation in the excited population from the local thermal equilibrium (LTE) or Boltzmann distribution. This reduces the IR emission intensity, an effect which is estimated to be significant for Jupiter (Melin et al. 2005). The reduction ratio of the H$_3^+$ density, $\eta(z) = n_{H_3^+,non\text{-}LTE}/n_{H_3^+,LTE}$, is a function of H$_2$ density and the temperature as shown in Fig. 12 (after Tao et al. 2011). As the temperature, i.e., the efficiency of the IR emission, increases and/or as the H$_2$ density decreases, this reduction (the non-LTE effect) becomes large. The non-LTE effect must therefore be considered when analysing IR spectra, and can be exploited to determine the incident electron energy, by considering the relationship between electron penetration depth and H$_2$ density (Tao et al. 2012).

Using the H$_3^+$ ion density, $N_{H_3^+}$, the LTE fraction η, and the atmospheric temperature T, the IR emission strength is estimated as follows,

$$I_{IR}(\omega_{if}, z) = N_{H_3^+}\eta(z)g(2J + 1)hc\omega_{if}A_{if}\exp(-E_f/k_BT)/Q(T) \qquad (24)$$

where, e.g., for the fundamental line, I_{IR} is the emission intensity; $\omega_{if} = 2529.5$ cm^{-1} is the wavenumber; $g = 4$ is the nuclear spin weight; $J = 1$ is the rotational quantum number of the upper level of transition; h is the Planck constant; c is the velocity of light; $A_{if} = 129$ s^{-1} is the Einstein coefficient; $E_f = 2616.5$ cm^{-1} is the energy of the upper level of the transition; k_B is the Boltzmann constant; $Q = \Sigma_i(2J + 1)g_i\exp(-E_i/k_BT)$ is the partition function.

2.3.3 Jupiter-Saturn and IR-UV Comparison

Tao et al. (2011) have developed a model of how the above UV and IR emissions respond to different auroral electron energy and flux, and the background atmospheric temperature.

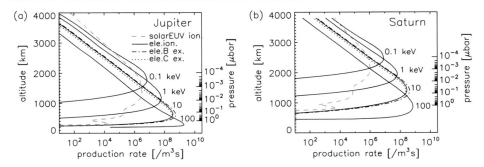

Fig. 13 Altitude profiles of ionization and excitation rates caused by auroral electrons (*black lines*) and solar EUV (*grey lines*) at (**a**) Jupiter and (**b**) Saturn. Ionization rates caused by auroral electrons increase with increasing electron initial energy, where energies of 0.1, 1, 10, and 100 keV with a flux of 0.15 μA m^{-2} have been shown. Excitation rates for B and C states caused by 10 keV electrons are shown by the *dot-dashed* and *dotted lines*, respectively. The *dashed grey lines* show the sum of production rates (H$_2^+$; H$^+$, and hydrocarbon ions) due to solar EUV

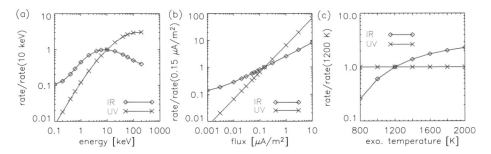

Fig. 14 Jupiter UV (*blue line with crosses*) and IR (*red line with diamonds*) dependences on (**a**) electron energy, (**b**) electron flux, and (**c**) temperature, after Tao et al. (2011)

This model accounts for UV absorption by hydrocarbons, ion chemistry, and H$_3^+$ non-LTE effects. Ionization and excitation profiles for auroral electrons at Jupiter and Saturn derived from the model are shown in Figs. 13(a) and 13(b), respectively. These can be compared with profiles given in Figs. 3 and 4. As remarked in section 'Electron Production Rate', higher energy electrons reach lower altitudes in the atmosphere before depositing their full energy.

The modelled dependences of altitude-integrated values of UV and IR emissions on electron energy ϵ_0, flux f_0, and exospheric temperature T_{ex} for Jupiter are shown by the diamonds and crosses in Fig. 14. The same is shown for Saturn in Fig. 15. The emission intensities are normalized to the conditions $\epsilon_0 = 10$ keV, $f_0 = 0.15$ μA m^{-2}, and $T_{ex} = 1200$ K for Jupiter and the same ϵ_0 and f_0 with $T_{ex} = 420$ K for Saturn. The normalized emission intensities of UV (lines in 117–174 nm range) and IR (Q(0,1-) line) are 38 kR, and 33 μW m^{-2} str^{-1} for Jupiter, and 37 kR and 0.80 μW m^{-2} str^{-1} for Saturn. Note that H$_2$O was included in the model by Tao et al. (2011) but not here for Saturn's high latitude regions.

The different dependence of emission rates on electron energy and temperature between the Jupiter and Saturn models can be summarized as two main points: (1) the temperature-dependence of IR emissions for Saturn covers three orders of magnitude (Fig. 15(c)), which is much larger than that for Jupiter (one order of magnitude, Fig. 14(c)), and (2) the electron

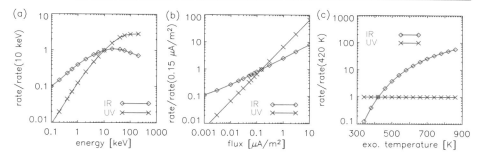

Fig. 15 Saturn UV (*blue line with crosses*) and IR (*red line with diamonds*) dependences on (**a**) electron energy, (**b**) electron flux, and (**c**) temperature, after Tao et al. (2011) but excluding the effects of H_2O

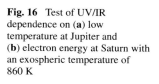

Fig. 16 Test of UV/IR dependence on (**a**) low temperature at Jupiter and (**b**) electron energy at Saturn with an exospheric temperature of 860 K

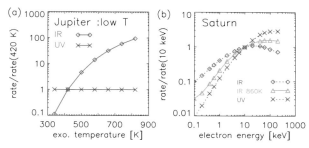

energy dependence of IR emission for Saturn (Fig. 15(a)) has a shallower slope in the energy range of 0.5–5 keV than that for Jupiter (Fig. 14(a)).

To explain the first of these differences, Fig. 16 shows the dependence of Jovian emission intensities across the same temperature range as initially considered for Saturn in Fig. 15(c), i.e. 300–820 K. The Jovian H_3^+ emission intensity now shows a large IR variation comparable to that of Saturn (Fig. 15(c)). To address the second difference, the electron energy dependence, Fig. 16(b) shows the dependence of Saturn's emission intensities on electron energy for a high temperature case. This profile now has a steeper slope in the range 0.5–5 keV. Therefore the differences between Saturn and Jupiter in both the rate of variation with temperature and the electron energy dependence of the IR emission are due to the lower prevailing temperature at Saturn.

2.3.4 Time Variation

Fig. 17 shows the processes from auroral electron precipitation to UV and IR emissions with their characteristic time scales, estimated by the above model. The UV aurora at Jupiter and Saturn is directly related to excitation by auroral electrons that impact molecular H_2, occurring over a time scale of 10^{-2} s. The IR auroral emission involves several time scales: while the auroral ionization process and IR transitions occur over $<10^{-2}$ s, the time scale for ion chemistry is much longer at 10^2–10^4 s. Associated atmospheric phenomena such as temperature variations and circulation are effective over time scales of $>10^4$ s. Tao et al. (2013) demonstrated the implications of these different timescales on the UV and IR emissions. They found that for events with a timescale of ~ 100 s, ion chemistry, which is present in the IR emission process but not the UV, could result in the production of different features between the two wavelength ranges. They also applied these results to observations

Fig. 17 Flowchart of UV and IR auroral emissions with the timescale for each process

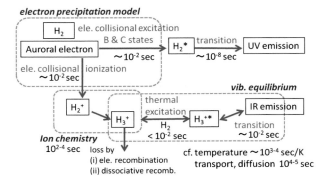

of the Jovian polar UV flashes identified by Bonfond et al. (2011) and the Io footprint aurora (Clarke et al. 2004) and showed that whether the IR intensity varies in correlation with the UV or not depends on the number flux of the auroral electrons and their characteristic energy. Section 3.4 summarises the comparison between observed UV and IR emissions.

2.4 Future Developments

Auroral particle energy deposition and transport models are critical for the assessment of the ionospheric state and the derivation of the auroral electron initial energy (and energy flux in the presence of ions) from UV analysis. A careful comparison between such models using clear and detailed information for input parameters, such as neutral profiles in altitude and incident electron intensity in energy and angle, should be carried out. Different methods adopted, such as Monte Carlo versus multi-stream, have been validated at Earth (e.g. Solomon 1993, 2001). There is therefore no reason, intrinsic to the methods adopted, to justify differences sometimes found between electron production rates induced by a given initial electron distribution in a given atmospheric model. Beside a careful comparison between suprathermal electron models, the contribution of energetic ions should be re-assessed, especially in the return current regions, and compared to the contribution from energetic electrons using realistic values for incident particle populations.

The estimate of the electron density by ionospheric models, critical for calculating ionospheric conductances, is limited by the large uncertainties in the k_1^* effective reaction rate. Detailed calculations of the amount of H_2 vibrational ($v \geq 4$), especially in the auroral regions, are required using the latest thermospheric density estimates and reaction rates. The k_1^* effective reaction rate is expected to change with (at least) season, latitude, and local time. Other limiting factors of the electron density assessment include: (1) the exact amount of water influx, which has significant effect at Saturn at low latitudes (Kliore et al. 2009; Moore et al. 2010), but whose contribution is not well known at Jupiter; and (2) the effect of dynamics, which could be large in the auroral regions (e.g. Smith and Aylward 2009; Tao et al. 2009; Galand et al. 2011; Müller-Wodarg et al. 2012). In addition, it seems highly relevant to try to characterize the properties and identify the origin of the sharp, narrow electron density peaks seen in Cassini/RSS profiles and most likely responsible for the diurnal distribution derived from SED analysis.

It is critical to improve the assessment of the ionospheric conductances, as they control the current density that flows through the ionosphere closing the global magnetospheric current system and strongly influence the Joule heating of the thermosphere in the auroral regions. To this aim we need to improve the estimate of input parameters (e.g., k_1^*, water influx, ionospheric electric field), which drive the ionospheric models and influence the assessment of the conductances. This requires detailed modeling efforts (e.g., of H_2 ($v \geq 4$)) combined with the analysis of a multi-instrumental dataset, including estimates from electron density (through radio occultations and SEDs (see section 'Observations of Electron Density'), H_3^+ density and temperature (from IR spectroscopic observations (see Sect. 3.3)), thermospheric densities and temperature (from UV occultations, Yelle and Miller 2004; Nagy et al. 2009), particle characteristics (from UV and X-ray spectroscopic observations (see Sects. 3.1.1 and 4)) and, in the near future through Juno at Jupiter, in situ particle and field measurements). As illustrated in Moore et al. (2010) and Müller-Wodarg et al. (2012), such information combined with self-consistent upper atmospheric models can be used to reduce the parameter space and improve our assessment of the ionospheric state, better constrain its drivers, and improve our understanding of the magnetosphere-ionosphere-thermosphere coupled system.

Such studies need to be carried out not only under steady-state auroral forcing but also for time variable auroral forcing, and considering the different timescales for different processes. While energy seems to be trapped at high latitudes for an imposed day-to-day variability (I. Müller-Wodarg, personal communication, 2012), higher frequency forcing as attested by auroral observations, from the X-rays and UV to the IR, may alter this picture and the system response at a global scale. Auroral emission models like that described above typically deal with a localised region including detailed collision processes. Therefore, the horizontal distributions (in longitude and latitude) of UV emission and IR spectra, which are affected by magnetospheric and thermospheric dynamics, are beyond the auroral emission model alone. In order to understand the relation between these auroral characteristics and their energy source using a modeling approach, coupling multiple models is essential, e.g., combining an auroral emission model with an atmospheric model to know the temperature variation and energy budget, and/or with a magnetosphere-ionosphere coupling model, a magnetosphere global model, or a magnetosphere chemical model to know where and when auroral electrons are energized and where H_2O precipitates. Finally, the use of low latitude observation- or model-based atmospheres results in discrepancies with auroral observations at high latitudes. Observation-model comparisons based on recent observations are also required to improve the atmosphere model at high altitudes as proposed by Gérard et al. (2009).

3 Ground- and Space-Based Observations of UV and IR Aurora

The largest outstanding gap in our understanding of the upper atmosphere of the gas giants is that they are all much hotter than solar input alone can produce. This 'missing' energy is very large: the temperatures observed in the thermosphere are several hundreds of Kelvin hotter than models can produce. The auroral process, whereby energetic particles impact the upper atmosphere at the intersection of magnetic field lines and the planetary atmosphere, is capable of injecting much energy, mainly in the form of Joule heating, into very localized regions on the planet. Therefore, the auroral process becomes a powerful source of energy for the upper atmosphere, heating it several hundreds of Kelvin above the temperature that solar heating alone can provide. It is not currently understood how the energy injected into the auroral regions could be redistributed—there is even evidence that injecting energy at the poles has the effect of cooling the equator (Smith et al. 2007a). In this section we discuss how remote sensing of the ultraviolet and infrared aurora can help us understand this energy transfer.

3.1 UV Observations

The UV aurorae can be divided into three regions by their characteristics: moon footprint aurora, main auroral emission, and high latitude ('polar') aurora.

Auroral footprints of the satellites Io, Europa, and Ganymede on Jupiter, and Enceladus on Saturn have been detected. The variation in the separation distance of the multiple Io footprint emissions (Bonfond et al. 2008), their intensity (Bonfond et al. 2012), tail length (Hill and Vasyliūnas 2002), and appearance and disappearance of the Enceladus footprint aurora (Pryor et al. 2011) have been attributed to variations in the plasma environments.

The main auroral emission is rather stable and encircles the magnetic poles. For Jupiter, the main emission surrounds a dark polar dawn region, while several features broaden in the dusk region (Grodent et al. 2003b). The main emission intensity decreases in the noon sector (Radioti et al. 2008a). Intense 'storms' of emission have been reported along the dawn main oval (Gustin et al. 2006).

An enhancement in the emission intensity of the Io plasma torus, observed in May 2007 (Yoneda et al. 2009), has been correlated with the following auroral behaviour: (i) a shift to lower latitude of the main auroral oval and a lesser shift of the Ganymede footprint, (ii) an increase in the main oval intensity, (iii) a decrease in the Io footprint auroral intensity (Bonfond et al. 2012), and (iv) a decrease in the (HOM) radio emission (Yoneda et al. 2013). An enhancement in the outward transport of heavy flux tubes, possibly caused by an increase in Io's volcanic activity, and their replacement by inward-moving hot, rare flux tubes (interchange) is suggested to decrease the plasma density around Io and the footprint aurora intensity (Bonfond et al. 2012; Hess et al. 2013). It is also proposed that movement of the main oval to lower latitude, also identified by comparing images from 2000 and 2005, could be caused by shrinking of the plasma corotation region in the middle magnetosphere, or enhancement of the azimuthal current which modifies the magnetic field mapping region, which could be caused by an increase in the mass outflow rate, or solar wind compression, as predicted by models (e.g. Hill 2001; Nichols and Cowley 2003, 2004; Grodent et al. 2008; Tao et al. 2010; Nichols 2011; Ray et al. 2012a). An external solar wind effect is also suggested for UV intensity variations (e.g. Clarke et al. 2009; Nichols et al. 2009b, see also Delamere et al. in this volume).

At Saturn, the observed shape of the main oval varies dynamically from a circle to a spiral shape (Clarke et al. 2005). The position of the centre of the auroral oval oscillates at a period close to that of planetary rotation (Nichols et al. 2010b), while the intensity of radio, infrared (H_3^+) and UV (H, H_2) emissions are modulated by the magnetospheric rotation periods separately in the northern and southern hemispheres (Sandel et al. 1982; Nichols et al. 2010a; Badman et al. 2012b; Carbary 2013). Saturn's main auroral emission also demonstrates a significant local time asymmetry, being generally more intense in the dawn-to-noon sector (Trauger et al. 1998; Lamy et al. 2009; Badman et al. 2012b; Carbary 2012). This asymmetry is related to the solar wind interaction with the outer magnetosphere, via a stronger flow shear between the solar wind flow and rotating magnetospheric plasma (anti-sunward v. sunward) on the dawnside than on the duskside (both anti-sunward).

At high latitudes, more local and shorter-time variations are often observed both for Jupiter and Saturn. At Jupiter the high latitude auroral emissions vary on timescales from several seconds (Waite et al. 2001; Bonfond et al. 2011) to a few days (Radioti et al. 2008b) in addition to persistent distributions characterized by dark or variably-bright regions (Grodent et al. 2003b) and sun-aligned arcs (Nichols et al. 2009a). At Saturn, the main emission sometimes intensifies and broadens toward high latitudes associated with solar wind compressions (Clarke et al. 2005) and small bifurcations have been reported close to the main aurora, and associated with the solar wind interaction (Radioti et al. 2011a; Badman et al. 2013).

3.1.1 UV Color Ratio Studies

Absorption of UV emission by hydrocarbons depends on wavelength with a large effect on short wavelengths <130 nm, as shown in Fig. 10. The UV color ratio, defined as the ratio of the intensity of a waveband unabsorbed by hydrocarbons (e.g. 155–162 nm) to that of an absorbed waveband (e.g. 123–130 nm), informs us how much hydrocarbon exists above the emission altitudes. Since the hydrocarbons exist at low altitudes, increases in the hydrocarbon column are an indicator of either enhanced penetration depth, and thus energy of the auroral primary particles, or of increases in the high-altitude hydrocarbon content caused by modification of the local atmosphere (e.g. Livengood and Moos 1990; Harris et al. 1996; Gérard et al. 2002, 2003; Gustin et al. 2004b).

The northern and southern aurora observed by the International Ultraviolet Explore (IUE) spacecraft showed that the attenuation by hydrocarbons varies in phase with intensity (Livengood and Moos 1990). Hubble Space Telescope (HST) observations also show a positive correlation between the energy flux deduced from the auroral brightness and the mean electron energy from the color ratio (Gustin et al. 2004b). Compared with their emission model, the electron energy producing the main emission lies between ∼30–200 keV, with a large enhancement around 08 LT possibly due to the occurrence of dawn storms. The energy flux varies between 2–30 mW m^{-2}. The observed relationship between the auroral electron energy fluxes and the electron energies in the main oval is compatible with that expected from Knight's theory of field-aligned currents (Knight 1973), taking source plasma parameters at the magnetospheric equator well within the observed range.

In addition, fitting of UV spectra can be used to determine the absorption by H_2, where the H_2 column density is also related to the auroral electron energy (e.g. Wolven and Feldman 1998; Gustin et al. 2009).

Colour ratio studies have also been performed for Saturn's H_2 aurora and reveal primary electron energies of 10–20 keV (Gustin et al. 2009; Lamy et al. 2013).

3.2 Visible Emission

Although intense reflection of solar radiation has so far prevented the detection of visible (VIS) emission on the dayside, nightside visible aurorae have been detected by Galileo at Jupiter (Vasavada et al. 1999; Ingersoll et al. 1998) and Cassini at Saturn (Kurth et al. 2009). At Jupiter, observations of the northern, nightside main emission and Io footprint made at visible wavelengths by Galileo showed they lined up well with their locations observed in the UV by HST on the dayside (Grodent et al. 2008). The visible aurora morphology changes with local time: from a multiple branch, latitudinally distributed pattern post-dusk to a single narrow arc before dawn. The power emitted at VIS wavelengths is 2–3 orders less than those at the UV/IR wavelengths. Detailed analysis of Saturn's visible auroral emissions has yet to be published.

3.3 Infrared Emission from H_3^+

We focus here on auroral IR emission, generated by transitions between rotational and/or vibrational states of H_2 and H_3^+ molecules. Some of the emission lines are at wavelengths that can be observed through the Earth's atmospheric window, i.e., around 2.1, 3.4 and 3.9

micron, by ground-based telescope facilities. Recently the Visual and Infrared Mapping Spectrometer (VIMS, Brown et al. 2004) instrument on Cassini has provided high spatial resolution observations of Saturn's IR aurora.

The observed IR emissions show both similarities and differences with the UV emissions in their spatial distribution. The IR emissions are also divided into the three characteristic regions: the moon footprint aurora, the main auroral emission, and high latitude aurorae. The Jovian IR main aurora and high latitude emission have both longitudinal and local time (LT) fixed features (Satoh and Connerney 1999). Baron et al. (1996) identified a positive correlation between Jupiter's spatially-unresolved IR auroral intensity and the solar wind dynamic pressure. Large-scale polar brightenings and multiple arc bifurcations have also been observed in Saturn's polar infrared emission, reflecting the solar wind interaction (Stallard et al. 2008a; Badman et al. 2011b, 2012a). Interestingly, the Io footprint was first observed in the infrared by Connerney et al. (1993) and remains the only moon footprint to have been observed at wavelengths outside the ultraviolet.

Observations of emissions from H_3^+ have probed the ionospheres of Jupiter, Saturn, and Uranus for over two decades. The parameters that can be derived from either imaging or spectral observations depend on the spectral resolution, wavelength coverage, and the signal-to-noise ratio (SNR). The parameters that can be derived are as follows:

1. The intensity of the observed H_3^+ emission reveals the morphology of the auroral deposition, which directly relates to where particle precipitation is sourced from in the magnetosphere (e.g. Connerney et al. 1998; Bunce et al. 2008; Badman et al. 2011a; O'Donoghue et al. 2013).

2. The H_3^+ temperature, which, when observed from environments in Local Thermal Equilibrium (LTE), is equal to the temperature of the surrounding neutrals (e.g. Stallard et al. 2002; Melin et al. 2007; O'Donoghue et al. 2014). The observed temperature stems from the energy that is injected into the upper atmosphere mainly via Joule heating (Achilleos et al. 1998; Bougher et al. 2005; Tao et al. 2009; Müller-Wodarg et al. 2006). Whilst the gas giants are generally thought to be in a state of *quasi*-LTE (Miller et al. 1990), this assumptions breaks down in the upper thermosphere of Jupiter (Melin et al. 2005), where higher ro-vibrational states are populated below the expected Boltzmann distributions, relative to lower states. Without accounting for this underpopulation, one may derive temperatures that are an under-estimation of the actual thermosphere temperature, weighted towards lower altitudes.

3. The ionospheric column integrated H_3^+ density, which is directly related to the conductivity, or the ability to drive currents through the upper atmosphere. As H_3^+ is formed via the ionization of molecular hydrogen, the number of ions present in the auroral ionosphere is a function of the particle precipitation energy and flux (Tao et al. 2011).

4. The total energy emitted by H_3^+ over all wavelengths (Lam et al. 1997; Stallard et al. 1999; Lamy et al. 2013). This is energy lost to the atmosphere via radiation to space, which has an overall cooling effect (Miller et al. 2010). This process becomes more effective as the temperature increases; at Jupiter, H_3^+ is said to behave as an effective thermostat, removing the atmosphere's ability to absorb any short-term injections of energy.

5. The line-of-sight ionospheric velocities derived from high resolution observations with $R = \lambda/\Delta\lambda > 25\,000$ (e.g. Stallard et al. 2001, 2007a, 2007b). The ionosphere is the conduit for vast magnetosphere-ionosphere coupling currents at Jupiter and Saturn, and is subjected to $\mathbf{j} \times \mathbf{B}$ forces to which the neutral atmosphere is oblivious. By measuring these ion winds, we are indirectly measuring the angular velocity of the regions in the

130

Fig. 18 A flowchart of the parameters that can be derived from H_3^+ spectral observations, which is a function of spectral resolution, wavelength coverage, and signal-to-nose

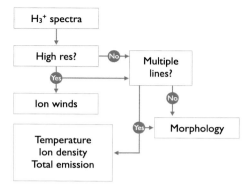

magnetosphere to and from which the ionospheric currents flow, assuming perfect coupling.

The flow-chart in Fig. 18 summarizes the spectral requirements needed to obtain these parameters. More recently, Tao et al. (2012) suggested that by comparing the intensity of different H_3^+ spectral lines, the flux and energy of the particle precipitation could be derived. Whilst this requires large SNRs, these developments have the potential to enable observations of H_3^+ to provide a near complete view of the energy terms of the upper atmosphere, without the need for simultaneous ultraviolet and infrared observations. As noted below, however, infrared observations are not sensitive to short term auroral variability, due to the relatively long life-time of the H_3^+ ion.

Jupiter is the giant planet closest to us and it has the strongest planetary magnetic field in our solar system. Consequently, it is the brightest source of H_3^+ in the night sky. Since the molecular ion was detected for the first time outside the laboratory by Drossart et al. (1989), there has been a plethora of both imaging and spectral studies, addressing a range of magnetospheric, ionospheric, and thermospheric questions. Here, we highlight a handful of studies that showcase the versatility of H_3^+ as a tool to study giant planets in general, and Jupiter in particular.

The ratio of emission line intensities reflects the temperature, which has been measured as high as ~ 1000 K at high latitudes (e.g., Stallard et al. 2001) and up to several hundred K at low-to-middle latitudes (Lam et al. 1997). Lam et al. (1997) analyzed medium resolution H_3^+ spectra obtained with CGS4 on the United Kingdom Infrared Telescope (UKIRT), covering a full rotation of the planet, thus deriving the H_3^+ temperature and density as a function of longitude. This provided a map of the noon ionosphere for all longitudes, showing that the temperature difference between the equator (which is cooler) and the auroral region is only a couple of hundreds of Kelvin. This highlights the complexity of the 'energy crisis'—in order to heat the entire atmosphere, this heating needs to be present at all latitudes.

The observed Doppler shift of the emission lines reveals the line-of-sight velocity of ion and neutral winds in the upper atmospheres of the giant planets. At Jupiter, the ion velocity reaches up to 3 km s^{-1} in the direction opposite to planetary rotation (Rego et al. 1999; Stallard et al. 2001), with spatial distributions associated with the emission intensity (Stallard et al. 2003). Raynaud et al. (2004) observed a H_3^+ 'hot-spot' on Jupiter's northern aurora using the Fourier Transform Spectrometer (FTS) mounted on the Canada France Hawaii Telescope (CFHT) in the 2 μm region. This region had a temperature 250 K higher than the rest of the auroral region, which had an average temperature of ~ 1150 K. Using the same data, Chaufray et al. (2011) were able to simultaneously derive, for the first time, the

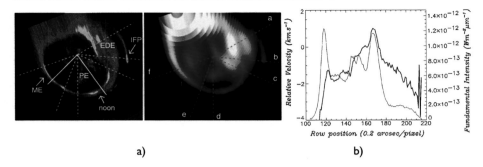

Fig. 19 (**a**) Simultaneous ultraviolet H and H_2 and infrared H_3^+ images of Jupiter's northern aurora from Radioti et al. (2013b) obtained in 2004. There are many similarities, but also significant differences, mainly in the polar emissions. (**b**) An example of the intensity (*light*) and velocity (*bold*) profiles of H_3^+ from 1998 observations of Jupiter's northern auroral oval from Stallard et al. (2001). This represents a cut through the oval, approximately connecting f and a in (**a**). The ionospheric plasma exhibits a strong lag from co-rotation

H_3^+ ion velocity and the neutral H_2 velocity, showing that the 'hot-spot' had an ion velocity of $3.1 \pm 0.4 \text{ km s}^{-1}$, whereas the neutral atmosphere rotated with an upper limit velocity of 1 km s^{-1}. The intensity (thin) and velocity (bold) profiles can be seen in Fig. 19(b), showing regions of corotation (steep gradients), sub-corotation (shallow gradients), and stagnation (flat). These latter regions are regions connecting directly to the solar-wind, whereas the former two connect to regions within the magnetosphere.

Using narrowband IRTF NSFCAM images of H_3^+ emission, including emission from the Io footprint aurora, Connerney et al. (1998) constrained the existing magnetic field models of Jupiter, adding important constraints to the morphology of the magnetic field at magnetic latitudes equatorward of the main auroral oval. This kind of imaging requires filters that are very narrow, and there are only three tuned to H_3^+ emission in existence today, all at the NASA IRTF.

Whilst limited by SNR and low spatial resolution, the techniques employed here highlight the versatility of the 2 μm region, observing through the telluric K window. In general, the telluric L and L' water absorption window between 3.4 and 4.1 μm offers brighter H_3^+ Q-branch transition (higher SNR), and less absorption.

Analysis of Cassini VIMS observations that were obtained during the Jupiter flyby at the end of 2000 by Stallard et al., (manuscript in preparation), revealed that the nightside (dusk to midnight) ionosphere was severely depleted in H_3^+, showing none of the mid- to low latitude emission observed by Rego et al. (2000) on the dayside at local noon. This absence indicates that there may not be a soft low-latitude component of particle precipitation away from the aurora.

At Saturn, IR spectral observations reveal anti-corotational convection (Stallard et al. 2007a) and a change in ionospheric velocities related to solar wind compression of the magnetosphere (Stallard et al. 2012a). Measurements of the temperature of the thermosphere vary significantly over time between $\sim$400–600 K (Melin et al. 2007, 2011; Stallard et al. 2012b; O'Donoghue et al. 2014).

3.4 Simultaneous Infrared and Ultraviolet Auroral Observations

By analyzing observations obtained in two or more wavelength bands that are both spatially overlapping and temporally simultaneous, it is possible to get a more complete view of the auroral processes. Because it has been very difficult to coordinate these multi-spectral campaigns of ground- and space-based observations, there are currently very few examples and many studies rely on statistical or average comparisons.

3.4.1 UV and IR Altitude Profiles

Grodent et al. (2001) developed a one dimensional model which couples a two-stream electron transport of electron energy deposition with a thermal conduction of Jupiter's atmosphere. The electron spectrum required is constrained by comparing the temperature predicted by the model with the observations. The characteristic energy varies from 100 eV for high altitude heating to 22 keV, while other non-particle heat sources are also required to balance the hydrocarbon cooling. This auroral electron spectrum produces UV emission with its peak at $\sim$200 km above the one bar pressure level, which is lower than the IR emission peak altitude $\sim$350 km (Clarke et al. 2004).

Limb imaging observations and comparison with spectroscopy of UV auroral emission provides a unique restriction for the high latitude atmosphere and auroral electron energy (Gérard et al. 2009). This study showed that the emission peak of Saturn's nightside is located 900–1300 km above the 1-bar level. In order to be coincident with the results given by the FUV and EUV spectra, a temperature enhancement is indicated compared to the low-latitudes. Comparing observations with the auroral profiles estimated by an auroral electron precipitation model that assumes a modified atmosphere profile, the characteristic energy of the precipitated electrons is found to be 5–30 keV.

The altitude profile of IR emission observed by Cassini shows the peak altitude lies at $\sim$1155 $\pm$ 25 km, almost the same as that of the UV, while the emission profile seems narrower in height than the UV profiles (Stallard et al. 2012c). This is explained by the large contribution to UV emission at higher altitude by Lyman α. Ambiguity in the H_2 profile assumed in the model study leads to a broader estimate of IR emission across altitudes.

3.4.2 Morphology and Time Variability

Clarke et al. (2004) presented the first study of simultaneous infrared and ultraviolet observations of Jupiter, comparing HST and ground based images. Having applied a 'best guess' de-convolution to the ground-based observations to remove the blurring effect of the Earth's atmosphere, they were able to compare images of auroral emissions in the two wavelength bands. They showed that on global scales, the morphology of the emission of H_2 and H_3^+ from the main oval is remarkably similar. There were, however, differences in the emission pole-ward of the main oval, regions connected to the solar wind (Vogt et al. 2011). Additionally the H_3^+ limb emission was brighter than predicted by a simple cosine function, likely to be the result of the bulk of the emission in the two wavelengths being produced at different altitudes.

Radioti et al. (2013a) analyzed the simultaneous Clarke et al. (2004) observations in greater detail and found that most of the main emission features (main oval, dark polar region, equator-ward emission, and the Io footprint), where all co-located in the infrared and the ultraviolet. However, polar emissions were not co-located in the two wavelengths. These emissions can be variable in the ultraviolet on time-scales of 2–3 minutes (Bonfond et al. 2011), which is much too short to be observed in the H_3^+ emission as discussed in Sect. 2.3.4. The excitation of H and H_2 by electron precipitation produces an almost instantaneous emission response, and the UV emission traces the instantaneous injection of energy into the

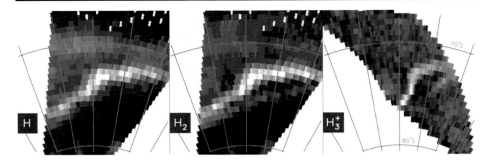

Fig. 20 The simultaneous H, H_2 (both in the ultraviolet), and H_3^+ (infrared) observations of Saturn's southern aurora of Melin et al. (2011), obtained with Cassini VIMS and UVIS. Whilst only a subset of these field-of-views are temporally simultaneous, there are significant differences between all three species

thermosphere. The IR picture is somewhat different. Because H_3^+ is formed via chemical reactions via the ionization of molecular hydrogen, and subsequently thermalizes to the surrounding neutral atmosphere, there is both a lag in response to precipitation and a life-time associated with the ion. The life-time is governed by the electron density, and is therefore a strong function of altitude. This means that H_3^+ does not respond to particle precipitation in the same manner as H and H_2 does in the ultraviolet. Given that H_3^+ life-times at Jupiter are 4 to 40 seconds (Radioti et al. 2013a) and up to 10 minutes at Saturn (Melin et al. 2011), the emitting ions can be subjected to significant horizontal transport within the thermosphere, and thus create a more diffuse view of the precipitation morphology (Tao et al. 2013; Radioti et al. 2013a). More relevant, however, is the fact that H_3^+ emissions will map the ion life-time average of the particle precipitation morphology. This stands in stark contrast to the ultraviolet, which represents an instantaneous view of the magnetospheric injection of energy.

Additionally, without a knowledge of the temperature variability of the thermosphere across the polar cap, it is difficult to disentangle the relationship between H_3^+ temperature and density, both of which drive intensity. Whilst hydrocarbons can feasibly be upwelled by energetic particle precipitation—destroying H_3^+—such an event could also act to increase the temperature, thus offsetting the decrease in intensity produced by the lost of ionospheric ions.

Using Cassini VIMS and UVIS data, Melin et al. (2011) analyzed simultaneous observations of Saturn's southern aurora, showing that over very small spatial scales, there can be significant differences between H, H_2, and H_3^+ emissions outside the main oval. This can be seen in Fig. 20. The broadness of H Lyman α is the result of multiple scattering within the thermosphere, whilst H_3^+ displays the same morphology as H_2 in the main oval.

Lamy et al. (2013) combined Cassini observations of Saturn's southern aurorae at radio, infrared, and ultraviolet wavelengths, while simultaneously monitoring the energetic neutral atom (ENA) intensification, which represents ion injections in the middle magnetosphere. These observations revealed three atmospheric auroral source regions: a main oval co-located with the bulk of the Saturn Kilometric Radiation (SKR) emission and lower intensity emissions poleward and equatorward of this. Sub-corotating features exist along the main oval, while overall the intensity is modulated in local time at the southern magnetospheric rotation period. The polar emissions from H_3^+ were more intense relative to the main oval than either of the UV emissions from H or H_2, but as the temperature was found to be

Fig. 21 Einstein Observatory
HRI (0.15–3 keV) image of
Jupiter, clearly displaying the
well separated emissions of the
aurorae. The *circle* outlines the
planet's disk, and the equatorial
plane is indicated by the two
linear segments (from Metzger
et al. 1983)

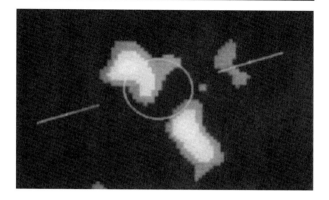

approximately constant across the whole auroral region, this is likely attributed to different electron energies or fluxes rather than a hot polar spot.

These studies, which combine observations from multiple instruments, have shown that orbiting space-based platforms provide excellent platforms from which to perform multi-spectral studies, revealing the conditions in both the magnetosphere (dynamics and electron energies and fluxes) and the atmosphere (temperatures and densities).

4 X-Ray Views of the Outer Planets

X-rays have become very relevant in the context of solar system observations more recently than other spectral bands, and mostly since the turn of the millennium, thanks to the high spatial resolution and the large collecting area of the Chandra and XMM-Newton observatories respectively. Planets, moons and comets have been detected and in some cases have been studied in detail. The significance of the Charge eXchange (CX) process has been realised (e.g. Dennerl 2010; Dennerl et al. 2012), with Solar Wind CX (SWCX) being responsible for the soft X-ray emission in many cases (Mars, Venus, Earth). Below is a review of what we have learnt so far about the X-ray emissions of Jupiter and Saturn, focusing on their aurorae, as well as a look at the many issues opened up by the discoveries, at the many questions still awaiting answers, ending up with some considerations about the other gas giants and future directions.

4.1 Jupiter

4.1.1 First Detection and Early Observations

The first X-ray detection of Jupiter takes us back to the Einstein Observatory: the planet was detected by the Imaging Proportional Counter (IPC) in 1979 and the High Resolution Imager (HRI) in 1981 (Metzger et al. 1983), and already then it was proposed that the emission may be related to energetic ion precipitation. Similarly good fits to the IPC spectra were obtained by a combination of line emission (oxygen, O, and in lower measure sulphur, S) and by electron bremsstrahlung. The luminosity of both aurorae combined (clearly distinguished in the HRI image, see Fig. 21) was estimated to be 4 GW (in the energy band 0.15–3 keV). The conclusion was that the electron input power would be too low and the spectral shape too soft for the X-ray emission to be explained by electron bremsstrahlung, and that it was

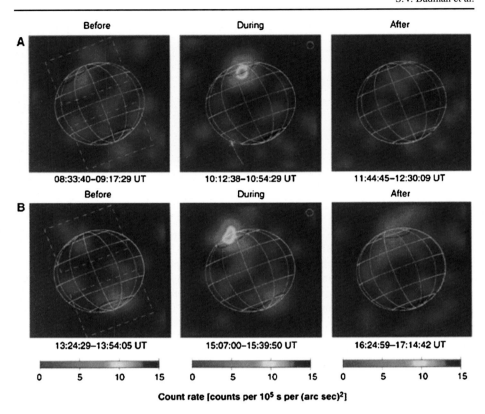

Fig. 22 ROSAT HRI images of Jupiter taken before, during and after the impacts of fragments K (*top*) and P2 (*bottom*) of comet Shoemaker-Levy 9 in July 1994. Brightenings of the Northern aurora are clearly seen during the impacts. The latitude-longitude grids show the planet's orientation at the *mid-point* of each exposure (from Waite et al. 1995)

more likely that heavy ion precipitation be the cause of the X-ray aurora (with the X-ray power being produced mostly in the O lines and 40 times less in S). Gehrels and Stone (1983) envisaged a scenario where O and S ions diffuse outward from the Io plasma torus to the middle magnetosphere, where they are accelerated, and then diffuse inward towards the planet. The ions are then scattered into the loss cone and ultimately precipitate into the upper atmosphere, where they slow and undergo radiative transitions—although the transitions were not yet attributed to CX. Waite et al. (1988) expanded this picture to include the results of IUE observations and suggested that, while ions are at the origin of the soft X-ray emission, 10–30 keV electrons are responsible for most of the UV emission, by the excitation of atmospheric hydrogen molecules.

ROSAT observations in the early 1990s confirmed this general picture (Waite et al. 1994). Of particular interest is the brightening of the X-ray aurora observed in the event of comet Shoemaker-Levy 9's plunge into Jupiter (see Fig. 22, Waite et al. 1995) possibly triggered by the impact itself or by comet fragments and dust transiting in the inner magnetosphere. Attention was also paid to the equatorial emission apparent in the ROSAT observations (Waite et al. 1997), which was then attributed to ion precipitation as well.

Fig. 23 Cartoon illustrating the CX process in the case of an O^{7+} ion from the solar wind encountering a water molecule of a cometary coma, acquiring an electron, being left in an excited state and emitting an X-ray line (from Dennerl 2009)

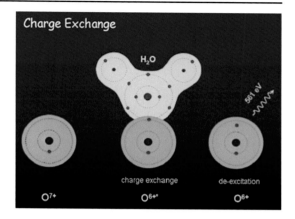

4.1.2 X-Ray Emission Processes

Cravens et al. (1995) were first to explain Jupiter's auroral soft X-ray emissions by Charge Transfer (or CX) of highly charged O ions, while Horanyi et al. (1988) had initially explained the auroral UV emissions by CX of lower charge states of O. A thorough review of the CX process is presented in Dennerl (2010). The process had been studied since the dawn of atomic physics, but was recognized as a very efficient mechanism of X-ray production only when invoked by Cravens (1997) to explain cometary X-ray emission. Basically, energetic highly charged ions (such as O^{7+} or O^{8+}) acquire an electron in the encounter with a neutral atom or a molecule, are left in an excited state and subsequently decay with the emission of characteristic soft X-ray lines (see the cartoon in Fig. 23, from Dennerl 2009). This process is now known to be ubiquitous in the universe, and is observed in our solar system, the local interstellar medium, supernova remnants, star forming regions, and starburst galaxies (see the review by Raymond 2012). Within the confines of the solar system SWCX is known to be responsible for the exospheric X-ray emissions of Venus and Mars (Dennerl 2008; Dennerl et al. 2006), and of our own Earth (e.g. Carter et al. 2010). In the case of the soft X-ray emission from Jupiter's aurorae, the origin of the ions has been matter of debate for some time, and currently a magnetospheric origin (i.e. from Io's volcanoes) is preferred, on spectroscopic grounds, over one from the solar wind (see Sect. 4.1.3).

In addition to CX, a variety of other processes are known to produce X-ray emission from planets and their moons. Recently electron bremsstrahlung has indeed been discovered in the spectra of Jupiter's aurorae, at energies (>2 keV) where CX is no longer the dominant emission mechanism (see Sect. 4.1.3). Line emission, following electron collisions, is observed, e.g. in the Earth's atmosphere. Elastic and fluorescent scattering of solar X-rays takes place in planetary atmospheres (and on moon surfaces), in such a way that the planetary disks are seen to mirror the solar X-ray variability, on short timescales and over the solar cycle (e.g. Bhardwaj et al. 2005a; Branduardi-Raymont et al. 2010).

4.1.3 Chandra and XMM-Newton Reveal Spatial, Spectral and Temporal Details

The first observations of Jupiter by Chandra in 2000 (Gladstone et al. 2002) returned very surprising results. While the ions producing X-rays by CX were originally thought to originate in the inner magnetosphere (i.e. Io), polar projections of the Chandra High Resolution Camera (HRC) X-ray photons indicate that the bright spot of auroral emission magnetically

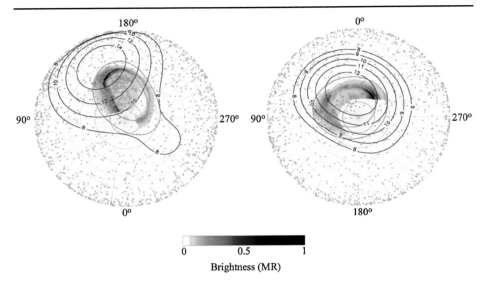

Fig. 24 Polar projections of the X-ray events (*light blue dots*) observed by the Chandra High Resolution Camera, superposed on the simultaneous UV images (*orange*) from HST STIS. The *green ovals* show the footprints of the magnetic lines that map out to 6 and 30 RJ. Clearly most of the Northern aurora X-rays are located well inside the *UV oval* (from Gladstone et al. 2002)

maps out to some $30 R_J$ (Fig. 24). This ignited the debate about where the ions are really coming from: inner or outer magnetosphere, Io (thus O and S ions) or solar wind (rich in carbon, C, ions)? Detailed modeling of the two options has been carried out by Cravens et al. (2003). For both scenarios, solar wind and magnetospheric origins, the ions need to undergo acceleration by electric potentials of at least 200 kV for the former, and 8 MV for the latter, in order to be stripped to high charge states, so as to produce sufficient X-ray flux to match the observations.

Even more surprisingly, Gladstone et al. (2002) reported that the flux in the bright spot was pulsating at a period of ~45 min (Fig. 25), with no correlations to e.g. Cassini upstream solar wind and energetic particle data at the time, although radio bursts, and associated electron bursts, of similar periodicity had been detected in 1992 during the Ulysses fly-by (Mac-Dowall et al. 1993). Such strict periodicity has never again been observed in Jupiter's X-ray aurorae, although chaotic variability, with power peaks in the 20–70 min range, was detected by Chandra in 2003 (Elsner et al. 2005). This change in the character of the variability, from organised to chaotic, may be explained by particle acceleration driven by pulsed reconnection at the dayside magnetopause between magnetospheric and magnetosheath field lines, as suggested by Bunce et al. (2004). The average potentials predicted in their case of solar wind 'fast flow', with high density, high field conditions, are of the order of 100 kV and 5 MV for electrons and ions, respectively.

Further simultaneous Chandra and HST STIS observations in 2003 revealed the interesting occurrence of a strong FUV flare in the north auroral region, temporally coincident with (and spatially adjacent to) a highly significant X-ray brightening (Elsner et al. 2005). This was taken to support the scenario where electrons and ions are simultaneously accelerated in the magnetosphere by strong field-aligned electric fields.

The large collecting area of the XMM-Newton telescopes allows the construction of spectral maps of Jupiter's X-ray emission in narrow energy bands and data obtained in 2003

Fig. 25 *Light curve* (*top*) and power spectrum (*bottom*) of the X-ray events from Jupiter's Northern auroral 'hot spot'. The ~45 min periodicity is clearly seen in the *light curve* and identified by the peak in the power spectrum (from Gladstone et al. 2002)

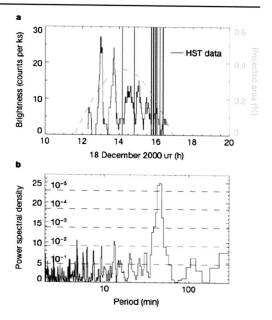

brought about more surprises (Branduardi-Raymont et al. 2007a). First, the spectral maps clearly display the planet's different X-ray morphology, dependent on the emission process involved. Those at the top of Fig. 26 show (left) CX emission concentrated in the aurorae (using the OVII band centred at 0.57 keV), and (right) scattered solar X-rays (bands centred on the Fe lines around 0.73 and 0.83 keV, typical of solar flares). Unexpected was the detection of auroral emission at higher energies (bottom panels) because CX lines are not present above ~2 keV, and at these energies the scattered disk emission has also died off. The spectra extracted for the two auroral regions and the low latitude disk are shown in Fig. 27. Below 2 keV the aurorae display the presence of the strong OVII line and evidence for other CX line emission at lower energies, while above ~2 keV the spectrum is a featureless continuum, consistent with electron bremsstrahlung. Interestingly, the bremsstrahlung component varied significantly in both flux and spectral shape between the two halves of the XMM-Newton observation, made in late November 2003. This coincided with a period of enhanced solar activity (the 'Halloween storm') when changes in solar wind dynamic pressure may have affected plasma acceleration in the Jovian magnetosphere. The ion CX line emission did not change at the time, possibly because of the much higher level of accelerating potentials required (Bunce et al. 2004).

The soft X-ray spectrum is well modeled by a combination of oxygen emission lines (OVII being the strongest) superposed on a bremsstrahlung continuum which is likely to represent the contribution of more CX line transitions below ~0.5 keV. One such line is resolved at ~0.32 keV, however, its attribution to C or S is uncertain because of the relatively large error on the fitted energy, although analysis of all the 2003 XMM-Newton datasets combined suggests a more likely interpretation as SXI (0.32 keV) or SXII (0.34 keV) rather than CVI (0.37 keV) (Branduardi-Raymont et al. 2007a). A preference for S over C line emission (and thus for a magnetospheric origin of the CX ions) is also indicated by Chandra Advanced CCD Imaging Spectrometer (ACIS) data (Hui et al. 2009, 2010b). However,

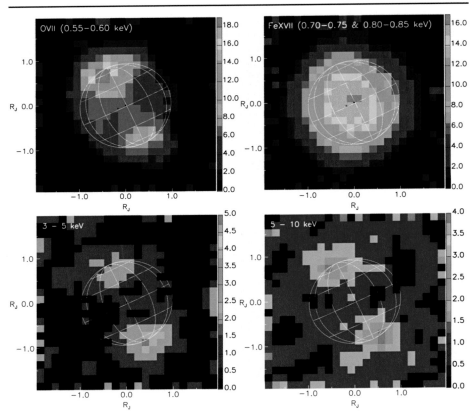

Fig. 26 Jupiter's differing morphology in narrow band X-ray spectral maps centred on the auroral CX OVII line (*top left*), on the Fe lines characteristic of solar flares (*top right*), and in higher energy bands where only the auroral emission is visible (XMM-Newton European Photon Imaging Camera data, from Branduardi-Raymont et al. 2007a)

a definitive conclusion on the origin of the ions has not been made. The Reflection Grating Spectrometer, which could easily resolve the S and C lines, does not have enough sensitivity at these energies to detect the lines above the noise. Through its high spectral resolving power, however, it is possible to measure the Doppler broadening of the CX OVII line, which gives an indication of the velocities and energies of the O ions. These are found to be of the order of 5000 km s^{-1}, or a few MeV, close to the levels predicted by Cravens et al. (2003) and Bunce et al. (2004). During the 2003 Chandra and XMM-Newton observations Jupiter's auroral power was measured to range between 0.4–0.7 GW (in the energy band 0.2–2 keV) and 40–90 MW (2–7 keV) (Branduardi-Raymont et al. 2007a).

The shape and flux level of the bremsstrahlung spectrum, dominating above 2 keV, in the XMM-Newton spectra from the 'quiet' part of the 2003 observation are in remarkable agreement with predictions by Singhal et al. (1992) for electron energies of few tens of keV (Branduardi-Raymont et al. 2007a). For completeness it is worth mentioning that the spectrum from the low latitude disk (see Fig. 27) is well represented by an optically thin coronal model with a temperature of ∼0.4 keV, which confirms the idea that the X-rays originate from scattered solar emission (see also Branduardi-Raymont et al. 2007b).

Fig. 27 XMM-Newton spectra of Jupiter's North and South aurorae, and of the low latitude disk: the OVII CX emission line at 0.57 keV is very prominent in the auroral spectra; the disk emission is harder and has the characteristics of an optically thin coronal spectrum; the electron bremsstrahlung component of the aurorae dominates above ∼2 keV (from Branduardi-Raymont et al. 2007a)

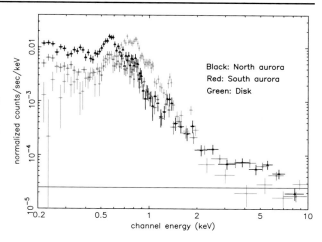

Fig. 28 Superposition of Chandra ACIS X-ray events (*large green dots*: >2 keV; *small green dots*: <2 keV) on the FUV emission (*orange*) observed with HST STIS. The footprints of the hard X-rays, expected to be of electron bremsstrahlung origin, coincide with the auroral oval and bright FUV features, indicating that the same electrons are most likely to produce both, X-ray and FUV emissions (from Branduardi-Raymont et al. 2008)

4.1.4 Auroral Morphology in Simultaneous Chandra and HST STIS Observations

The great value of truly simultaneous observations in different energy bands was demonstrated by a study of the different morphology of the Jovian X-ray and FUV auroral emissions (Branduardi-Raymont et al. 2008). Figure 28 shows the superposition of X-ray events detected by Chandra ACIS (each small green dot corresponds to a <2 keV photon, and each big dot to a photon of >2 keV energy) over the FUV emission (in orange) detected in simultaneous HST STIS observations. Note that the exposure time for the FUV image shown was 100 s while the X-ray photons shown were accumulated over approximately one Jovian rotation. Clearly the >2 keV X-rays (from electron bremsstrahlung) fall coincident with the bright auroral oval and regions of enhanced FUV emission, while those of <2 keV energy (ionic CX) fall inside the oval (as we knew already from the HRC, which maps them out to >30R_J away from the planet). Given that the FUV emission is expected to originate from excitation of atmospheric H_2 molecules and H atoms by 10–100 keV electrons, it is natural to make the connection that the same electron population is responsible for both the hard X-ray and FUV emissions. Also, the fluxes in the two bands are in line with this picture, being within a factor of 10 of the ratio of 10^{-5} predicted by Singhal et al. (1992).

4.1.5 The Galilean Satellites, the Io Plasma Torus and Jupiter's Radiation Belts

The detection by Chandra of Jupiter's Galilean satellites Io, Europa and possibly Ganymede (Elsner et al. 2002) has been interpreted, on flux grounds, as evidence for fluorescence scattering on their surfaces of energetic H, O and S ions, probably originating from the Io Plasma Torus (IPT). The X-ray emission of the IPT itself, also clearly detected by Chandra, is made up of a very soft continuum, a large fraction of which could be due to non-thermal electron bremsstrahlung, and a single spectral feature, a line at $\sim$0.57 keV; the origin of this is unclear because neither fluorescence of solar X-rays nor CX can produce the observed flux.

Finally, diffuse hard (1–5 keV) X-ray emission from around Jupiter, reported recently on the basis of a deep Suzaku observation (Ezoe et al. 2010), has been attributed to non-thermal electrons in the radiation belts and the IPT. However, synchrotron and bremsstrahlung processes cannot explain it on energetic and spectral grounds, and the energetic electron density required to produce the X-rays by inverse Compton scattering of solar photons is an order of magnitude larger than that estimated from an empirical model of the charge particle distribution around Jupiter (assuming the emission is truly diffuse and not the integrated emission of background sources).

4.1.6 Open Questions

While in the last decade of Chandra and XMM-Newton observations we have learnt a great amount about the X-ray properties of the Jovian system, many questions have also been raised, as it always happens when we open a new exploration window on the Universe.

Despite the spectral evidence in favour of a magnetospheric origin of the ions undergoing CX and producing the soft X-rays in Jupiter's aurorae, it is worth considering whether there is still a role for SWCX, and if so, what fraction of the emission may be due to it. On the one hand, while mentioning that SWCX may contribute, Cravens et al. (2003) point out that, in this case, bright UV proton auroral emission would also be expected, but is not seen, thus excluding a pure SWCX scenario. On the other hand, since the ion fluxes in the outer magnetosphere are insufficient to explain the observed auroral X-ray emission, another ion source (possibly the solar wind) may be contributing. If the $\sim$45 min periodicity observed by Chandra is an analogue of the quasi-periodic radio bursts reported by MacDowall et al. (1993), the phenomenon may be under solar wind control, as the bursts were reported to be. Bunce et al. (2004) suggest that pulsed reconnection phenomena should be more intense under high density solar wind conditions, when the magnetosphere is compressed, so this could be used as evidence for the ions origin. Alternatively, Cravens et al. (2003) note that if the pulsations have a 'particle bounce' origin this would imply a magnetospheric scenario, unrelated to the solar wind. The rare occurrence of the pulsations may hold a clue and if a new detection were to be made, correlation with the solar wind conditions at the time would add decisive information. Could the ions be precipitating directly from the solar wind? The X-ray hot spot location (Gladstone et al. 2002) lying in the vicinity of the Jovian cusps would support this possibility, although acceleration is still required to explain the X-ray fluxes observed. How do the timescales of ion and electron precipitation compare? Only further simultaneous studies of the UV and X-ray emissions can help to take this further.

An opportunity to advance this quest is offered by the JAXA Sprint-A mission, launched in September 2013 with the Hisaki EUV spectrograph on-board, and dedicated to the study

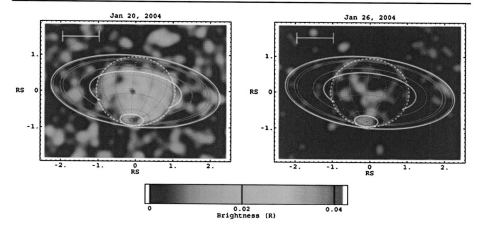

Fig. 29 Chandra images of Saturn taken one week apart in Jan 2004, showing how the disk X-ray emission was brighter in the first observation, at the time when X-rays from a strong solar flare reached the planet. The enhancement at the south pole shows the same spectral character of the rest of the planet's disk and is not evidence of an X-ray aurora (from Bhardwaj et al. 2005c)

of the tenuous plasma surrounding planets in our solar system. The primary target is Jupiter, and the emission from the IPT in particular. The aim is to explore the possible links between the IPT emission distribution, the strength and character of Jupiter's auroral emissions and the conditions of the solar wind. Concurrently with the Hisaki observations, a large multi-wavelength campaign, including X-ray observations, has been organised to gather important diagnostic data on the complex array of physical processes that operate in Jupiter's environment. Some answers may well come from this endeavour, for example, examination of the energetics of the particles in the IPT should help establish what mechanism leads to its OVII line emission.

Other questions, still wide open, concern the comparison between north and south X-ray aurorae in flux, temporal variability and spectrum, and also Io and its footprint: is there an X-ray equivalent, which has not yet been detected? Finally, given the known presence of a magnetosphere and auroral emissions associated with Ganymede, one can also speculate whether there may be a magnetospheric component in the X-ray emissions of the Galilean moons. Only more sensitive and higher duty cycle observations will be able to shed light on this. However, the difficulty of realizing them with an Earth orbiting X-ray observatory leads to the conclusion that a much more effective option is to have X-ray observations in situ at the planets, incorporating X-ray instrumentation in future planetary missions.

4.2 Saturn

4.2.1 Disk X-Ray Emission Under Solar Control: No X-Ray Aurorae?

By analogy with Jupiter, X-ray aurorae powered by CX could also be expected on Saturn, yet none have been observed so far. The disk and polar cap have similar coronal-type spectra ($kT \sim 0.5$ keV averaged over the years) and the disk flux variability strictly correlates with that of solar X-rays, demonstrating that the planet's X-ray emission is controlled by the Sun (Bhardwaj et al. 2005c; Branduardi-Raymont et al. 2010). This is clearly illustrated by Fig. 29, which compares the view of Saturn in two Chandra observations separated by about a week; the planet was brighter by a factor of 3 during the first observation, coincident (after correction for light travel times) with a strong flare going off on the Sun (Bhardwaj et al. 2005c).

UV and radio brightenings of Saturn's aurora have been found to correlate with the arrival of solar wind shocks at the planet (Clarke et al. 2009), suggesting that solar wind ions may also have a role in producing X-ray aurorae by CX. In this case, as shown by Cravens (2000), the emitted power is proportional to both the density and speed of the solar wind, thus the passage of a solar wind shock at the planet may produce an X-ray auroral brightening. Theoretical estimates of Saturn's auroral fluxes (Hui et al. 2010a) for un-accelerated solar wind ions are within a factor of a few of the sensitivity of current instrumentation, although acceleration of the ions in the planet's magnetic field would raise these estimates. Chandra observations were obtained in 2011 (Branduardi-Raymont et al. 2013), triggered by the expected arrival of solar wind shocks that had been propagated from measurements at 1 AU using the 1-D MHD code mSWiM (Zieger and Hansen 2008). Variability in Saturn's X-ray emission was observed, but once again it was due to a flare in the solar X-rays scattered by the planet's atmosphere. Stringent upper limits of 2 MW (photon energies of 0.3–2 keV) and 17 MW (2–8 keV) were derived on Saturn's auroral emissions, excluding the presence of accelerating potentials down to $\sim$10 kV. Upper limits of 4 MW were also set on X-rays from each of Titan and Enceladus.

A by-product of these triggered Chandra observations was also a validation of the solar wind propagation technique. At the time, Cassini was crossing Saturn's magnetopause and bow shock as identified in the Cassini magnetometer and electron data. The standoff distances of the boundaries inferred from the in situ measurements were compared with those derived from the propagations. Measurements and propagations were matched by shifting the propagations by +1.9 days, which is consistent in magnitude and direction with the shifts established by Clarke et al. (2009). During the period covered by the 2011 Chandra observations Cassini radio data (RPWS) also showed a strong enhancement, indicating a compression of the magnetosphere (Branduardi-Raymont et al. 2013).

4.2.2 X-Rays from Saturn's Rings

Chandra has also revealed X-ray emission from Saturn's rings. The spectrum is dominated by a single line centred at 0.53 keV, indicative of atomic O Kα fluorescence, most likely the result of excitation of the oxygen trapped in the icy water particles making up the rings. Bhardwaj et al. (2005b) suggested that this may be due to solar X-ray illumination, however, the apparent lack of correlation with solar activity over the years may point to an alternative explanation, such as electron injections linked to the planet's thunderstorms (Branduardi-Raymont et al. 2010). More observations at high angular resolution (only Chandra can spatially separate ring from disk emission) are needed to search systematically for correlations with solar activity and/or planetary seasons.

4.2.3 Open Questions

The X-ray exploration of Saturn, and thus our understanding of its workings, is clearly less advanced than that of Jupiter. Is there really no X-ray aurora on Saturn? The conclusion from the searches made so far is that much more dramatic solar wind enhancements than those used to trigger the Chandra observations in 2011 may be needed if we are to make a detection, and/or much more sensitive instrumentation. Could an alternative ion source, internal to the Kronian system, e.g. Enceladus, contribute an element of CX? Are Titan, as it moves in and out of the solar wind, and Enceladus, with its active cryo-volcanoes, X-ray sources, and by which mechanism? Their environment and physical conditions would be favourable to ionic CX, particle precipitation and fluorescence. Is there a link between the emissions from Saturn and its rings? These are all fascinating issues, which unfortunately are most likely to remain unsolved until we take a major step up in our experimental capability.

4.3 Uranus and Neptune

Detection of X-ray aurorae at Uranus and Neptune is hampered by their vast distances and by the conditions of their environments. By making a comparison with Jupiter and assuming a simple scaling law for the planetary parameters most relevant to X-ray auroral production, such as dipole magnetic moment and magnetospheric particle density, it is clear that any emissions would be well below detectability with current instrumentation, unless some other physical factor were to provide an unexpected contribution. For example, the very large tilt angle (59°) between the magnetic dipole and rotation axes of Uranus, and the 30 % offset of the dipole from the centre of the planet, with the consequent order of magnitude difference in surface magnetic field between day- and night-side, might have the effect of enhancing the auroral power above that expected from the simple extrapolation (Branduardi-Raymont et al. 2010).

4.4 Conclusions

With XMM-Newton and Chandra, planetary science has acquired a new observing regime which has revealed many unexpected sides of our solar system, and has led to many new questions. Planetary X-ray astronomy has come of age, and now unexpected discoveries must be turned into fully understood physics. Real progress can only be made by recognizing the high potential of X-ray observing, and by offering X-ray instrumentation the same platform as more traditional wavebands have enjoyed for decades, that which allows in situ measurements. This will bring about higher sensitivity and spatial resolution, together with the improved spectral resolving power of the most modern imaging devices. Recent developments in lightweight optics show that a low-requirement (mass, power, data rates) X-ray telescope for planetary exploration is a feasible proposition. It would also work in great synergy with in situ UV and particle instrumentation, contributing to establish the dynamics and energetics of the particles populating planetary environments, and would validate and test the consistency of models developed from more 'traditional' measurement techniques. On the other hand, remote global X-ray observations at much higher sensitivity and spectral resolving power than afforded by XMM-Newton and Chandra are now forthcoming following ESA's recent selection of the science theme 'The Hot and Energetic Universe' for its next large mission, which would be addressed by the proposed Athena mission. The non-dispersive character of the planned cryogenic spectrometer will enable Jupiter's auroral and scattered solar emissions, as well as the Io Plasma Torus, to be individually mapped spatially and spectrally at high resolution. The search for auroral X-ray emissions on Saturn, as well as attempts to detect Uranus and Neptune, will be pushed to much fainter flux limits than currently possible. X-ray spectra of the Galilean satellites, and (speculatively) Saturn's moons, will enable the search for magnetospheric emission components, as well as allowing surface composition analysis by fluorescence.

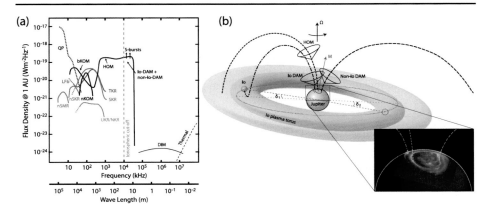

Fig. 30 (**a**) Average spectra of known planetary radio emissions, adapted from Cecconi (2010). (**b**) Expected locus of Jovian radio sources, adapted from Cecconi et al. (2012)

5 Jupiter and Saturn Magnetospheric Dynamics: A Diagnosis from Radio Emissions

In this section, we will characterise the spectral and spatial properties of Jovian and Kronian auroral radio emissions and detail the rich diagnosis they bring on internally-driven (magnetodisc, planet-satellite interactions) compared to externally-driven (solar wind) magnetospheric dynamics.

5.1 Spectral and Spatial Properties of Auroral Radio Emissions

5.1.1 Historical Context

All the explored magnetized planets are powerful radio sources at frequencies ranging from a few kHz to a few tens of MHz (Fig. 30). Among these, only the decametric emission (DAM) of Jupiter, the frequency of which exceeds the terrestrial ionospheric cutoff (~10 MHz), can be observed from the ground. Jovian DAM emissions were first detected in 1955 (Burke and Franklin 1955), while the terrestrial kilometric radiation (TKR) was later discovered by observations from space in the 1960s, and investigated in detail by numerous in situ auroral orbiters in the following decades (such as Freja, Viking, FAST). The exploration of the solar system by the Voyager (1980s) and Ulysses (1990s) spacecraft, completed by that of the Jovian magnetosphere with Galileo (2000s), later revealed hectometric (HOM) and kilometric (KOM) components of emissions at Jupiter, and kilometric emissions at Saturn (SKR), Uranus (UKR) and Neptune (NKR). The reader is referred to post-Galileo comparative reviews for more information (Zarka 1998, 2004, and references therein).

Hereafter, we focus on radio emissions radiated by the auroral regions and planet-moon flux tubes of Jupiter and Saturn's magnetospheres, which are the brightest radio emitters of the solar system. These emissions reduce to free-space electromagnetic waves propagating on extraordinary (X) and ordinary (O) modes. We therefore exclude other types of emissions such as low frequency continuum, trapped Z-mode or whistler-mode radiation (narrowband emissions, auroral hiss), electrostatic waves or atmospheric emissions (lightning).

5.1.2 Properties of Radiated Waves

Jupiter and Saturn display characteristic remote properties, more generally common to all auroral planetary radio emissions, which can be summarised as:

- very intense non-thermal radiation ($\sim 10^{11}$ W radiated by Jupiter, $\sim 10^9$ W by Saturn), predominantly in the X mode;
- instantaneous emission at $f \sim f_{ce}$ (f_{ce} is the electron gyrofrequency);
- sources along high latitude magnetic field lines, hosting energetic (keV) electrons, and co-located with atmospheric aurorae, where $f_{pe} \ll f_{ce}$ (f_{pe} is the electron plasma frequency);
- 100 % circular or elliptical polarization (the sign of which indicates the magnetic hemisphere of origin);
- very anisotropic beaming (thin conical sheet) leading to strong visibility effects;
- sensitivity to magnetospheric dynamics at relevant timescales (e.g. planetary rotation, orbit of moons, solar wind activity).

In situ measurements within the terrestrial auroral regions led to the identification of the Cyclotron Maser Instability (CMI) as the driver of the TKR emission (Wu and Lee 1979; Wu 1985; Treumann 2000, 2006, and references therein). This mechanism operates in regions where $f_{pe} \ll f_{ce}$ as a resonant wave-particle interaction between non-maxwellian electrons gyrating around magnetic field lines and a background of radio waves. These are amplified close to f_{ce} at the expense of the electron (perpendicular) energy. This free energy may come from loss-cone, ring or shell electron distributions, which all yield positive growth rates (Roux et al. 1993; Louarn and Le Quéau 1996; Delory et al. 1998; Ergun et al. 2000), with slightly different wave properties in terms of emission frequency (above/below f_{ce}) or emission angle relative to the magnetic field (i.e. the wave beaming). Efficient amplification additionally requires the size of the source region to exceed several times the wavelength. The local conversion efficiency from the total (perpendicular) electron free energy to X mode wave energy can reach ~ 1 % (Benson and Calvert 1979).

As Jupiter and Saturn auroral radio waves display remote properties consistent with CMI, this mechanism was postulated to be a universal generation process common to all magnetized planets (Zarka 1992). This hypothesis could recently be validated for Saturn with Cassini in situ measurements within the SKR source region (Lamy et al. 2010; Mutel et al. 2010; Schippers et al. 2011; Menietti et al. 2011), yielding a 1 % (2 % peak) electron-to-wave energy conversion efficiency (Lamy et al. 2011). The JUNO mission will specifically investigate this and other properties of the Jovian auroral regions in the coming decade.

5.1.3 Diagnosis

The understanding of the generation mechanism and the subsequent remote properties of radiated waves provide a powerful diagnosis of the nature and the dynamics of the underlying coupling between the solar wind, the magnetosphere, the moons, and the ionosphere at the origin of these emissions.

Spectral and spatial properties are intrinsically related because the emission frequency f is limited by the electron gyrofrequency f_{ce}, itself linearly proportional to the magnetic field. The emission frequency f therefore directly indicates the altitude of the source above the ionosphere. This allows one to instantaneously locate the radio sources and to track possible motions throughout the auroral regions. The detected emission also indicates a source region fulfilling CMI requirements with $f_{pe} \ll f_{ce}$ and energetic electrons whose distribution is unstable (shell, ring, loss cone). The main advantage of radio observations relies on the capability for long-term, quasi-continuous, remote measurements at high spectral and temporal resolution. The Poynting flux, organized in time-frequency (dynamic) spectra, provides essential information on the auroral activity. Beyond pioneering analysis of the most obvious variations (rotational or moon-induced modulation, solar wind forcing), a refined interpretation of dynamic spectra, focused on shorter sub-structures, has been the subject of recent modeling studies.

In parallel, higher level observables, such as the wave polarization (Stokes parameters), and/or the position of radio sources, can be retrieved with sophisticated instrumentation and data processing techniques, either space-based (goniopolarimetry, Cecconi 2010) or ground-based (LOFAR phased array). The wave polarization reveals several important parameters: the sense of circular polarization depends on the hemisphere of origin, while the quantitative degree of polarization depends on the magneto-ionic propagation mode and the degree of wave-plasma coupling along the ray path. The position of radio sources enables one to perform radio imaging and to map active field lines in real time.

The use of such observables to investigate the Jovian and Kronian magnetospheric dynamics are illustrated through several examples below.

5.2 Jupiter

Jupiter's auroral radio emissions are divided between Io (the most intense) and non-Io emissions, regularly observed from the ground above 10 MHz since the 1950s, and at low frequencies from space with Voyager/Ulysses/Galileo, or more recently with Cassini and STEREO. All these emissions are strongly modulated at the planet rotation period (9 h 55 min), as a result of the magnetic dipole tilt.

5.2.1 Io-Jupiter: The Case for Moon-Planet Interactions

The Io-Jupiter interaction is due to the motion of Io relative to the Jovian magnetic field, which generates an electric current closing in the Jovian ionosphere (Neubauer 1980). This electrodynamic coupling was the first discovered and is the most powerful case of satellite-magnetosphere interactions. It thus stands as the archetype of such interactions (4 cases confirmed so far: Io, Europa, Ganymede and Enceladus), and, by a small extension, rapidly moving interacting regions (Hess et al. 2011b). The most prominent feature of the Io-related radio emissions is their well-defined arc shape observed in the time-frequency plane at timescales of hours. Sub-structures include the well known Jovian S-bursts at timescales of milliseconds.

Radio Arcs   Arc-shaped emissions are primarily due to the small spatial size of the interaction region (i.e. Io). A more extended interaction region would generate a continuous suite of arcs and form a continuum of radio emission. The arc shape is then a direct consequence of the anisotropy of the emission pattern of individual radio sources, which is a thin ($\sim 1°$ wide) conical sheet with a wide opening angle relative to the magnetic field vector. The source is detected only when the observer crosses the cone sheet, so it can be seen at most twice by a fixed observer as the source rotates with the planet, even though it emits continuously. The time delay between these two observations depends on the cone opening angle and on the observer's motion relative to the source.

Although the arc shape was explained a long time ago, its use as a diagnostic of the interaction parameters is quite recent. Hess et al. (2008) computed the theoretical opening of the emission cone and showed that, apart from the altitude, it mostly depended on (1) the electron distribution function, (2) the emitting electron energy and (3) the plasma parameters determining the refraction index (Ray and Hess 2008; Mottez et al. 2010). The morphology of radio arcs permits the diagnosis of the current system powering the emission, as different current systems lead to different beamings. Shell driven emission with nearly constant emission angles are obtained in auroral cavities for steady-state systems, whereas loss-cone driven CMI with an emission angle rapidly decreasing close to the planet are obtained for transient currents (Mottez et al. 2010).

The Io-related current system is Alfvénic (Crary 1997; Hess et al. 2008, 2010), in accordance with our knowledge of the Io-Jupiter interaction (Neubauer 1980). A numerical model called ExPRES (Hess et al. 2008) has been developed for extended simulation studies. It takes into account both the physical parameters of the interaction, to compute the beaming angle, and the geometry of an observation, to ultimately compute simulated dynamic spectra of the emissions. The fit of the simulated dynamic spectra to observational data allowed Hess et al. (2010) to measure the variation of the electron energy with Io's System III longitude.

Fine Structures Fine structures, called millisecond or short (S-)bursts, are very common. They have a short duration ($\sim$10 ms at a given frequency) and drift in frequency with time. The emission frequency is close to f_{ce}, and thus relates to the source altitude, therefore this drift is the result of the source motion along magnetic field lines.

Zarka et al. (1996) and Hess et al. (2007b) measured the drift rate versus frequency and showed that the source motion was generally consistent with the adiabatic motion of electrons moving away from Jupiter, which allows one to measure the emitting electron energy from the measurements of the drift rate. These authors showed that the electrons have an energy between 2 and 5 keV which appears to vary as a function of Io's system III longitude.

Hess et al. (2007b) also detected localized jumps of the electron kinetic energy, interpreted as a localized electron acceleration due to localized electric potential drops. Hess et al. (2009b) showed that these potential drops are actually moving away at the local ion acoustic velocity, and thus are probably solitary ion acoustic waves.

Finally, Hess et al. (2007a) simulated dynamic spectra of the radio emissions induced by electrons accelerated by periodic Alfvén waves (see Fig. 31(b)). These are similar to the observed dynamic spectra of the S-bursts, validating Alfvén waves as the primary source of electron acceleration in the Io-Jupiter interaction.

5.2.2 Non-Io Emissions and Rotational Dynamics

The origin of most of the non-Io emissions often remains a mystery as only the narrowband kilometric (nKOM) emission sources have been clearly identified as being plasma wave generated on the borders of the Io plasma torus (Reiner et al. 1993b).

Direction-finding studies using Ulysses observations (Reiner et al. 1993a; Ladreiter et al. 1994) concluded that part of the hectometric (HOM) emissions occurs along field lines mapping to regions between 4 and $10R_J$, i.e. in the extended Io torus. In the decameter range, Panchenko et al. (2013) observed radio arcs resembling Io's in sub-corotation with the same period as that observed by Steffl et al. (2006) in the torus, and interpreted as the beating of the System III (internal) and IV (torus perturbations) periods. The beating originates from a peak of the hot electron population density near 290° of longitude (Steffl et al. 2008) caused by an Io-like interaction powered by empty flux tubes moving inward in the torus (Hess et al. 2011b). The Io-like decameter arcs observed by Panchenko et al. (2013) could be related to an Io-like interaction powered by the interchange instability in the torus and the HOM emissions located by Ulysses may be their lower frequency counterpart.

The dynamic spectra of the Jovian emissions often exhibit slowly drifting bands in which the background emissions are alternately enhanced and dimmed. These bands (so-called

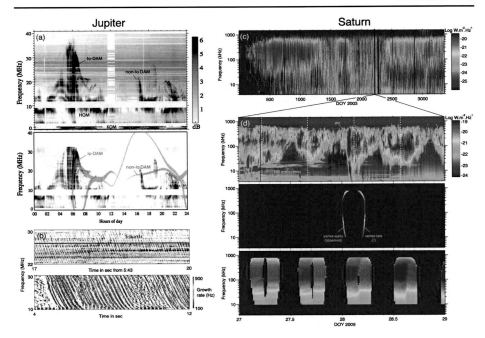

Fig. 31 Observed and modelled radio emissions of (*left*) Jupiter and (*right*) Saturn. (**a**) Voyager 2/PRA observations of the Jovian auroral radio emissions on 16 July 1979 (*top*) and simulations of Io and non-Io DAM arcs (*bottom*), adapted from (Cecconi et al. 2012). (**b**) Nançay/NDA observations of Jovian S-bursts over a few seconds of April 1995 (*top*) and associated simulations (*bottom*), adapted from Hess et al. (2009a). (**c**) Cassini/RPWS observations of SKR from 2003 to 2012, adapted from Kimura et al. (2013). (**d**) Cassini/RPWS observations of SKR from 27 to 28 January (*top panel*) and simulations of the rotational modulation (*middle panel*) and of a sub-corotating arc (*bottom panel*), adapted from Lamy et al. (2013)

modulation lanes) are due to flux tubes with densities differing from their environment and acting as a diffracting grating (Imai et al. 1997). Io-related (localized) background emissions permit one to locate the position of these flux tubes in a region between the orbits of Amalthea and Europa (Arkhypov and Rucker 2007). Refraction undergone in the equatorial torus is responsible for attenuated lanes in the HOM range (Gurnett et al. 1998; Menietti et al. 2003).

Intense non-Io arcs have also been observed in the Jovian magnetosphere by STEREO (Lamy et al. 2012) and simultaneously by the Nançay decametric array, Cassini, and Galileo (Hess et al. 2014). Their corotation rate, close to ∼100 %, indicates that their source is in the inner or middle magnetosphere, although their intensity seems to be modulated by the solar wind conditions.

Less structured emissions exist for which the decametric emissions mostly originate from the dusk side (from the Voyager 1 flyby Barrow 1981; Hess et al. 2012), and extend deeply into the hectometric and kilometric ranges (e.g. Cassini observations in Hess et al. 2014). The low sub-corotation rates of these radio sources (down to 50 %) indicates that the interaction powering these emissions occurs in the outer magnetosphere. UV and infrared observations show that the poleward (probably solar wind related) aurorae are also mostly emitted on the dusk side, thus the latter radio emissions may also be caused by the magnetosphere-solar wind interaction (Grodent et al. 2003b). Quasi-periodic kilometric bursts are also observed pulsating with a quasi-period of about 40 minutes. Their origin seems to be among the most poleward of all radio emissions (Kimura et al. 2011).

5.2.3 Solar Wind Control

Jupiter's outer magnetosphere is sensitive to variations of the solar wind dynamic pressure (e.g. Smith et al. 1978), but the relation between Jupiter auroral emissions and the solar wind pressure is complex and only some of the radio components are sensitive to it. Several observations have shown that auroral emissions are enhanced during times of higher solar wind pressure (Barrow 1978; Zarka and Genova 1983; Genova et al. 1987) or are triggered by interplanetary shocks (Barrow 1979; Terasawa et al. 1978; Prangé et al. 1996, 2004; Gurnett et al. 2002; Clarke et al. 2009).

Hess et al. (2013) compared Nançay observations of non-Io emissions to solar wind parameters (magnetic field, velocity and density) propagated from Earth to Jupiter. They found that fast reverse shocks generated dawn and dusk radio emissions, whereas forward shocks generate emissions from dusk only which may later move toward the dawn side. A multi-spacecraft study of the Jovian radio emissions during Cassini's approach to the planet (Hess et al. 2014) confirmed those results and additionally showed that the corotation rate of these radio source (usually about 50 %) increased to more than 80 % for strong magnetospheric compressions.

5.3 Saturn

Saturn's kilometric radiation, discovered during the Voyager flyby of Saturn in 1980 (Kaiser et al. 1980, 1984b), has been observed by only two spacecraft since then, Ulysses in the 1990s and Cassini, in orbit since mid-2004. Further details on SKR average properties can be found in Kaiser et al. (1984b), Lamy et al. (2008b) and references therein. These observations revealed that SKR is subject to significant variations at various timescales.

5.3.1 Rotational Dynamics

Rotational Modulation The most obvious SKR temporal variation is its strong, regular, rotational modulation at ~11h, which reveals the prominent role of the fast planetary rotation on magnetospheric dynamics. The modulation of northern SKR discovered by Voyager was interpreted as the result of strobe-like intense flashes emitted by radio sources fixed in local time (Desch and Kaiser 1981; Kaiser et al. 1981). The dawnside location of the latter was indirectly inferred from visibility considerations (Galopeau et al. 1995, and references therein). Similarly to other planets, this radio period was taken as a direct measurement of the inner rotation period. However, in the absence of any measurable tilt between the magnetic and rotation axis, the origin of the modulation itself remained unexplained.

Further distant observations by Ulysses surprisingly revealed that the SKR period, measured alternately from the southern and northern hemispheres, varies with time, at a level of ~1 % over several years (Galopeau and Lecacheux 2000). This result definitely precluded the observed radio period from providing the internal rotation rate, and raised the additional question of the origin of a period varying over long timescales.

The quasi-continuous observations by Cassini since 2004, equipped with a sophisticated radio experiment, brought a set of important results to light. Thanks to long-term time series, Zarka et al. (2007) identified weekly modulations of the southern SKR period during the pre-orbit insertion interval (subject to little visibility effect). A positive correlation with variations of the solar wind speed suggested an external control of the period's variation. Investigating yearly variations, (Gurnett et al. 2009, 2010a, and references therein) showed that SKR is modulated at (slightly) different periods in southern and northern hemispheres, both varying by ∼1 % over years and reaching each other after the equinox of 2009. The existence and the variation of SKR periods was proposed to result from a seasonal forcing of the magnetosphere-ionosphere coupling at the origin of these radio emissions. Further insights were provided by higher level observables. Statistical studies of the position of SKR sources (Cecconi et al. 2009; Lamy et al. 2009) showed (i) that they lie on magnetic field lines colocated with the atmospheric auroral oval, with a strong local time variation of their intensity maximising at dawn (Lamy et al. 2009), and (ii) that the southern modulation is produced by an active region, extended in longitude, and rotating at the southern SKR period (Lamy 2011). The latter result, supported by an independent analysis of phases built from radio and magnetic field data (Andrews et al. 2011) and validated by another independent modeling study (Lamy et al. 2013), changed the simple strobe-like picture derived from Voyager to an intrinsic search-light phenomenon, which displays strobe-like characteristics when the observer is in view of the most intense, dawnside sources. This feature is consistent with the ubiquitous search-light modulation of various other magnetospheric observables at both SKR periods (modulation of particles and magnetic field, oscillations of magnetospheric boundaries and the auroral oval). The sum of these observations is proposed to result from two co-existing field-aligned current (FAC) systems, rotating at southern and northern radio periods (see e.g. Andrews et al. 2010). The ultimate driver of these FAC systems, though, is still unknown.

Source Regions in Sub-corotation At timescales shorter than the ∼11h periods (referred to as the 'rigid' corotation period), SKR dynamic spectra often display arcs lasting for a few minutes to a few hours, either vertex-early and/or vertex-late shaped, similar to Jovian DAM arcs (Boischot et al. 1981; Thieman and Goldstein 1981). These structures, together with signal disappearance close to the planet, were quantitatively modeled as the result of visibility effects owing to the relative motion of radio sources with respect to the observer (Lamy et al. 2008a). More precisely, the correct modeling required oblique beaming angles with a steep decrease at high frequency, and active field lines moving in sub-corotation (here 90 %).

Oblique beaming angles were theoretically obtained by assuming loss cone-driven CMI with 20 keV electrons. However, while the observed SKR beaming is indeed oblique with a decrease at high frequency, it is significantly variable (Cecconi et al. 2009; Lamy et al. 2009), although whether it varies with time, source position, and/or with wave direction is an open question. In addition, in situ measurements revealed shell distributions with 6–9 keV electrons within the SKR source region (Lamy et al. 2010; Schippers et al. 2011), shown to be an efficient CMI-driver able to produce the observed SKR intensities (Mutel et al. 2010). A possible way to account for oblique and variable beaming from shell-driven (quasi-perpendicular) emission relies on refraction close and far from the source.

Atmospheric auroral sources in sub-corotation have long been observed along the main auroral oval. Therefore, as the bulk of SKR is emitted on field lines co-located with the main oval, it is not surprising to observe sub-corotating sources at radio and optical wavelengths, as illustrated with recent simultaneous observations of a single auroral hot spot moving at 65 % of corotation (Lamy et al. 2013). The range of sub-corotating velocities additionally matches the velocity of the ambient cold plasma populating auroral field lines (Thomsen et al. 2010). The co-existence of rotational and sub-corotational dynamics on adjacent field lines, likely relating to the intrinsic nature of the rotating FAC systems, remains to be further investigated.

5.3.2 Longer-Term Variations

SKR also exhibits variations on timescales longer than a planetary rotation, ranging from days to years.

Since Voyager, the solar wind has been known to be a key ingredient for driving SKR emissions, the most striking evidence of which was revealed by the sudden drop off of SKR intensity when Saturn was immersed in Jupiter's magnetotail for several intervals in 1981 (Desch 1983). Precisely, the level of SKR emission was found to be highly correlated with the solar wind dynamic (ram) pressure (Desch and Rucker 1983, 1985), later confirmed by Cassini (Rucker et al. 2008), rather than with the geometry of the magnetic field, which controls Earth substorms. Such a correspondence was tracked with long-term time series from approximately fixed spacecraft locations, which limits the visibility effects discussed in section 'Rotational Dynamics' above. This property was related to acceleration processes specific to Saturn, most efficient on the dawn sector, such as the Kelvin-Helmholtz instability on the flank of the magnetopause (Galopeau et al. 1995), field-aligned currents initiated by the shear between open and closed field lines (Cowley et al. 2004b), or the shear of swept back closed field lines (Southwood and Kivelson 2009).

The effect of interplanetary shocks on auroral emissions was recently investigated in more detail with coordinated observations (Kurth et al. 2005; Badman et al. 2008; Clarke et al. 2009). These authors showed an overall brightening of the SKR spectrum lasting for several planetary rotations, matching a dawnside intensification of the auroral oval, and with a characteristic extension toward low frequencies. Importantly, the southern SKR phase was shown to be unaltered by solar wind compressions. Investigating such auroral intensifications from the magnetotail, Jackman et al. (2009, 2010) showed that SKR low frequency extensions coincide with plasmoid ejections.

Most recently, Kimura et al. (2013) investigated very long-term variations of northern and southern SKR spectra, separated by polarization, spanning six years of measurements. This study confirms the prominent role of solar wind pressure over one solar cycle, and additionally identifies a seasonal dependence of the SKR activity, maximising in summer.

5.4 Summary

Observations at radio wavelengths, either acquired from the ground (high temporal and spectral sampling, interferometry) or from space-based probes (observations below 10 MHz, goniopolarimetry), have provided a wealth of information on the auroral processes at work in the magnetospheres of Jupiter and Saturn. Following the first analyses of radio emission flux, spectra, and time-variability, recent developments include accurate modeling of CMI-driven radiation, and the study of higher level observables (wave polarization, the source location). Such analyses have been illustrated with a few examples of internally-driven processes at Jupiter (Io and non-Io DAM visibility, millisecond S-bursts) and Saturn (rotational and sub-corotational modulation, seasonal effects) and compared to externally-driven processes (solar wind influences). The diagnosis provided by low frequency radio observations is not only of interest for the further study of giant planet magnetospheres (with JUNO and JUICE), but also more generally for all planetary and possible exoplanetary radio sources (Zarka 2007), with LOFAR, Bepi-Colombo and future missions toward the outer heliosphere.

6 Auroral Signatures of Magnetospheric Dynamics and Boundaries at Jupiter and Saturn

In this section the interpretation of auroral emissions in terms of magnetospheric dynamics is described, focussing in particular on the signatures of open-closed field line boundaries as evidence of the solar wind interaction, and their differences from magnetodisk-related processes and emissions.

The concept of an 'open' magnetosphere was first described for the terrestrial magnetosphere by Dungey (1963). He described how dayside planetary magnetic field lines can become open to the solar wind via magnetic reconnection with the interplanetary field at the magnetopause. The open field lines are then dragged anti-sunward by the magnetosheath flow to form the magnetotail lobes. The lobe field lines drift to the tail current sheet where reconnection occurs again to close the field lines. The newly-closed field lines are accelerated back toward the planet and circulate around to the dayside to complete the 'Dungey cycle' of flux circulation. The disconnected portion of open flux is lost downtail. Cowley et al. (2003, 2004b) have applied this concept to the rapidly-rotating magnetospheres of Jupiter and Saturn to illustrate the nature of plasma flow in different regions of the magnetospheres (see also Delamere et al., this issue). In addition to the Dungey-cycle model of flux circulation, it has been proposed that the open field regions in the magnetospheres of Jupiter and Saturn could be maintained by viscous processes at the boundaries allowing for flux and plasma exchange, such as reconnection within Kelvin-Helmholtz vortices on the dusk flank (Delamere and Bagenal 2010; Desroche et al. 2013).

The characteristics of the open field region are mainly defined by the change in plasma population, i.e., the loss of previously-trapped magnetospheric plasma and entry of magnetosheath plasma, and the anti-sunward convection of the open field lines. In contrast, as explained in Sect. 1, the magnetodisk regions of the giant planet magnetospheres are characterised by a trapped, warm plasma population, including heavy ions originating from the moons, sub- or co-rotating with the planet. In this section we focus on the auroral signatures of open and closed field regions, and their boundaries, as these can be remotely monitored and provide a more global picture than the restricted spatial sampling of an in situ spacecraft.

6.1 Open-Closed Boundaries in Jupiter's Magnetosphere

The extent of an open field region and the existence of a Dungey-cycle in Jupiter's magnetosphere have been debated because of the large size of the magnetosphere, and hence long transport times (Cowley et al. 2003, 2008; Badman and Cowley 2007; McComas and Bagenal 2007; Delamere and Bagenal 2010). Although, as mentioned above, the processes leading to the replenishment of open flux have not been conclusively identified (i.e. large-scale Dungey-cycle circulation versus localised viscous interactions), in situ plasma measurements and remote sensing of ionospheric flows have indicated the presence of a persistent open field region, as discussed below.

6.1.1 Evidence for an Open Field Region

In situ Measurements Ulysses was, so far, the only spacecraft to sample Jupiter's high latitude region. It reached latitudes of ~45° during its encounter in Feb 1992. Simpson et al. (1992) identified a region of Jupiter's magnetosphere analogous to the Earth's polar cap, where the fluxes of MeV particles decreased, indicating their loss to the interplanetary

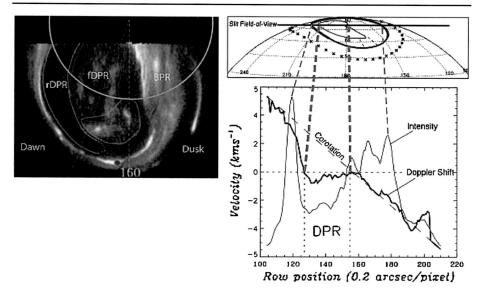

Fig. 32 (*Left*) polar projection of the northern UV auroral region showing the shape and position of the dark region (*yellow contour*), the swirl region (*red contour*), and the active region (*green contour*) as they appear at CML = 160° (marked with a *vertical green dashed line*). Longitude 180° is highlighted with a *red dashed line*. The *red dot* locates the magnetic footprint of Ganymede (VIP4 model) as the orbital longitude of the satellite matches the CML and therefore indicates the direction of magnetic noon at $15R_J$ (Grodent et al. 2003b). The *purple circle* is latitude 74°, the projected location of the slit field of view of the data on the right. (*Top right*) viewing geometry and (*bottom right*) Doppler shifted H_3^+ IR emission profile from Stallard et al. (2003), showing the stagnated flows in the dark polar region (DPR). This DPR corresponds to the swirl region on the UV image *on the left*. Both images illustrate the dawn-dusk asymmetry of the polar auroral emission intensity (from Delamere and Bagenal 2010)

medium. Supporting evidence was provided by a decrease of the proton/helium abundance ratio to values typical of interplanetary space, disappearance of the anisotropy in the corotational direction for ~MeV protons, disappearance of the hot magnetospheric electrons, the detection of auroral hiss, and anti-sunward ion flow (Bame et al. 1992; Simpson et al. 1992; Stone et al. 1992; Cowley et al. 1993). The field-aligned current detected at the boundary as a perturbation in the magnetic field had only a weak signature (~1 nT) implying that the ionospheric conductivity at the magnetic footprint of the spacecraft was low (Cowley et al. 1993). However, the size and dynamics of this polar cap region could not be determined by the single spacecraft encounter.

Measurements of the magnetic field in the dawnside magnetotail lobes by Voyager showed they exhibit very low fluctuations compared to the plasma sheet, consistent with the extremely low electron densities detected ($<10^{-5}$ cm^{-3}) (Gurnett et al. 1980; Acuña et al. 1983).

Auroral Observations Jupiter's main auroral ovals have been shown to be generated by currents associated with the breakdown of corotation at 20–$30R_J$ at the inner edge of the equatorial middle magnetosphere (Cowley and Bunce 2001; Hill 2001). However, it is not yet certain which of the various auroral features seen at higher latitudes maps to the open lobes or their boundary in the ionosphere. Figure 32 illustrates the dark (yellow contour), swirl (red contour), and active (green contour) regions of the Jovian northern UV polar

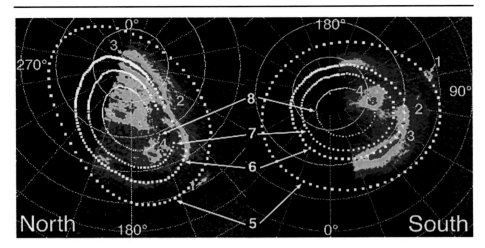

Fig. 33 Two consecutive color-coded polar projected maps of Jupiter's aurora on 15 August 1999 using HST-STIS (1180–1530 Å FWHM). *Dark blue*: the faint solar reflected flux, *light blue-green*: moderate emissions including (2) the 'low-latitude belt', and *red*: brightest auroral features, (1) Io footprint, (3) main oval, (4) conjugate polar cusps. The *dotted curve on top* of (3) is the Pallier-Prangé (PP) reference main oval. (7, 8) are the PP derived inner reference ovals, and (5) is the VIP4 model Io footprint (from Pallier and Prangé 2004)

aurora (Grodent et al. 2003b). The light purple circle indicates the projected location of the slit used to obtain the data shown in the right hand panel. This is the slit used at NASA IRTF to observe the infrared H_3^+ intensity and line-of-sight velocities (as described in Sect. 3.3), shown in the lower right panel. Stallard et al. (2003) and Cowley et al. (2003) suggested that the dark polar region (DPR), which was held fixed relative to the planetary rotation, could be the footprint of the open field lines. However, as shown in Fig. 32 there is significant UV 'swirl' emission at 0–200 kR above background in this region, which requires collisions with electrons with energies greater than the H_2 ionization energy, 15.4 eV. It is not clear how the required fluxes and energies of electrons would be present on supposedly open, plasma-depleted field lines, particularly as Cowley et al. (2003) suggested this would be a region of downward current, i.e. upward-moving electrons.

Vogt et al. (2011) attempted to address this issue by applying a flux equivalency mapping between Jupiter's equatorial magnetosphere and the ionosphere. They considered any region mapping beyond the magnetopause or beyond $150R_J$ downtail to be open flux, and found that corresponded to a region of approximately 40° longitude by 20° latitude in the ionosphere. The amount of open flux was estimated to be ∼700 GWb in each hemisphere, in agreement with estimates based on the size of the magnetotail lobes and the average field strength (Acuña et al. 1983; Joy et al. 2002). In terms of the observed auroral features, Vogt et al. (2011) estimated that the active, swirl, and part of the dark regions all mapped to open field lines (see Fig. 32), but the question remains as to what could produce the swirl emission on open field lines.

Pallier and Prangé (2001) and Pallier and Prangé (2004) determined a reference main oval from UV observations and scaled it to higher latitudes in a search for a persistent open-closed field line boundary aurora. They identified an arc of aurora in the northern hemisphere, surrounding a dark area of radius ∼10°, as a possible signature of the open-closed boundary. They also identified diffuse spots near local noon as signatures of the magnetospheric cusps in both the northern and southern hemispheres (labelled 4 in Fig. 33).

The color ratio (see section 'Comparison Between Electron and Ion Energy Deposition' and Sect. 3.1.1) of these spots was particularly large, leading to an estimate of the characteristic electron energy (assuming a pure electron beam) of $\sim$200 keV, at the upper end of the range of values usually measured for different components of the aurora (Gustin et al. 2004b).

6.1.2 Auroral Signatures of Reconnection at the Open-Closed Boundary

In addition to the search for a persistent auroral signature of the open-closed boundary and cusps, other more transient features, possibly associated with reconnection events, have been identified. Waite et al. (2001) showed a localised flare in the UV aurora reaching 37 MR (total H_2 and H emission) on a timescale of $\sim$70 s. Bonfond et al. (2011) identified intensifications in the UV polar emission with a 2 min periodicity. They magnetically mapped the location of these flares to the vicinity of the dayside magnetopause, and pointed out that their periodicity is similar to that identified for flux transfer events, which are bursts of reconnection with the interplanetary magnetic field (Walker and Russell 1985).

Signatures of Dungey-cycle tail reconnection, i.e. the closure of open magnetic field lines, have not been identified in Jupiter's aurora. Small spots have been observed inside the dawn arc of the main aurora in both UV and IR images, and related to Vasyliunas-cycle tail reconnection, which involves reconfiguration of closed field lines (Grodent et al. 2004; Radioti et al. 2008b, 2011b).

6.1.3 Comparison to Magnetodisk-Related Emissions

In addition to the polar aurorae described above, at least some of which are related to the solar wind interaction, features have been identified at latitudes lower than Jupiter's main oval. The most obvious of these are the spots and downstream tails associated with the moons Io, Europa, and Ganymede (Connerney et al. 1993; Clarke et al. 2002). The moons orbit at radial distances of 5.9, 9.4, and 15.1 R_J and their auroral footprints provide valuable constraints for magnetic field models which seek to map magnetic field lines between the magnetosphere and the ionosphere (e.g. Connerney et al. 1998; Grodent et al. 2008; Hess et al. 2011a). The variability of the main oval and moon footprints attributed to increased mass-loading and hot plasma injection was described in Sect. 3.1.

One further auroral feature related to magnetodisk processes is the variable emission located equatorward of the main oval. Tomás et al. (2004) investigated electron pitch angle distributions measured by Galileo and identified a persistent, sharp transition between inner, trapped (maximum fluxes field-perpendicular) and outer, bidirectional populations at 10–17 R_J in the equatorial plane. They suggest that this transition could be caused by whistler waves scattering the electrons into a more field-aligned distribution, which then precipitate into the ionosphere to produce a relatively discrete auroral arc observed equatorward of the main oval (Grodent et al. 2003a). Transient, diffuse equatorward emissions of varying spatial extent have also been identified and related to injections of hotter plasma from larger radial distances, either via wave scattering, or field-aligned currents at the edges of the high pressure injected cloud (e.g. Mauk et al. 2002).

In conclusion, some high latitude auroral features have been suggested to be signatures of an open field region or its boundary, including magnetopause reconnection events. A polar cap-like region has been detected in situ at high latitudes and in the magnetotail, but its extent is not well constrained. The extent and replenishment of Jupiter's open field lines, which represent the transfer of plasma and momentum with the solar wind, are not well understood.

6.2 Open-Closed Boundaries in Saturn's Magnetosphere

The efficiency of reconnection between Saturn's planetary field and the interplanetary magnetic field, which creates the open field region, has been questioned because of the supposed low efficiency of reconnection at the magnetopause in the high plasma beta regime (Scurry and Russell 1991; Masters et al. 2012), although this has been contradicted by observations at the Earth (Grocott et al. 2009). However, in situ measurements of the magnetopause have shown evidence for reconnection, from changes in the component of the magnetic field normal to the magnetopause, and/or the detection of heated or mixed plasma populations (Huddleston et al. 1997; McAndrews et al. 2008; Lai et al. 2012; Badman et al. 2013). These latter studies have concluded that reconnection is able to proceed at a sufficient rate at different locations across the magnetopause to produce a persistent open flux region at Saturn.

6.2.1 Characteristics of the Open Field Region

In situ Measurements   As at Jupiter, the observations made by in situ spacecraft have provided evidence of an open field region in Saturn's magnetosphere. Ness et al. (1981) identified a tail lobe from Voyager-1 magnetic field data with a diameter of $80R_S$ and likened it to the terrestrial magnetotail. The high latitude orbits made by Cassini have since provided a wealth of in situ measurements of the lobe and polar cap structure. For example, a decrease in electron flux by several orders of magnitude was observed by Cassini as it passed from the dayside magnetosphere to the higher latitudes over the southern polar cap (Bunce et al. 2008). This was interpreted as a crossing from closed to open field lines. Gurnett et al. (2010b) identified a plasma density boundary at high latitudes in Saturn's magnetosphere using Langmuir Probe measurements of electron density. They also related the decrease in density to the appearance of auroral hiss (broadband whistler mode waves observed at frequencies below the plasma frequency). Examination of the high energy ($\sim$200 keV) electron data indicated an upward electron anisotropy, suggesting that no electrons were returning from magnetic mirror points in the opposite hemisphere and hence that the high latitude field lines were open. An example of these measurements is shown in Fig. 34.

Auroral Observations   The auroral field-aligned currents in Saturn's high latitude magnetosphere are also identified in Fig. 34 as perturbations in the azimuthal (B_ϕ) component. These field aligned currents are responsible for Saturn's main auroral emission and lie close to the boundary between open and closed field lines, driven by the flow shear between antisunward open and outer magnetosphere flux tubes, and the sub-corotating middle and inner magnetospheric flux tubes (Cowley et al. 2004b, 2004a; Bunce et al. 2008). The auroral oval maps to the outer magnetosphere, beyond the ring current, with the poleward boundary of the aurora mapping to the vicinity of the magnetopause on the dayside (Carbary et al. 2008; Belenkaya et al. 2011).

The auroral oval is therefore observed to change its size and power in response to solar wind conditions as open flux is created and destroyed (Clarke et al. 2005, 2009; Crary et al. 2005; Bunce et al. 2005a; Badman et al. 2005). A selection of images acquired by HST demonstrating the variability of the southern UV aurorae is shown in Fig. 35. Badman et al. (2006) showed from these images that the southern auroral oval varies in position from 2–20° co-latitude. If the poleward boundary of the auroral oval is used as a proxy for the open-closed boundary (this is likely to be an upper limit following the discussion above), the amount of open flux threading the high latitude polar cap is estimated to be 15–50 GWb (Badman et al. 2005, 2014).

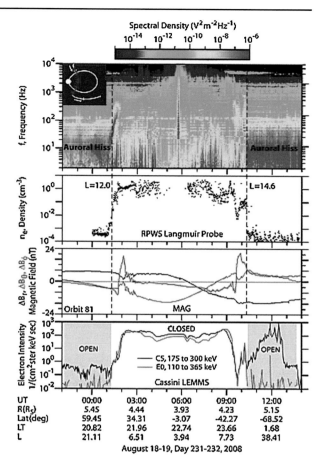

Fig. 34 A multi-plot comparison of (*top*) the electric field spectrum of auroral hiss, (*top middle*) the electron density from the Langmuir probe, (*bottom middle*) three magnetic field components from the magnetometer (MAG), and (*bottom*) the electron flux from the MIMI-LEMMS energetic electron detector (from Gurnett et al. 2010b)

6.2.2 Auroral Signatures of Reconnection at the OCB

Several localised auroral features have been identified and related to reconnection processes close to the open-closed field line boundary. Gérard et al. (2004) identified an auroral spot poleward of the noon main auroral arc and suggested it was the signature of precipitation in the magnetospheric cusps. Bunce et al. (2005a) modelled the ionospheric response to flow vortices produced by magnetopause reconnection events under different IMF conditions including auroral field aligned currents and related emission intensities. If plasma conditions are favourable then under northward IMF, reconnection is expected to proceed at the sub-solar magnetopause, resulting in anti-sunward ionospheric flows and currents close to the open-closed boundary (main oval), the opening of dayside magnetic field lines and subsequent expansion of the dayside auroral oval to lower latitudes. Conversely, high-latitude lobe reconnection would occur under prolonged southward IMF and result in reversed vortical flows and currents poleward of the open-closed boundary in the ionosphere. Sub-solar reconnection is therefore related to the intensification of the main auroral arc in the noon region, while high-latitude reconnection is related to localised auroral emission poleward of the main oval (Bunce et al. 2005a; Gérard et al. 2005).

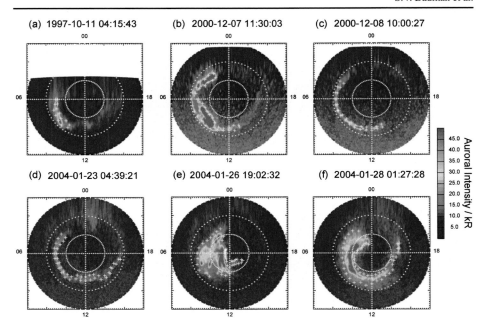

Fig. 35 Selection of six UV images of Saturn's southern aurora obtained during the interval 11 October 1997 to 30 January 2004, with the date and start time of each image shown at the *top* of each plot. The images are projected onto a polar grid from the pole to 30° co-latitude, viewed as though looking through the planet onto the southern pole. Noon is *at the bottom of each plot*, and *dawn to the left*, as indicated. The UV auroral intensity is plotted according to the colour scale shown on the *right-hand side* of the figure. The *white crosses mark* the poleward and equatorward boundaries of the auroral emissions (from Badman et al. 2006)

Using high-sensitivity Cassini instruments, the signatures of transient magnetopause reconnection events have been identified in the noon and post-noon sectors. These appear as bifurcations of the main auroral arc that have been observed to travel poleward while the end connected to the main oval sub-corotates (Radioti et al. 2011a; Badman et al. 2012a, 2013). Similar dusk sector features have been shown to be non-conjugate with their appearance in each hemisphere related to the direction of the IMF; specifically, they will be favoured in the northern hemisphere for $B_Y < 0$ and in the southern hemisphere for $B_Y > 0$ because of the different sense of the associated field aligned current patterns and the source plasma populations (Meredith et al. 2013).

The signatures of nightside reconnection events have also been identified. Broad infilling of the polar cap region has been interpreted as a large tail reconnection event in response to a solar wind compression of the magnetosphere (Cowley et al. 2005; Badman et al. 2005; Stallard et al. 2012a). Smaller-scale UV spots and blobs have been attributed to tail energisation events, likely driven by reconnection (Mitchell et al. 2009b; Jackman et al. 2013), but it has not yet been possible to conclude whether they are associated with reconfiguration of stretched, closed field lines (the Vasyliunas cycle) or the closure of open lobe field lines.

6.2.3 Interpretation and Differences from Magnetodisk Processes

While the above discussion has concentrated on the solar wind-related emissions in the vicinity of the open-closed field line boundary, the auroral signatures of internally-driven processes have also been detected at Saturn. A relatively broad mid-latitude auroral oval has been observed in ground-based measurements of H_3^+ emission, with the peak emission at 62°N and 58°S (Stallard et al. 2008b, 2010). Stallard et al. (2010) suggest that this corresponds to the location where plasma flow initially departs from rigid corotation at radial distances of $3–4R_S$ in the magnetosphere (Wilson et al. 2009), and invoke a system of corotation-enforcement currents flowing between the ionosphere and inner magnetosphere. The small radial distance of this corotation-breakdown region is somewhat unexpected, given that the region of maximum ion formation is further out, closer to $\sim 6R_S$ (Sittler et al. 2008).

A similar feature was observed in HST UV images but the emission is very faint compared to the main oval (1.7 kR) such that it can only be observed on the nightside when the tilt angle is large so that the emission is limb-brightened (Grodent et al. 2005). Grodent et al. (2010) suggested that this emission could be driven by precipitation of keV electrons identified in the magnetosphere at $4–11R_S$. They demonstrated that these electrons, scattered by whistler waves into the loss cone, would have sufficient energy flux to produce the level of UV emission observed, such that the field-aligned currents suggested by Stallard et al. (2010) are not required. One further possible generation mechanism is the precipitation of hot protons from the ring current (e.g. Mitchell et al. 2009b). Interestingly, simultaneous observations of the UV and IR emissions have revealed instances of an equatorward arc at 70°S present at all wavelengths (Lamy et al. 2013), and present only in H and H_2, but not H_3^+ (Melin et al. 2011) (see also Sect. 3.4.2).

More localised diffuse emission features have been identified in the dayside UV aurora by Radioti et al. (2013b), who related them to ENA emissions in the same local time sectors of the magnetosphere. They suggested that both the auroral and ENA emissions are the signatures of injections of hot plasma in the magnetosphere although the origin of the injections is unclear.

In addition to these diffuse auroral features, the auroral footprint of Enceladus has been observed in a small number of the UVIS images (Pryor et al. 2011). The reason for its variable intensity is most likely to be the time-variability of the cryogenic plume activity on the moon affecting the local plasma conditions. So far the Enceladus auroral footprint has only been identified in the UV.

In conclusion, while Saturn's auroral emissions vary strongly with the solar wind interaction, their precise relationship to the open-closed boundary, as revealed by different instrumentation, is not yet determined. Monitoring the size and shape of the auroral emissions provides a valuable tool for describing the extent to which the solar wind and interplanetary magnetic field are controlling Saturn's magnetosphere, relative to magnetospheric dynamics.

7 Future Observations and Outstanding Issues

Much of our understanding of auroral processes at the giant planets has come from dedicated ground- and space-based telescope observing campaigns. Long-term sequences of observations are also provided by high inclination views of the polar regions by orbiters. In 2016 there will be the opportunity for simultaneous polar observations of two different environments with the NASA polar orbiter Juno at Jupiter, and Cassini's high inclination orbits at Saturn. These types of observation are valuable for imaging both hemispheres independent of the planet's season (when one hemisphere is preferentially observed from the Earth), and for obtaining a good view of the nightside aurora, which is difficult to observe from the

Earth. These orbits also provide invaluable simultaneous in situ detections of auroral plasma and currents with imaging or spectra of the conjugate aurora.

Unfortunately, beyond late 2017 there will be no orbital spacecraft at any of the giant planets until 2030 when the European Space Agency (ESA) Juice mission arrives at Jupiter. Secondly, with the ageing HST facing retirement within the not too distant future, there will be no facilities capable of observing in the ultraviolet, with no replacement yet in the pipeline. These facts mean, by necessity, that we are entering an era where ground-based infrared observations of H_3^+ will be the main tool with which to study the magnetosphere-ionosphere-thermosphere interaction at the gas giants. This is not to say, of course, that this upcoming era is entirely bleak—ground-based telescopes are getting larger, with much improved instrumentation, with many facilities developing the capability of removing the influence of the Earth's atmosphere via adaptive optics (AO). The planetary observing capabilities of the James Webb Space Telescope, scheduled for launch in 2018, are under investigation. The limitation of infrared observations, however, remains the long lifetime of H_3^+, which precludes the study of short term auroral variability observed in the UV. Ongoing and future observations of Jupiter's radio emissions will be provided by the Low Frequency Array (LOFAR), Nançay Decametric Array (NDA) and Stereo spacecraft.

The outstanding questions for different scientific targets are given at the end of each section above. A common idea is for coordinated observations at different wavelengths, and with in situ measurements of solar wind or magnetospheric field and plasma conditions. Coordinated observations are required to study the full thermosphere-ionosphere-magnetosphere coupled system, including Io (or Enceladus at Saturn) activity and solar wind conditions. Observations on different timescales are also required. For Io-related variations at Jupiter, the observing interval should cover several months. To study the solar wind variation, the observing interval should be at least one week to resolve the time scale for magnetospheric compression and the following expansion phase. The observable parameters required are Io's volcano activity, the Io torus, the IR and UV aurora, radio and X-ray emissions, and the solar wind (ideally in-situ near the planet or at least propagated from near-Earth measurements). Since solar wind-driven variation causes compression of the magnetosphere followed by expansion, and different response processes should occur in each phase, the temporal variation is important. At Saturn, combined studies using Cassini remote-multi-wavelength and in-situ observations are an ongoing approach (see Sect. 3.4). A coordinated observation campaign at Jupiter was carried out in early 2014 when EUV spectral observations of the Io torus and Jupiter's polar region were taken by the JAXA Sprint-A/Hisaki mission. The results of such campaigns will provide significant advances in our understanding of the relative contributions of solar wind and magnetodisk driving processes at the giant planets.

Acknowledgements The authors acknowledge the support of EUROPLANET RI project (Grant agreement no.: 228319) funded by EU; and also the support of the International Space Science Institute (Bern). SVB was supported by a Royal Astronomical Society Research Fellowship. MG was partially supported by the Science and Technology Facilities Council (STFC) through the Consolidated Grant to Imperial College London. C. Tao was supported by a JSPS Postdoctoral Fellowship for Research Abroad. The Editor thanks the work of two anonymous referees.

References

N. Achilleos, S. Miller, J. Tennyson, A.D. Aylward, I. Müller-Wodarg, D. Rees, JIM: A time-dependent, three-dimensional model of Jupiter's thermosphere and ionosphere. J. Geophys. Res. **103**, 20089–20112 (1998). doi:10.1029/98JE00947

M.H. Acuña, K.W. Behannon, J.E.P. Connerney, Jupiter's magnetic field and magnetosphere, in *Physics of the Jovian Magnetosphere*, ed. by A.J. Dessler, (1983), pp. 1–50

J.M. Ajello, W. Pryor, L. Esposito, I. Stewart, W. McClintock, J. Gustin, D. Grodent, J.-C. Gérard, J.T. Clarke, The Cassini campaign observations of the Jupiter aurora by the ultraviolet imaging spectrograph and the space telescope imaging spectrograph. Icarus **178**, 327–345 (2005). doi:10.1016/j.icarus.2005.01.023

D.J. Andrews, A.J. Coates, S.W.H. Cowley, M.K. Dougherty, L. Lamy, G. Provan, P. Zarka, Magnetospheric period oscillations at Saturn: Comparison of equatorial and high-latitude magnetic field periods with north and south Saturn kilometric radiation periods. J. Geophys. Res. **115**, 12252 (2010). doi:10.1029/2010JA015666

D.J. Andrews, B. Cecconi, S.W.H. Cowley, M.K. Dougherty, L. Lamy P. G, P. Zarka, Planetary period oscillations in Saturn's magnetosphere: Evidence in magnetic field phase data for rotational modulation of Saturn kilometric radiation emissions. J. Geophys. Res. **116** (2011). doi:10.1029/2011JA016636

O.V. Arkhypov, H.O. Rucker, Amalthea's modulation of Jovian decametric radio emission. Astron. Astrophys. **467**, 353–358 (2007). doi:10.1051/0004-6361:20066505

S.V. Badman, S.W.H. Cowley, Significance of Dungey-cycle flows in Jupiter's and Saturn's magnetospheres, and their identification on closed equatorial field lines. Ann. Geophys. **25**, 941–951 (2007). doi:10.5194/angeo-25-941-2007

S.V. Badman, E.J. Bunce, J.T. Clarke, S.W.H. Cowley, J.-C. Gérard, D. Grodent, S.E. Milan, Open flux estimates in Saturn's magnetosphere during the January 2004 Cassini-HST campaign, and implications for reconnection rates. J. Geophys. Res. **110** (2005). doi:10.1029/2005JA011240

S.V. Badman, S.W.H. Cowley, J.-C. Gérard, D. Grodent, A statistical analysis of the location and width of Saturn's southern auroras. Ann. Geophys. **24**(12), 3533–3545 (2006)

S.V. Badman, S.W.H. Cowley, L. Lamy, B. Cecconi, P. Zarka, Relationship between solar wind corotating interaction regions and the phasing and intensity of Saturn kilometric radiation bursts. Ann. Geophys. **26**(12), 3641–3651 (2008)

S.V. Badman, N. Achilleos, K.H. Baines, R.H. Brown, E.J. Bunce, M.K. Dougherty, H. Melin, J.D. Nichols, T. Stallard, Location of Saturn's northern infrared aurora determined from Cassini VIMS images. Geophys. Res. Lett. **38** (2011a). doi:10.1029/2010GL046193

S.V. Badman, C. Tao, A. Grocott, S. Kasahara, H. Melin, R.H. Brown, K.H. Baines, M. Fujimoto, T. Stallard, Cassini VIMS observations of latitudinal and hemispheric variations in Saturn's infrared auroral intensity. Icarus **216**, 367–375 (2011b). doi:10.1016/j.icarus.2011.09.031

S.V. Badman, N. Achilleos, C.S. Arridge, K.H. Baines, R.H. Brown, E.J. Bunce, A.J. Coates, S.W.H. Cowley, M.K. Dougherty, M. Fujimoto, G. Hospodarsky, S. Kasahara, T. Kimura, H. Melin, D.G. Mitchell, T. Stallard, C. Tao, Cassini observations of ion and electron beams at Saturn and their relationship to infrared auroral arcs. J. Geophys. Res. **117** (2012a). doi:10.1029/2011JA017222

S.V. Badman, D.J. Andrews, S.W.H. Cowley, L. Lamy, G. Provan, C. Tao, S. Kasahara, T. Kimura, M. Fujimoto, H. Melin, T. Stallard, R.H. Brown, K.H. Baines, Rotational modulation and local time dependence of Saturn's infrared H_3^+ auroral intensity. J. Geophys. Res. **117**(A9), 09228 (2012b)

S.V. Badman, A. Masters, H. Hasegawa, M. Fujimoto, A. Radioti, D. Grodent, N. Sergis, M.K. Dougherty, A.J. Coates, Bursty magnetic reconnection at Saturn's magnetopause. Geophys. Res. Lett. **40**, 1027–1031 (2013). doi:10.1002/grl.50199

S.V. Badman, C.M. Jackman, J.D. Nichols, J.-C. Gérard, Open flux in Saturn's magnetosphere. Icarus **231**, 137–145 (2014). doi:10.1016/j.icarus.2013.12.004

F. Bagenal, P.A. Delamere, Flow of mass and energy in the magnetospheres of Jupiter and Saturn. J. Geophys. Res. **116**, 5209 (2011). doi:10.1029/2010JA016294

F. Bagenal, T.E. Dowling, W.B. McKinnon, *Jupiter* (Cambridge University Press, Cambridge, 2004)

S.J. Bame, B.L. Barraclough, W.C. Feldman, G.R. Gisler, J.T. Gosling, D.J. McComas, J.L. Phillips, M.F. Thomsen, B.E. Goldstein, M. Neugebauer, Jupiter's magnetosphere: Plasma description from the ulysses flyby. Science **257**, 1539–1543 (1992). doi:10.1126/science.257.5076.1539

R.L. Baron, T. Owen, J.E.P. Connerney, T. Satoh, J. Harrington, Solar wind control of Jupiter's H_3^+ auroras. Icarus **120**, 437–442 (1996). doi:10.1006/icar.1996.0063

C.H. Barrow, Jupiter's decametric radio emission and solar activity. Planet. Space Sci. **26**, 1193–1199 (1978). doi:10.1016/0032-0633(78)90059-4

C.H. Barrow, Association of corotating magnetic sector structure with Jupiter's decameter-wave radio emission. J. Geophys. Res. **84**, 5366–5372 (1979). doi:10.1029/JA084iA09p05366

C.H. Barrow, Latitudinal beaming and local time effects in the decametre-wave radiation from Jupiter observed at the Earth and from Voyager. Astron. Astrophys. **101**, 142–149 (1981)

D. Barrow, K.I. Matcheva, Impact of atmospheric gravity waves on the Jovian ionosphere. Icarus **211**, 609–622 (2011). doi:10.1016/j.icarus.2010.10.017

D.J. Barrow, K.I. Matcheva, Modeling the effect of atmospheric gravity waves on Saturn's ionosphere. Icarus **224**(1), 32–42 (2013). doi:10.1016/j.icarus.2013.01.027

E.S. Belenkaya, S.W.H. Cowley, J.D. Nichols, M.S. Blokhina, V.V. Kalegaev, Magnetospheric mapping of the dayside UV auroral oval at Saturn using simultaneous HST images, Cassini IMF data, and a global magnetic field model. Ann. Geophys. **29**, 1233–1246 (2011). doi:10.5194/angeo-29-1233-2011

R.F. Benson, W. Calvert, Isis 1 observations at the source of auroral kilometric radiation. Geophys. Res. Lett. **6**, 479–482 (1979). doi:10.1029/GL006i006p00479

A. Bhardwaj, G.R. Gladstone, Auroral emissions of the giant planets. Rev. Geophys. **38**, 295–354 (2000). doi:10.1029/1998RG000046

A. Bhardwaj, G. Branduardi-Raymont, R.F. Elsner, G.R. Gladstone, G. Ramsay, P. Rodriguez, R. Soria, J.H. Waite, T.E. Cravens, Solar control on Jupiter's equatorial X-ray emissions: 26-29 November 2003 XMM-Newton observation. Geophys. Res. Lett. **32**, 3 (2005a). doi:10.1029/2004GL021497

A. Bhardwaj, R.F. Elsner, J.H. Waite Jr., G.R. Gladstone, T.E. Cravens, P.G. Ford, The discovery of oxygen Kα X-ray emission from the rings of Saturn. Astrophys. J. Lett. **627**, 73–76 (2005b). doi:10.1086/431933

A. Bhardwaj, R.F. Elsner, J.H. Waite Jr., G.R. Gladstone, T.E. Cravens, P.G. Ford Chandra, Observation of an X-ray flare at Saturn: Evidence of direct solar control on Saturn's disk X-ray emissions. Astrophys. J. Lett. **624**, 121–124 (2005c). doi:10.1086/430521

A. Boischot, Y. Leblanc, A. Lecacheux, B.M. Pedersen, M.L. Kaiser, Arc structure in Saturn's radio dynamic spectra. Nature **292**, 727 (1981). doi:10.1038/292727a0

B. Bonfond, D. Grodent, J.-C. Gérard, A. Radioti, J. Saur, S. Jacobsen, UV Io footprint leading spot: A key feature for understanding the UV Io footprint multiplicity? Geophys. Res. Lett. **35**(5) (2008). doi:10.1029/2007GL032418

B. Bonfond, M.F. Vogt, J.-C. Gérard, D. Grodent, A. Radioti, V. Coumans, Quasi-periodic polar flares at Jupiter: A signature of pulsed dayside reconnections? Geophys. Res. Lett. **38**, 2104 (2011). doi:10.1029/2010GL045981

B. Bonfond, D. Grodent, J.-C. Gérard, T. Stallard, J.T. Clarke, M. Yoneda, A. Radioti, J. Gustin, Auroral evidence of Io's control over the magnetosphere of Jupiter. Geophys. Res. Lett. **39** (2012). doi:10.1029/2011GL050253

B. Bonfond, S. Hess, F. Bagenal, J.-C. Gérard, D. Grodent, A. Radioti, J. Gustin, J.T. Clarke, The multiple spots of the Ganymede auroral footprint. Geophys. Res. Lett. **40**, 4977–4981 (2013). doi:10.1002/grl.50989

S.W. Bougher, J.H. Waite, T. Majeed, G.R. Gladstone, Jupiter thermospheric general circulation model (JTGCM): Global structure and dynamics driven by auroral and Joule heating. J. Geophys. Res. **110**, 4008 (2005). doi:10.1029/2003JE002230

G. Branduardi-Raymont, A. Bhardwaj, R.F. Elsner, G.R. Gladstone, G. Ramsay, P. Rodriguez, R. Soria, J.H. Waite Jr., T.E. Cravens, A study of Jupiter's aurorae with XMM-Newton. Astron. Astrophys. **463**, 761–774 (2007a). doi:10.1051/0004-6361:20066406

G. Branduardi-Raymont, A. Bhardwaj, R.F. Elsner, G.R. Gladstone, G. Ramsay, P. Rodriguez, R. Soria, J.H. Waite, T.E. Cravens, Latest results on Jovian disk X-rays from XMM-Newton. Planet. Space Sci. **55**, 1126–1134 (2007b). doi:10.1016/j.pss.2006.11.017

G. Branduardi-Raymont, R.F. Elsner, M. Galand, D. Grodent, T.E. Cravens, P. Ford, G.R. Gladstone, J.H. Waite, Spectral morphology of the X-ray emission from Jupiter's aurorae. J. Geophys. Res. **113**, 2202 (2008). doi:10.1029/2007JA012600

G. Branduardi-Raymont, A. Bhardwaj, R.F. Elsner, P. Rodriguez, X-rays from Saturn: A study with XMM-Newton and Chandra over the years 2002-05. Astron. Astrophys. **510**, 73 (2010). doi:10.1051/0004-6361/200913110

G. Branduardi-Raymont, P.G. Ford, K.C. Hansen, L. Lamy, A. Masters, B. Cecconi, A.J. Coates, M.K. Dougherty, G.R. Gladstone, P. Zarka, Search for Saturn's X-ray aurorae at the arrival of a solar wind shock. J. Geophys. Res. **118** (2013). doi:10.1002/jgra.50112

R.H. Brown, K.H. Baines, G. Bellucci, J.P. Bibring, B.J. Buratti, F. Capaccioni, P. Cerroni, R.N. Clark, A. Coradini, D.P. Cruikshank, P. Drossart, V. Formisano, R. Jaumann, Y. Langevin, D.L. Matson, T.B. McCord, V. Mennella, E. Miller, R.M. Nelson, P.D. Nicholson, B. Sicardy, C. Sotin, The Cassini visual and infrared mapping spectrometer (VIMS) investigation. Space Sci. Rev. **115**(1–4), 111–168 (2004). doi:10.1007/s11214-004-1453-x

E.J. Bunce, S.W.H. Cowley, J.A. Wild, Azimuthal magnetic fields in Saturn's magnetosphere: Effects associated with plasma sub-corotation and the magnetopause-tail current system. Ann. Geophys. **21**, 1709–1722 (2003). doi:10.5194/angeo-21-1709-2003

E.J. Bunce, S.W.H. Cowley, T.K. Yeoman, Jovian cusp processes: Implications for the polar aurora. J. Geophys. Res. **109**, 9 (2004). doi:10.1029/2003JA010280

E.J. Bunce, S.W.H. Cowley, S.E. Milan, Interplanetary magnetic field control of Saturn's polar cusp aurora. Ann. Geophys. **23**, 1405–1431 (2005a). doi:10.5194/angeo-23-1405-2005

E.J. Bunce, S.W.H. Cowley, D.M. Wright, A.J. Coates, M.K. Dougherty, N. Krupp, W.S. Kurth, A.M. Rymer, In situ observations of a solar wind compression-induced hot plasma injection in Saturn's tail. Geophys. Res. Lett. **322**, L20S04 (2005b). doi:10.1029/2005GL022888

E.J. Bunce, C.S. Arridge, J.T. Clarke, A.J. Coates, S.W.H. Cowley, M.K. Dougherty, J.-C. Gérard, D. Grodent, K.C. Hansen, J.D. Nichols, D.J. Southwood, D.L. Talboys, Origin of Saturn's aurora: Simultaneous observations by Cassini and the Hubble space telescope. J. Geophys. Res. **113** (2008). doi:10.1029/2008JA013257

B.F. Burke, K.L. Franklin, Observations of a variable radio source associated with the planet Jupiter. J. Geophys. Res. **60**, 213–217 (1955). doi:10.1029/JZ060i002p00213

J.F. Carbary, The morphology of Saturn's ultraviolet aurora. J. Geophys. Res. **117** (2012). doi:10.1029/2012JA017670

J.F. Carbary, Longitude dependences of Saturn's ultraviolet aurora. Geophys. Res. Lett. **40**(10), 1902–1906 (2013). doi:10.1002/grl.50430

J.F. Carbary, D.G. Mitchell, P. Brandt, E.C. Roelof, S.M. Krimigis, Statistical morphology of ENA emissions at Saturn. J. Geophys. Res. **113**, 5210 (2008). doi:10.1029/2007JA012873

J.A. Carter, S. Sembay, A.M. Read, A high charge state coronal mass ejection seen through solar wind charge exchange emission as detected by XMM-Newton. Mon. Not. R. Astron. Soc. **402**, 867–878 (2010). doi:10.1111/j.1365-2966.2009.15985.x

B. Cecconi, Goniopolarimetric techniques for low-frequency radio astronomy in space, in *Observing Photons in Space*, vol. 9, (2010), pp. 263–277

B. Cecconi, L. Lamy, P. Zarka, R. Prangé, W.S. Kurth, P. Louarn, Goniopolarimetric study of the revolution 29 perikrone using the Cassini radio and plasma wave science instrument high-frequency radio receiver. J. Geophys. Res. **114**, 3215 (2009). doi:10.1029/2008JA013830

B. Cecconi, S. Hess, A. Hérique, M.R. Santovito, D. Santos-Costa, P. Zarka, G. Alberti, D. Blankenship, J.-L. Bougeret, L. Bruzzone, W. Kofman, Natural radio emission of Jupiter as interferences for radar investigations of the icy satellites of Jupiter. Planet. Space Sci. **61**, 32–45 (2012). doi:10.1016/j.pss.2011.06.012

J.-Y. Chaufray, T.K. Greathouse, G.R. Gladstone, J.H. Waite, J.-P. Maillard, T. Majeed, S.W. Bougher, E. Lellouch, P. Drossart, Spectro-imaging observations of Jupiter's 2 μm auroral emission. II: Thermospheric winds. Icarus **211**, 1233–1241 (2011). doi:10.1016/j.icarus.2010.11.021

J.T. Clarke, J. Ajello, G. Ballester, L. Ben Jaffel, J. Connerney, J.-C. Gérard, G.R. Gladstone, D. Grodent, W. Pryor, J. Trauger, J.H. Waite, Ultraviolet emissions from the magnetic footprints of Io, Ganymede and Europa on Jupiter. Nature **415**(6875), 997–1000 (2002)

J.T. Clarke, D. Grodent, S.W.H. Cowley, E.J. Bunce, P. Zarka, J.E.P. Connerney, T. Satoh, Jupiter's aurora, in *Jupiter. The Planet, Satellites and Magnetosphere*, ed. by F. Bagenal, T.E. Dowling, W.B. McKinnon, (2004), pp. 639–670

J.T. Clarke, J.-C. Gérard, D. Grodent, S. Wannawichian, J. Gustin, J. Connerney, F. Crary, M. Dougherty, W. Kurth, S.W.H. Cowley, E.J. Bunce, T. Hill, J. Kim, Morphological differences between Saturn's ultraviolet aurorae and those of Earth and Jupiter. Nature **433**(7027), 717–719 (2005). doi:10.1038/nature03331

J.T. Clarke, J. Nichols, J.-C. Gerard, D. Grodent, K.C. Hansen, W. Kurth, G.R. Gladstone, J. Duval, S. Wannawichian, E. Bunce, S.W.H. Cowley, F. Crary, M. Dougherty, L. Lamy, D. Mitchell, W. Pryor, K. Retherford, T. Stallard, B. Zieger, P. Zarka, B. Cecconi, Response of Jupiter's and Saturn's auroral activity to the solar wind. J. Geophys. Res. **114** (2009). doi:10.1029/2008JA013694

J.E.P. Connerney, J.H. Waite, New model of Saturn's ionosphere with an influx of water from the rings. Nature **312**(5990), 136–138 (1984)

J.E.P. Connerney, R. Baron, T. Satoh, T. Owen, Images of excited H_3^+ at the foot of the Io flux tube in Jupiter's atmosphere. Science **262**, 1035–1038 (1993). doi:10.1126/science.262.5136.1035

J.E.P. Connerney, M.H. Acuña, N.F. Ness, T. Satoh, New models of Jupiter's magnetic field constrained by the Io flux tube footprint. J. Geophys. Res. **103**, 11929–11940 (1998). doi:10.1029/97JA03726

S.W.H. Cowley, E.J. Bunce, Origin of the main auroral oval in Jupiter's coupled magnetosphere-ionosphere system. Planet. Space Sci. **49**, 1067–1088 (2001). doi:10.1016/S0032-0633(00)00167-7

S.W.H. Cowley, A. Balogh, M.K. Dougherty, T.M. Edwards, R.J. Forsyth, R.J. Hynds, K. Staines, Ulysses observations of anti-sunward flow on Jovian polar cap field lines. Planet. Space Sci. **41**, 987–998 (1993). doi:10.1016/0032-0633(93)90103-9

S.W.H. Cowley, E.J. Bunce, T.S. Stallard, S. Miller, Jupiter's polar ionospheric flows: Theoretical interpretation. Geophys. Res. Lett. **30**, 1220 (2003). doi:10.1029/2002GL016030

S.W.H. Cowley, E.J. Bunce, J.M. O'Rourke, A simple quantitative model of plasma flows and currents in Saturn's polar ionosphere. J. Geophys. Res. **109** (2004a). doi:10.1029/2003JA010375

S.W.H. Cowley, E.J. Bunce, R. Prange, Saturn's polar ionospheric flows and their relation to the main auroral oval. Ann. Geophys. **22**(4), 1379–1394 (2004b)

S.W.H. Cowley, S.V. Badman, E.J. Bunce, J.T. Clarke, J.-C. Gérard, D. Grodent, C.M. Jackman, S.E. Milan, T.K. Yeoman, Reconnection in a rotation-dominated magnetosphere and its relation to Saturn's auroral dynamics. J. Geophys. Res. **110**(A2) (2005). doi:10.1029/2004JA010796

S.W.H. Cowley, C.S. Arridge, E.J. Bunce, J.T. Clarke, A.J. Coates, M.K. Dougherty, J.-C. Gérard, D. Grodent, J.D. Nichols, D.L. Talboys, Auroral current systems in Saturn's magnetosphere: Comparison of theoretical models with Cassini and HST observations. Ann. Geophys. **26**(9), 2613–2630 (2008)

F.J. Crary, On the generation of an electron beam by Io. J. Geophys. Res. **102**, 37–50 (1997). doi:10.1029/96JA02409

F. Crary, J. Clarke, M. Dougherty, P. Hanlon, K. Hansen, J. Steinberg, B. Barraclough, A. Coates, J. Gerard, D. Grodent, W. Kurth, D. Mitchell, A. Rymer, D. Young, Solar wind dynamic pressure and electric field as the main factors controlling Saturn's aurorae. Nature **433**(7027), 720–722 (2005). doi:10.1038/nature03333

T.E. Cravens, Vibrationally excited molecular hydrogen in the upper atmosphere of Jupiter. J. Geophys. Res. **92**, 11083–11100 (1987). doi:10.1029/JA092iA10p11083

T.E. Cravens, Comet Hyakutake x-ray source: Charge transfer of solar wind heavy ions. Geophys. Res. Lett. **24**, 105–108 (1997). doi:10.1029/96GL03780

T.E. Cravens, Heliospheric X-ray emission associated with charge transfer of the solar wind with interstellar neutrals. Astrophys. J. Lett. **532**, 153–156 (2000). doi:10.1086/312574

T.E. Cravens, N. Ozak, *Auroral Ion Precipitation and Acceleration at the Outer Planets*. Washington DC American Geophysical Union Geophysical Monograph Series, vol. 197 (2012), pp. 287–294. doi:10.1029/2011GM001159

T.E. Cravens, E. Howell, J.H. Waite, G.R. Gladstone, Auroral oxygen precipitation at Jupiter. J. Geophys. Res. **100**, 17153–17162 (1995). doi:10.1029/95JA00970

T.E. Cravens, J.H. Waite, T.I. Gombosi, N. Lugaz, G.R. Gladstone, B.H. Mauk, R.J. MacDowall, Implications of Jovian X-ray emission for magnetosphere-ionosphere coupling. J. Geophys. Res. **108**, 1465 (2003). doi:10.1029/2003JA010050

A. Dalgarno, M. Yan, W. Liu, Electron energy deposition in a gas mixture of atomic and molecular hydrogen and helium. Astrophys. J. Suppl. Ser. **125**, 237–256 (1999). doi:10.1086/313267

P.A. Delamere, F. Bagenal, Solar wind interaction with Jupiter's magnetosphere. J. Geophys. Res. **115**, 10201 (2010). doi:10.1029/2010JA015347

G.T. Delory, R.E. Ergun, C.W. Carlson, L. Muschietti, C.C. Chaston, W. Peria, J.P. McFadden, R. Strangeway, FAST observations of electron distributions within AKR source regions. Geophys. Res. Lett. **25**, 2069–2072 (1998). doi:10.1029/98GL00705

K. Dennerl, X-rays from Venus observed with Chandra. Planet. Space Sci. **56**, 1414–1423 (2008). doi:10.1016/j.pss.2008.03.008

K. Dennerl, High resolution X-ray spectroscopy of comets with Xmm-newton/rgs (2009). http://www.mssl.ucl.ac.uk/~gbr/workshop3/papers/comets_mssl_2009_kd.pdf

K. Dennerl, Charge transfer reactions. Space Sci. Rev. **157**, 57–91 (2010). doi:10.1007/s11214-010-9720-5

K. Dennerl, C.M. Lisse, A. Bhardwaj, V. Burwitz, J. Englhauser, H. Gunell, M. Holmström, F. Jansen, V. Kharchenko, P.M. Rodríguez-Pascual, First observation of Mars with XMM-Newton. High resolution X-ray spectroscopy with RGS. Astron. Astrophys. **451**, 709–722 (2006). doi:10.1051/0004-6361:20054253

K. Dennerl, C.M. Lisse, A. Bhardwaj, D.J. Christian, S.J. Wolk, D. Bodewits, T.H. Zurbuchen, M. Combi, S. Lepri, Solar system X-rays from charge exchange processes. Astron. Nachr. **333**, 324 (2012). doi:10.1002/asna.201211663

M.D. Desch, Radio emission signature of Saturn immersions in Jupiter's magnetic tail. J. Geophys. Res. **88**, 6904–6910 (1983). doi:10.1029/JA088iA09p06904

M.D. Desch, M.L. Kaiser, Voyager measurement of the rotation period of Saturn's magnetic field. Geophys. Res. Lett. **8**, 253–256 (1981). doi:10.1029/GL008i003p00253

M.D. Desch, H.O. Rucker, The relationship between Saturn kilometric radiation and the solar wind. J. Geophys. Res. **88**, 8999–9006 (1983). doi:10.1029/JA088iA11p08999

M.D. Desch, H.O. Rucker, Saturn radio emission and the solar wind—Voyager-2 studies. Adv. Space Res. **5**, 333–336 (1985). doi:10.1016/0273-1177(85)90159-0

M. Desroche, F. Bagenal, P.A. Delamere, N. Erkaev, Conditions at the magnetopause of Saturn and implications for the solar wind interaction. J. Geophys. Res. **118**, 3087–3095 (2013). doi:10.1002/jgra.50294

A.J. Dessler, *Physics of the Jovian Magnetosphere* (Cambridge University Press, Cambridge, 1983)

M.K. Dougherty, K.K. Khurana, F.M. Neubauer, C.T. Russell, J. Saur, J.S. Leisner, M.E. Burton, Identification of a dynamic atmosphere at Enceladus with the Cassini magnetometer. Science **311**(5766), 1406–1409 (2006). doi:10.1126/science.1120985. http://www.sciencemag.org/content/311/5766/1406.abstract

M.K. Dougherty, L.W. Esposito, S.M. Krimigis, *Saturn from Cassini-Huygens* (Springer, Dordrecht Heidelberg London New York, 2009). doi:10.1007/978-1-4020-9217-6

P. Drossart, J.P. Maillard, J. Caldwell, S.J. Kim, J.K.G. Watson, W.A. Majewski, J. Tennyson, S. Miller, S.K. Atreya, J.T. Clarke, J.H. Waite, R. Wagener, Detection of H_3^+ on Jupiter. Nature **340**, 539–541 (1989)

J.W. Dungey, The structure of the exosphere or adventures in velocity space, in *Geophysics, the Earth's Environment*, ed. by C. De Witt, J. Hieblot, L. Le Beau, (1963), p. 503

U.A. Dyudina, A.P. Ingersoll, S.P. Ewald, C.C. Porco, G. Fischer, W.S. Kurth, R.A. West, Detection of visible lightning on Saturn. Geophys. Res. Lett. **37**, 9205 (2010). doi:10.1029/2010GL043188

R.F. Elsner, G.R. Gladstone, J.H. Waite, F.J. Crary, R.R. Howell, R.E. Johnson, P.G. Ford, A.E. Metzger, K.C. Hurley, E.D. Feigelson, G.P. Garmire, A. Bhardwaj, D.C. Grodent, T. Majeed, A.F. Tennant, M.C. Weisskopf, Discovery of soft X-ray emission from Io, Europa, and the Io plasma torus. Astrophys. J. **572**, 1077–1082 (2002). doi:10.1086/340434

R.F. Elsner, N. Lugaz, J.H. Waite, T.E. Cravens, G.R. Gladstone, P. Ford, D. Grodent, A. Bhardwaj, R.J. MacDowall, M.D. Desch, T. Majeed, Simultaneous Chandra X ray, Hubble space telescope ultraviolet, and Ulysses radio observations of Jupiter's aurora. J. Geophys. Res. **110**, 1207 (2005). doi:10.1029/2004JA010717

R.E. Ergun, C.W. Carlson, J.P. McFadden, G.T. Delory, R.J. Strangeway, P.L. Pritchett, Electron-cyclotron maser driven by charged-particle acceleration from magnetic field-aligned electric fields. Astrophys. J. **538**, 456–466 (2000). doi:10.1086/309094

Y. Ezoe, K. Ishikawa, T. Ohashi, Y. Miyoshi, N. Terada, Y. Uchiyama, H. Negoro, Discovery of diffuse hard X-ray emission around Jupiter with Suzaku. Astrophys. J. Lett. **709**, 178–182 (2010). doi:10.1088/2041-8205/709/2/L178

W.M. Farrell, M.L. Kaiser, M.D. Desch, A model of the lightning discharge at Jupiter. Geophys. Res. Lett. **26**, 2601–2604 (1999). doi:10.1029/1999GL900527

G. Fischer, M.D. Desch, P. Zarka, M.L. Kaiser, D.A. Gurnett, W.S. Kurth, W. Macher, H.O. Rucker, A. Lecacheux, W.M. Farrell, B. Cecconi, Saturn lightning recorded by Cassini/RPWS in 2004. Icarus **183**, 135–152 (2006). doi:10.1016/j.icarus.2006.02.010

G. Fischer, W.S. Kurth, U.A. Dyudina, M.L. Kaiser, P. Zarka, A. Lecacheux, A.P. Ingersoll, D.A. Gurnett, Analysis of a giant lightning storm on Saturn. Icarus **190**, 528–544 (2007). doi:10.1016/j.icarus.2007.04.002

G. Fischer, D.A. Gurnett, W.S. Kurth, F. Akalin, P. Zarka, U.A. Dyudina, W.M. Farrell, M.L. Kaiser, Atmospheric electricity at Saturn. Space Sci. Rev. **137**, 271–285 (2008). doi:10.1007/s11214-008-9370-z

G. Fischer, D.A. Gurnett, P. Zarka, L. Moore, U.A. Dyudina, Peak electron densities in Saturn's ionosphere derived from the low-frequency cutoff of Saturn lightning. J. Geophys. Res. **116**, 4315 (2011). doi:10.1029/2010JA016187

B.L. Fleshman, P.A. Delamere, F. Bagenal, T. Cassidy, The roles of charge exchange and dissociation in spreading Saturn's neutral clouds. J. Geophys. Res. **117**, 5007 (2012). doi:10.1029/2011JE003996

D. Flower, *Molecular Collisions in the Interstellar Medium* (Cambridge University Press, Cambridge, 1990)

J.L. Fox, M.I. Galand, R.E. Johnson, Energy deposition in planetary atmospheres by charged particles and solar photons. Space Sci. Rev. **139**, 3–62 (2008). doi:10.1007/s11214-008-9403-7

K. Fukazawa, T. Ogino, R.J. Walker, A simulation study of dynamics in the distant Jovian magnetotail. J. Geophys. Res. **115** (2010). doi:10.1029/2009JA015228

M. Galand, S. Chakrabarti, Auroral processes in the solar system. Washington DC American Geophysical Union Geophysical Monograph Series **130**, 55 (2002)

M. Galand, S. Chakrabarti, Proton aurora observed from the ground. J. Atmos. Terr. Phys. **68**, 1488–1501 (2006). doi:10.1016/j.jastp.2005.04.013

M. Galand, D. Lummerzheim, Contribution of proton precipitation to space-based auroral FUV observations. J. Geophys. Res. **109**, 3307 (2004). doi:10.1029/2003JA010321

M. Galand, L. Moore, B. Charnay, I. Müller-Wodarg, M. Mendillo, Solar primary and secondary ionization at Saturn. J. Geophys. Res. **114**, 6313 (2009). doi:10.1029/2008JA013981

M. Galand, L. Moore, I. Müller-Wodarg, M. Mendillo, S. Miller, Response of Saturn's auroral ionosphere to electron precipitation: Electron density, electron temperature, and electrical conductivity. J. Geophys. Res. **116**, 9306 (2011). doi:10.1029/2010JA016412

P.H.M. Galopeau, A. Lecacheux, Variations of Saturn's radio rotation period measured at kilometer wavelengths. J. Geophys. Res. **105**, 13089–13102 (2000). doi:10.1029/1999JA005089

P.H.M. Galopeau, P. Zarka, D.L. Quéau, Source location of Saturn's kilometric radiation: The Kelvin-Helmholtz instability hypothesis. J. Geophys. Res. **1002**, 26397–26410 (1995). doi:10.1029/95JE02132

N. Gehrels, E.C. Stone, Energetic oxygen and sulfur ions in the Jovian magnetosphere and their contribution to the auroral excitation. J. Geophys. Res. **88**, 5537–5550 (1983). doi:10.1029/JA088iA07p05537

F. Genova, P. Zarka, C.H. Barrow, Voyager and Nancay observations of the Jovian radio-emission at different frequencies—Solar wind effect and source extent. Astron. Astrophys. **182**, 159–162 (1987)

J.-C. Gérard, V. Singh, A model of energy deposition of energetic electrons and EUV emission in the Jovian and Saturnian atmospheres and implications. J. Geophys. Res. **87**, 4525–4532 (1982). doi:10.1029/JA087iA06p04525

J.-C. Gérard, J. Gustin, D. Grodent, P. Delamere, J.T. Clarke, Excitation of the FUV Io tail on Jupiter: Characterization of the electron precipitation. J. Geophys. Res. **107**, 1394 (2002). doi:10.1029/2002JA009410

J.-C. Gérard, J. Gustin, D. Grodent, J.T. Clarke, A. Grard, Spectral observations of transient features in the FUV Jovian polar aurora. J. Geophys. Res. **108**, 1319 (2003). doi:10.1029/2003JA009901

J.-C. Gérard, D. Grodent, J. Gustin, A. Saglam, J.T. Clarke, J.T. Trauger, Characteristics of Saturn's FUV aurora observed with the space telescope imaging spectrograph. J. Geophys. Res. **109**(A9) (2004). doi:10.1029/2004JA010513

J.-C. Gérard, E.J. Bunce, D. Grodent, S.W.H. Cowley, J.T. Clarke, S.V. Badman, Signature of Saturn's auroral cusp: Simultaneous Hubble space telescope FUV observations and upstream solar wind monitoring. J. Geophys. Res. **110** (2005). doi:10.1029/2005JA011094

J.-C. Gérard, B. Bonfond, J. Gustin, D. Grodent, J.T. Clarke, D. Bisikalo, V. Shematovich, Altitude of Saturn's aurora and its implications for the characteristic energy of precipitated electrons. Geophys. Res. Lett. **36** (2009). doi:10.1029/2008GL036554

J.-C. Gérard, J. Gustin, W.R. Pryor, D. Grodent, B. Bonfond, A. Radioti, G.R. Gladstone, J.T. Clarke, J.D. Nichols, Remote sensing of the energy of auroral electrons in Saturn's atmosphere: Hubble and Cassini spectral observations. Icarus **223**(1) (2013). doi:10.1016/j.icarus.2012.11.033

G.R. Gladstone, J.H. Waite, D. Grodent, W.S. Lewis, F.J. Crary, R.F. Elsner, M.C. Weisskopf, T. Majeed, J.-M. Jahn, A. Bhardwaj, J.T. Clarke, D.T. Young, M.K. Dougherty, S.A. Espinosa, T.E. Cravens, A pulsating auroral X-ray hot spot on Jupiter. Nature **415**, 1000–1003 (2002)

T.I. Gombosi, T.P. Armstrong, C.S. Arridge, K.K. Khurana, S.M. Krimigis, N. Krupp, A.M. Persoon, M.F. Thomsen, Saturn's magnetospheric configuration, in *Saturn from Cassini-Huygens*, ed. by M.K. Dougherty, L.W. Esposito, S.M. Krimigis (Springer, Dordrecht Heidelberg London New York, 2009), pp. 203–255

A. Grocott, S.V. Badman, S.W.H. Cowley, S.E. Milan, J.D. Nichols, T.K. Yeoman, Magnetosonic Mach number dependence of the efficiency of reconnection between planetary and interplanetary magnetic fields. J. Geophys. Res. **114** (2009). doi:10.1029/2009JA014330

D. Grodent, J.H. Waite Jr., J.-C. Gérard, A self-consistent model of the Jovian auroral thermal structure. J. Geophys. Res. **106**, 12933–12952 (2001). doi:10.1029/2000JA900129

D. Grodent, J.T. Clarke, J. Kim, J.H. Waite, S.W.H. Cowley, Jupiter's main auroral oval observed with HST-STIS. J. Geophys. Res. **108**, 1389 (2003a). doi:10.1029/2003JA009921

D. Grodent, J.T. Clarke, J.H. Waite, S.W.H. Cowley, J.-C. Gérard, J. Kim, Jupiter's polar auroral emissions. J. Geophys. Res. **108**, 1366 (2003b). doi:10.1029/2003JA010017

D. Grodent, J.-C. Gérard, J.T. Clarke, G.R. Gladstone, J.H. Waite, A possible auroral signature of a magnetotail reconnection process on Jupiter. J. Geophys. Res. **109**, 5201 (2004). doi:10.1029/2003JA010341

D. Grodent, J.-C. Gérard, S.W.H. Cowley, E.J. Bunce, J.T. Clarke, Variable morphology of Saturn's southern ultraviolet aurora. J. Geophys. Res. **110** (2005). doi:10.1029/2004JA010983

D. Grodent, B. Bonfond, J.-C. GéRard, A. Radioti, J. Gustin, J.T. Clarke, J. Nichols, J.E.P. Connerney, Auroral evidence of a localized magnetic anomaly in Jupiter's northern hemisphere. J. Geophys. Res. **113**, 9201 (2008). doi:10.1029/2008JA013185

D. Grodent, A. Radioti, B. Bonfond, J.-C. Gérard, On the origin of Saturn's outer auroral emission. J. Geophys. Res. **115**, 8219 (2010). doi:10.1029/2009JA014901

D.A. Gurnett, W.S. Kurth, F.L. Scarf, The structure of the Jovian magnetotail from plasma wave observations. Geophys. Res. Lett. **7**, 53–56 (1980). doi:10.1029/GL007i001p00053

D.A. Gurnett, W.S. Kurth, J.D. Menietti, A.M. Persoon, An unusual rotationally modulated attenuation band in the Jovian hectometric radio emission spectrum. Geophys. Res. Lett. **25**, 1841–1844 (1998). doi:10.1029/98GL01400

D.A. Gurnett, W.S. Kurth, G.B. Hospodarsky, A.M. Persoon, P. Zarka, A. Lecacheux, S.J. Bolton, M.D. Desch, W.M. Farrell, M.L. Kaiser, H.-P. Ladreiter, H.O. Rucker, P. Galopeau, P. Louarn, D.T. Young, W.R. Pryor, M.K. Dougherty, Control of Jupiter's radio emission and aurorae by the solar wind. Nature **415**, 985–987 (2002)

D.A. Gurnett, W.S. Kurth, G.B. Hospodarsky, A.M. Persoon, T.F. Averkamp, B. Cecconi, A. Lecacheux, P. Zarka, P. Canu, N. Cornilleau-Wehrlin, P. Galopeau, A. Roux, C. Harvey, P. Louarn, R. Bostrom, G. Gustafsson, J.-E. Wahlund, M.D. Desch, W.M. Farrell, M.L. Kaiser, K. Goetz, P.J. Kellogg, G. Fischer, H.-P. Ladreiter, H. Rucker, H. Alleyne, A. Pedersen, Radio and plasma wave observations at Saturn from Cassini's approach and first orbit. Science **307**, 1255–1259 (2005). doi:10.1126/science.1105356

D.A. Gurnett, A. Lecacheux, W.S. Kurth, A.M. Persoon, J.B. Groene, L. Lamy, P. Zarka, J.F. Carbary, Discovery of a north-south asymmetry in Saturn's radio rotation period. Geophys. Res. Lett. **36**, 16102 (2009). doi:10.1029/2009GL039621

D.A. Gurnett, J.B. Groene, A.M. Persoon, J.D. Menietti, S.-Y. Ye, W.S. Kurth, R.J. MacDowall, A. Lecacheux, The reversal of the rotational modulation rates of the north and south components of Saturn kilometric radiation near equinox. Geophys. Res. Lett. **37**, 24101 (2010a). doi:10.1029/2010GL045796

D.A. Gurnett, A.M. Persoon, A.J. Kopf, W.S. Kurth, M.W. Morooka, J.-E. Wahlund, K.K. Khurana, M.K. Dougherty, D.G. Mitchell, S.M. Krimigis, N. Krupp, A plasmapause-like density boundary at high latitudes in Saturn's magnetosphere. Geophys. Res. Lett. **37**, 16806 (2010b). doi:10.1029/2010GL044466

J. Gustin, P.D. Feldman, J.-C. Gérard, D. Grodent, A. Vidal-Madjar, L. Ben Jaffel, J.-M. Desert, H.W. Moos, D.J. Sahnow, H.A. Weaver, B.C. Wolven, J.M. Ajello, J.H. Waite, E. Roueff, H. Abgrall, Jovian auroral spectroscopy with FUSE: Analysis of self-absorption and implications for electron precipitation. Icarus **171**, 336–355 (2004a). doi:10.1016/j.icarus.2004.06.005

J. Gustin, J.-C. Gérard, D. Grodent, S.W.H. Cowley, J.T. Clarke, A. Grard, Energy-flux relationship in the FUV Jovian aurora deduced from HST-STIS spectral observations. J. Geophys. Res. **109**, 10205 (2004b). doi:10.1029/2003JA010365

J. Gustin, J.-C. Gérard, G.R. Gladstone, D. Grodent, J.T. Clarke, Characteristics of Jovian morning bright FUV aurora from Hubble space Telescope/space telescope imaging spectrograph imaging and spectral observations. J. Geophys. Res. **111**, 9220 (2006). doi:10.1029/2006JA011730

J. Gustin, J.-C. Gérard, W.R. Pryor, P.D. Feldman, D. Grodent, G. Holsclaw, Characteristics of Saturn's polar atmosphere and auroral electrons derived from HST/STIS, FUSE and Cassini/UVIS spectra. Icarus **200**(1), 176–187 (2009). doi:10.1016/j.icarus.2008.11.013

J. Gustin, B. Bonfond, D. Grodent, J.-C. Gérard, Conversion from HST ACS and STIS auroral counts into brightness, precipitated power, and radiated power for H_2 giant planets. J. Geophys. Res. **117**, 7316 (2012). doi:10.1029/2012JA017607

J. Gustin, J.-C. Gérard, D. Grodent, G.R. Gladstone, J.T. Clarke, W.R. Pryor, V. Dols, B. Bonfond, A. Radioti, L. Lamy, J.M. Ajello, Effects of methane on giant planet's UV emissions and implications for the auroral characteristics. J. Mol. Spectrosc. **291**, 108–117 (2013). doi:10.1016/j.jms.2013.03.010

C.J. Hansen, L. Esposito, A.I.F. Stewart, J. Colwell, A. Hendrix, W. Pryor, D. Shemansky, R. West, Enceladus' water vapor plume. Science **311**, 1422–1425 (2006). doi:10.1126/science.1121254

W. Harris, J.T. Clarke, M.A. McGrath, G.E. Ballester, Analysis of Jovian auroral H Ly-alpha emission (1981–1991). Icarus **123**, 350–365 (1996). doi:10.1006/icar.1996.0164

S. Hess, F. Mottez, P. Zarka, Jovian S burst generation by Alfvén waves. J. Geophys. Res. **112**, 11212 (2007a). doi:10.1029/2006JA012191

S. Hess, P. Zarka, F. Mottez, Io Jupiter interaction, millisecond bursts and field-aligned potentials. Planet. Space Sci. **55**, 89–99 (2007b). doi:10.1016/j.pss.2006.05.016

S. Hess, B. Cecconi, P. Zarka, Modeling of Io-Jupiter decameter arcs, emission beaming and energy source. Geophys. Res. Lett. **35**, 13107 (2008). doi:10.1029/2008GL033656

S. Hess, F. Mottez, P. Zarka, Effect of electric potential structures on Jovian S-burst morphology. Geophys. Res. Lett. **36** (2009a). doi:10.1029/2009GL039084

S. Hess, P. Zarka, F. Mottez, V.B. Ryabov, Electric potential jumps in the Io-Jupiter flux tube. Planet. Space Sci. **57**(1), 23–33 (2009b). doi:10.1016/j.pss.2008.10.006

S.L.G. Hess, A. Petin, P. Zarka, B. Bonfond, B. Cecconi, Lead angles and emitting electron energies of Io-controlled decameter radio arcs. Planet. Space Sci. **58**(10), 1188–1198 (2010). doi:10.1016/j.pss.2010.04.011

S.L.G. Hess, B. Bonfond, P. Zarka, D. Grodent, Model of the Jovian magnetic field topology constrained by the Io auroral emissions. J. Geophys. Res. **116**, 5217 (2011a). doi:10.1029/2010JA016262

S.L.G. Hess, P.A. Delamere, F. Bagenal, N.M. Schneider, A.J. Steffl, Longitudinal modulation of hot electrons in the Io plasma torus. J. Geophys. Res. **116** (2011b). doi:10.1029/2011JA016918

S.L.G. Hess, E. Echer, P. Zarka, Solar wind pressure effects on Jupiter decametric radio emissions independent of Io. Planet. Space Sci. **70**, 114–125 (2012). doi:10.1016/j.pss.2012.05.011

S.L.G. Hess, B. Bonfond, P.A. Delamere, How could the Io footprint disappear? Planet. Space Sci. **89**, 102–110 (2013). doi:10.1016/j.pss.2013.08.014

S.L.G. Hess, E. Echer, P. Zarka, L. Lamy, P. Delamere, Multi-instrument study of the Jovian radio emissions triggered by solar wind shocks and inferred magnetospheric subcorotation rates. Planet. Space Sci. (2014, submitted)

T.W. Hill, Inertial limit on corotation. J. Geophys. Res. **84**, 6554–6558 (1979). doi:10.1029/JA084iA11p06554

T.W. Hill, The Jovian auroral oval. J. Geophys. Res. **106**, 8101–8108 (2001). doi:10.1029/2000JA000302

T.W. Hill, V.M. Vasyliūnas, Jovian auroral signature of Io's corotational wake. J. Geophys. Res. **107**, 1464 (2002). doi:10.1029/2002JA009514

M.E. Hill, D.K. Haggerty, R.L. McNutt, C.P. Paranicas, Energetic particle evidence for magnetic filaments in Jupiter's magnetotail. J. Geophys. Res. **114** (2009). doi:10.1029/2009JA014374

Y. Hiraki, C. Tao, Parameterization of ionization rate by auroral electron precipitation in Jupiter. Ann. Geophys. **26**, 77–86 (2008). doi:10.5194/angeo-26-77-2008

M. Horanyi, T.E. Cravens, J.H. Waite Jr., The precipitation of energetic heavy ions into the upper atmosphere of Jupiter. J. Geophys. Res. **93**, 7251–7271 (1988). doi:10.1029/JA093iA07p07251

T.S. Huang, T.W. Hill, Corotation lag of the Jovian atmosphere, ionosphere, and magnetosphere. J. Geophys. Res. **94**, 3761–3765 (1989). doi:10.1029/JA094iA04p03761

D.E. Huddleston, C.T. Russell, G. Le, A. Szabo, Magnetopause structure and the role of reconnection at the outer planets. J. Geophys. Res. **102**(A11), 24289–24302 (1997)

D.L. Huestis, Hydrogen collisions in planetary atmospheres, ionospheres, and magnetospheres. Planet. Space Sci. **56**, 1733–1743 (2008). doi:10.1016/j.pss.2008.07.012

Y. Hui, D.R. Schultz, V.A. Kharchenko, P.C. Stancil, T.E. Cravens, C.M. Lisse, A. Dalgarno, The Ion-induced charge-exchange X-ray emission of the Jovian auroras: Magnetospheric or solar wind origin? Astrophys. J. **702**, 158–162 (2009). doi:10.1088/0004-637X/702/2/L158

Y. Hui, T.E. Cravens, N. Ozak, D.R. Schultz, What can be learned from the absence of auroral X-ray emission from Saturn? J. Geophys. Res. **115**, 10239 (2010a). doi:10.1029/2010JA015639

Y. Hui, D.R. Schultz, V.A. Kharchenko, A. Bhardwaj, G. Branduardi-Raymont, P.C. Stancil, T.E. Cravens, C.M. Lisse, A. Dalgarno, Comparative analysis and variability of the Jovian X-ray spectra detected by the Chandra and XMM-Newton observatories. J. Geophys. Res. **115**, 7102 (2010b). doi:10.1029/2009JA014854

K. Imai, L. Wang, T.D. Carr, Modeling Jupiter's decametric modulation lanes. J. Geophys. Res. **102**, 7127–7136 (1997). doi:10.1029/96JA03960

A.P. Ingersoll, A.R. Vasavada, B. Little, C.D. Anger, S.J. Bolton, C. Alexander, K.P. Klaasen, W.K. Tobiska, Imaging Jupiter's aurora at visible wavelengths. Icarus **135**, 251–264 (1998). doi:10.1006/icar.1998.5971

M. Ishimoto, M.R. Torr, Energetic He(+) precipitation in a mid-latitude aurora. J. Geophys. Res. **92**, 3284–3292 (1987). doi:10.1029/JA092iA04p03284

C.M. Jackman, L. Lamy, M.P. Freeman, P. Zarka, B. Cecconi, W.S. Kurth, S.W.H. Cowley, M.K. Dougherty, On the character and distribution of lower-frequency radio emissions at Saturn and their relationship to substorm-like events. J. Geophys. Res. **114**, 8211 (2009). doi:10.1029/2008JA013997

C.M. Jackman, C.S. Arridge, J.A. Slavin, S.E. Milan, L. Lamy, M.K. Dougherty, A.J. Coates, In situ observations of the effect of a solar wind compression on Saturn's magnetotail. J. Geophys. Res. **115**, 10240 (2010). doi:10.1029/2010JA015312

C.M. Jackman, J.A. Slavin, S.W.H. Cowley, Cassini observations of plasmoid structure and dynamics: Implications for the role of magnetic reconnection in magnetospheric circulation at Saturn. J. Geophys. Res. **116** (2011). doi:10.1029/2011JA016682

C.M. Jackman, N. Achilleos, S.W.H. Cowley, E.J. Bunce, A. Radioti, D. Grodent, S.V. Badman, M.K. Dougherty, W. Pryor, Auroral counterpart of magnetic field dipolarizations in Saturn's tail. Planet. Space Sci. **82**, 34–42 (2013)

S.P. Joy, M.G. Kivelson, R.J. Walker, K.K. Khurana, C.T. Russell, T. Ogino, Probabilistic models of the Jovian magnetopause and bow shock locations. J. Geophys. Res. **107**, 1309 (2002). doi:10.1029/2001JA009146

S. Jurac, J.D. Richardson, A self-consistent model of plasma and neutrals at Saturn: Neutral cloud morphology. J. Geophys. Res. **110**, 9220 (2005). doi:10.1029/2004JA010635

M.L. Kaiser, M.D. Desch, J.W. Warwick, J.B. Pearce, Voyager detection of nonthermal radio emission from Saturn. Science **209**, 1238–1240 (1980). doi:10.1126/science.209.4462.1238

M.L. Kaiser, M.D. Desch, A. Lecacheux, Saturnian kilometric radiation—Statistical properties and beam geometry. Nature **292**, 731–733 (1981). doi:10.1038/292731a0

M.L. Kaiser, M.D. Desch, J.E.P. Connerney, Saturn's ionosphere—Inferred electron densities. J. Geophys. Res. **89**, 2371–2376 (1984a). doi:10.1029/JA089iA04p02371

M.L. Kaiser, M.D. Desch, W.S. Kurth, A. Lecacheux, F. Genova, B.M. Pedersen, D.R. Evans, Saturn as a radio source, in *Saturn*, ed. by T. Gehrels, M.S. Matthews, (1984b), pp. 378–415

S. Kasahara, E.A. Kronberg, N. Krupp, T. Kimura, C. Tao, S.V. Badman, A. Retinò, M. Fujimoto, Magnetic reconnection in the Jovian tail: X-line evolution and consequent plasma sheet structures. J. Geophys. Res. **116**, 11219 (2011). doi:10.1029/2011JA016892

V. Kharchenko, W. Liu, A. Dalgarno, X ray and EUV emission spectra of oxygen ions precipitating into the Jovian atmosphere. J. Geophys. Res. **103**, 26687–26698 (1998). doi:10.1029/98JA02395

V. Kharchenko, A. Dalgarno, D.R. Schultz, P.C. Stancil, Ion emission spectra in the Jovian X-ray aurora. Geophys. Res. Lett. **33**, 11105 (2006). doi:10.1029/2006GL026039

V. Kharchenko, A. Bhardwaj, A. Dalgarno, D.R. Schultz, P.C. Stancil, Modeling spectra of the north and south Jovian X-ray auroras. J. Geophys. Res. **113**, 8229 (2008). doi:10.1029/2008JA013062

K.K. Khurana, M.G. Kivelson, V.M. Vasyliunas, N. Krupp, J. Woch, A. Lagg, B.H. Mauk, W.S. Kurth, The configuration of Jupiter's magnetosphere, in *Jupiter. The Planet, Satellites and Magnetosphere*, ed. by F. Bagenal, T.E. Dowling, W.B. McKinnon, (2004), pp. 593–616

Y.H. Kim, J.L. Fox, The Jovian ionospheric E region. Geophys. Res. Lett. **18**, 123–126 (1991). doi:10.1029/90GL02587

Y.H. Kim, J.L. Fox, The chemistry of hydrocarbon ions in the Jovian ionosphere. Icarus **112**, 310–325 (1994). doi:10.1006/icar.1994.1186

Y.H. Kim, J.L. Fox, H.S. Porter, Densities and vibrational distribution of H(3+) in the Jovian auroral ionosphere. J. Geophys. Res. **97**, 6093–6101 (1992). doi:10.1029/92JE00454

Y.H. Kim, W.D. Pesnell, J.M. Grebowsky, J.L. Fox, Meteoric ions in the ionosphere of Jupiter. Icarus **150**, 261–278 (2001). doi:10.1006/icar.2001.6590

T. Kimura, F. Tsuchiya, H. Misawa, A. Morioka, H. Nozawa, M. Fujimoto, Periodicity analysis of Jovian quasi-periodic radio bursts based on Lomb-Scargle periodograms. J. Geophys. Res. **116**, 3204 (2011). doi:10.1029/2010JA016076

T. Kimura, L. Lamy, C. Tao, S.V. Badman, S. Kasahara, B. Cecconi, P. Zarka, A. Morioka, Y. Miyoshi, D. Maruno, Y. Kasaba, M. Fujimoto, Long-term modulations of Saturn's auroral radio emissions by the solar wind and seasonal variations controlled by the solar ultraviolet flux. J. Geophys. Res. **118**(11), 7019–7035 (2013). doi:10.1002/2013JA018833

M.G. Kivelson, Moon-magnetosphere interactions: A tutorial. Adv. Space Res. **33**, 2061 (2004). doi:10.1016/j.asr.2003.08.042

A.J. Kliore, I.R. Patel, G.F. Lindal, D.N. Sweetnam, H.B. Hotz, J.H. Waite, T. McDonough, Structure of the ionosphere and atmosphere of Saturn from Pioneer 11 Saturn radio occultation. J. Geophys. Res. **85**, 5857–5870 (1980). doi:10.1029/JA085iA11p05857

A.J. Kliore, A.F. Nagy, E.A. Marouf, A. Anabtawi, E. Barbinis, D.U. Fleischman, D.S. Kahan, Midlatitude and high-latitude electron density profiles in the ionosphere of Saturn obtained by Cassini radio occultation observations. J. Geophys. Res. **114**, 4315 (2009). doi:10.1029/2008JA013900

S. Knight, Parallel electric fields. Planet. Space Sci. **21**, 741–750 (1973). doi:10.1016/0032-0633(73)90093-7

P.S. Krstić, Inelastic processes from vibrationally excited states in slow $H^+ + H2$ and $H + H2^+$ collisions: Excitations and charge transfer. Phys. Rev. A **66**, 042717 (2002). doi:10.1103/PhysRevA.66.042717

N. Krupp, J. Woch, A. Lagg, B. Wilken, S. Livi, D.J. Williams, Energetic particle bursts in the predawn Jovian magnetotail. Geophys. Res. Lett. **25**, 1249–1252 (1998). doi:10.1029/98GL00863

N. Krupp, A. Lagg, S. Livi, B. Wilken, J. Woch, E.C. Roelof, D.J. Williams, Global flows of energetic ions in Jupiter's equatorial plane: First-order approximation. J. Geophys. Res. **106**, 26017–26032 (2001). doi:10.1029/2000JA900138

W.S. Kurth, D.A. Gurnett, J.T. Clarke, P. Zarka, M.D. Desch, M.L. Kaiser, B. Cecconi, A. Lecacheux, W.M. Farrell, P. Galopeau, J.-C. Gérard, D. Grodent, R. Prangé, M.K. Dougherty, F.J. Crary, An Earth-like correspondence between Saturn's auroral features and radio emission. Nature **433**, 722–725 (2005). doi:10.1038/nature03334

W.S. Kurth, E.J. Bunce, J.T. Clarke, F.J. Crary, D.C. Grodent, A.P. Ingersoll, U.A. Dyudina, L. Lamy, D.G. Mitchell, A.M. Persoon, W.R. Pryor, J. Saur, T. Stallard, Auroral processes, in *Saturn from Cassini-Huygens*, ed. by M.K. Dougherty, L.W. Esposito, S.M. Krimigis (Springer, Dordrecht Heidelberg London New York, 2009)

H.P. Ladreiter, P. Zarka, A. Lecacheux, Direction finding study of Jovian hectometric and broadband kilometric radio emissions: Evidence for their auroral origin. Planet. Space Sci. **42**, 919–931 (1994). doi:10.1016/0032-0633(94)90052-3

H.R. Lai, H.Y. Wei, C.T. Russell, C.S. Arridge, M.K. Dougherty, Reconnection at the magnetopause of Saturn: Perspective from FTE occurrence and magnetosphere size. J. Geophys. Res. **117** (2012). doi:10.1029/2011JA017263

H.A. Lam, N. Achilleos, S. Miller, J. Tennyson, L.M. Trafton, T.R. Geballe, G. Ballester, A baseline spectroscopic study of the infrared auroras of Jupiter. Icarus **127** (1997). doi:10.1006/icar.1997.5698

L. Lamy, Variability of southern and northern periodicities of Saturn kilometric radiation, in *Planetary Radio Emissions*, ed. by H.O. Rucker (Austrian Acad. Sci. Press, Vienna, 2011), pp. 39–50. doi:10.1553/PRE7s39

L. Lamy, P. Zarka, B. Cecconi, S. Hess, R. Prangé, Modeling of Saturn kilometric radiation arcs and equatorial shadow zone. J. Geophys. Res. **113**, 10213 (2008a). doi:10.1029/2008JA013464

L. Lamy, P. Zarka, B. Cecconi, R. Prangé, W.S. Kurth, D.A. Gurnett, Saturn kilometric radiation: Average and statistical properties. J. Geophys. Res. **113**, 7201 (2008b). doi:10.1029/2007JA012900

L. Lamy, B. Cecconi, R. Prangé, P. Zarka, J.D. Nichols, J.T. Clarke, An auroral oval at the footprint of Saturn's kilometric radio sources, colocated with the UV aurorae. J. Geophys. Res. **114**, 10212 (2009). doi:10.1029/2009JA014401

L. Lamy, P. Schippers, P. Zarka, B. Cecconi, C.S. Arridge, M.K. Dougherty, P. Louarn, N. André, W.S. Kurth, R.L. Mutel, D.A. Gurnett, A.J. Coates, Properties of Saturn kilometric radiation measured within its source region. Geophys. Res. Lett. **37**, 12104 (2010). doi:10.1029/2010GL043415

L. Lamy, B. Cecconi, P. Zarka, P. Canu, P. Schippers, W.S. Kurth, R.L. Mutel, D.A. Gurnett, D. Menietti, P. Louarn, Emission and propagation of Saturn kilometric radiation: Magnetoionic modes, beaming pattern, and polarization state. J. Geophys. Res. **116**, 4212 (2011). doi:10.1029/2010JA016195

L. Lamy, R. Prangé, K.C. Hansen, J.T. Clarke, P. Zarka, B. Cecconi, J. Aboudarham, N. André, G. Branduardi-Raymont, R. Gladstone, M. Barthélémy, N. Achilleos, P. Guio, M.K. Dougherty, H. Melin, S.W.H. Cowley, T.S. Stallard, J.D. Nichols, G. Ballester, Earth-based detection of Uranus' aurorae. Geophys. Res. Lett. **39**, 7105 (2012). doi:10.1029/2012GL051312

L. Lamy, R. Prangé, W. Pryor, J. Gustin, S.V. Badman, H. Melin, T. Stallard, D.G. Mitchell, P.C. Brandt, Multi-spectral simultaneous diagnosis of Saturn's aurorae throughout a planetary rotation. J. Geophys. Res. **118**, 1–27 (2013). doi:10.1002/jgra.50404

L.J. Lanzerotti, T.P. Armstrong, R.E. Gold, K.A. Anderson, S.M. Krimigis, R.P. Lin, M. Pick, E.C. Roelof, E.T. Sarris, G.M. Simnett, The hot plasma environment at Jupiter—ULYSSES results. Science **257**, 1518–1524 (1992). doi:10.1126/science.257.5076.1518

G.F. Lindal, D.N. Sweetnam, V.R. Eshleman, The atmosphere of Saturn—An analysis of the Voyager radio occultation measurements. Astron. J. **90**, 1136–1146 (1985)

T.A. Livengood, H.W. Moos, Jupiter's north and south polar aurorae with IUE data. Geophys. Res. Lett. **17**, 2265–2268 (1990). doi:10.1029/GL017i012p02265

D.A. Lorentzen, Latitudinal and longitudinal dispersion of energetic auroral protons. Ann. Geophys. **18**, 81–89 (2000). doi:10.1007/s00585-000-0081-3

P. Louarn, D. Le Quéau, Generation of the auroral kilometric radiation in plasma cavities—II. The cyclotron maser instability in small size sources. Planet. Space Sci. **44**, 211–224 (1996). doi:10.1016/0032-0633(95)00122-0

P. Louarn, A. Roux, S. Perraut, W. Kurth, D. Gurnett, A study of the large-scale dynamics of the Jovian magnetosphere using the Galileo plasma wave experiment. Geophys. Res. Lett. **25**, 2905–2908 (1998). doi:10.1029/98GL01774

R.J. MacDowall, M.L. Kaiser, M.D. Desch, W.M. Farrell, R.A. Hess, R.G. Stone, Quasiperiodic Jovian radio bursts: Observations from the ulysses radio and plasma wave experiment. Planet. Space Sci. **41**, 1059–1072 (1993). doi:10.1016/0032-0633(93)90109-F

T. Majeed, J.C. McConnell, The upper ionospheres of Jupiter and Saturn. Planet. Space Sci. **39**, 1715–1732 (1991). doi:10.1016/0032-0633(91)90031-5

A. Masters, J.P. Eastwood, M. Swisdak, M.F. Thomsen, C.T. Russell, N. Sergis, F.J. Crary, M.K. Dougherty, A.J. Coates, S.M. Krimigis, The importance of plasma β conditions for magnetic reconnection at Saturn's magnetopause. Geophys. Res. Lett. **39** (2012). doi:10.1029/2012GL051372

K.I. Matcheva, D.J. Barrow, Small-scale variability in Saturn's lower ionosphere. Icarus **221**, 525–543 (2012). doi:10.1016/j.icarus.2012.08.022

B.H. Mauk, J.T. Clarke, D. Grodent, J.H. Waite, C.P. Paranicas, D.J. Williams, Transient aurora on Jupiter from injections of magnetospheric electrons. Nature **415**, 1003–1005 (2002)

B.H. Mauk, D.C. Hamilton, T.W. Hill, G.B. Hospodarsky, R.E. Johnson, C. Paranicas, E. Roussos, C.T. Russell, D.E. Shemansky, E.C. Sittler, R.M. Thorne, Fundamental plasma processes in Saturn's magnetosphere, in *Saturn from Cassini-Huygens*, ed. by M.K. Dougherty, L.W. Esposito, S.M. Krimigis, (2009). Chap. Fundamental plasma processes in Saturn's magnetosphere. doi:10.1007/978-1-4020-9217-6

H.J. McAndrews, C.J. Owen, M.F. Thomsen, B. Lavraud, A.J. Coates, M.K. Dougherty, D.T. Young, Evidence for reconnection at Saturn's magnetopause. J. Geophys. Res. **113**(A4) (2008). doi:10.1029/2007JA012581

D.J. McComas, F. Bagenal, Jupiter: A fundamentally different magnetospheric interaction with the solar wind. Geophys. Res. Lett. **34** (2007). doi:10.1029/2007GL031078

J.C. McConnell, J.B. Holberg, G.R. Smith, B.R. Sandel, D.E. Shemansky, A.L. Broadfoot, A new look at the ionosphere of Jupiter in light of the UVS occultation results. Planet. Space Sci. **30**, 151–167 (1982). doi:10.1016/0032-0633(82)90086-1

M.B. McElroy, The ionospheres of the Major planets. Space Sci. Rev. **14**, 460–473 (1973). doi:10.1007/BF00214756

H. Melin, S. Miller, T. Stallard, D. Grodent, Non-LTE effects on H_3^+ emission in the Jovian upper atmosphere. Icarus **178**, 97–103 (2005). doi:10.1016/j.icarus.2005.04.016

H. Melin, S. Miller, T. Stallard, L.M. Trafton, T.R. Geballe, Variability in the H_3^+ emission of Saturn: Consequences for ionisation rates and temperature. Icarus **186**(1), 234–241 (2007). doi:10.1016/j.icarus.2006.08.014

H. Melin, T. Stallard, S. Miller, J. Gustin G. M, S.V. Badman, W.R. Pryor, J. O'Donoghue, R.H. Brown, K.H. Baines, Simultaneous Cassini VIMS and UVIS observations of Saturn's southern aurora: Comparing

⚛ Springer

emissions from H, H_2 and H_3^+ at a high spatial resolution. Geophys. Res. Lett. **38** (2011). doi:10.1029/2011GL048457

H. Menager, M. Barthélemy, J. Lilensten, H. Lyman, α line in Jovian aurorae: Electron transport and radiative transfer coupled modelling. Astron. Astrophys. **509**, 56 (2010). doi:10.1051/0004-6361/200912952

J.D. Menietti, D.A. Gurnett, G.B. Hospodarsky, C.A. Higgins, W.S. Kurth, P. Zarka, Modeling radio emission attenuation lanes observed by the Galileo and Cassini spacecraft. Planet. Space Sci. **51**, 533–540 (2003). doi:10.1016/S0032-0633(03)00078-3

J.D. Menietti, R.L. Mutel, P. Schippers, S.-Y. Ye, D.A. Gurnett, L. Lamy, Analysis of Saturn kilometric radiation near a source center. J. Geophys. Res. **116**, 12222 (2011). doi:10.1029/2011JA017056

C.J. Meredith, S.W.H. Cowley, K.C. Hansen, J.D. Nichols, T.K. Yeoman, Simultaneous conjugate observations of small-scale structures in Saturn's dayside ultraviolet auroras—Implications for physical origins. J. Geophys. Res. **118**(5), 2244–2266 (2013). doi:10.1002/jgra.50270

A.E. Metzger, D.A. Gilman, J.L. Luthey, K.C. Hurley, H.W. Schnopper, F.D. Seward, J.D. Sullivan, The detection of X rays from Jupiter. J. Geophys. Res. **88**, 7731–7741 (1983). doi:10.1029/JA088iA10p07731

S. Miller, R.D. Joseph, J. Tennyson, Infrared emissions of H_3^+ in the atmosphere of Jupiter in the 2.1 and 4.0 micron region. Astrophys. J. Lett. **360**, 55–58 (1990). doi:10.1086/185811

S. Miller, A. Aylward, G. Millward, Giant planet ionospheres and thermospheres: The importance of ion-neutral coupling. Space Sci. Rev. **116**, 319–343 (2005). doi:10.1007/s11214-005-1960-4

S. Miller, T. Stallard, H. Melin, J. Tennyson, H_3^+ cooling in planetary atmospheres. Faraday Discuss. **147**, 283 (2010). doi:10.1039/c004152c

G. Millward, S. Miller, T. Stallard, A.D. Aylward, N. Achilleos, On the dynamics of the Jovian ionosphere and Thermosphere. III. The modelling of auroral conductivity. Icarus **160**, 95–107 (2002). doi:10.1006/icar.2002.6951

G. Millward, S. Miller, T. Stallard, N. Achilleos, A.D. Aylward, On the dynamics of the Jovian ionosphere and thermosphere. Icarus **173**, 200–211 (2005). doi:10.1016/j.icarus.2004.07.027

D.G. Mitchell, J.F. Carbary, S.W.H. Cowley, T.W. Hill, P. Zarka, The dynamics of Saturn's magnetosphere, in *Saturn from Cassini-Huygens*, ed. by M.K. Dougherty, L.W. Esposito, S.M. Krimigis (Springer, Dordrecht Heidelberg London New York, 2009a). Chap. The dynamics of Saturn's magnetosphere. doi:10.1007/978-1-4020-9217-6

D.G. Mitchell, S.M. Krimigis, C. Paranicas, P.C. Brandt, J.F. Carbary, E.C. Roelof, W.S. Kurth, D.A. Gurnett, J.T. Clarke, J.D. Nichols, J.-C. Gérard, D.C. Grodent, M.K. Dougherty, W.R. Pryor, Recurrent energization of plasma in the midnight-to-dawn quadrant of Saturn's magnetosphere, and its relationship to auroral UV and radio emissions. Planet. Space Sci. **57**, 1732–1742 (2009b). doi:10.1016/j.pss.2009.04.002

L.E. Moore, M. Mendillo, Are plasma depletions in Saturn's ionosphere a signature of time-dependent water input? Geophys. Res. Lett. **34**(12) (2007). doi:10.1029/2007GL029381

L.E. Moore, M. Mendillo, I.C.F. Müller-Wodarg, D.L. Murr, Modeling of global variations and ring shadowing in Saturn's ionosphere. Icarus **172**, 503–520 (2004). doi:10.1016/j.icarus.2004.07.007

L. Moore, A.F. Nagy, A.J. Kliore, I. Müller-Wodarg, J.D. Richardson, M. Mendillo, Cassini radio occultations of Saturn's ionosphere: Model comparisons using a constant water flux. Geophys. Res. Lett. **33**, 22202 (2006). doi:10.1029/2006GL027375

L. Moore, M. Galand, I. Müller-Wodarg, R. Yelle, M. Mendillo, Plasma temperatures in Saturn's ionosphere. J. Geophys. Res. **113**, 10306 (2008). doi:10.1029/2008JA013373

L. Moore, I. Müller-Wodarg, M. Galand, A. Kliore, M. Mendillo, Latitudinal variations in Saturn's ionosphere: Cassini measurements and model comparisons. J. Geophys. Res. **115**, 11317 (2010). doi:10.1029/2010JA015692

L. Moore, G. Fischer, I. Müller-Wodarg, M. Galand, M. Mendillo, Diurnal variation of electron density in Saturn's ionosphere: Model comparisons with Saturn electrostatic discharge (SED) observations. Icarus **221**, 508–516 (2012). doi:10.1016/j.icarus.2012.08.010

J.I. Moses, S.F. Bass, The effects of external material on the chemistry and structure of Saturn's ionosphere. J. Geophys. Res. **105**, 7013–7052 (2000). doi:10.1029/1999JE001172

F. Mottez, S. Hess, P. Zarka, Explanation of dominant oblique radio emission at Jupiter and comparison to the terrestrial case. Planet. Space Sci. **58**, 1414–1422 (2010). doi:10.1016/j.pss.2010.05.012

I.C.F. Müller-Wodarg, M. Mendillo, R.V. Yelle, A.D. Aylward, A global circulation model of Saturn's thermosphere. Icarus **180**, 147–160 (2006). doi:10.1016/j.icarus.2005.09.002

I.C.F. Müller-Wodarg, L. Moore G. M, M. Mendillo, Magnetosphere–atmosphere coupling at Saturn: 1—Response of thermosphere and ionosphere to steady state polar forcing. Icarus **221**(2) (2012). doi:10.1016/j.icarus.2012.08.034

R.L. Mutel, J.D. Menietti, D.A. Gurnett, W. Kurth, P. Schippers, C. Lynch, L. Lamy, C. Arridge, B. Cecconi, CMI growth rates for Saturnian kilometric radiation. Geophys. Res. Lett. **37**, 19105 (2010). doi:10.1029/2010GL044940

A.F. Nagy, A.J. Kliore, E. Marouf, R. French, M. Flasar, N.J. Rappaport, A. Anabtawi, S.W. Asmar, D. Johnston, E. Barbinis, G. Goltz, D. Fleischman, First results from the ionospheric radio occultations of Saturn by the Cassini spacecraft. J. Geophys. Res. **111**, 6310 (2006). doi:10.1029/2005JA011519

A.F. Nagy, A.J. Kliore, M. Mendillo, S. Miller, L. Moore, J.I. Moses, I. Müller-Wodarg, D.E. Shemansky, Upper atmosphere and ionosphere of Saturn, in *Saturn from Cassini-Huygens*, ed. by M.K. Dougherty, L.W. Esposito, S.M. Krimigis, (2009)

N.F. Ness, M.H. Acuna, R.P. Lepping, J.E.P. Connerney, K.W. Behannon, L.F. Burlaga, F.M. Neubauer, Magnetic field studies by Voyager 1—Preliminary results at Saturn. Science **212**, 211–217 (1981). doi:10.1126/science.212.4491.211

F.M. Neubauer, Nonlinear standing Alfven wave current system at Io—Theory. J. Geophys. Res. **85**, 1171–1178 (1980). doi:10.1029/JA085iA03p01171

J.D. Nichols, Magnetosphere-ionosphere coupling at Jupiter-like exoplanets with internal plasma sources: Implications for detectability of auroral radio emissions. Mon. Not. R. Astron. Soc. **414**, 2125–2138 (2011). doi:10.1111/j.1365-2966.2011.18528.x

J.D. Nichols, S.W.H. Cowley, Magnetosphere-ionosphere coupling currents in Jupiter's middle magnetosphere: Dependence on the effective ionospheric Pedersen conductivity and iogenic plasma mass outflow rate. Ann. Geophys. **21**, 1419–1441 (2003). doi:10.5194/angeo-21-1419-2003

J. Nichols, S. Cowley, Magnetosphere-ionosphere coupling currents in Jupiter's middle magnetosphere: Effect of precipitation-induced enhancement of the ionospheric Pedersen conductivity. Ann. Geophys. **22**, 1799–1827 (2004). doi:10.5194/angeo-22-1799-2004

J.D. Nichols, J.T. Clarke, J.C. Gérard, D. Grodent, Observations of Jovian polar auroral filaments. Geophys. Res. Lett. **36** (2009a). doi:10.1029/2009GL037578

J.D. Nichols, J.T. Clarke, J.C. Gérard, D. Grodent, K.C. Hansen, Variation of different components of Jupiter's auroral emission. J. Geophys. Res. **114**, 6210 (2009b). doi:10.1029/2009JA014051

J.D. Nichols, B. Cecconi, J.T. Clarke, S.W.H. Cowley, J.-C. Gérard, A. Grocott, D. Grodent, L. Lamy, P. Zarka, Variation of Saturn's UV aurora with SKR phase. Geophys. Res. Lett. **37** (2010a). doi:10.1029/2010GL044057

J.D. Nichols, S.W.H. Cowley, L. Lamy, Dawn-dusk oscillation of Saturn's conjugate auroral ovals. Geophys. Res. Lett. **372**, 24102 (2010b). doi:10.1029/2010GL045818

J. O'Donoghue, T.S. Stallard, H. Melin, G.H. Jones, S.W.H. Cowley, S. Miller, K.H. Baines, J.S.D. Blake, The domination of Saturn's low-latitude ionosphere by ring 'rain'. Nature **496**(7444), 193–195 (2013). doi:10.1038/nature12049

J. O'Donoghue, T.S. Stallard, H. Melin, S.W.H. Cowley, S.V. Badman, L. Moore, S. Miller, C. Tao, K.H. Baines, J.S.D. Blake, Conjugate observations of Saturn's northern and southern H_3^+ aurorae. Icarus **229**, 214–220 (2014). doi:10.1016/j.icarus.2013.11.009

N. Ozak, D.R. Schultz, T.E. Cravens, V. Kharchenko, Y.-W. Hui, Auroral X-ray emission at Jupiter: Depth effects. J. Geophys. Res. **115**, 11306 (2010). doi:10.1029/2010JA015635

L. Pallier, R. Prangé, More about the structure of the high latitude Jovian aurorae. Planet. Space Sci. **49**, 1159–1173 (2001). doi:10.1016/S0032-0633(01)00023-X

L. Pallier, R. Prangé, Detection of the southern counterpart of the Jovian northern polar cusp: Shared properties. Geophys. Res. Lett. **31**, 6701 (2004). doi:10.1029/2003GL018041

M. Panchenko, H. Rucker, W. Farrell, Periodic bursts of Jovian non-Io decametric radio emission. Planet. Space Sci. **77**, 3–11 (2013)

J.D. Patterson, T.P. Armstrong, C.M. Laird, D.L. Detrick, A.T. Weatherwax, Correlation of solar energetic protons and polar cap absorption. J. Geophys. Res. **106**, 149–164 (2001). doi:10.1029/2000JA002006

J.J. Perry, Y.H. Kim, J.L. Fox, H.S. Porter, Chemistry of the Jovian auroral ionosphere. J. Geophys. Res. **104**, 16541–16566 (1999). doi:10.1029/1999JE900022

R. Prangé, D. Rego, J.-C. Gerard, Auroral Lyman alpha and H2 bands from the giant planets. 2: Effect of the anisotropy of the precipitating particles on the interpretation of the 'color ratio'. J. Geophys. Res. **100**, 7513–7521 (1995). doi:10.1029/94JE03176

R. Prangé, D. Rego, D. Southwood, P. Zarka, S. Miller, W. Ip, Rapid energy dissipation and variability of the Io-Jupiter electrodynamic circuit. Nature **379**, 323–325 (1996). doi:10.1038/379323a0

R. Prangé, L. Pallier, K.C. Hansen, R. Howard, A. Vourlidas, R. Courtin, C. Parkinson, An interplanetary shock traced by planetary auroral storms from the Sun to Saturn. Nature **432**, 78–81 (2004). doi:10.1038/nature02986

W.R. Pryor, A.M. Rymer, D.G. Mitchell, T.W. Hill, D.T. Young, J. Saur, G.H. Jones, S. Jacobsen, S.W.H. Cowley, B.H. Mauk, A.J. Coates, J. Gustin, D. Grodent, J.-C. Gérard, L. Lamy, J.D. Nichols, S.M. Krimigis, L.W. Esposito, M.K. Dougherty, A.J. Jouchoux, A.I.F. Stewart, W.E. McClintock, G.M. Holsclaw, J.M. Ajello, J.E. Colwell, A.R. Hendrix, F.J. Crary, J.T. Clarke, X. Zhou, The auroral footprint of Enceladus on Saturn. Nature **472**, 331–333 (2011). doi:10.1038/nature09928

A. Radioti, J.-C. Gérard, D. Grodent, B. Bonfond, N. Krupp, J. Woch, Discontinuity in Jupiter's main auroral oval. J. Geophys. Res. **113**, 1215 (2008a). doi:10.1029/2007JA012610

A. Radioti, D. Grodent, J.-C. Gérard, B. Bonfond, J.T. Clarke, Auroral polar dawn spots: Signatures of internally driven reconnection processes at Jupiter's magnetotail. Geophys. Res. Lett. **35**, 3104 (2008b). doi:10.1029/2007GL032460

A. Radioti, D. Grodent, J.-C. Gérard, E. Roussos, C. Paranicas, B. Bonfond, D.G. Mitchell, N. Krupp, S. Krimigis, J.T. Clarke, Transient auroral features at Saturn: Signatures of energetic particle injections in the magnetosphere. J. Geophys. Res. **114**, 3210 (2009). doi:10.1029/2008JA013632

A. Radioti, D. Grodent, J.-C. Gérard, S.E. Milan, B. Bonfond, J. Gustin, W.R. Pryor, Bifurcations of the main auroral ring at Saturn: Ionospheric signatures of consecutive reconnection events at the magnetopause. J. Geophys. Res. **116** (2011a). doi:10.1029/2011JA016661

A. Radioti, D. Grodent, J.-C. Gérard, M.F. Vogt, M. Lystrup, B. Bonfond, Nightside reconnection at Jupiter: Auroral and magnetic field observations from 26 July 1998. J. Geophys. Res. **116**, 3221 (2011b). doi:10.1029/2010JA016200

A. Radioti, M. Lystrup, B. Bonfond, J.-C. Gérard, Jupiter's aurora in ultraviolet and infrared: Simultaneous observations with the Hubble space telescope and the NASA infrared telescope facility. J. Geophys. Res. **118**(5), 2286–2295 (2013a). doi:10.1002/jgra.50245

A. Radioti, E. Roussos, D. Grodent, J.-C. Gérard, N. Krupp, D.G. Mitchell, J. Gustin, B. Bonfond, W. Pryor, Signatures of magnetospheric injections in Saturn's aurora. J. Geophys. Res. **118**, 1922–1933 (2013b). doi:10.1002/jgra.50161

L.C. Ray, S. Hess, Modelling the Io-related DAM emission by modifying the beaming angle. J. Geophys. Res. **113**, 11218 (2008). doi:10.1029/2008JA013669

L.C. Ray, R.E. Ergun, P.A. Delamere, F. Bagenal, Magnetosphere-ionosphere coupling at Jupiter: Effect of field-aligned potentials on angular momentum transport. J. Geophys. Res. **115**, 9211 (2010). doi:10.1029/2010JA015423

L.C. Ray, R.E. Ergun, P.A. Delamere, F. Bagenal, Magnetosphere-ionosphere coupling at Jupiter: A parameter space study. J. Geophys. Res. **117**, 1205 (2012a). doi:10.1029/2011JA016899

L.C. Ray, M. Galand, L.E. Moore, B.L. Fleshman, Characterizing the limitations to the coupling between Saturn's ionosphere and middle magnetosphere. J. Geophys. Res. **117**, 7210 (2012b). doi:10.1029/2012JA017735

J.C. Raymond, X-rays from charge transfer in astrophysics: Overview. Astron. Nachr. **333**, 290 (2012). doi:10.1002/asna.201211677

E. Raynaud, E. Lellouch, J.-P. Maillard, G.R. Gladstone, J.H. Waite, B. Bézard, P. Drossart, T. Fouchet, Spectro-imaging observations of Jupiter's 2-μm auroral emission. I. H_3^+ distribution and temperature. Icarus **171**, 133–152 (2004). doi:10.1016/j.icarus.2004.04.020

M.H. Rees, *Physics and Chemistry of the Upper Atmosphere* (Cambridge University Press, Cambridge, 1989)

D. Rego, R. Prange, J.-C. Gerard, Auroral Lyman α and H_2 bands from the giant planets: 1. Excitation by proton precipitation in the Jovian atmosphere. J. Geophys. Res. **99**, 17075–17094 (1994). doi:10.1029/93JE03432

D. Rego, R. Prangé, L. Ben Jaffel, Auroral Lyman α and H_2 bands from the giant planets 3. Lyman α spectral profile including charge exchange and radiative transfer effects and H_2 color ratios. J. Geophys. Res. **104**, 5939–5954 (1999). doi:10.1029/1998JE900048

D. Rego, S. Miller, N. Achilleos, R. Prangé, R.D. Joseph, Latitudinal profiles of the Jovian IR emissions of H_3^+ at 4 μm with the NASA infrared telescope facility: Energy inputs and thermal balance. Icarus **147**, 366–385 (2000). doi:10.1006/icar.2000.6444

M.J. Reiner, J. Fainberg, R.G. Stone, Source characteristics of Jovian hectometric radio emissions. J. Geophys. Res. **98**, 18767–18777 (1993a). doi:10.1029/93JE01779

M.J. Reiner, J. Fainberg, R.G. Stone, M.L. Kaiser, M.D. Desch, R. Manning, P. Zarka, B.-M. Pedersen, Source characteristics of Jovian narrow-band kilometric radio emissions. J. Geophys. Res. **98**, 13163 (1993b). doi:10.1029/93JE00536

A. Roux, A. Hilgers, H. de Féraudy, D. Le Quéau, P. Louarn, S. Perraut, A. Bahnsen, M. Jespersen, E. Ungstrup, M. André, Auroral kilometric radiation sources—In situ and remote observations from Viking. J. Geophys. Res. **98**, 11657 (1993). doi:10.1029/92JA02309

H.O. Rucker, M. Panchenko, K.C. Hansen, U. Taubenschuss, M.Y. Boudjada, W.S. Kurth, M.K. Dougherty, J.T. Steinberg, P. Zarka, P.H.M. Galopeau, D.J. McComas, C.H. Barrow, Saturn kilometric radiation as a monitor for the solar wind? Adv. Space Res. **42**, 40–47 (2008). doi:10.1016/j.asr.2008.02.008

B.R. Sandel, D.E. Shemansky, A.L. Broadfoot, J.B. Holberg, G.R. Smith, J.C. McConnell, D.F. Strobel, S.K. Atreya, T.M. Donahue, H.W. Moos, D.M. Hunten, R.B. Pomphrey, S. Linick, Extreme ultraviolet observations from the Voyager 2 encounter with Saturn. Science **215**, 548–553 (1982). doi:10.1126/science.215.4532.548

T. Satoh, J.E.P. Connerney, Jupiter's H_3^+ emissions viewed in corrected Jovimagnetic coordinates. Icarus **141**, 236–252 (1999). doi:10.1006/icar.1999.6173

P. Schippers, C.S. Arridge, J.D. Menietti, D.A. Gurnett, L. Lamy, B. Cecconi, D.G. Mitchell, N. André, W.S. Kurth, S. Grimald, M.K. Dougherty, A.J. Coates, N. Krupp, D.T. Young, Auroral electron distributions within and close to the Saturn kilometric radiation source region. J. Geophys. Res. **116**, 05203 (2011). doi:10.1029/2011JA016461

L. Scurry, C.T. Russell, Proxy studies of energy transfer to the magnetosphere. J. Geophys. Res. **96**, 9541–9548 (1991). doi:10.1029/91JA00569

J.A. Simpson, J.D. Anglin, A. Balogh, J.R. Burrows, S.W.H. Cowley, P. Ferrando, B. Heber, R.J. Hynds, H. Kunow, R.G. Marsden, Energetic charged-particle phenomena in the Jovian magnetosphere—First results from the ULYSSES COSPIN collaboration. Science **257**, 1543–1550 (1992). doi:10.1126/science.257.5076.1543

R.P. Singhal, S.C. Chakravarty, A. Bhardwaj, B. Prasad, Energetic electron precipitation in Jupiter's upper atmosphere. J. Geophys. Res. **97**, 18245 (1992). doi:10.1029/92JE01894

E.C. Sittler, N. Andre, M. Blanc, M. Burger, R.E. Johnson, A. Coates, A. Rymer, D. Reisenfeld, M.F. Thomsen, A. Persoon, M. Dougherty, H.T. Smith, R.A. Baragiola, R.E. Hartle, D. Chornay, M.D. Shappirio, D. Simpson, D.J. McComas, D.T. Young, Ion and neutral sources and sinks within Saturn's inner magnetosphere: Cassini results. Planet. Space Sci. **56**, 3–18 (2008). doi:10.1016/j.pss.2007.06.006

T.G. Slanger, T.E. Cravens, J. Crovisier, S. Miller, D.F. Strobel, Photoemission phenomena in the solar system. Space Sci. Rev. **139**, 267–310 (2008). doi:10.1007/s11214-008-9387-3

C.G.A. Smith, A.D. Aylward, Coupled rotational dynamics of Saturn's thermosphere and magnetosphere: A thermospheric modelling study. Ann. Geophys. **26**, 1007–1027 (2008). doi:10.5194/angeo-26-1007-2008

C.G.A. Smith, A.D. Aylward, Coupled rotational dynamics of Jupiter's thermosphere and magnetosphere. Ann. Geophys. **27**, 199–230 (2009). doi:10.5194/angeo-27-199-2009

E.J. Smith, R.W. Fillius, J.H. Wolfe, Compression of Jupiter's magnetosphere by the solar wind. J. Geophys. Res. **83**, 4733–4742 (1978). doi:10.1029/JA083iA10p04733

C.G.A. Smith, S. Miller, A.D. Aylward, Magnetospheric energy inputs into the upper atmospheres of the giant planets. Ann. Geophys. **23**, 1943–1947 (2005). doi:10.5194/angeo-23-1943-2005

C.G.A. Smith, A.D. Aylward, G.H. Millward, S. Miller, L.E. Moore, An unexpected cooling effect in Saturn's upper atmosphere. Nature **445**, 399–401 (2007a). doi:10.1038/nature05518

H.T. Smith, R.E. Johnson, E.C. Sittler, M. Shappirio, D. Reisenfeld, O.J. Tucker, M. Burger, F.J. Crary, D.J. McComas, D.T. Young, Enceladus: The likely dominant nitrogen source in Saturn's magnetosphere. Icarus **188**, 356–366 (2007b). doi:10.1016/j.icarus.2006.12.007

S.C. Solomon, Auroral electron transport using the Monte Carlo method. Geophys. Res. Lett. **20**, 185–188 (1993). doi:10.1029/93GL00081

S.C. Solomon, Auroral particle transport using Monte Carlo and hybrid methods. J. Geophys. Res. **106**, 107–116 (2001). doi:10.1029/2000JA002011

D.J. Southwood, M.G. Kivelson, The source of Saturn's periodic radio emission. J. Geophys. Res. **114**, 9201 (2009). doi:10.1029/2008JA013800

T. Stallard, S. Miller, G.E. Ballester, D. Rego, R.D. Joseph, L.M. Trafton, The H_3^+ latitudinal profile of Saturn. Astrophys. J. Lett. **521**, 149–152 (1999). doi:10.1086/312189

T. Stallard, S. Miller, G. Millward, R.D. Joseph, On the dynamics of the Jovian ionosphere and thermosphere. I. The measurement of ion winds. Icarus **154**, 475–491 (2001). doi:10.1006/icar.2001.6681

T. Stallard, S. Miller, G. Millward, R.D. Joseph, On the dynamics of the Jovian ionosphere and thermosphere. II. The measurement of H_3^+ vibrational temperature, column density, and total emission. Icarus **156**, 498–514 (2002). doi:10.1006/icar.2001.6793

T.S. Stallard, S. Miller, S.W.H. Cowley, E.J. Bunce, Jupiter's polar ionospheric flows: Measured intensity and velocity variations poleward of the main auroral oval. Geophys. Res. Lett. **30**, 1221 (2003). doi:10.1029/2002GL016031

T. Stallard, S. Miller, H. Melin, M. Lystrup, M.K. Dougherty, N. Achilleos, Saturn's auroral/polar H_3^+ infrared emission I. General morphology and ion velocity structure. Icarus **189**(1), 1–13 (2007a). doi:10.1016/j.icarus.2006.12.027

T. Stallard, C. Smith, S. Miller, H. Melin, M. Lystrup, A. Aylward, N. Achilleos, M.K. Dougherty, Saturn's auroral/polar H_3^+ infrared emission—II. A comparison with plasma flow models. Icarus **191**(2), 678–690 (2007b). doi:10.1016/j.icarus.2007.05.016

T. Stallard, S. Miller, M. Lystrup, N. Achilleos, E.J. Bunce, C.S. Arridge, M.K. Dougherty, S.W.H. Cowley, S.V. Badman, D.L. Talboys, R.H. Brown, K.H. Baines, B.J. Buratti, R.N. Clark, C. Sotin, P.D. Nicholson, P. Drossart, Complex structure within Saturn's infrared aurora. Nature **456**(7219), 214–217 (2008a). doi:10.1038/nature07440

T. Stallard, S. Miller, H. Melin, M. Lystrup, S.W.H. Cowley, E.J. Bunce, N. Achilleos, M. Dougherty, Jovian-like aurorae on Saturn. Nature **453**(7198), 1083–1085 (2008b). doi:10.1038/nature07077

T. Stallard, H. Melin, S.W.H. Cowley, S. Miller, M.B. Lystrup, Location and magnetospheric mapping of Saturn's mid-latitude infrared auroral oval. Astrophys. J. Lett. **722**, 85–89 (2010). doi:10.1088/2041-8205/722/1/L85

T.S. Stallard, A. Masters, S. Miller, H. Melin, E.J. Bunce, C.S. Arridge, N. Achilleos, M.K. Dougherty, S.W.H. Cowley, Saturn's auroral/polar H_3^+ infrared emission: The effect of solar wind compression. J. Geophys. Res. **117**, 12302 (2012a). doi:10.1029/2012JA018201

T.S. Stallard, H. Melin, S. Miller, J. O'Donoghue, S.W.H. Cowley, S.V. Badman, A. Adriani, R.H. Brown, K.H. Baines, Temperature changes and energy inputs in giant planet atmospheres: What we are learning from H_3^+. Philos. Trans. R. Soc. Lond. A **370**, 5213–5224 (2012b). doi:10.1098/rsta.2012.0028

T.S. Stallard, H. Melin, S. Miller, S.V. Badman, R.H. Brown, K.H. Baines, Peak emission altitude of Saturn's H_3^+ aurora. Geophys. Res. Lett. **39**(15), L15103 (2012c). doi:10.1029/2012GL052806

A.J. Steffl, P.A. Delamere, F. Bagenal, Cassini UVIS observations of the Io plasma torus. III. Observations of temporal and azimuthal variability. Icarus **180**, 124–140 (2006). doi:10.1016/j.icarus.2005.07.013

A.J. Steffl, P.A. Delamere, F. Bagenal, Cassini UVIS observations of the Io plasma torus. IV. Modeling temporal and azimuthal variability. Icarus **194**, 153–165 (2008). doi:10.1016/j.icarus.2007.09.019

R.G. Stone, B.M. Pedersen, C.C. Harvey, P. Canu, N. Cornilleau-Wehrlin, M.D. Desch, C. de Villedary, J. Fainberg, W.M. Farrell, K. Goetz, ULYSSES radio and plasma wave observations in the Jupiter environment. Science **257**, 1524–1531 (1992). doi:10.1126/science.257.5076.1524

C. Tao, H. Fujiwara, Y. Kasaba, Neutral wind control of the Jovian magnetosphere-ionosphere current system. J. Geophys. Res. **114**, 8307 (2009). doi:10.1029/2008JA013966

C. Tao, H. Fujiwara, Y. Kasaba, Jovian magnetosphere-ionosphere current system characterized by diurnal variation of ionospheric conductance. Planet. Space Sci. **58**, 351–364 (2010). doi:10.1016/j.pss.2009.10.005

C. Tao, S.V. Badman, M. Fujimoto, UV and IR auroral emission model for the outer planets: Jupiter and Saturn comparison. Icarus **213**, 581–592 (2011). doi:10.1016/j.icarus.2011.04.001

C. Tao, S.V. Badman, T. Uno, M. Fujimoto, On the feasibility of characterising Jovian auroral electrons via H_3^+ infrared line emission analysis. Icarus **221**, 236–247 (2012). doi:10.1016/j.icarus.2012.07.015

C. Tao, S.V. Badman, M. Fujimoto, Characteristic time scales of Uv and Ir auroral emissions at Jupiter and Saturn and their possible observable effects, in *Proc. of the 12th Symposium on Planetary Science* (TERRAPUB, Japan, 2013)

T. Terasawa, K. Maezawa, S. Machida, Solar wind effect on Jupiter's non-Io-related radio emission. Nature **273**, 131 (1978). doi:10.1038/273131a0

J.R. Thieman, M.L. Goldstein, Arcs in Saturn's radio spectra. Nature **292**, 728–731 (1981). doi:10.1038/292728a0

M.F. Thomsen, D.B. Reisenfeld, D.M. Delapp, R.L. Tokar, D.T. Young, F.J. Crary, E.C. Sittler, M.A. Mc-Graw, J.D. Williams, Survey of ion plasma parameters in Saturn's magnetosphere. J. Geophys. Res. **115**, 10220 (2010). doi:10.1029/2010JA015267

A.T. Tomás, J. Woch, N. Krupp, A. Lagg, K.-H. Glassmeier, W.S. Kurth, Energetic electrons in the inner part of the Jovian magnetosphere and their relation to auroral emissions. J. Geophys. Res. **109**, 6203 (2004). doi:10.1029/2004JA010405

J.T. Trauger, J.T. Clarke, G.E. Ballester, R.W. Evans, C.J. Burrows, D. Crisp, J.S. Gallagher, R.E. Griffiths, J.J. Hester, J.G. Hoessel, J.A. Holtzman, J.E. Krist, J.R. Mould, R. Sahai, P.A. Scowen, K.R. Stapelfeldt, A.M. Watson, Saturn's hydrogen aurora: Wide field and planetary camera 2 imaging from the Hubble space telescope. J. Geophys. Res. **103**(E9), 20237–20244 (1998). doi:10.1029/98JE01324

R.A. Treumann, Planetary radio emission mechanisms: A tutorial, in *Radio Astronomy at Long Wavelengths*, ed. by R.G. Stone, K.W. Weiler, M.L. Goldstein, J.-L. Bougeret. Washington DC American Geophysical Union Geophysical Monograph Series, vol. 119, (2000)

R.A. Treumann, The electron-cyclotron maser for astrophysical application. Astron. Astrophys. Rev. **13**, 229–315 (2006). doi:10.1007/s00159-006-0001-y

A.R. Vasavada, A.H. Bouchez, A.P. Ingersoll, B. Little, C.D. Anger (Galileo SSI Team), Jupiter's visible aurora and Io footprint. J. Geophys. Res. **104**, 27133–27142 (1999). doi:10.1029/1999JE001055

V.M. Vasyliunas, Plasma distribution and flow, in *Physics of the Jovian Magnetosphere*, ed. by A.J. Dessler (Cambridge University Press, Cambridge, 1983), pp. 395–453

M.F. Vogt, M.G. Kivelson, K.K. Khurana, S.P. Joy, R.J. Walker, Reconnection and flows in the Jovian magnetotail as inferred from magnetometer observations. J. Geophys. Res. **115** (2010). doi:10.1029/2009JA015098

M.F. Vogt, M.G. Kivelson, K.K. Khurana, R.J. Walker, B. Bonfond, D. Grodent, A. Radioti, Improved mapping of Jupiter's auroral features to magnetospheric sources. J. Geophys. Res. **116**, 3220 (2011). doi:10.1029/2010JA016148

J.H. Waite Jr., The ionosphere of Saturn. Ph.D. thesis, Michigan Univ., Ann Arbor (1981)

J.H. Waite Jr., J.T. Clarke, T.E. Cravens, C.M. Hammond, The Jovian aurora—Electron or ion precipitation? J. Geophys. Res. **93**, 7244–7250 (1988). doi:10.1029/JA093iA07p07244

J.H. Waite Jr., F. Bagenal, F. Seward, C. Na, G.R. Gladstone, T.E. Cravens, K.C. Hurley, J.T. Clarke, R. Elsner, S.A. Stern, ROSAT observations of the Jupiter aurora. J. Geophys. Res. **99**, 14799 (1994). doi:10.1029/94JA01005

J.H. Waite Jr., G.R. Gladstone, K. Franke, W.S. Lewis, A.C. Fabian, W.N. Brandt, C. Na, F. Haberl, J.T. Clarke, K.C. Hurley, M. Sommer, S. Bolton, ROSAT observations of X-ray emissions from Jupiter during the impact of comet Shoemaker-Levy 9. Science **268**, 1598–1601 (1995). doi:10.1126/science.268.5217.1598

J.H. Waite Jr., W.S. Lewis, G.R. Gladstone, T.E. Cravens, A.N. Maurellis, P. Drossart, J.E.P. Connerney, S. Miller, H.A. Lam, Outer planet ionospheres—A review of recent research and a look toward the future. Adv. Space Res. **20**, 243 (1997). doi:10.1016/S0273-1177(97)00542-5

J.H. Waite, T.E. Cravens, J. Kozyra, A.F. Nagy, S.K. Atreya, R.H. Chen, Electron precipitation and related aeronomy of the Jovian thermosphere and ionosphere. J. Geophys. Res. **88**, 6143–6163 (1983). doi:10.1029/JA088iA08p06143

J.H. Waite, G.R. Gladstone, W.S. Lewis, P. Drossart, T.E. Cravens, A.N. Maurellis, B.H. Mauk, S. Miller, Equatorial X-ray emissions: Implications for Jupiter's high exospheric temperatures. Science **276**, 104–108 (1997). doi:10.1126/science.276.5309.104

J.H. Waite, G.R. Gladstone, W.S. Lewis, R. Goldstein, D.J. McComas, P. Riley, R.J. Walker, P. Robertson, S. Desai, J.T. Clarke, D.T. Young, An auroral flare at Jupiter. Nature **410**, 787–789 (2001)

R.J. Walker, C.T. Russell, Flux transfer events at the Jovian magnetopause. J. Geophys. Res. **90**, 7397–7404 (1985). doi:10.1029/JA090iA08p07397

J.W. Warwick, J.B. Pearce, D.R. Evans, T.D. Carr, J.J. Schauble, J.K. Alexander, M.L. Kaiser, M.D. Desch, M. Pedersen, A. Lecacheux, G. Daigne, A. Boischot, C.H. Barrow, Planetary radio astronomy observations from Voyager 1 near Saturn. Science **212**, 239–243 (1981). doi:10.1126/science.212.4491.239

J.W. Warwick, D.R. Evans, J.H. Romig, J.K. Alexander, M.D. Desch, M.L. Kaiser, M.G. Aubier, Y. Leblanc, A. Lecacheux, B.M. Pedersen, Planetary radio astronomy observations from Voyager 2 near Saturn. Science **215**, 582–587 (1982). doi:10.1126/science.215.4532.582

R.J. Wilson, R.L. Tokar, M.G. Henderson, Thermal ion flow in Saturn's inner magnetosphere measured by the Cassini plasma spectrometer: A signature of the Enceladus torus? Geophys. Res. Lett. **36**, 23104 (2009). doi:10.1029/2009GL040225

J. Woch, N. Krupp, A. Lagg, B. Wilken, S. Livi, D.J. Williams, Quasi-periodic modulations of the Jovian magnetotail. Geophys. Res. Lett. **25**, 1253–1256 (1998). doi:10.1029/98GL00861

J. Woch, N. Krupp, A. Lagg, A. Tomás, The structure and dynamics of the Jovian energetic particle distribution. Adv. Space Res. **33**, 2030–2038 (2004). doi:10.1016/j.asr.2003.04.050

B.C. Wolven, P.D. Feldman, Self-absorption by vibrationally excited H_2 in the Astro-2 Hopkins ultraviolet telescope spectrum of the Jovian aurora. Geophys. Res. Lett. **25**, 1537–1540 (1998). doi:10.1029/98GL01063

C.S. Wu, Kinetic cyclotron and synchrotron maser instabilities—Radio emission processes by direct amplification of radiation. Space Sci. Rev. **41**, 215–298 (1985). doi:10.1007/BF00190653

C.S. Wu, L.C. Lee, A theory of the terrestrial kilometric radiation. Astrophys. J. **230**, 621–626 (1979). doi:10.1086/157120

R.V. Yelle, S. Miller, Jupiter's thermosphere and ionosphere, in *Jupiter. The Planet, Satellites and Magnetosphere*, ed. by F. Bagenal, T.E. Dowling, W.B. McKinnon, (2004), pp. 185–218

M. Yoneda, M. Kagitani, S. Okano, Short-term variability of Jupiter's extended sodium nebula. Icarus **204**(2), 589–596 (2009). doi:10.1016/j.icarus.2009.07.023

M. Yoneda, F. Tsuchiya, H. Misawa, B. Bonfond, C. Tao, M. Kagitani, S. Okano, Io's volcanism controls Jupiter's radio emissions. Geophys. Res. Lett. **40**(4), 671–675 (2013)

P. Zarka, On detection of radio bursts associated with Jovian and Saturnian lightning. Astron. Astrophys. **146**, 15–18 (1985)

P. Zarka, The auroral radio emissions from planetary magnetospheres—What do we know, what don't we know, what do we learn from them? Adv. Space Res. **12**, 99–115 (1992). doi:10.1016/0273-1177(92)90383-9

P. Zarka, Auroral radio emissions at the outer planets: Observations and theories. J. Geophys. Res. **103**, 20159–20194 (1998). doi:10.1029/98JE01323

P. Zarka, Radio and plasma waves at the outer planets. Adv. Space Res. **33**, 2045–2060 (2004). doi:10.1016/j.asr.2003.07.055

P. Zarka, Plasma interactions of exoplanets with their parent star and associated radio emissions. Planet. Space Sci. **55**, 598–617 (2007). doi:10.1016/j.pss.2006.05.045

P. Zarka, F. Genova, Low-frequency Jovian emission and solar wind magnetic sector structure. Nature **306**, 767–768 (1983). doi:10.1038/306767a0

P. Zarka, T. Farges, B.P. Ryabov, M. Abada-Simon, L. Denis, A scenario for Jovian S-bursts. Geophys. Res. Lett. **23**, 125–128 (1996). doi:10.1029/95GL03780

P. Zarka, L. Lamy, B. Cecconi, R. Prangé, H.O. Rucker, Modulation of Saturn's radio clock by solar wind speed. Nature **450**, 265–267 (2007). doi:10.1038/nature06237

B. Zieger, K.C. Hansen, Statistical validation of a solar wind propagation model from 1 to 10 au. J. Geophys. Res. **113**(A8) (2008). doi:10.1029/2008JA013046

DOI 10.1007/978-1-4939-3395-2_6
Reprinted from *Space Science Reviews* Journal, DOI 10.1007/s11214-014-0047-5

Magnetic Reconnection and Associated Transient Phenomena Within the Magnetospheres of Jupiter and Saturn

Philippe Louarn · Nicolas Andre ·
Caitriona M. Jackman · Satoshi Kasahara ·
Elena A. Kronberg · Marissa F. Vogt

Received: 10 January 2014 / Accepted: 19 April 2014 / Published online: 15 May 2014
© Springer Science+Business Media Dordrecht 2014

Abstract We review in situ observations made in Jupiter and Saturn's magnetosphere that illustrate the possible roles of magnetic reconnection in rapidly-rotating magnetospheres. In the Earth's solar wind-driven magnetosphere, the magnetospheric convection is classically described as a cycle of dayside opening and tail closing reconnection (the Dungey cycle). For the rapidly-rotating Jovian and Kronian magnetospheres, heavily populated by internal plasma sources, the classical concept (the Vasyliunas cycle) is that the magnetic reconnection plays a key role in the final stage of the radial plasma transport across the disk. By cutting and closing flux tubes that have been elongated by the rotational stress, the reconnection process would lead to the formation of plasmoids that propagate down the tail, contributing to the final evacuation of the internally produced plasma and allowing the return of the magnetic flux toward the planet. This process has been studied by inspecting possible 'local' signatures of the reconnection, as magnetic field reversals, plasma flow anisotropies, energetic particle bursts, and more global consequences on the magnetospheric activity.

The investigations made at Jupiter support the concept of an 'average' X-line, extended in the dawn/dusk direction and located at 90–120 Jovian radius (R_J) on the night side. The existence of a similar average X-line has not yet been established at Saturn, perhaps by lack of statistics. Both at Jupiter and Saturn, the reconfiguration signatures are consistent with magnetospheric dipolarizations and formation of plasmoids and flux ropes. In several

P. Louarn (✉) · N. Andre
Institut d'Astrophysique et de Planétologie (IRAP), CNRS/UPS, Toulouse, France
e-mail: philippe.louarn@irap.omp.eu

C.M. Jackman
University of Southampton, Southampton, UK

S. Kasahara
Institute of Space and Astronautical Science JAXA, Sagamihara, Kanagawa, Japan

E.A. Kronberg
Max Planck Institute for Solar System Research, Göttingen, Germany

M.F. Vogt
University of Leicester, Leicester, UK

 Springer

cases, the reconfigurations also appear to be closely associated with large scale activations of the magnetosphere, seen from the radio and auroral emissions. Nevertheless, the statistical study also suggests that the reconnection events and the associated plasmoids are not frequent enough to explain a plasma evacuation that matches the mass input rate from the satellites and the rings. Different forms of transport should thus act together to evacuate the plasma, which still needs to be established. Investigations of reconnection signatures at the magnetopause and other processes as the Kelvin-Helmholtz instability are also reviewed. A provisional conclusion would be that the dayside reconnection is unlikely a crucial process in the overall dynamics. On the small scales, the detailed analysis of one reconnection event at Jupiter shows that the local plasma signatures (field-aligned flows, energetic particle bursts...) are very similar to those observed at Earth, with likely a similar scaling with respect to characteristic kinetic lengths (Larmor radius and inertial scales).

Keywords Magnetosphere · Giant planet

1 Introduction

In plasma physics, magnetic reconnection designates the process by which magnetic field lines are broken and merged to form new field lines. It is often considered as a key driver of the dynamics of magnetic structures and the most likely explanation of their explosive energy releases, by a chain of phenomena occurring at many different spatial and temporal scales.

Magnetic reconnection is a subject of active researches (see books by Biskamp 2005; Priest and Forbes 2007; Birn and Priest 2007) addressing a large range of topics, from the analysis of global consequences of magnetic topology evolutions to the detailed investigation of the kinetic mechanisms that may provide dissipation in collisionless plasmas. The reconnection can be interpreted as a process that allows a magnetic system to reach a new equilibrium at smaller potential energy, with the released magnetic energy being converted into mechanical energy or radiated away. It can also break existing plasma confinements so that more efficient plasma transport can occur over large distances. Important open questions concern the identification of the triggers and drivers of reconnection, the effects of the pre-existing geometry, its temporal behavior (is it more a continuous or a sporadic process?) and the associated acceleration and transport mechanisms. Furthermore, as the reconnection requires non-ideal effects to relax the magnetic 'frozen-in' constraint, another mystery is to identify the mechanisms that provide a form of dissipation in the plasma and control the reconnection rate. Works in this domain include the detailed exploration of the basic kinetic processes using spacecraft data (see Vaivads et al. 2004, 2006; Phan et al. 2007; Mozer et al. 2011, and references therein), simulations studies, including extended 3-D particle-in-cell simulations (Drake and Shay 2007; Pritchett and Mozer 2009; Karimabadi et al. 2005; Daughton et al. 2011; Olshevsky et al. 2013) and analytical models (Boozer 2013).

Reconnection in natural plasmas has been mostly studied in the context of Earth's magnetosphere and solar corona. The magnetospheres of giant planets, especially Jupiter and Saturn which combine a fast rotation and prolific internal plasma sources, offer a different and complementary perspective. They are indeed characterized by the presence of plasma tori and magneto-disks that can be considered as examples of magnetic structures having some kind of cylindrical symmetry (especially in the inner magnetosphere), organized by the rotational stress, with a dynamics largely controlled by the radial plasma and flux transport. How important is the reconnection in driving their dynamics? Are there specific characteristics of reconnection processes in rotating systems that inherently control the overall dynamics? What are the transient mechanisms associated to reconnection events occurring in disks? Are there specificities in the microscopic dissipation processes due to the complex plasma composition? These are questions of potential wide astrophysical interest that can be addressed by in-situ investigations in the Jovian and Kronian disks.

It should be stressed that answering these questions is not easy and requires sophisticated instrumentation. Other mechanisms than the reconnection may indeed explain or play a role in transient magnetospheric phenomena. For example, even in the well-studied Earth case, the debate is vivid between the supporters of the reconnection model of substorms and authors proposing alternatives as medium scale current-driven and pressure-driven (ballooning) instabilities (Roux et al. 1991; Lui et al. 1991; Liu et al. 2012, see also the general review by Kennel 1995 and the scientific presentation of the THEMIS mission Angelopoulos 2008). At Jupiter and Saturn, important roles are attributed to mechanisms such as ionosphere-magnetosphere coupling to drive the plasma rotation, diffusion processes, interchange, ballooning instabilities, not excluding solar wind interactions and Kelvin-Helmholtz instabilities, in explaining the plasma transport and the magnetospheric dynamics. The undisputable characterization of these different processes, including the reconnection, requires a full characterization of the electric and magnetic fields, as well as measurements of the ion and electron distribution functions. Ideally, multi-spacecraft measurements would be also needed to investigate the magnetic topology evolutions. This is beyond the capabilities of the instruments that have yet explored Jupiter and Saturn and one should consider that there are still ambiguities in the identification of the key magnetospheric processes that operate at Jupiter and Saturn.

This being stated, we review Galileo and Cassini observations of phenomena that may be related to reconnection. More generally, these observations document how fast rotating and heavily populated magnetospheres behave. Some emphasis will be put on the Jovian case simply because the Jupiter's magnetodisk is often considered as archetypical with, perhaps, more studies of the disk dynamics than for the Kronian case. With Cassini observations at Saturn, the tail and the magnetopause reconnection will also be reviewed. In Sect. 2, we discuss what could be the specificities of the reconnection in magnetodisks. We sketch the classical Dungey and Vasyliunas models of magnetospheric convection; they indeed serve as the conceptual frameworks in which many investigations of reconnection and related phenomena are performed. We also list reasons for which the reconnection might proceed differently in rotating plasma than at Earth. The topic of the dayside reconnection is reviewed in Sect. 2.3. In Sects. 3 and 4, the emphasis is put on observations made in the disks and the tails. We treat the cases of Jupiter and Saturn separately, starting from the description of local signatures then discussing global aspects of the dynamics. A general discussion with prospective directions of future investigations is proposed in Sect. 5.

2 Possible Specificities of Reconnection at Jupiter and Saturn

2.1 Reconnection in Solar-Driven System

To introduce the role that the reconnection may play in magnetospheric dynamics, it is useful to consider the classical picture proposed by Dungey (1961) that describes the dynamics of a solar-driven magnetosphere, as is the case at the Earth (Fig. 1). This seminal model attributes a central role to the reconnection, first to 'open' the magnetosphere and allow its energy coupling with the solar wind, second to 'close' the tail magnetic field, relax the accumulated magnetic energy and allow a return of the magnetic flux towards Earth, then to the dayside, which ultimately closes the cycle.

Fig. 1 Dynamics of a solar-driven magnetosphere (Dungey cycle). The sketch describes how the magnetosphere is opened when the Interplanetary Magnetic Field becomes anti-parallel to the planetary dipole, which enhances the reconnection at the magnetopause, leads to an accumulation of magnetic flux in the tail and, finally, triggers reconnection events and sporadic magnetic energy releases in the magnetotail (substorms). Solar-driven magnetospheres are characterized by a bursty activity

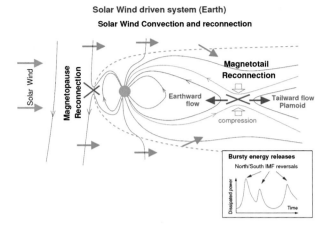

The intuitive idea that underlies the use of magnetic reconnection in space physics is that the current density and, thus, the magnetic reversal across a current sheet ($\Delta B_x/L$, for a model sheet of thickness L, lying in the X/Y plane with current flowing in Y) cannot indefinitely increase without triggering magnetic merging. As the sheet thickness becomes comparable to kinetic length scales such as the Larmor radius or the skin depth, magnetic field lines initially located on both sides of the sheet may merge. This changes the magnetic topology, relaxes the magnetic constraint and may contribute to the transport of magnetic flux, heats and accelerates particles.

In Dungey cycle, this principle is first applied to explain how the interplanetary magnetic field (IMF), when turned southward, merges with the Earth's northward-directed dipole field at the dayside magnetopause. The newly opened flux tubes are then dragged in the anti-sunward direction by their 'root' in the solar wind. This contributes to add open magnetic flux in the magnetotail. The reconnection is again invoked for the second stage of the cycle. The accumulated magnetic flux progressively stretches the tail current sheet and finally, when its thickness becomes comparable to kinetic scales, triggers the merging across the sheet. The closed flux tubes return back to the inner magnetosphere and a plasmoid is evacuated in the antisolar direction. This picture is also closely related to the near-Earth X-line model of substorms (see review by Kennel 1995). In situations of strong inputs of the solar wind energy, the tail reconnection would take place close to the Earth (15–25 R_E) and trigger major energy releases in the magnetosphere. Overall, the dissipated power in a solar-driven system would then appear sporadic, with phases of slow energy accumulations (few hour time scales at Earth) and intense much shorter energy releases (over a few ten minute scales at Earth), largely controlled by the variations of the solar wind dynamic pressure and IMF direction.

This picture largely relies on the standard 2D model sketched in Fig. 2. The reconnection is supposed to occur in thin planar structures, organized in a quasi-1D equilibrium by the magnetic and plasma pressures. The relaxation of the magnetic stress leads to particle accelerations and the formation of plasma jets that propagate symmetrically on both sides of the reconnection points. Figure 2 is adapted to the tail reconnection with, thus, a high symmetry in Z (the normal to the sheet) but different boundary conditions in $+/-X$. The picture for the reconnection at the magnetopause would be different, with a strong asymmetry along the normal to the sheet (magnetosheath/magnetosphere) and a leading role played by external

Fig. 2 Standard model of the 2D reconnection (adapted to the tail configuration). This sketch is commonly used to predict the signatures of the reconnection. *Cases 1, 2, and 3* describe observations made by spacecraft located planetward, close to, and anti-planetward of the reconnection site

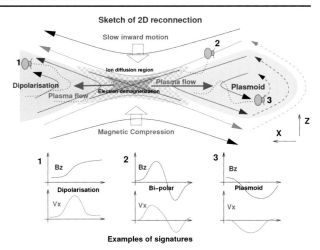

driving processes, linked to the continuous incident flux of energy and momentum from the solar wind.

The reconnection does not necessarily require an anti-parallel configuration. Reconnection in sheared configurations can also be considered (the 'component' reconnection corresponding to a non-null B_Y in our model); this notion is important at the magnetopause. In general, the plasma conditions are expected to influence the occurrence of reconnection (Quest and Coroniti 1981). In particular, a high value of the plasma beta parameter (the ratio of plasma to magnetic pressure) in the magnetosheath could inhibit component reconnection (Swisdak et al. 2003; Phan et al. 2010; Masters et al. 2012). As discussed in Sect. 2.3, this may have an impact on the role of the dayside reconnection at Jupiter and Saturn.

This 2D model is almost systematically used to predict the magnetotail signatures of reconnection. They are sketched in Fig. 2 as a function of the spacecraft position with respect to the reconnection site. On the planet side of the event (case 1), the magnetic relaxation leads to a more dipolar configuration with a larger vertical magnetic component (dipolarization); often accompanied by a planetward flow. On the anti-planet side (case 3), a magnetic structure of closed field lines—a plasmoid—is formed and evacuated. The signature is a reversal of the normal field direction (southward turn at Earth, northward at Jupiter and Saturn) and an anti-planetward flow. A related signature is the flux rope, a helical structure characterized by a strong guide field (typically in the east-west, or GSM y component at the Earth). Finally, if the spacecraft crosses the reconnection point (case 2), a combination of both signatures—the bipolar pulse (double South-North peak) accompanied by a flow reversal—is observed.

As mentioned in the introduction, these signatures are not without ambiguities. Recognizing the precise role played by the reconnection or another process in a transient magnetospheric phenomenon is difficult. The typical example is that a ballooning instability or a current-driven instability may also lead to a dipolarization. To try to avoid too hasty interpretation, we will thus often prefer the terms 'dipolarization' or 'reconfiguration' to the more specific, process-oriented, 'reconnection', except for bipolar pulses accompanied with flow reversals that can be considered as more reliable signatures of reconnection. As shown later, the statistical survey of the signatures may also help to settle a coherent view and define 'average' locations where reconnection events occur.

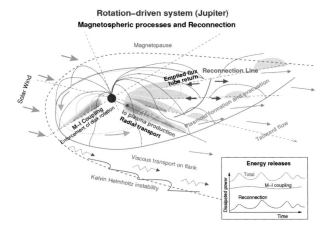

Rotation–driven system (Jupiter)
Magnetospheric processes and Reconnection

Fig. 3 Dynamics of a rotation-driven magnetosphere. The sketch illustrates the possible location of the reconnection X-line from which plasmoids are formed and evacuated. The *continuous green line* sketches the external edge the disk, with its 'detached' part resulting from the reconnection. Other forms of transport processes are indicated: as the radial plasma diffusion in the inner disk and the plasma 'drizzle', related to viscous interaction of the flank of the magnetosphere. The typical time profile of the dissipated power is shown. It emphasizes the importance of the continuous dissipation linked to the M-I coupling. The part of energy dissipation due to sporadic reconnections is not expected to dominate the overall energy release

2.2 Reconnection in Rotation-Driven System

Compared to the Earth, Jupiter and Saturn have a stronger magnetic dipole, a faster rotation and magnetospheres which are populated by abundant internal sources of particles. The particles expelled by Io and Enceladus lead to the formation of relatively dense plasma torus at typical distances of 6 R_J and 4 R_S, respectively. Locally, the plasma density peak to values of ~ 100 cm^{-3} (Saturn) and $\sim$ a few 1000 cm^{-3} (Jupiter), thus, factors 100 to 1000 larger than at Earth. Both magnetospheres are then largely organized by the way the plasma, produced at a rate of ~ 1000 kg/s by Io (Bagenal and Sullivan 1981) and ~ 8–250 kg/s by Enceladus and the rings (Gombosi et al. 2009; Bagenal and Delamere 2011), is driven in rotation by its magnetic connection to the ionosphere, confined towards equatorial/centrifugal plane, transported radially, heated and accelerated, to finally be expelled down tail (Belcher 1983; Krimigis and Roelof 1983; Vasyliunas 1983). The plasma convection in these rotation-driven magnetosphere is classically sketched by the Vasyliunas cycle (Vasyliunas 1983), an analog of the Dungey cycle, adapted to rotating and internally populated system.

In short, Vasyliunas's model assumes that the reconnection is the process that underlies the last stage of radial plasma transport in the disk. The reconnection allows the transfer of the internally produced plasma to the tail and the return of the emptied flux tubes to inner regions (Fig. 3). In the night side and at large distances from Jupiter (more than 60–80 R_J), the low pressure in the tail cannot compete with the rotational stress, the flux tubes then progressively stretch out and thin, leading ultimately to a reconnection and the formation of an X line. Plasmoids are then formed and expelled out. The emptied flux tubes, connected to Jupiter, then return back for a new rotation. In this model, the reconnection is not the primary mechanism that drives the magnetospheric activity. According to the most accepted conception (Hill 1979), the Jovian magnetospheric dynamics and the related energy release are indeed dominated by the magnetic coupling that enforces the rotation of the newly created plasma populations. This coupling is the result of a current system that links

Fig. 4 Sketch illustrating possible specificities of the reconnection in disk. The *upper panel* shows the general geometry and the various forms of reconnection that possibly act at the outer edge of the disk. The reconnection can be organized in sporadic events with formation of plasma jets and plasmoids. The X-line can also be the location of a more continuous cross-field plasma diffusion. The *lower panel* sketches some particularities of disk plasma sheets. They may differ from more classical magnetotails by: (1) the different height scales of the various plasma species, (2) the ambipolar field, with a parallel-to-B component, (3) the significant vertical field, (4) the importance of the radial pressure gradient that counter-balances the rotational stress

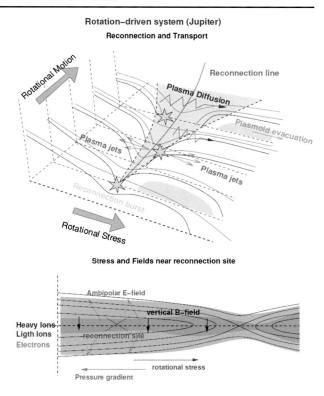

the disk to the ionosphere. Via $j \times B$ forces, it continuously extracts rotation energy from the ionosphere to the disk, at a rate that is mostly controlled by the mass input and the radial transport. Overall, this internal coupling leads to much more regular power dissipation than the external solar wind coupling.

There are many open questions concerning the specificities of reconnection in rotation-driven system. For example, the way the reconnection helps the radial plasma transport is not elucidated (see Fig. 4). By analogy with terrestrial substorms, a possibility is that the reconnection occurs sporadically, with formation of plasma jets and plasmoids, the difference being that the dissipated energy is the rotational energy progressively accumulated in the disk. However, as presented later, estimates of the mass and the frequency of the plasmoids suggest that the sporadic reconnection is not sufficient to explain an evacuation equivalent to the internal production rate, both at Jupiter and Saturn (Sects. 3.2, 3.3, 4.2, 4.3). Other mechanism need to be considered as, for example, forms of continuous cross-field transport across the X-line and possible viscous-like interactions resulting from the interaction with the solar wind.

About the details of the plasma processes involved in the reconnection, it would also not be surprising that reconnection in disks differs from the standard 2D model. In disks, the main stress is linked to the rotation and, thus, oriented radially (X direction, in the plane of the current sheet), which is different from the magnetic compression, exerted vertically (Z direction, perpendicular to the plane of the current sheet) in standard plasmasheet (see Fig. 2). A significant vertical magnetic field and a planetward oriented pressure gradient are also needed to construct the equilibrium solution. The disk reconnection should thus operate on closed field lines, in the presence of a non-negligible vertical field, when at Earth the merging occurs on open and almost antiparallel field lines. At small scales, the rotational stress and the presence of multiple charged heavy ions likely make the disk situation more complex. For example, since heavy ions are concentrated in the vicinity of the central plane by the centrifugal force, an ambipolar electric field has to be taken into account to restore the quasi-neutrality.

This discussion shows that reconnection in disks may differ from what is sketched in the classical 2D model for many reasons. There are surprisingly very few theoretical/simulation studies of disk situations, at the exception (at the author's knowledge) of the analysis proposed by Yin et al. (2000) which remains limited to a rather simple 2D model. This can be stressed already: the conceptual tools that would permit the detailed study of plasma reconnection processes in rotating plasma are still lacking.

2.3 Interplay Between Internally and Externally Driven Processes

Obviously, as the Jovian and Kronian magnetospheres interact with the solar wind, they are not immune to external drivers. The role of the solar wind as a trigger of Jupiter and Saturn's magnetospheric processes has been demonstrated a long time ago from the radio emissions (Barrow et al. 1986; Ladreiter and Leblanc 1989; Kaiser 1993; Prangé et al. 1993; Zarka 1998; Rucker et al. 2008). There are also many evidences that the solar wind effects may drive a spectacular auroral activity (Gurnett et al. 2002; Prangé et al. 2004; Tao et al. 2005) with, for example, clear indications that the Jovian's auroral emissions may brighten at times of solar wind disturbances (Pryor et al. 2005; Nichols et al. 2007; see also the review in Clarke et al. 2009). The physical nature of these correlations is still not firmly established. They can indeed be direct consequences of an enhanced reconnection at the magnetopause, leading possibly to a more active Dungey cycle. On the other hand, the variations of the solar wind pressure may also directly impact the activity of the internal magnetospheric engine.

A first way to envisage these topics is to investigate how the Dungey and the Vasyliunas cycles may interact and, thus, to analyze how the dynamics of the rotationally ejected plasma can be combined with cycles of dayside opening and tail closing reconnection (Cowley and Bunce 2001, 2003; Cowley et al. 2003, 2005; Badman et al. 2005; Badman and Cowley 2007; Cowley et al. 2007). Regarding the role of external drivers in regulating the internal, rotationally-dominated, dynamics, it has been proposed that an increased solar wind pressure could oppose to the outflow rate of magnetospheric plasma and, thus, weakens the intensity of the rotationally-enforced current system (Southwood and Kivelson 2001). Combining MHD and kinetic concepts, the same authors have discussed the role of the forces externally imposed at the magnetopause on the dynamics and the energetics of rotating flux tubes (Kivelson and Southwood 2005, see also Vogt et al. 2014). The existence of an operant Dungey cycle at Jupiter has even been challenged. McComas and Bagenal (2007) have suggested that it might be incomplete; raising the question of the effectiveness of the return flow resulting from the tail field closure given the huge scale of the Jovian magnetosphere, this statement being nevertheless later criticized by Cowley et al. (2008). Other forms of interactions have also been investigated. A quite radically different conception is that Kelvin-Helmholtz (KH) instability developing at the magnetopause transfers an anti-solar stress to the rotationally driven plasma (Delamere and Bagenal 2010, 2013; Delamere et al. 2011, 2013, note also the early work by Galopeau et al. 1995, about the KH instability at Saturn). This would imply small-scale, intermittent reconnections on the flanks of the magnetopause, resulting in a viscous-like interaction that ultimately leads to a diffusive 'drizzle' of the plasma across the magnetic field.

The interaction between internally and externally driven processes is likely of particular importance at Saturn. The rings and Enceladus are prolific internal sources of particles and may even lead to larger supply rate than Io at Jupiter when normalized to intrinsic planetary properties (Vasyliunas 2008). However, the smaller magnetic dipole less efficiently couples the new plasma populations to the ionosphere so that the rate of rotational energy that can be transferred to the Kronian torus/disk is reduced compared to Jupiter: $\sim$50–200 GW instead of 5–10 TW (Bagenal and Delamere 2011). The part of the magnetosphere internally driven by the rotation is also reduced compared to Jupiter with, comparatively, a larger role that may be attributed to external processes, in the form of more effective Dungey cycle (Badman et al. 2005; Badman and Cowley 2007) or viscous-like interactions linked to the development of KH instabilities (Desroche et al. 2012, 2013; Delamere et al. 2013).

What do we know about solar wind interactions at Jupiter and Saturn and, more specifically, about reconnection at the magnetopause from the observation?

At Jupiter, possible signatures of dayside reconnection have been observed with Pioneer and Voyager data (Walker and Russell 1985); these signatures seem less frequent and weaker than observed at Earth, possibly indicating a lower overall reconnection rate. Other evidences for dayside reconnections come from the observations of the highest latitude aurorae (the polar aurorae). The intensity of these transient and localized emissions, magnetically connected to the dayside magnetosphere (Pallier and Prangé 2001, 2004), can increase by factors of a few tens on minute time scales (Waite et al. 2001). More recent HST observations have revealed polar flares that are pulsed on timescales of 2–3 minutes. A refined magnetic mapping (Vogt et al. 2011) shows that they most probably originate from the dayside magnetopause, at 50–100 R_J, and then likely constitutes evidences of pulsed reconnection at the magnetopause (Bonfond et al. 2011).

At Saturn, several auroral signatures of dayside reconnection have also been reported. Their location may give indications on the position of the reconnection events. A low-latitude reconnection would lead to an intensification of the near-noon auroral arc, while lobe reconnection could be manifest as a distinct spot poleward of the noon auroral arc (Bunce et al. 2005; Gérard et al. 2005). Analysis of Cassini auroral images has also revealed poleward arc bifurcations in the high-latitude noon sector, which were interpreted as the signatures of bursty reconnection events (Milan 2009; Radioti et al. 2011, 2013). It has also been suggested that similar forms of auroral spots could be linked to Kelvin-Helmholtz vortices (Grodent et al. 2011). Recently, Badman et al. (2013) have tracked the evolution of auroral signatures. For the studied event, the reconnection activity was first manifest as a cusp spot, explained by high-latitude lobe reconnection, which then evolve toward auroral bifurcations at the noon oval. This was interpreted as bursty reconnection under conditions of strong solar wind compression of the magnetosphere. Direct field and particle signatures of reconnection at Saturn's magnetopause were first reported by Huddleston et al. (1997), showing evidence for rotational discontinuity from Voyager 1 data. With Cassini magnetic and field measurements, it has been even possible to distinguish different form of reconnection with evidences for low-latitude anti-parallel reconnection and more distant high-latitude events, presenting features similar to those observed at Earth (McAndrews et al. 2008).

Interestingly, the extent to which dayside reconnection operates is a topic of intense debate (Masters et al. 2014 and references therein). The reconnection rate is modulated in some manner by the orientation and magnitude of the IMF, as well as the value of the plasma β. Conflicting early studies have indicated that reconnection is (i) feasible and can be important (e.g. Huddleston et al. 1997; Grocott et al. 2009), or (ii) that reconnection can be suppressed by the high Mach number regime at Saturn (e.g. Scurry and Russell 1991). While evidences of dayside reconnection have been presented (see above), recent studies claim its role is rather limited (Lai et al. 2012). More recently, work has focused on the factors that may govern the reconnection rate, such as the plasma β (Masters et al. 2012;

Desroche et al. 2013). Fuselier et al. (2014) have analyzed an extended set of reconnection events, suggesting that there is an effect of high Mach number regime in inhibiting component reconnection at Saturn. Using also an extended set of reconnection events, Masters et al. (2014), have estimated the reconnection electric field and shown that it is generally small (0.01 mV/m). They thus concluded that magnetopause reconnection is likely not a major driver of the dynamics at Saturn.

The review of the literature thus provides several examples of bursts in Jupiter and Saturn's activity that are likely triggered by solar wind perturbations. In addition, there is in overall a clear correlation between solar wind effects and various proxies of the magnetospheric activity. That this operates mainly through reconnection at the magnetopause is challenged. There are possibilities of direct actions of the solar wind pressure on the internal magnetospheric engine, powered by the rotation; KH instability and forms of viscous-like interactions may also play a significant role.

3 Observations in Jupiter's Disk

3.1 Local Reconnection Signatures

Most of the *in situ* measurements from Jupiter's magnetosphere were made by the Galileo spacecraft, which orbited Jupiter from late 1995 to 2003, though data are also available from seven spacecraft flybys. The Galileo spacecraft included a magnetometer (Kivelson et al. 1992), which measured the direction and magnitude of the magnetic field, and an energetic particle detector (EPD), which measured the temporal and spatial distribution of ions (20 keV–55 MeV) and electrons (15 keV–11 MeV) (Williams et al. 1992). The EPD measurements provide particle anisotropies which can be used to infer flow. Generally, it can be noted that the Galileo coverage (up to 150 R_J), when scaled to dipole strength and standoff distance, is analogous to $\sim$30 R_E at Earth (Jackman and Arridge 2011) so that, generally speaking, very little is known about distant magnetospheric processes. The same remark applies to Cassini and Saturn.

The analysis of the magnetic field and energetic particle data has established the typical quiet-time field configuration and global flow patterns. In the inner magnetosphere, inside of 20 R_J (Jovian radii), the magnetic field is primarily dipolar and the plasma in the Jovian magnetodisk rotates with the same angular velocity as the planet. Beyond this distance, the flow of the plasma lags behind the rotation of the planet but remains predominantly in corotational direction (Krupp et al. 2001). In the middle ($\sim$20–40 R_J) and outer (>40 R_J) magnetosphere the magnetic field is stretched in the radial direction by the current sheet, and is swept back azimuthally as the magnetodisk plasma is no longer corotating with the planet (Smith et al. 1974; Khurana and Kivelson 1993). The equatorial field is southward. Both the plasma flow pattern and the magnetic field configuration display strong local time asymmetries (Krupp et al. 2001; Kivelson and Khurana 2002; Khurana et al. 2004). In the post-midnight to dawn local time sector the azimuthal plasma flows are stronger, the field lines are most radially stretched and are most bent back due to interaction with the solar wind. By comparison, near dusk the plasma flows are slowest and the magnetic field is more dipolar and may be swept forward.

As presented in Sect. 2, possible signatures of reconnection events are identified in magnetic field data by transient reconfigurations to a more dipolar field geometry, and in particle data by deviations in the typical corotational flow pattern in favor of increased flow in the radial direction. The assumption is that the magnetic relaxation transfers radial momentum to the plasma, which is added to the rotation flow. The first observational evidence of a signature of reconnection at Jupiter was found in Voyager particle and magnetometer data by Nishida (1983). They reported strong outward radial flows associated with a reversal of the magnetic field to a northward direction, and attributed these changes to magnetic reconnection. More recently, case studies (Krupp et al. 1998; Russell et al. 1998) and surveys of the Galileo EPD data and magnetometer data have identified hundreds of disturbed intervals that have been interpreted as possible reconnection signatures.

Kronberg et al. (2005) used Galileo EPD data to identify 34 reconfiguration events that were characterized by strong radial flows occurring at the same time as increases to the north-south component of the magnetic field (B_θ component). Vogt et al. (2010) surveyed magnetometer data from Galileo and several other spacecraft and found 249 reconfiguration events in which B_θ, the north-south component of the magnetic field, was significantly enhanced over background levels, indicating reconfiguration to a more dipolar field geometry. Depending on the nature of the observed B_θ signature, the change of the magnetic field configuration may be associated with: plasmoids (regions defined by closed field lines and associated with quasi-bipolar fluctuations in B_θ), traveling compression regions (TCRs, the regions where the field strength is increased above and below the plasmoids), post-plasmoid plasma sheet formations, dipolarization and flux ropes (Kasahara et al. 2011, 2013; Kronberg et al. 2012). The disparity between the number of events identified from the particle data and the number of events identified in the magnetometer data can partially be explained by the fact that the magnetometer data have a higher time resolution (typically 24 seconds per vector) than the EPD data (typically 3–11 minutes), making it easier to identify transient features. Additionally, the intervals studied by Kronberg et al. are generally long in duration (~10–20 hours) and include more than one of the Vogt et al. events.

In general, there is good agreement between the particle and magnetometer data, as most of the intervals of high radial anisotropy identified in the particle data were accompanied by field dipolarizations or reversals (Kronberg et al. 2005). Additionally, there is a good correlation between the direction of radial flow and the direction of B_θ, the meridional or north-south component of the magnetic field (Kronberg et al. 2008a; Vogt et al. 2010), as was seen in the Nishida (1983) events. Though the magnetometer does not directly measure flow, the radial flow direction can be inferred through the sign of B_θ, the north-south component of the magnetic field, and the flow direction inferred from the anisotropies is frequently consistent with the changes seen in the magnetic field. Planetward (tailward) of the x-line, it is most likely that flow will be radially inward (outward) and B_θ will be positive (negative), so outward (inward) radial anisotropies are expected to be associated with a negative (positive) B_θ.

Magnetic field and particle data from a reconfiguration event studied by Kronberg et al. (2005) are shown in Fig. 5. During the quiet period before and after the disturbed interval, the flow in the corotational direction (azimuthal component of the ion anisotropy, denoted by a green line) dominates the flow in the radial direction (red line). Despite the large distance from Jupiter (100 R_J), this suggests that Galileo is still located in a plasma that flows in the azimuthal direction, as expected if rotational effects dominate. The particle intensity, shown in the top panel, smoothly changes on periodic, 10-hour time intervals. These changes are associated with the spacecraft crossing periodically through the current sheet center due to the 10° tilt of the magnetic field axis with respect to the rotational axis. During the quiet time, Galileo has not encountered the distinct plasma sheet/lobe boundaries. The magnetic field during the quiet time is largely in the radial direction, with a small (~1–2 nT) positive

Fig. 5 Energetic particle (EPD) and magnetic field (MAG) observations on Galileo orbit G2 from DOY 269, 05:00 to DOY 272, 05:00 in 1996. From top to bottom are displayed: omnidirectional ion intensities (0.042–3.2 MeV) (*first panel*); first order ion anisotropies in the radial (positive is outward) and corotational direction (*second panel*); the magnetic field components (*third to fifth panels*) in SIII coordinates (the radial component is positive in the outward direction, the azimuthal component positive in the direction of Jupiter's rotation, the south-north component positive southward) and its magnitude (*sixth panel*); *continuous horizontal lines* outline "quiet" and disturbed periods, *dashed lines* labeled (*Q* and *D*) to specific times referred to some representative current sheet crossings; in the text and a sketch below. *Sketch*: Meridional cut through the magnetodisk configuration. Jupiter is from the *left*. At the *right side* the profiles of ion pressure P_i and of the magnetic pressure P_b are shown. From Kronberg et al. (2005)

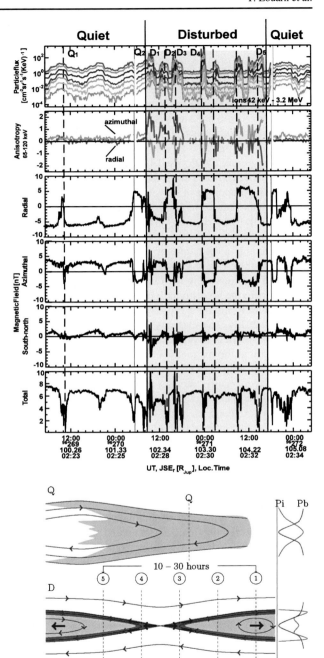

B_θ component, compared to the radial component (4–5 nT). The disturbed interval begins roughly at 08:00 on day 270 of 1996, lasts until roughly 17:00 on day 271, and includes two of the reconnection signatures identified by Vogt et al. (2010) from the magnetometer data.

The start of the disturbed interval is marked by changes in the flow pattern and magnetic field configuration. Most noticeably, the flow in the radial direction dominates over the azimuthal component, while the azimuthal component points predominantly in the anticorotational direction. During the disturbed interval the strong plasma flows are initially in the outward radial direction, then after $\sim$10 hours we observe strong negative radial anisotropies coupled with the super-corotational flow, which indicates sunward-moving plasma. Such clear flow reversals are observed in 53 % of the Jovian reconfiguration events. They indicate that the X-line propagates tailward (Kronberg et al. 2005). The ion intensity, measured by the particle flux in the top panel, changes rapidly during the disturbed interval. This indicates that Galileo intermittently entered the lobes where the plasma density is low. The magnetic field data from the disturbed interval show at least two periods of relatively brief ($\sim$tens of minutes) field reversals ($B_\theta < 0$) in which $|B_\theta|$ is larger than background values, indicating reconfiguration to a more dipolar field.

This observation, made at $\sim$100 R_J in the post midnight sector $\sim$2:30 MLT, combined both magnetic and particle signatures of the reconnection. It can be considered as a clear indication of reconnection events occurring sporadically at the edge of the disk.

3.2 Event Properties

Statistics studies of the signatures found in energetic particle and magnetic field data have established their average properties such as spatial distribution, scale size, recurrence period, and mass lost, and have also identified the location of a statistical separatrix separating inward and outward flow. In this section we review these properties and discuss what the results suggest for Jupiter's magnetospheric dynamics.

The spatial distribution of 249 reconfiguration events identified from magnetometer data (Vogt et al. 2010) is shown in Fig. 6. The solid black lines in the figure show the spacecraft orbits for times when there were few data gaps and when the spacecraft was within $15°$ latitude of the equatorial plane. The color of each event symbol indicates the dominant sign of B_θ—positive, negative, or bipolar—during the event, as this signature can be used to infer flow direction. Events are indicated in the figure by a triangle symbol if they induced an acceleration of energetic ions and electrons as inferred from the flow-anisotropy of energetic ions in the Galileo EPD data at energies between 65 and 120 keV (Kronberg et al. 2005). While many more events are found at post-midnight local times, more data are available post-midnight than pre-midnight, and when this is taken into account one finds that events are roughly evenly distributed on either side of midnight. However, it is worth noting that the Kronberg et al. flow anisotropy events are much more spatially restricted, being located mainly post-midnight and at distances larger than $\sim$60 R_J. The most distant of the flow anisotropy events were observed at $\sim$2500 R_J by New Horizons (e.g. McComas et al. 2007). Recently, Kasahara et al. (2013) analyzed particle data from all of the 249 events found in the magnetometer data, and showed that the nature of the particle signature during the event depends on local time. They found that large density changes, indicating lobe reconnection, and large radial flows were observed during the dawn sector events only, whereas the flow direction remained corotational, and little or no density changes were observed for the dusk sector events. Therefore Kasahara et al. (2013) proposed that a process other than tail reconnection may be responsible for the duskside field dipolarization events.

Also shown by the purple lines in Fig. 6 are the locations of two statistical x-lines separating inward and outward flow. The dashed purple line indicates the x-line drawn by Woch et al. (2002) using energetic particle observations, which indicate the flow direction, and the solid purple line indicates the x-line derived by Vogt et al. (2010), who inferred the flow

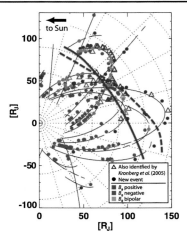

Fig. 6 The location of reconnection signatures observed in magnetometer data (*circles*) and in both magnetometer and Galileo EPD data (*triangles*). This is an equatorial plane view, and the Sun is to the left. Events are colored *red*, *blue*, or *green* according to the sign of B_θ, the north-south component of the magnetic field, which provides a good proxy for the radial flow direction. The *purple lines* show the location of the statistical x-line separating inward and outward flow according to Woch et al. (2002) (*dashed line*) and Vogt et al. (2010) (*solid line*). *Black solid lines* indicate spacecraft orbits. Modified from Vogt et al. (2010), Fig. 8

direction from magnetic field data using the sign of B_θ. The two x-lines show mutual agreement and confirm that the magnetodisk at certain distance becomes unstable and that this distance depends on the local time, being ~ 90 R$_J$ near dawn and farther out at earlier local times.

Many of the observed reconnection events are indicative of plasmoids, which are characterized by a quasi-symmetric bipolar B_θ signatures. Roughly 17 percent of the reconnection signatures observed in the magnetometer data can be identified as tailward-moving plasmoids based on the nature of the B_θ signature (Vogt et al. 2014). An example of such a plasmoid event is given in Fig. 5 at about 11:00 on day 270. Such plasmoid-induced events last on average between 10 and 20 min, not considering the duration of the transition of Galileo through the post-plasmoid plasma sheet, which usually is substantially longer. The average speed of the plasmoid is about 460 km s^{-1}. This implies that the radial length of the plasmoids range from 5 to 10 R$_J$, with the mean value being 9 ± 2.6 R$_J$. Their speed is Alfvénic (Kronberg et al. 2008b). The mass loss through one plasmoid is estimated by Kronberg et al. (2008a) to be about 8×10^5 kg (the particle number density is 0.025 cm^{-3}, the average ion mass is assumed to be 16 proton mass, the length of the plasmoid is 9 R$_J$, its thickness is 2 R$_J$ and the azimuthal scale is 200 R$_J$), which is consistent with the estimates of Vogt et al. (2014) with masses from 2.8×10^4 kg to $\sim 2 \times 10^6$ kg. It is interesting to estimate the role of these different signatures in the mass transport. Given a recurrence rate of 2–3 events per day (which can be considered as a higher value), the estimates mass loss rate is ~ 1–125 kg/s, much smaller than the 500–1000 kg/s of Iogenic plasma (Vogt et al. 2014), especially considering that an azimuth extension of 200 R$_J$ for the plasmoids is likely an upper value.

Another transient feature frequently associated with a reconfiguration event is the presence of field-aligned energetic particle beams, observed in the lobes. An example is given by the reconfiguration event shown in Fig. 7, where two field-aligned beams (1) and (3) precede bursty bulk flows (2) and (4). Field-aligned beam (3) is associated with a TCR signature and

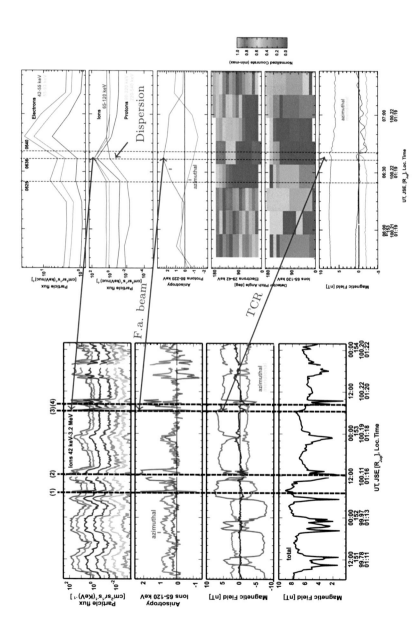

Fig. 7 Energetic particle (EPD) and magnetic field (MAG) observations on Galileo orbit G8 from DOY 151, 06:00 to DOY 154, 02:00 in 1997, from left in the same format as Fig. 1, only 3 magnetic field components are put in one panel. The *vertical lines* denote field-aligned beams and bursty bulk flows mainly at the boundary layers. The *zoomed part* from the right shows from top to the bottom: the electron intensities, ion and proton intensities, the first order ion anisotropies, detector pitch-angle distributions for electrons at 29–42 keV and for ions at 65–120 keV and the magnetic field

a flux rope. Most of these beams are ion streams associated with field-aligned electron flows, in support of the X-line formation scenario by Grigorenko et al. (2009). The field-aligned ion beams show a time dispersion which implies that its origin is at one point or line. The dispersion is seen also for energetic ions of different species by New Horizons by Hill et al. (2009). The total mass loss during these events is estimated to be about 10^5 kg, which is less than the loss through plasmoids (Kronberg et al. 2012). The rate of kinetic energy release during field-aligned beams is on the order of 9 TW, which is enough to supply, e.g., the Jovian polar auroral emissions observed by Radioti et al. (2010) (see Sect. 3.4).

Interestingly, the measurements shown in Fig. 5 suggest that a part of the mass is released not through plasmoids themselves but at the plasma sheet boundary layers. It is striking that the strongest increases of the ion intensity are observed at the plasma sheet boundary layers (see sketch in Fig. 5), namely when the modulus of the radial magnetic field component starts to decrease from the lobe values (see panel 3 from the top). It is consistent with the fact that the ion intensity (pressure) is lower at the plasma sheet center and higher in the transition region between the plasma sheet and the lobe (in contrast with the quiet time, which is characterized by a peak at the current sheet center), see sketch in Fig. 5. If it is confirmed by additional cases, this observation would suggest that a continuous form of reconnection might exist with the formation of a continuous X-line that would ease the radial plasma evacuation (see Fig. 4).

3.3 The Large Scale Perspective

It is also interesting to look at the preconditions and the occurrence frequency of large reconfiguration events. When Galileo was located Jupiterward of the X-line, the magnetic field data showed a recurrent stretching of the plasma sheet (Kronberg et al. 2007). Before the onset of the reconfiguration event, the radial magnetic field component in the lobe grows gradually together with a decrease of the polar magnetic field in the current sheet center (Fig. 8). This is interpreted as a signature of the current sheet stretching. At the same time, an increase of the spectral energy index γ, $I \sim E^{-\gamma}$, is observed (spectral softening). The change is associated with the fact that ions at higher energies (larger gyroradii) are no longer trapped in the current sheet and therefore not seen by the detector. Similar events of plasma sheet thinning were observed by Louarn et al. (2000). Eventually, the thin current sheet is disrupted by explosive reconfiguration events characterized by a bipolar magnetic field signature (dipolarization). The dipolarization is associated with a hardening of the spectral energy index gamma (see Fig. 8). Reconnection can be triggered by micro instabilities in the thin current sheets when its thickness is comparable to the gyroradius of the ion and the magnetic field is so distended that $|B_\theta / B_r| \leq 0.025$ (Zimbardo 1993). This condition is satisfied just before the reconfiguration event has happened, see Fig. 8.

Based on EPD data, the reconfiguration of the Jovian magnetosphere appears to be periodic, at least for certain time intervals as during the G2 and G7 orbits, with 2–4 day cadence (Fig. 9). The signature of periodic stretching/thinning of the current sheet and the following dipolarization is well seen in the spectral index gamma (Woch et al. 1998; Kronberg et al. 2007, 2008a, 2009), especially planetward of the X-line. Tailward of the X-line, the periodic signatures of the mass-release on several days scale are rather seen in the first order anisotropy (Krupp et al. 1998; Kronberg et al. 2005, 2007, 2009) and associated with bipolar magnetic field fluctuations. A periodic plasma sheet thinning and thickening were discussed by Vasyliunas et al. (1997). Also observations of ion velocity dispersions, anisotropies and compositional variations by New Horizons show typical repetition time scale of several days

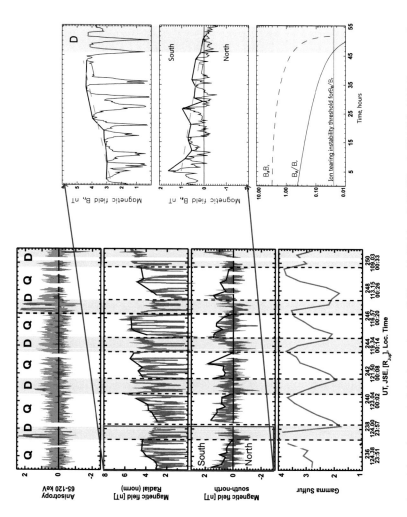

Fig. 8 From the left: Energetic particle and magnetic field observations on Galileo orbit E16 from DOY 235 to DOY 251 in 1998. From top to bottom: first order anisotropies; the absolute value of the radial magnetic field component, the *black line* presents the rough envelope of the lobe values (transient events are neglected); the south-north magnetic field component with the rough envelope of the values in the vicinity of the current sheet; the sulfur energy spectral index γ, over 10 hour-averages. From the right: zoomed part of the marked by *blue line* the time period, the corresponding magnetic field components and their linear fits; at the bottom by *solid line* the ratio of these linear fits is shown which becomes equal to the ion tearing instability threshold right at the beginning of the disturbed time

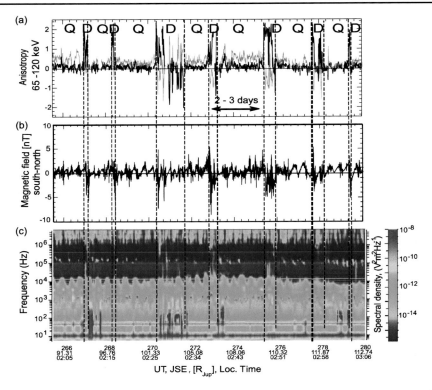

Fig. 9 (**a**) First order anisotropies in the radial and corotational direction; (**b**) the south-north magnetic field component; (**c**) a time frequency spectrogram from PWS data: The observations have been taken in the time interval DOY 265, 1200 to DOY 280, 0000 in 1996 (Galileo orbit G2). Details on the origin of the various radio emissions seen by PWS are described in Fig. 10

(McComas et al. 2007; McNutt et al. 2007). Note, however, that the periodicity is not systematic. There are also time periods of several months during which the magnetosphere does not present signs of periodic activity (Louarn et al. 2000).

A large scale perspective of these reconfigurations can be given by investigating how they are related to intensifications of the radio emissions. In Fig. 10, we present observations of radio emissions made with the Galileo PWS instrument (Gurnett et al. 1992), during the G2 orbit. This figure details the radio data that are shown in Fig. 9. The data are organized in three frequency ranges (respectively: (1) a high frequency band, from 50 kHz to 5.2 MHz, (2) a medium one, from 50 kHz to 250 kHz, and (3) a low one, below 20 kHz), each of them corresponding to emissions coming from different magnetospheric regions: (1) the auroral region (the so-called HOM and b-KOM), (2) the Io torus (n-KOM) and, (3) the continuum emission that filled the whole magnetosphere. The low frequency cut-off of this last emission gives the local upper-hybrid frequency which is, in practice, very close to the plasma frequency.

As described in Louarn et al. (1998, 2000), these dynamic spectra show the occurrence of repetitive radio events, indicated by red ticks in Fig. 10. They are called 'energetic events' in Louarn et al. (1998, 2000, 2001). They are characterized by sudden increases of the intensity of the auroral emissions, a change in the morphology of the n-KOM and strong perturbations in the low frequency cut-off of the continuum. This is indicative of the occurrence of global

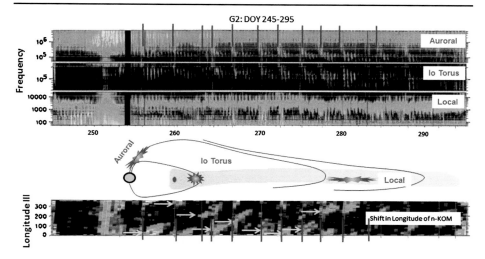

Fig. 10 Details of the radio emissions observed by PWS during the time interval DOY245-1996 to DOY295-1996 (G2 orbit). The *upper panel* shows the dynamical spectra of the 3 main types of Jovian radio emissions. The *middle panel* indicates the location of their respective sources. The *red ticks* mark the occurrence of the 'energetic events'. They coincide with magnetic perturbations and flow anisotropies (see Fig. 9). The *lower panel* is a plot of the nKOM intensity as a function of Galileo system III longitude. Each event corresponds to a shift in longitude of the source (*yellow arrow*). This observation illustrates the global aspect of the 'energetic events' and the related reconnection signatures (from Louarn et al. 2000)

dynamical processes that affect, simultaneously, the auroral zone (2–4 R_J on high latitude field lines), the external part of the Io torus ($\sim$8–10 R_J in the equatorial plane) and the local plasma density, at about 100 R_J. A fascinating feature is also that each new 'event' is linked to the formation of a new source in the torus, located at a different system III longitude than the previous one. They are also related to variations of the density and, most likely, of the total mass content of the disc (Louarn et al. 2000), which may indicate a link with variations in the radial mass transport. Finally, one case studied in Louarn et al. (2001) suggests that these intensifications are linked to injections of energetic particles at 10–15 R_J. The analysis of the whole Galileo data set shows that this is general: all 'energetic events' that have been observed when Galileo is close to Jupiter were correlated with injections of particles (Louarn et al. 2013). This constitutes a further evidence of the global aspect of these phenomena.

These radio events were discovered independently of the quasi-periodic bursts and anisotropies in the energetic particles reported by Woch et al. (1998) and Krupp et al. (1998). It is now clear that all these signatures are various aspects of the same global magnetospheric phenomena.

3.4 Possible Model of the Periodicity

One of possible scenarios in order to explain the periodicity is that the magnetosphere is internally driven. Starting from an initial mass-loaded state further mass-loading of rapidly-rotating flux tubes will lead to the increase of the inertial moment of the tail plasma and the centrifugal force. The magnetotail will continuously stretch and at a certain moment the reconnection conditions will be satisfied, upon which the x-line/point will be formed and the explosive plasma release occurs. As mass-loading is continuous this process will be repeated again and again. The feasibility of this scenario is estimated in the model by Kronberg et al. (2007):

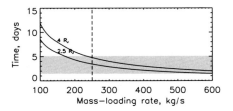

Fig. 11 The intrinsic time constant (in days) of the Jovian magnetosphere needed for mass-loading of the magnetodisc until it becomes unstable and reconnection starts versus the mass-loading rate of the magnetodisc. *Two lines* correspond to the initial thickness of the magnetodisc. The *shaded area* shows the observed periodicities of the plasma sheet topology change from e.g. Woch et al. (1998), Kronberg et al. (2009) and shown in Figs. 4, 5. The *vertical dashed line* show the most probable mass-loading rate

The additional forces will change the stress balance in the Jovian magnetotail. The local stress balance in the co-rotating system, following (Vasyliunas 1983), can be written as:

$$\dot{\rho}\boldsymbol{\Omega} \times \boldsymbol{r} + 2\rho\boldsymbol{\Omega} \times \dot{\boldsymbol{r}} + \rho\boldsymbol{\Omega} \times (\boldsymbol{\Omega} \times \boldsymbol{r}) + \nabla \cdot \boldsymbol{P} = \boldsymbol{j} \times \boldsymbol{B}, \tag{1}$$

where r is the radial distance, $\boldsymbol{P}$ the plasma pressure tensor, $\boldsymbol{j}$ the electric current density, $\boldsymbol{B}$ the magnetic field strength. The mass-loading is allowed in Eq. (1) by including the local time derivative of the mass density. The centrifugal force and mass-loading lead to the change of the magnetic field configuration. The azimuthal component of the local stress balance in the equatorial plane responsible for thinning of the current sheet is

$$\rho\Omega^2 r - \frac{\partial P}{\partial r} \cong -\frac{B_r B_\theta}{\mu_0 d}. \tag{2}$$

We can define the ratio α, which is related to the current sheet topology, by assumption $\frac{\partial P}{\partial r} \cong -\rho\Omega^2 r$ based on observations (Kronberg et al. 2007), and its corresponding temporal variation which depends on the mass-loading rate and the centrifugal force

$$\alpha = \left| \frac{B_\theta B_r}{\mu_0 d} \right| \cong 2\rho\Omega^2 r, \qquad \frac{d\alpha}{dt} = 2\dot{\rho}\Omega^2 r. \tag{3}$$

From this formula the estimation of the intrinsic magnetospheric time constant for the stretching process is derived:

$$\tau \cong \frac{\alpha^{rec} - \alpha^0}{2\dot{\rho}\Omega^2 r} \cong \frac{\frac{B_r^{rec} B_\theta^{rec}}{d^{rec}} - \frac{B_r^0 B_\theta^0}{d^0}}{2\dot{\rho}\mu_0\Omega^2 r}, \tag{4}$$

where the terms with "0" correspond to the initial state of the magnetotail configuration (see Fig. 9) and the terms with "rec" related to values taken just before the disturbed period starts.

The results are seen in the Fig. 11, which is in reasonable agreement with observations.

The formula (3) can be explained more qualitatively. Apparently the result of this simple conceptual model is basically the same as the result of the "minimal substorm model" for explaining the waiting time between two successive substorms by Freeman and Morley (2004). The difference is only in the energy sources. The parameter α_0 characterizes the initial state of magnetospheric configuration and can be influenced by the solar wind conditions. The parameter α_{rec} is a critical energy threshold determined by the conditions for the current sheet configuration favoring reconnection (i.e., the conditions at which the current sheet thickness is of the order of the ion gyroradius). This parameter can be also influenced by the solar wind environment. The parameter $d\alpha/dt = 2\dot{\rho}\Omega^2 r$ (see also Eq. (2)) is the energy power: The energy power in the terrestrial case comes from the solar wind, and in the

Jovian case it depends on the mass-loading power and the centrifugal force, thus it describes an internal energy source. As periodicities are frequently observed in the Jovian magnetosphere (Kronberg et al. 2009) and confirmed by modeling to be feasible, we suggest that the main driving mechanism of the Jovian reconfiguration events is internal. A similar model was developed for Saturn by Rymer et al. (2013).

The models which describe the circulation in the Jovian magnetosphere, e.g. the Vasyliunas cycle that was already presented, the model by Cowley et al. (2003) who combined Vasyliunas cycle with the solar wind interaction Dungey-cycle, the model by Kivelson and Southwood (2005) which outlined the dynamics of the plasma sheet as a function of local time, do not include the transient nature of the magnetotail dynamics described above. It would be an important task for the future to incorporate these transient features discussed above in a more complex model.

It is also natural to compare the reconfiguration events with the terrestrial substorm processes, as suggested by Russell et al. (2000). A phenomenology of individual Jovian substorm-like events was also considered by Woch et al. (1999) and Kronberg et al. (2005). More specifically, the main phenomena of a magnetospheric substorm at Earth can be classified in three groups (see Vasyliunas, abstract for ICS-9): (1) enhanced energy dissipation (auroral emissions) accompanied by (2) changes of the magnetic field configuration from stretched to nearly dipolar, (3) fast (order of Alfven speed or more) plasma bulk flows in the magnetotail. This phenomenological grouping appears to be applicable to Jupiter. The phenomena of group (1) were observed by the Hubble Space Telescope (HST) and related to ground observations, see e.g. Radioti et al. (2010, 2011), Ge et al. (2010). Louarn et al. (1998) reported on periodic intensifications of the auroral emissions using in situ PWS data. These intensifications were associated with the change of the magnetic field topology and with particle bursty bulk flows by Woch et al. (1999). Signatures from group (2) are seen as a change of the magnetic field configuration from a tail-like to a more dipolar configuration (Russell et al. 1998, 2000; Kronberg et al. 2007, 2008a; Vogt et al. 2010; Kasahara et al. 2011; Ge et al. 2007). Eventually the fast plasma bulk flows associated with phenomena of two previous groups were reported by Krupp et al. (2001), McComas et al. (2007), Kronberg et al. (2008b). Transient dispersive ion events were observed by New Horizons at distance from about 880 to 2200 R_J in the Jovian magnetotail by Hill et al. (2009). The source of these events was estimated to be at 150 R_J in the magnetotail, where the Jovian near X-line is located. Therefore, the signatures from all 3 groups are observed in the magnetosphere and are consistent with each other. If one replaces the solar wind stress as in terrestrial case by stress from mass loading and rotationally driven iogenic ion outflow, then the substorm concept might well be adapted to the Jovian magnetosphere.

3.5 Auroral Signatures of Tail Reconnection

Remote observations of Jupiter's ultraviolet and infrared aurora show polar spots which are thought to be associated with inward moving flow from tail reconnection (Grodent et al. 2004), based on their emitted power, mapped location, and typical recurrence time. These spots are located $\sim 1°$ poleward of the main auroral emission, which is associated with a corotation enforcement current system that maps to a radial distance of 20–30 R_J in the magnetosphere (Hill 1979, 2001; Cowley and Bunce 2001). Most of the spots are observed near dawn, and hence are commonly termed "polar dawn spots", though spots have also been reported pre-midnight (Radioti et al. 2011). Figure 12 shows one such nightside spot, which maps to ~ 50–90 R_J and ~ 2100–2400 LT, and a polar dawn spot that maps to ~ 50–80 R_J and ~ 0200–0400 LT (Vogt et al. 2011). The figure also shows the mapping of the

Fig. 12 Polar view of Jupiter's UV aurora as imaged by HST. A polar dawn spot and nightside spot, both thought to be associated with inward flow from tail reconnection, are *circled in pink*. The *white line* shows the location of the statistical x-line (Woch et al. 2002; Vogt et al. 2010) mapped into the ionosphere using the method of Vogt et al. (2011). The *colored lines* delineate three different regions of the polar aurora. Modified from Grodent et al. (2003)

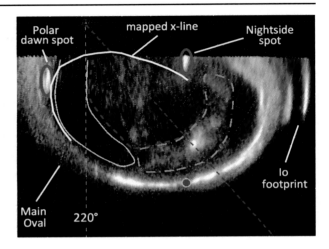

statistical x-line from Woch et al. (2002) and Vogt et al. (2010); most spots lie equatorward of the mapped x-line position in the ionosphere, meaning they map to distances inside of the statistical x-line, which is consistent with their association with inward moving flow released during tail reconnection (Radioti et al. 2010).

In a recent study, Radioti et al. (2011) presented observations of a nightside spot in the UV auroral data taken at nearly the same time that the Galileo magnetometer recorded a reconfiguration signature in the pre-midnight magnetotail (Vogt et al. 2010). The spot mapped close to the position of Galileo and the emitted power derived from the flow bubble in the magnetic field measurements closely matched the emitted power of the nightside spot seen with HST. In general, the emitted power of the polar spots is consistent with the field-aligned currents estimated from typical tail flow bursts (Radioti et al. 2010). It can also be mentioned that HST images collected intermittently from February to June 1997 showed that the polar spots are quite common, appearing on nearly half of the days with observations, and have a typical recurrence period of ~2–3 days (Radioti et al. 2008). This recurrence time is similar to the ~2–3 day periodicity that has been observed in flow bursts (see above).

The auroral observations also nicely complement the available *in situ* magnetic field and particle data by providing a way to study magnetospheric dynamics on a global scale. Without multi-point spacecraft measurements it is difficult to distinguish between spatially and temporally varying features, but the auroral observations enable remote sensing of the magnetosphere over a large range of radial distances and local times. For example, the fact that the polar spots are relatively limited in azimuthal extent suggests that reconnection at Jupiter occurs in relatively narrow channels, consistent with calculations of the flow channel width based on the *in situ* magnetic field data, which estimated a ~15–20 R_J azimuthal extent, or a few percent of the magnetotail width (Vogt et al. 2010).

The auroral observations also show that reconnection at Jupiter may occur simultaneously, in narrow channels, at multiple points across the magnetotail, with spots mapping to both the dawn and pre-midnight local time sectors. Radioti et al. (2011) reported several other cases of simultaneous dawn and nightside spots, though they concluded the total number of observations was too small to conclusively determine whether nightside and dawn spots are typically observed together. Future observational campaigns may help resolve this issue.

3.6 Fine Scale Structures in Relation with Jovian Tail Transient Reconnection Events

So far we have reviewed the large-scale (a few to hundreds of R_J) perspectives of Jovian tail reconnection. They may be compared to the single-fluid global MHD simulations as has been done previously (e.g., Krupp et al. 2001). On the other hand, a speculation from the observations in the Earth's magnetosphere suggests that finer scale structures, in which ions and electrons decouple and/or multi-ion species separate, are embedded in the large scale structures. Resolving such structures would help understanding of key physical processes (e.g., force balance/imbalance and energy dissipation). Unfortunately, however, the large distance between Jupiter and Earth limits the mission data rate, resulting in poor time resolution (and thus coarse spatial resolution). In the case of Galileo, which is a unique Jovian orbiter at the time of this review paper, the failure of the high-gain antenna further hampered transmission of high-resolution data. Nonetheless, the Jovian magnetosphere contains significant amount of heavy ions, and it results in the larger ion scale compared to proton-electron plasma in the Earth's case (note that ion inertial length and gyro radius are larger for heavier ions). Relatively lower plasma sheet density in the tail ($>50\ R_J$) also leads to the larger ion inertial length than the commonly observed values in the near-Earth tail. These facts allow us (at least in some cases) to get a brief glimpse of ion-scale structures of the plasma sheet during reconnection events. Below we review some observations which exhibit reconnection-related fine structures in/around the Jovian nightside plasma sheet, despite moderate time resolution of charged particle data ($\sim$ten minutes) and magnetic field data (a few tens of seconds) of the Galileo spacecraft.

Recent multi-spacecraft missions in the Earth's magnetosphere have revealed earthward propagation of magnetic pulses, which are considered to be a jet front of transient reconnection. They are characterized by a sharp rise followed by a gradual decay of B_θ, and the front (vertical current sheet) thickness is typically of the order of ion gyro-radius and inertial length (Sergeev et al. 2009; Runov et al. 2011; Fu et al. 2012). The propagation speed is close to the local Alfven velocity, and particle energisation and significant density depletion often follow that. Several numerical studies have also reproduced such structures and motions in reconnection simulations.

Similar anomalous magnetic field pulse in the Jovian tail was reported by Russell et al. (1998), and they argued that it is generated by transient reconnection as well. More recently, by using multiple instruments, Kasahara et al. (2011) showed that the magnetic structure was coincident with Alfvénic ion flow as well as an appreciable density decrease, which are similar to the Earth's case. Figure 13 compares the Jovian event to the Earth's one (likely spacecraft path of Galileo is schematically drawn in Fig. 14). In both cases, significant planetward/sunward velocity increase, density depletion, and plasma heating (energetic particle flux enhancement) were similarly observed. The estimated Alfvén speed in this Jovian event was $\sim$650 km/s (derived from the observed magnetic field, electron density, and assumed mass-per-charge of $10m_p/e$, where m_p and e are the proton mass and charge, respectively). Hence the observed flow of 450 km/s is sub-Alfvénic, in conformity with the Earth's case.

In the Jovian case, the duration of the steep increase of B_θ was 36 ± 12 s (the large uncertainty is due to the sampling time of 24 s), and then the front thickness is estimated to be $\sim$10000–20000 km (a few tenth of R_J), assuming the front structure moved with the ion flow velocity (ion velocity was calculated from energetic particle anisotropy). This is apparently much thicker than the Earth's case (300–1000 km or 0.05–0.16 R_E, where R_E is the Earth's radius (Sergeev et al. 2009; Runov et al. 2011; Fu et al. 2012); see also Table 1). Despite such an apparently-large thickness compared to Earth case scales, however, the observed jet front still holds the ion-scale structure. For

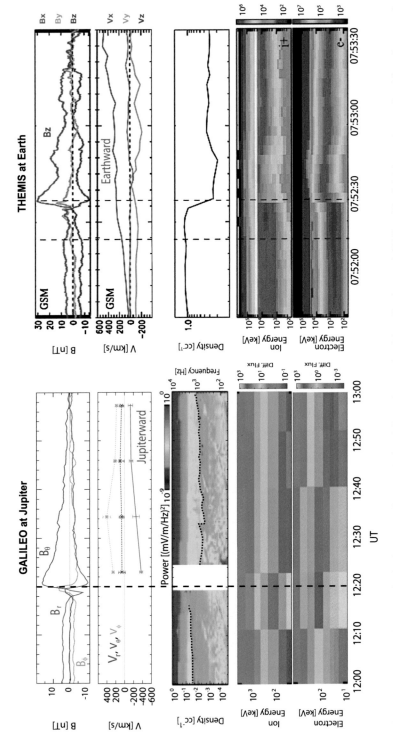

Fig. 13 Comparison of reconnection jet front events between Jupiter (*left panels*, reproduced from Kasahara et al. 2011) and Earth (*right panels*, reproduced from Runov et al. 2009)

Fig. 14 Jet front and Galileo's relative motion during the observation. *Arrows* indicate flows of electrons and ions. From Kasahara et al. (2011)

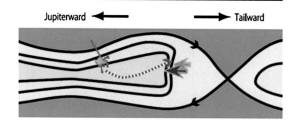

Jupiterward ◄—————   ————► Tailward

Table 1 Comparison of scale length between Jupiter and Earth

Parameters	Jupiter	Earth
Jet front thickness	10000–20000 km	300–1000 km
Ion inertial length	13000–20000 km	300–500 km
Ion gyro radius	6000–8000 km	500–600 km

instance, the ion inertial length c/ω_{pi} is $\sim$13000 km behind the front for O^{2+} or S^{3+} ($\sim$20000 km for O^+), which is comparable to the front thickness (c is the light speed, $\omega_{pi} = (4\pi neq/m_i)^{1/2}$, $m_i/q \sim 10m_p/e$, $n \sim 0.003$ cc^{-1}). Such a large inertial length compared to the Earth's case is due to the larger ion mass and lower electron density.

Additionally, the front thickness is not much larger than the gyroradii of streaming ions ahead of the front. The gyroradii of thermal ions can also be compared with the front thickness (for a temperature Ti of 20 keV/q and adopting $B \sim 10$ nT, the thermal gyroradii are $\sim$6000 km for O^{2+} and S^{3+}, and $\sim$8000 km for O^+, which are not much smaller than the front thickness.

Based on the list of magnetic field disturbance events by Vogt et al. (2010), a survey of reconnection jet front events was conducted (Kasahara et al. 2013). Figure 15a shows all the events with circle sizes indicating the magnitude of the peak magnetic field during events and colors illustrating the signs of the peak value (red and blue represent southward and northward peaks, respectively). The authors found prominent jet fronts mainly on the dawnside. Figure 15b shows only high-peak events (the peak magnitude is >8 nT) with vectors exhibiting ion flows. In addition to the sunward/Jupiterward propagating events as shown in Fig. 13, similar but tailward propagating events were also found, in conformity with previous anisotropy analyses (Woch et al. 2002). As can be seen in the superposed epoch plots in Fig. 16, prominent events exhibit significant density depletion down to <0.01 cc^{-1}. Such a low density is suggestive of low-density plasma intrusion due to the initiation of lobe reconnection, although it is not easy to unambiguously identify lobe reconnection.

Not all but in several moderate events on the dawnside, similar velocity deflection/enhancement and density depletion were also identified (Figs. 15c and 15d). On the other hand, on the duskside, no appreciable density decrease was found although some flow deflection events are identified. This fact indicates lobe reconnection can rarely be seen inside 150 R$_J$ on the duskside, in sharp contrast with the situation on the dawnside.

Reconnection jet front structures described above can be detected when a spacecraft is close to the plasma sheet center. On the other hand, when a spacecraft pass by the separatrixes, which locate on both sides of the plasma sheet center, field-aligned energetic particle beams are frequently observed (e.g., Kronberg et al. 2012). The energetic particles are launched from the reconnection diffusion region and stream away along the magnetic field. Since the magnetic field lines convect inward (equatorward and away from

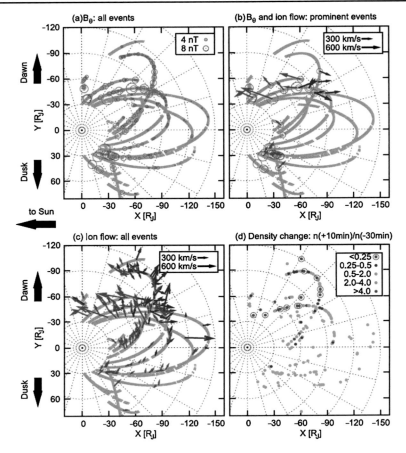

Fig. 15 (**a**) Distribution of B_θ events. The *size of circle* indicates the peak strength of B_θ (reference size is indicated at the *top right*). *Red* and *blue* indicate those events with $B_\theta > 0$ (southward) and $B_\theta < 0$ (northward), respectively. (**b**) Same as Panel (**a**) but only for prominent B_θ events ($B_\theta > 8$ nT). Ion flows perpendicular to the magnetic field are also plotted when they are available. The length of arrow displays the flow speed and reference length is indicated at the *top right*. (**c**) Ion flows perpendicular to the magnetic field for all events. (**d**) Density ratio between 10 min after the B_θ peak and 30 min before that. The value below unity indicates the density decrease after the B_θ peak, and vice versa. In Panels (**a**)–(**c**), the Galileo trajectory is shown in *grey* only for the period in which magnetic field data with sufficient time resolution (60 s/sample or better) are available

the X-line), the energetic particles are also transported inward. The inward convection of the energetic particles proceeds with a common speed, whilst the energetic particles stream along the field line with the speed which is dependent on the energy, mass, and the pitch angle. As a result, layering structures of energetic particles in which higher (lower)-velocity particles stream in the outer (inner) layer are formed, and in fact have been observed in the Earth's magnetosphere (Scholer et al. 1986; Onsager et al. 1991; Sarafopoulos et al. 1997).

Figure 17 displays the velocity layer structure associated with Jovian tail reconnection. The energetic particle species were electrons, protons, oxygen ions, and sulphur ions. These observations are similar to those reported by Sarafopoulos et al. (1997) for the terrestrial

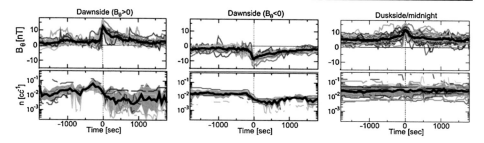

Fig. 16 Superposed epoch plots of plasma and field parameters for each category. B_θ and the electron density n are shown. The average values and standard deviations are shown by *thick black lines* and *grey shades*, respectively. Each *coloured line* corresponds to a particular event throughout a column

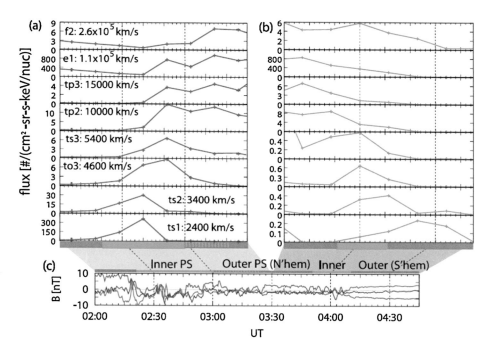

Fig. 17 Omni-directional particle fluxes (**a**) for 0200–0330 UT and (**b**) for 0330–0415 UT and (**c**) the magnetic field data on 17/06/1998. Channels f2 and e1 are of electrons (*top* and *second panels*), and others are of ions. Galileo's location inferred from the magnetic field is illustrated by *purple* and *green bars*. The inner plasma sheet (*inner PS*), the outer plasma sheet (*outer PS*) in the northern hemisphere (*N'hem*), and the outer plasma sheet in the southern hemisphere (*S'hem*) are shown

distant tail. Interestingly, they also noted that in some cases in the Earth's magnetosphere the electron flux peaked at the central plasma sheet rather than at the separatrix, and argued that such a signature would be detected when the field line is closed and the electrons are trapped in a loop or flux-rope structure. According to this interpretation, the present Jovian observations which represent the highest electron flux in the outer plasma sheet/separatrix (and thus the lesser flux in the inner plasma sheet) are consistent with an open field line topology (i.e., lobe reconnection).

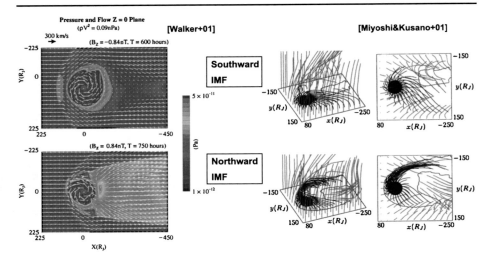

Fig. 18 Global simulations to explore the effect of IMF. For Southward IMF, X-line is formed on the dawn-side in the simulation by Miyoshi and Kusano (2001) (*right top*, illustrated by the *black line* indicating the place where the meridional magnetic field is zero), whilst no X-line is formed in the simulation by Walker et al. (2001). For northward IMF, the X-line extends over the midnight in both simulations

3.7 Simulation and Theory on the Jovian Disc/Tail Dynamics

We now review the existing theoretical works including simulations/modelling which treat Jovian disk/tail reconnection. Of remarkable interest is the comparison of the importance of solar wind effects and internal processes on the transient events, including periodic phenomena. Microscopic view of tail reconnection in the multiple/heavy ion environment is another important problem which is worth studying with numerical simulations.

It is well known that the solar wind dynamic pressure significantly affects auroral morphologies and hence magnetospheric dynamics. However, a link between the solar wind pressure and nightside reconnection has not been well established observationally. Based on the model calculation of solar wind propagation and Galileo observations, Tao et al. (2005) argued that the plasma sheet thinning by the vertical compression are caused by enhanced dynamic pressure.

For southward IMF, no tail X-line is formed in the simulation by Walker et al. (2001), whilst plasma sheet reconnection operates on the dawnside in the case of Miyoshi and Kusano (2001). For null IMF, plasma sheet/lobe reconnection is identified by Ogino et al. (1998). For typical solar wind dynamic pressure (>0.02 nPa) under null IMF, X-line does not appreciably extend to the dusk sector (Fukazawa et al. 2006). For northward IMF, lobe reconnection occurs and an X-line can extend well to the dusk sector. There is a general trend that stronger northward IMF leads to an X-line closer to Jupiter (Fukazawa et al. 2006).

Figure 18 shows the results of southward (top panels) and northward (bottom panels) IMF taken from Walker et al. (2001) (left panels) and Miyoshi and Kusano (2001) (right panels). It should be noted that these results are not necessarily identical. For instance, no X-line is formed in the top left panel, whilst plasma sheet reconnection occurs in the top right panel. It is not clear if this difference is due to the different solar wind parameters or due to the different boundary conditions and numerical methods. If simulation results from several different groups (for which boundary conditions are different) are compared with the same solar wind parameters, they may provide some clue to understand what controls magnetospheric responses.

Present global simulations, however, mainly focuses on the interaction with the solar wind, and have not addressed the internal driving of disturbance events such as transient tail reconnection and associated (or independent) hot-plasma injection back into the inner magnetosphere. The tail disturbance of 2–3 days period is one of the issues which await theoretical approach. There is no consensus on if the observed periodic bursts are explained purely by the internal process or it requires specific solar wind conditions (cf. Kronberg et al. 2007; Fukazawa et al. 2010). Further studies are needed to reveal if transient events occur only under some specific solar wind condition or not.

Another interesting aspect is the effect of multiple ions on the reconnection. For instance, multiple scale structure is expected around the center of reconnection sites (i.e., diffusion region and separatrixes), as illustrated by Shay and Swisdak (2004) and Markidis et al. (2011). However, we already mentioned in Sect. 2 that detailed simulations of reconnection processes in plasma conditions that are relevant for the study of plasma disks, as proposed by Yin et al. (2000), are very rare. This is a largely open field and, clearly, new calculations would greatly help the interpretation of future observations in the Jovian magnetosphere.

4 Observations in Saturn's Disk and Tail

4.1 Local Reconnection Signatures

The first spacecraft to sample Saturn's magnetosphere were Pioneer 11 and Voyager 1 and 2. However they merely glimpsed the nightside of the planet and revealed little about dynamics of the magnetotail. With the arrival of the Cassini spacecraft in 2004, we had a chance to explore Saturn's magnetotail in detail for the first time. Indeed the first hint at the existence of the reconnection process in Saturn's tail came during the Saturn Orbit Insertion manoeuvre. Bunce et al. (2005) reported a field disturbance which they attributed to a dipolarization, accompanied by hot plasma injection. They interpreted this as solar wind compression-induced reconnection in the tail followed by injection of hot plasma into and around the inner magnetosphere.

Since orbit insertion the Cassini spacecraft subsequently explored many regions of Saturn's magnetosphere. The best nightside coverage came during 2006 when the spacecraft executed its deepest tail orbits, reaching maximum downtail distances of ~ 68 R_S (1 $R_S = 60268$ km), albeit with a strong dawn-dusk asymmetry, with little exploration beyond ~ 25 R_S on the dusk-side. Figure 19 shows a trajectory plot of the coverage of the first few years of the Cassini mission, including the location of 5 possible signatures of reconnection from 2006 discussed in Jackman et al. (2008). Since this study, a resurvey of the Cassini magnetometer data from 2006 uncovered many more events, and Jackman et al. (2011) reported 50 possible signatures, 34 plasmoids and 16 travelling compression regions (TCRs). At the time of writing, significant work is ongoing to categorize a number of new candidate plasmoid and TCRs at Saturn, to decipher their average properties, to probe their interior structure, and to calculate their likely recurrence rate (Jackman et al. 2014).

How are these structures identified in the data? The principles are basically the same as those used with Galileo data. Reconfigurations and possible signatures of reconnection are identified in magnetometer data through looking for a change from a radially stretched field

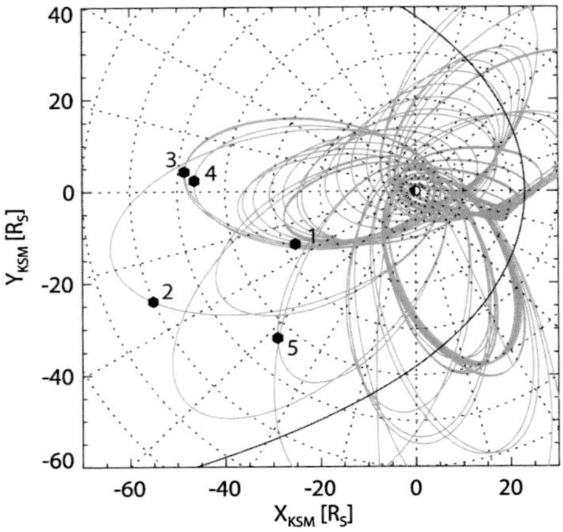

Fig. 19 Orbit of the Cassini spacecraft around Saturn shown in an equatorial projection in Kronocentric Solar Magnetospheric (KSM) coordinates, where X points from Saturn to the Sun and Z is north in the plane containing X and the Saturn rotation axis. The Sun is to the right of the diagram, and the magnetopause with subsolar standoff distance of 26 R$_S$ is shown in *black* (Arridge et al. 2006). The location of five tail reconnection events is shown on the plot. From Jackman et al. (2008)

Fig. 20 Schematic figure of the magnetic signatures of various magnetotail structure: (**a**) Plasmoid, (**b**) Plasmoid with a flux rope core, (**c**) Flux rope and (**d**) TCR. This figure is for the terrestrial magnetosphere, but is also applicable at Jupiter and Saturn where field directions are reversed due to the oppositely-directed planetary dipoles. From Zong et al. (2004)

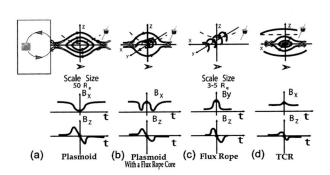

line to a dipolarized field line, or a plasmoid signature. These are manifested as changes in the north-south component of the field. Figure 20 from Zong et al. (2004) illustrates the various configurations of the tail field that lead to various signatures. It completes the sketches shown in Fig. 2. This figure is for the case of the Earth, but it is applicable at Jupiter and Saturn also. Specifically plasmoids may be characterized in terms of their interior structure, yielding either loop-like or flux rope-like signatures at the centre of the structures. Meanwhile travelling compression regions (TCRs) manifest themselves as a localized compression in the field, signifying the warping of lobe field lines around plasmoids as they move downtail.

Figure 21 shows an example of a tailward-moving plasmoid as observed by the Cassini magnetometer on August 4th 2006. The data are in Kronocentric Radial Theta Phi (KRTP) co-ordinates, where the radial component is positive outward from Saturn, the theta component is positive southward, and the azimuthal component is positive in the direction of corotation (in a prograde direction). This spherical polar system referenced to the northern spin and magnetic axis of the planet has been showed to be the best system for unambiguously detecting reconnection events in Saturn's tail. Changes in the north-south component indicate plasmoids/TCR and/or dipolarizations, while the radial and azimuthal components can also be used to elucidate the degree of corotation of the plasma. A thorough discussion of the unambiguous detection of reconnection in Saturn's tail was presented by Jackman et al. (2009a). The main signature of interest in Fig. 21 is clear to see in the second panel; a strong northward turning of the field followed by a slow recovery of the field back to

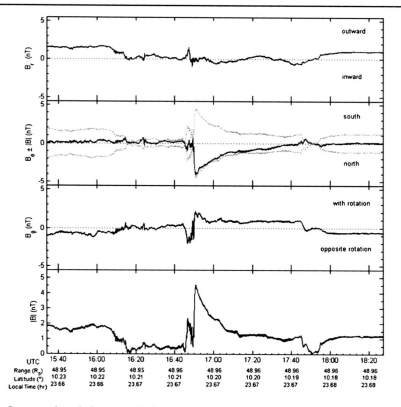

Fig. 21 One second resolution magnetic field measurements from Cassini in Saturn's magnetotail for an interval from August 4, 2006. The *panels* show radial, theta, and azimuthal components in KRTP co-ordinates, and the total magnetic field strength. In the *second panel*, the magnitude of the field is superimposed in *red* (plus and minus) on the theta component, to illustrate when the values become comparable. Information detailing the radial distance, latitude, and local time of the spacecraft with respect to Saturn is given. From Jackman et al. (2007)

pre-event values. This event was observed while the spacecraft was near the centre of the current sheet (as indicated by the near-zero radial field component). It is also worth noting that for several minutes after the main plasmoid passage, the radial and azimuthal components have the same sign, which may be interpreted as the plasma being sped up from corotation.

Of course this is just one example of a plasmoid at Saturn and it is of interest to explore not just case studies but also to understand the average properties of reconnection products. Thus in Fig. 22 we show a superposed epoch analysis of 34 plasmoids at Saturn. This plot shows that the average tailward-moving plasmoid at Saturn is represented by a ~1 nT field deflection, although the signature itself is highly asymmetric, with a ~0.3 nT southward deflection and a ~0.7 nT northward turning. The duration of this central field change, as defined by the local south/north extrema in the field, is ~8 minutes. However, there is a long (~58 min) interval after plasmoid release where the field remains northward. This interval is analogous to the post-plasmoid-plasma-sheet (PPPS), originally proposed for the Earth (Richardson et al. 1987). It corresponds to an interval post-plasmoid release where previously open flux is being closed through ongoing lobe reconnection. In the next section

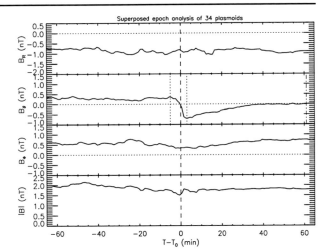

Fig. 22 The radial, north-south and azimuthal components of the magnetic field together with the magnetic field strength for a superposed epoch analysis of 34 plasmoids at Saturn. The zero epoch is the central event time (marked by the *vertical dashed line*) and the plot range spans 65 min either side of the events. *Vertical dotted* and *dot-dashed lines* mark the start/end times of the average plasmoid and the end of the PPPS. From Jackman et al. (2011)

we discuss the PPPS and the role of reconnection in flux closure in more detail. We also note that the total field strength dips slightly coincident with the passage of a plasmoid. This lack of a "core field" signature hints that the interior structure of plasmoids at Saturn is loop-like as opposed to flux rope-like, although this is currently being intensively studied (Jackman et al. 2014), and has been shown to be highly trajectory-dependent (e.g. Borg et al. 2012).

While signatures of plasmoids and TCRs can be most easily observed using magnetometer data in the first instance, multiple other data sets have been used to provide a complete picture of the effects of reconnection on magnetospheric dynamics.

Figure 23 shows data from the Cassini magnetometer (MAG) and Cassini Plasma Spectrometer (CAPS) instruments during a reconnection event on March 4th 2006. The largest plasmoid is visible at ∼2300 UT in the form of a large (∼3 nT) northward turning of the field. This is accompanied by a strong change in the plasma flows in Saturn's tail. The flows prior to the plasmoid release were primarily in the azimuthal direction, whereas after plasmoid release there is a strong spike in the radial component of the velocity. The ion composition was generally (but not always) dominated by W^+, which raises interesting questions about the initiation process for reconnection and its location, topics which we tackle in Sect. 4.2.

A further example of using plasma data to study the effects of reconnection on the magnetotail is shown in Fig. 24. McAndrews et al. (2008) employed data from the CAPS ion mass spectrometer (IMS) to extract flow velocities. In this flow map the effect of plasmoid release is clear to see, in particular for the March 4th event shown in Fig. 23. During "steady state", plasma flow in the tail tends to be predominantly in the azimuthal direction, whereas plasmoid release can result in dramatic change of plasma flow direction to nearly radial.

Reconfiguration also results in changes in energetic neutral atom (ENA) fluxes. The relationship between substorm-like events at Saturn and ENA was first shown by Mitchell et al. (2005) and has subsequently been discussed in a number of papers (e.g. Hill et al. 2008; Mitchell et al. 2009). Figure 25 shows an example of an ENA intensification from day 129 of 2008. Dynamic ENA emissions have been found to peak in the 20–30 R_S region, leading to the suggestion that this is the region where Saturn's reconnection X-line may lie. We discuss reconnection onset in more detail in Sect. 4.2.

Fig. 23 MAG and CAPS data for the plasmoid event on March 4th 2006. The *top panel* shows the magnetic field components, the *second panel* shows number densities of electrons, water-group ions W^+, and protons H^+. The *next panel* shows temperatures for the same species. The *bottom three panels* show ion velocity components in the same (r, θ, φ) coordinate system. Rigid corotation would imply $v_\varphi = 430$ km/s, off the top of the v_φ scale. From Hill et al. (2008)

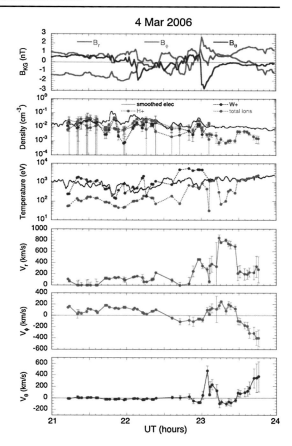

4.2 Event Properties

The question of where reconnection begins is an important one. As mentioned above for Jupiter, the statistical analysis of the change of the north-south component of the field, combined with signatures of flow moving tailward and planetward, can be used to derive a statistical separatrix. However, we have not been able to do this at Saturn yet due to the comparatively smaller number of events observed (total of 50 reported by Jackman et al. 2011) and due to the fact that the small number of planetward-moving events observed are spread over a relatively wide range of radial distances in an unsystematic way.

Thus, in the absence of a statistical x-line based on observation, we look to modeling to provide a sense of where reconnection may be likely to begin in Saturn's tail. Figure 26 shows the output of a MHD model illustrating how the position of Saturn's reconnection x-line can change depending on solar wind conditions. The radial range of x-line positions in their model is from ∼25–40 R_S, with the x-line moving closer to the planet and narrowing following solar wind compression of the magnetosphere.

What is the recurrence rate of reconnection at Saturn? Despite the almost perfect alignment of Saturn's magnetic and spin axes, many phenomena in Saturn's magnetosphere display distinctly periodic behavior at or near the expected planetary period. Thus it is natural that the question has been raised as to whether tail reconnection (and plasmoid release) also

Fig. 24 Equatorial flow pattern of ions in Saturn's magnetosphere. The velocity vector of each position was determined using the spacecraft position as the arrow's origin. Arrow length represents velocity and colour the total ion density. Scale arrows for strict corotation at radial distances of 5, 15, 25, 35 and 45 R_S have been indicated along with the magnitude of the corotation speed at that distance. The velocities are in a Saturn-centered equatorial system with the Sun toward the *right* of the figure. A sample magnetopause surface from Arridge et al. (2006) has been included for context. The flows detected during two plasmoid events on March 4, 2006 and August 4, 2006 have been included. Note that the flow speed observed on March 4, 2006 has been scaled down by ~50 % relative to the rest of the flows for clarity. From McAndrews et al. (2008)

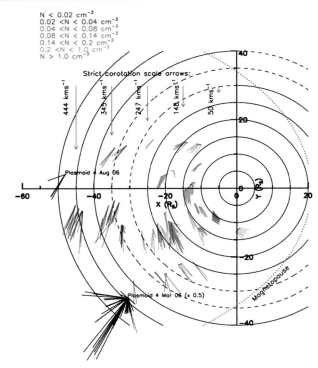

occurs with regularity. Several model outputs hint at quasi-periodic plasmoid release (e.g. Zieger et al. 2010; Jia et al. 2012) with a higher recurrence rate with higher solar wind dynamic pressure. However, despite extensive sampling of the magnetotail by the Cassini spacecraft in 2006, no evidence has been found for plasmoid release every planetary rotation from in situ field and plasma data sets. Note, however, that the interpretation of an absence of observational evidence could be ambiguous: it cannot be excluded that Cassini misses plasmoid signatures in case they do not span the entire tail. Jackman et al. (2009a) caution that the use of appropriate co-ordinate systems is crucial when analyzing magnetometer data to differentiate between observations of plasmoids and observations of Saturn's wavy current sheet. However, while it does not appear from rigorous data surveys that plasmoids are present every ~10.8 hours, there is a curious relationship between the occurrence times of plasmoid release and the phase of the planetary rotation. Jackman et al. (2009b) studied nine reconnection events and found that 8 out of 9 occurred during a particular sector of Saturn Kilometric Radiation phase, a sector where the SKR power would be expected to be rising with time. This implied that there may be a preferred longitude for plasmoid release. Since the discovery of many more plasmoid and TCR examples in Saturn's tail since then, work is underway to explore this relationship (and its physical interpretation) in much more detail (Jackman et al. 2014).

4.3 The Large Scale Perspective

We now consider the larger role that tail reconnection can play in magnetospheric dynamics. Figure 27 shows the large-scale cycles of plasma flow at work in Saturn's magnetosphere, showing the principle of a co-existence between the Dungey's and the Vasyliunas's cycle

Day 129, 2008 (First Intensification at 0945 UT)

Hydrogen 20-50 keV	Hydrogen 50-80 keV	Oxygen 64-144 keV

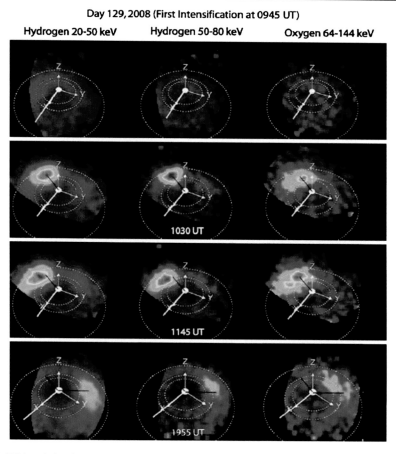

Fig. 25 ENA emission from Saturn's magnetosphere on Day 129, 2008. *Heavy white line* indicates direction toward sun in every image. The *light white arrow* points to dusk, and the *pink coordinate axes* rotate with the planet, indicating the SLS3 system. *Light white circles* are orbits of Dione (6.5 R_S), Rhea (8.5 R_S), and Titan (20 R_S). *Each column* contains data for the species and energy indicated, and *each row* is from the time labeled in *white* in the center column. The images are 1 hour integrations centered at the time indicated. The *black diagonal line* in the images in the *second row* is drawn from the center of Saturn to the peak emission in the image. This line is repeated each column, showing the initial enhancement in emission at the same location for each species and energy. The same line is drawn in the third row, to illustrate the rotation of the peak emission over the intervening 1.25 hours. In the *bottom row*, the line is repeated for reference, and a new line is drawn from the center of Saturn to the peak 20–50 keV H emission (*left column*). The same line is repeated in the *center* and *right columns*, illustrating the additional rotation experienced by the higher energy source ions responsible for the ENA emission in those images. From Mitchell et al. (2009)

(see Sect. 2.3). One of the key differences between the Dungey and Vasyliunas cycles is that the former is driven by the solar wind and involves the closing of previously open flux, whereas the latter is internally driven and involves closed field lines only. Hence, as discussed in Sect. 2, we might expect the signatures of reconnection associated with these cycles to have characteristic differences.

An interesting question concerns the location of the reconnection X-line (Dungey's or Vasyliunas's cycle X-line) as a function of local time across the magnetotail. The theoreti-

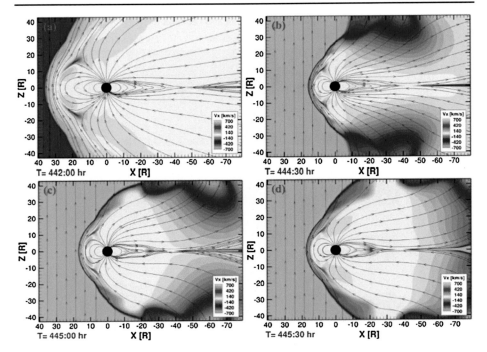

Fig. 26 Magnetospheric configuration prior to and after the shock compression around $T = 442$ h. Color contours of Vx (km/s) and magnetic field lines in the XZ plane at $Y = 0$ (the noon-midnight meridian) are shown in each panel. (**a**) Model results at $T = 442$ h, prior to the shock arrival; (**b**)–(**d**) model results from subsequent times after the shock arrival. From Jia et al. (2012)

cal picture put forward by Cowley et al. (2004) suggested that reconnection on closed field lines occurs predominantly in the dusk sector, with Dungey-cycle open field line reconnection dominating towards dawn. Thomsen et al. (2013) surveyed the dusk orbits of Cassini in 2010, and found no evidence for outward flow in this region. They interpreted this to mean that Vasyliunas-style reconnection may have occurred on the dusk flank but that these plasmoids are still trapped within outer closed field lines and thus not free to escape downtail until they reach the post-midnight sector.

Figure 28 is a schematic picture of the expected field signatures associated with Dungey- and Vasyliunas-cycle plasmoid release. A key difference is the presence (or absence) of a PPPS. As discussed in Sect. 4.1 above, the superposed epoch analysis of plasmoids at Saturn indicates the presence of a significant PPPS, indicating that reconnection in Saturn's tail may result in the closure of significant flux. This does not preclude Vasyliunas cycle activity, as it is entirely possible that closed field line reconnection is ongoing prior to the involvement of open field lines. Jackman et al. (2011) estimated that the average flux closed during a reconnection event in Saturn's tail is $\sim$3 GWb. Although this involves making assumptions about the azimuthal extent of the reconnection region (with an upper limit of $\sim$90 R$_S$ as the full tail width), and the length of region (obtained from the duration of observation multiplied by the plasma velocity), this agrees with the model predictions of Jia et al. (2012) who estimated $\sim$3.5 GWb. In the next section we will discuss how this correlates with auroral images.

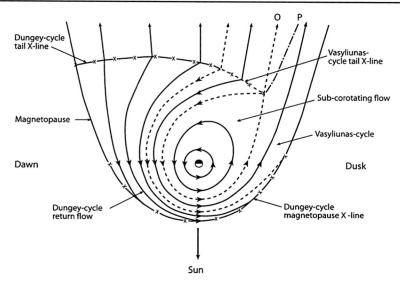

Fig. 27 Sketch of the plasma flow in the equatorial plane of Saturn's magnetosphere, where the direction to the Sun is at the *bottom* of the diagram, dusk is to the *right*, and dawn is to the *left*. *Solid curves with arrows* show plasma streamlines, *short-dashed curves with arrows* show the boundaries between flow regimes (also streamlines), the *solid curves joined by crosses* show the reconnection lines associated with the Dungey cycle, and the *dashed curves with crosses* show the tail reconnection line associated with the Vasyliunas cycle. The *curve indicated by the O* marks the path of the plasmoid O line in the Vasyliunas cycle flow (also a streamline), while *P* marks the outer limit of the plasmoid field lines, which eventually asymptotes to the dusk tail magnetopause. From Cowley et al. (2004)

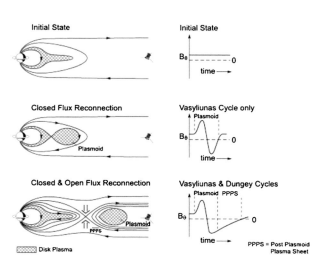

Fig. 28 Schematic of three possible states of the magnetotail and the associated B_θ signatures as measured by a spacecraft situated down the tail. From Jackman et al. (2011)

As already mentioned, a potential important role of the plasmoid ejection is to contribute to the final evacuation of the plasma of internal origin (mostly rings and Enceladus at Saturn).

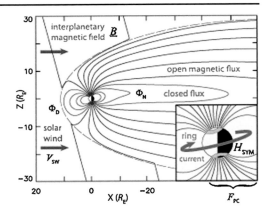

Fig. 29 A schematic diagram of the magnetosphere illustrating regions of closed (*red*) and open (*blue*) field lines. The *inset panel* shows the relationship between the footprints of the open flux and the size of the polar cap, the dim ionospheric regions encircled by the auroral ovals. From Milan (2009)

4.4 Auroral Signatures

While in situ measurements in the tail undoubtedly provide excellent information of reconnection products, auroral imaging is a fantastic complementary tool to help us understand the global effects of magnetic reconnection.

Figure 29 shows how regions of open and closed magnetic flux map into the auroral zone. At Saturn, the main auroral oval has been shown to be formed at the boundary between open and closed magnetic field lines (e.g. Bunce et al. 2008). Badman et al. (2005) studied a sequence of Hubble Space Telescope (HST) images which showed the expansion and contraction of the main auroral oval at Saturn. They measured the size of the polar cap and calculated the flux content of the magnetotail from this, showing that it varied from ∼15–50 GWb, changing constantly in response to addition of flux at the dayside and removal of flux at the nightside via reconnection. Thus sharp changes in the size of the main auroral oval at Saturn can be used as a remote proxy for understanding changing flux content globally.

There are also smaller-scale changes in Saturn's auroral emission which can be linked to tail reconnection. Figure 30 from Jackman et al. (2013) shows examples of small spots post-midnight, just poleward of Saturn's main auroral oval. These are suggested to be linked to reconnection and dipolarization of the field in Saturn's tail, resulting in a diversion of the cross-tail current, and the production of field-aligned currents which in turn yield this distinct ionospheric counterpart.

5 Concluding Remarks

There are many challenges to studying reconnection processes at the outer planets, including the lack of available upstream solar wind measurements, the difficulty in separating temporal and spatial effects with single spacecraft measurements, and incomplete field/particle measurements. These difficulties should always be considered when interpreting Galileo and Cassini observations. This caution being expressed, the set of work presented in this review illustrates that possible signatures of reconnection are relatively commonly observed in the Jovian and the Kronian disks and tails. At Jupiter, reconnection signatures in the magnetic field data (nearly 250 events) and energetic particle measurements (34 events) have been recorded in the nightside middle and outer magnetosphere. They are detected at typical distances of ∼70–100 R_J. Statistically, these signatures define an average 'X-line' region near

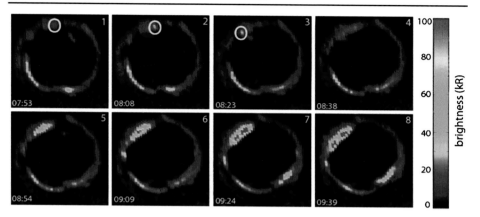

Fig. 30 UVIS pseudo-images of Saturn's northern polar region for 2008 day 129. All images are aligned with noon to the *bottom*, dawn to the *left* and dusk to the *right*. The start time of each 12-minute slew is marked on each image. The spot suggested to be linked to a dipolarization of the tail is circled in *yellow*. The *color bar* gives a correspondence between the color table and the emission brightness in kiloRayleighs (kR) of H_2, where $1\,\text{kR} = 10^9$ photons cm^{-2} s^{-1} emitted in 4π sr by H_2 molecules in the EUV + FUV range (excluding Ly-α) and assuming no absorption by methane (Gustin et al. 2009). The polar projection procedure does not preserve photometry; therefore, the color table may only be used as a proxy for the projected emission brightness. From Jackman et al. (2013)

90 R_J, with evidence of (i) inward flows and dipolarizations on its planet side and (ii) outward flows and plasmoids on its anti-planet side. At Saturn, the number of reported events is about 50, seen at distances larger than 25 R_S. At both Jupiter and Saturn, the magnetic signatures of tail reconnection are similar to those observed at Earth, however with a dominant proportion of loop-like plasmoids rather than flux ropes.

In general, the reconfiguration events are sporadic, both at Jupiter and Saturn. They are associated with relaxation of magnetic stress (dipolarization), they are accompanied by fast flow bursts and, in several cases, their radio and auroral signatures have been detected, showing that they may be associated to large scale magnetospheric disturbances. However, an important distinction concerns the origin of the relaxed energy. At Earth, the relaxation is a response to the transfer of magnetic flux from the dayside to the nightside resulting from the 'opening' of the magnetosphere due to reconnection triggered by solar wind interactions. At Jupiter, the relaxed energy is likely of internal origin. It is linked to the internal plasma supply, the mass loading of the disk and the resulting progressive accumulation of rotational stress. The case of Saturn is still not clarified. The most recent studies nevertheless suggest that reconnection at the magnetopause is unlikely to be a major process, so that Saturn could be more Jupiter-like than thought some years ago.

Many new questions and directions of investigations have also been opened by Galileo and Cassini observations. They concern, for example, (1) the role of the reconnection in the radial plasma transport and (2) in the final evacuation from the disk, (3) the dawn/dusk asymmetry and the interaction between Dungey and Vasyliunas cycles, (4) the global picture of the reconfigurations, (5) their possible periodicity, and (6) the details of the plasma processes that are able to trigger reconnection in a disk, with questions on their spatial and temporal scales. What is the place of reconnection in the general pictures of the Jovian and Kronian magnetospheric dynamics is a largely open question. Equally important is the following problem: are there fundamental characteristics of magnetic reconnection that inherently determine the main dynamical modes of the giant magnetospheres? What are they and how do they regulate the dynamics?

Point 1: Reconnection and radial transport. The role played by reconnection in the mechanism of radial plasma transport across the disk would need clarification. On the one hand, evidences of dipolarization have been seen in the inner/middle magnetosphere, as close as 30 R_J in the Jovian case, as well as reversals in B_θ at distance of 44 R_J (Vogt et al. 2010), which indicates that reconnection may act in the middle part of the disc. On the other hand, although the statistical analysis is perhaps still incomplete, these signatures seem rare. The total time spent by Galileo and Cassini at distances within 60 R_J and 25 R_S is at least several hundreds of hours, so that the small number of reported cases of reconnection in the inner/middle disc is likely meaningful. If this is confirmed, this would mean that the reconnection is not the prime mechanism that contributes to the radial plasma transport in the major portion of the disk (from typically the Io torus to 60–80 R_J, at Jupiter). The reconnection would be rather the mechanism that allows the final plasma evacuation at the edge of the disk and its transition to the tail. However, it is also not excluded that the reconnection in a disk might take a different form than the classic tail signatures and hence has not yet been identified from the available observations.

Point 2: Reconnection and plasma evacuation. The key question concerns the effective role of plasmoid formation in the final plasma evacuation. At Jupiter, the current estimates of the mass of a single plasmoid ($\sim 3 \times 10^4$–2×10^6 kg) and their recurrence rate ($\sim$hours to days) suggest that they would not play a major role in the plasma evacuation. The sporadic reconnection is estimated to result in a mass loss rate of ~ 1–125 kg/s, which would thus only explain a minor fraction of the iogenic plasma input to the magnetosphere (~ 500–1000 kg/s). A similar conclusion was obtained by Bagenal and Delamere (2011), in the case of Saturn. Even with generous number, as plasmoids with a volume of $(10 R_J)^3$ and density of 0.01 cm^{-3}, an ejection rate of 200 per day would be needed to evacuate 100 kg/s. Given that the production rate is 8–250 kg/s, tens of plasmoids should be evacuated per day, which has not been observed. The observation of a continuous flow at the boundary of the plasmasheet, with flux that could reach $\sim 10^3$ kg/s, at Alfvenic speed, could offer a solution. If this is confirmed by further observations, one interesting question would be to investigate whether this flow is related to a continuous form of reconnection, leading to a permanent cross-field plasma transport. This flow may be completed by the transport resulting from viscous interaction with the solar wind on the flanks of the magnetosphere and, possibly, microscopic processes as the loss by charge exchange (see estimates in Bagenal and Delamere 2011).

Point 3: Dawn/dusk asymmetry. Is the reconnection seen in the dawn sector similar to the one occurring in the pre-midnight and midnight sectors? This question is addresses the relative importance of internal/external effects and their organization in local time. At Jupiter, it seems clear that more signatures of reconnection seen closer to the planet on the dawn sector (50 R_J) than on the midnight sector (100 R_J) (Vogt et al. 2010; Kasahara et al. 2013). At Saturn, a recent work by Thomsen et al. (2013) raises the hypothesis of plasmoid production at dusk and their trapping until they reach the post-midnight sector. The origin of these asymmetries is not elucidated. This thematic is also related to the interaction between Dungey and Vasyliunas cycles, which is a particular important question at Saturn. The statistical distribution of reconnection at Saturn does not show a specific organization, with for example an 'average' X-line that would mark a transition from the disk to the tail as at Jupiter. This opens the question on how the Kronian disk is connected to the tail, and on how variable is this transition region.

Point 4: Global picture of reconfiguration events. At Jupiter, the radio observations reveal that the sporadic reconnection events, identified from magnetic field dipolarizations and reversals and flow anisotropies are related to intensifications of the radio emissions coming from the auroral regions and the Io torus. The radio emission intensifications are also associated with significant changes in the vertical density profile of the disk (Louarn et al. 2000) and, thus, most likely, changes in its mass content. Furthermore, in one case occurring when Galileo is in the inner magnetosphere, it can be shown that the radio event correlates with an injection of energetic particles in the Io plasma torus. In overall, this leads to a large scale picture of the dynamics of the disk, in which the sporadic reconnections are one element among others. More investigations would be needed to characterize these global phenomena, including their auroral signatures. A similar picture should well apply at Saturn, with low frequency extension in the SKR linked to reconnections events (Jackman et al. 2009b), and radio intensifications observed at the time of recurrent plasma energizations (Mitchell et al. 2009). The possible link between these energization and changes in the disk plasma contents is an important topic that would deserve future analysis

Point 5: Periodicity. The periodicity is another topic of great interest. There are time intervals of several weeks during which the Jovian magnetosphere presents a quasi-periodic activity, with typical period of 2–3 days. This quasi-periodic behavior, however, is not systematic and there are also long intervals during which no periodicity can be reported. These apparently different states in the dynamics of the Jovian magnetosphere are unexplained.

Point 6: Reconnection process in rotating plasma. We have listed many reasons for which reconnection in a disk may differ from the standard 2D model. To date, the most detailed investigation of a reconnection event occurring in the Jovian disk suggests that the processes scale as the typical ion Larmor radius, meaning scales about a factor 10–20 larger at Jupiter than at Earth (Kasahara et al. 2011). This is a first attempt to analyze the meso and small scale processes that may trigger reconnection in a rotating multi-species plasma. This topic is largely unexplored, even by numerical means. Performing theoretical analysis and simulations of reconnection in plasma configurations that are relevant for the analysis of the dynamics of magneto-disk should be a priority for the interpretation of future observations.

Acknowledgements   The authors acknowledge the support of EUROPLANET RI project (Grant agreement no.: 228319) funded by EU; and also the support of the International Space Science Institute (Bern). The Editor thanks the work of two anonymous referees.

References

V. Angelopoulos, The THEMIS mission. Space Sci. Rev. **141**, 5–34 (2008). doi:10.1007/s11214-008-9336-1

C.S. Arridge, N. Achilleos, M.K. Dougherty, K.K. Khurana, C.T. Russell, Modeling the size and shape of Saturn's magnetopause with variable dynamic pressure. J. Geophys. Res. **111**, A11227 (2006). doi:10.1029/2005JA011574

S.V. Badman, S.W.H. Cowley, Significance of Dungey-cycle flows in Jupiter's and Saturn's magnetospheres, and their identification on closed equatorial field lines. Ann. Geophys. **25**, 941–951 (2007)

S.V. Badman, E.J. Bunce, J.T. Clarke, S.W.H. Cowley, J.-C. Gérard, D. Grodent, S.E. Milan, Open flux estimates in Saturn's magnetosphere during the January 2004 Cassini-HST campaign, and implications for reconnection rates. J. Geophys. Res. **110**, A11216 (2005). doi:10.1029/2005JA011240

S.V. Badman, A. Masters, H. Hasegawa, M. Fujimoto, A. Radioti, D. Grodent, N. Sergis, M.K. Dougherty, A.J. Coates, Bursty magnetic reconnection at Saturn's magnetopause. Geophys. Res. Lett. **40**, 1027–1031 (2013). doi:10.1002/grl.50199

F. Bagenal, P.A. Delamere, Flow of mass and energy in the magnetospheres of Jupiter and Saturn. J. Geophys. Res. **116**, A05209 (2011)

F. Bagenal, J.D. Sullivan, Direct plasma measurements in the Io torus and inner magnetosphere of Jupiter. J. Geophys. Res. **86**, 8447–8466 (1981). doi:10.1029/JA096iA10p08447

C.H. Barrow, F. Genova, M.D. Desch, Solar wind control of Jupiter's decametric radio emission. Astron. Astrophys. **165**, 244–250 (1986)

J.W. Belcher, The low-energy plasma in the jovian magnetosphere, in *Physics of the Jovian Magnetosphere*, ed. by A.J. Dessler (Cambridge University Press, Cambridge, 1983), pp. 68–105. doi:10.1029/CBO9780511564574.005

J. Birn, E.R. Priest (eds.), *MHD and Collisionless Theory and Observations* (Cambridge University Press, Cambridge, 2007)

D. Biskamp, *Magnetic Reconnection in Plasmas*. Cambridge Monographs on Plasma Physics (2005)

B. Bonfond, M.F. Vogt, J.-C. Gerard, D. Grodent, A. Radioti, V. Coumans, Quasi-periodic polar flares at Jupiter: a signature of pulsed dayside reconnections? Geophys. Res. Lett. **38**, L02104 (2011). doi:10.1029/2010GL045981

A.H. Boozer, Model of magnetic reconnection in space and astrophysical plasmas. Phys. Plasmas **20**, 032903 (2013)

A.L. Borg, M.G.G.T. Taylor, J.P. Eastwood, Observations of magnetic flux ropes during magnetic reconnection in the Earth's magnetotail. Ann. Geophys. **30**, 761–773 (2012). doi:10.5194/angeo-30-761-2012

E.J. Bunce, S.W.H. Cowley, D.M. Wright, A.J. Coates, M.K. Dougherty, N. Krupp, W.S. Kurth, A.M. Rymer, In situ observations of a solar wind compression-induced hot plasma injection in Saturn's tail. Geophys. Res. Lett. **32**, L20S04 (2005). doi:10.1029/2005GL022888

E.J. Bunce et al., Origin of Saturn's aurora: simultaneous observations by Cassini and the Hubble Space Telescope. J. Geophys. Res. **113**, A09209 (2008). doi:10.1029/2008JA013257

J.T. Clarke et al., Response of Jupiter's and Saturn's auroral activity to the solar wind. J. Geophys. Res. **114**, A05210 (2009). doi:10.1029/2008JA013694

S.W.H. Cowley, E.J. Bunce, Origin of the main auroral oval in Jupiter's coupled magnetosphere-ionosphere system. Planet. Space Sci. **49**, 1067–1088 (2001)

S.W.H. Cowley, E.J. Bunce, Corotation-driven magnetosphere-ionosphere coupling currents in Saturn's magnetosphere and their relation to the auroras. Ann. Geophys. **21**, 1691–1707 (2003)

S.W.H. Cowley, E.J. Bunce, T.S. Stallard, S. Miller, Jupiter's polar ionospheric flows: theoretical interpretation. Geophys. Res. Lett. **30**, 1220 (2003)

S.W.H. Cowley, E.J. Bunce, J.M. O'Rourke, A simple quantitative model of plasma flows and currents in Saturn's polar ionosphere. J. Geophys. Res. **109**, A05212 (2004). doi:10.1029/2003JA010375

S.W.H. Cowley, I.I. Alexeev, E.S. Belenkaya, E.J. Bunce, C.E. Cottis, V.V. Kalegaev, J.D. Nichols, R. Prange, F.J. Wilson, A simple axisymmetric model of magnetosphere-ionosphere coupling currents in Jupiter's polar ionosphere. J. Geophys. Res. **110**, 11209 (2005)

S.W.H. Cowley, J.D. Nichols, D.J. Andrews, Modulation of Jupiter's plasma flow, polar currents, and auroral precipitation by solar wind-induced compressions and expansions of the magnetosphere: a simple theoretical model. Ann. Geophys. **25**, 1433–1463 (2007)

S.W.H. Cowley, S.V. Badman, S.M. Imber, S.E. Milan, Comment on "Jupiter: a fundamentally different magnetospheric interaction with the solar wind" by D.J. McComas and F. Bagenal. Geophys. Res. Lett. **35**, 10101 (2008)

W. Daughton, V. Roytershteyn, H. Karimabadi, L. Yin, B.J. Albright, B. Bergen, K.J. Bowers, Role of electron physics in the development of turbulent magnetic reconnection in collisionless plasmas. Nat. Phys. **7**(7), 539–542 (2011)

P.A. Delamere, F. Bagenal, Solar wind interaction with Jupiter's magnetosphere. J. Geophys. Res. **115**, 110201 (2010). doi:10.1029/2010JA015347

P.A. Delamere, F. Bagenal, Magnetotail structure of the giant magnetospheres: implications of the viscous interaction with the solar wind. J. Geophys. Res. **118**, 1–9 (2013)

P.A. Delamere, R.J. Wilson, A. Masters, Kelvin-Helmholtz instability at Saturn's magnetopause: hybrid simulations. J. Geophys. Res. **116**, 10222 (2011)

P.A. Delamere, R.J. Wilson, S. Eriksson, F. Bagenal, Magnetic signatures of Kelvin-Helmholtz vortices on Saturn's magnetopause: global survey. J. Geophys. Res. **118**, 393–404 (2013)

M. Desroche, F. Bagenal, P.A. Delamere, N. Erkaev, Conditions at the expanded Jovian magnetopause and implications for the solar wind interaction. J. Geophys. Res. **117**, A07202 (2012). doi:10.1029/2012JA017621

M. Desroche, F. Bagenal, P.A. Delamere, N. Erkaev, Conditions at the magnetopause of Saturn and implications for the solar wind interaction. J. Geophys. Res. **118**, 3087–3095 (2013). doi:10.1002/jgra.50294

J.F. Drake, M.A. Shay, Fundamentals of collisionless reconnection, in *Reconnection of Magnetic Fields: MHD and Collisionless Theory and Observations*, ed. by J. Birn, E.R. Priest (Cambridge University Press, Cambridge, 2007), p. 87

J.W. Dungey, Interplanetary field and the auroral zones. Phys. Rev. Lett. **6**, 47–48 (1961)

M.P. Freeman, S.K. Morley, A minimal substorm model that explains the observed statistical distribution of times between substorms. Geophys. Res. Lett. **31**(12), 807 (2004)

H.S. Fu, Y.V. Khotyaintsev, A. Vaivads, M. André, S.Y. Huang, Occurrence rate of earthward-propagating dipolarization fronts. Geophys. Res. Lett. **39**, L10101 (2012). doi:10.1029/2012GL051784

K. Fukazawa, T. Ogino, R.J. Walker, Configuration and dynamics of the Jovian magnetosphere. J. Geophys. Res. **111**, 10207 (2006). doi:10.1029/2006JA011874

K. Fukazawa, T. Ogino, R.J. Walker, A simulation study of dynamics in the distant Jovian magnetotail. J. Geophys. Res. **115**, A09219 (2010). doi:10.1029/2009JA015228

S.A. Fuselier, R. Frahm, W.S. Lewis, A. Masters, J. Mukherjee, S.M. Petrinec, I.J. Sillanpaa, The location of magnetic reconnection at Saturn's magnetopause: a comparison with Earth. J. Geophys. Res. **119** (2014). doi:10.1002/2013JA019684

P.H.M. Galopeau, P. Zarka, D. Le Quéau, Source location of Saturn's kilometric radiation: the Kelvin-Helmholtz instability hypothesis. J. Geophys. Res. **100**(E12), 26397–26410 (1995). doi:10.1029/95JE02132

Y.S. Ge, L.K. Jian, C.T. Russell, Growth phase of Jovian substorms. Geophys. Res. Lett. **34**, L23106 (2007)

Y.S. Ge, C.T. Russell, K.K. Khurana, Reconnection sites in Jupiter's magnetotail and relation to Jovian auroras. Planet. Space Sci. **58**, 1455–1469 (2010)

J.-C. Gérard, E.J. Bunce, D. Grodent, S.W.H. Cowley, J.T. Clarke, S.V. Badman, Signature of Saturn's auroral cusp: simultaneous Hubble Space Telescope FUV observations and upstream solar wind monitoring. J. Geophys. Res. **110**, A11201 (2005). doi:10.1029/2005JA011094

T.I. Gombosi, T.P. Armstrong, C.S. Arridge, K.K. Khurana, S.M. Krimigris, N. Krupp, A.M. Persoon, M.F. Thomsen, Saturn's magnetospheric configuration, in *Saturn from Cassini-Huygens*, ed. by M. Dougherty et al. (Springer, Berlin, 2009), p. 257

E.E. Grigorenko, M. Hoshino, M. Hirai, T. Mukai, L.M. Zelenyi, "Geography" of ion acceleration in the magnetotail: X-line versus current sheet effects. J. Geophys. Res. **114**, A03203 (2009)

A. Grocott, S.V. Badman, S.W.H. Cowley, S.E. Milan, J.D. Nichols, T.K. Yeoman, Magnetosonic Mach number dependence of the efficiency of reconnection between planetary and interplanetary magnetic fields. J. Geophys. Res. **114**, A07219 (2009). doi:10.1029/2009JA014330

D. Grodent, J.T. Clarke, J.H. Waite Jr., S.W.H. Cowley, J.-C. Gérard, J. Kim, Jupiter's polar auroral emissions. J. Geophys. Res. **108**, 1366 (2003). doi:10.1029/2003JA010017

D. Grodent, J.-C. Gérard, J.T. Clarke, G.R. Gladstone, J.H. Waite, A possible auroral signature of a magnetotail reconnection process on Jupiter. J. Geophys. Res. **109**, A05201 (2004). doi:10.1029/2003JA010341

D. Grodent, J. Gustin, J.-C. Gérard, A. Radioti, B. Bonfond, W.R. Pryor, Small-scale structures in Saturn's ultraviolet aurora. J. Geophys. Res. **116**, A09225 (2011). doi:10.1029/2011JA016818

D.A. Gurnett, W.S. Kurth, R.R. Shaw, A. Roux, R. Gendrin, C.F. Kennel, F.L. Scarf, S.D. Shawhan, The Galileo plasma waves investigation. Space Sci. Rev. **60**, 341 (1992)

D.A. Gurnett et al., Control of Jupiter's radio emission and aurorae by the solar wind. Nature **415**(6875), 985–987 (2002)

J. Gustin, J.-C. Gérard, W. Pryor, P.D. Feldman, D. Grodent, G. Holsclaw, Characteristics of Saturn's polar atmosphere and auroral electrons derived from HST/STIS, FUSE and Cassini/UVIS spectra. Icarus **200**, 176–187 (2009)

T.W. Hill, Inertial limit on corotation. J. Geophys. Res. **84**, 6554–6558 (1979). doi:10.1029/JA084iA11p06554

T.W. Hill, The Jovian auroral oval. J. Geophys. Res. **106**, 8101–8107 (2001)

T.W. Hill et al., Plasmoids in Saturn's magnetotail. J. Geophys. Res. **113**, A01214 (2008). doi:10.1029/2007JA012626

M.E. Hill, D.K. Haggerty, R.L. McNutt, C.P. Paranicas, Energetic particle evidence for magnetic filaments in Jupiter's magnetotail. J. Geophys. Res. **114**, A11201 (2009)

D.E. Huddleston, C.T. Russell, G. Le, A. Szabo, Magnetopause structure and the role of reconnection at the outer planets. J. Geophys. Res. **102**, 24289–24302 (1997)

C.M. Jackman, C.S. Arridge, Statistical properties of the magnetic field in the Kronian magnetotail lobes and current sheet. J. Geophys. Res. **116**, A05224 (2011). doi:10.1029/2010JA015973

C.M. Jackman, C.T. Russell, D.J. Southwood, C.S. Arridge, N. Achilleos, M.K. Dougherty, Strong rapid dipolarizations in Saturn's magnetotail: in situ evidence of reconnection. Geophys. Res. Lett. **34**, L11203 (2007). doi:10.1029/2007GL029764

C.M. Jackman et al., A multi-instrument view of tail reconnection at Saturn. J. Geophys. Res. **113**, A11213 (2008). doi:10.1029/2008JA013592

C.M. Jackman, C.S. Arridge, H.J. McAndrews, M.G. Henderson, R.J. Wilson, Northward field excursions in Saturn's magnetotail and their relationship to magnetospheric periodicities. Geophys. Res. Lett. **36**, L16101 (2009a). doi:10.1029/2009GL039149

C.M. Jackman, L. Lamy, M.P. Freeman, P. Zarka, B. Cecconi, W.S. Kurth, S.W.H. Cowley, M.K. Dougherty, On the character and distribution of lower-frequency radio emissions at Saturn and their relationship to substorm-like events. J. Geophys. Res. **114**, A08211 (2009b). doi:10.1029/2008JA013997

C.M. Jackman, J.A. Slavin, S.W.H. Cowley, Cassini observations of plasmoid structure and dynamics: implications for the role of magnetic reconnection in magnetospheric circulation at Saturn. J. Geophys. Res. **116**, A10212 (2011). doi:10.1029/2011JA016682

C.M. Jackman, N. Achilleos, S.W.H. Cowley, E.J. Bunce, A. Radioti, D. Grodent, S.V. Badman, M.K. Dougherty, W. Pryor, Auroral counterpart of magnetic field dipolarizations in Saturn's tail. Planet. Space Sci. **82–83**, 34–42 (2013). doi:10.1016/j.pss.2013.03.010

C.M. Jackman, J.A. Slavin, M.G. Kivelson, D.J. Southwood, N. Achilleos, M.F. Thomsen, G.A. DiBraccio, J.P. Eastwood, M.P. Freeman, M.K. Dougherty, M.F. Vogt, Saturn's dynamic magnetotail: a comprehensive magnetic field and plasma survey of plasmoids and travelling compression regions, and their role in global magnetospheric dynamics. J. Geophys. Res. (2014, submitted)

X. Jia, K.C. Hansen, T.I. Gombosi, M.G. Kivelson, G. Tóth, D.L. DeZeeuw, A.J. Ridley, Magnetospheric configuration and dynamics of Saturn's magnetosphere: a global MHD simulation. J. Geophys. Res. **117**, A05225 (2012). doi:10.1029/2012JA017575

M.L. Kaiser, Time-variable magnetospheric radio emissions from Jupiter. J. Geophys. Res. **98**, 18757 (1993)

H. Karimabadi, W. Daughton, K.B. Quest, Physics of saturation of collisionless tearing mode as a function of guide field. J. Geophys. Res. **110**, A03214 (2005). doi:10.1029/2004JA010749

S. Kasahara, E.A. Kronberg, N. Krupp, T. Kimura, C. Tao, S.V. Badman, A. Retino, M. Fujimoto, Magnetic reconnection in the Jovian tail: X-line evolution and consequent plasma sheet structures. J. Geophys. Res. **116**, A11219 (2011). doi:10.1029/2011JA016892

S. Kasahara, E.A. Kronberg, N. Krupp, T. Kimura, C. Tao, S.V. Badman, M. Fujimoto, Asymmetric distribution of reconnection jet fronts in the Jovian nightside magnetosphere. J. Geophys. Res. **118**, 375–384 (2013). doi:10.1029/2012JA018130

C.F. Kennel, *Convection and Substorms: Paradigms of Magnetospheric Phenomenology* (Oxford University Press, London, 1995)

K.K. Khurana, M.G. Kivelson, Inference of the angular velocity of plasma in the Jovian magnetosphere from the sweepback of magnetic field. J. Geophys. Res. **98**, 67 (1993)

K.K. Khurana, M.G. Kivelson, V.M. Vasyliunas, N. Krupp, J. Woch, A. Lagg, B.H. Mauk, W.S. Kurth, The configuration of Jupiter's magnetosphere, in *Jupiter. The Planet, Satellites and Magnetosphere*, ed. by F. Bagenal, T.E. Dowling, W.B. McKinnon. Cambridge Planetary Science, vol. 1 (Cambridge University Press, Cambridge, 2004), pp. 593–616. ISBN 0-521-81808-7

M.G. Kivelson, K.K. Khurana, Properties of the magnetic field in the Jovian magnetotail. J. Geophys. Res. **107**, 1196 (2002)

M.G. Kivelson, D.J. Southwood, Dynamical consequences of two modes of centrifugal instability in Jupiter's outer magnetosphere. J. Geophys. Res. **110**, A12209 (2005)

M.G. Kivelson, K.K. Khurana, J.D. Means, C.T. Russell, R.C. Snare, The Galileo magnetic field investigation. Space Sci. Rev. **60**, 357–383 (1992)

S.M. Krimigis, E.C. Roelof, Low energy particle population, in *Physics of the Jovian Magnetosphere*, ed. by A.J. Dessler (Cambridge University Press, Cambridge, 1983), pp. 106–156. doi:10.1029/CBO9780511564574.005

E.A. Kronberg, J. Woch, N. Krupp, A. Lagg, K.K. Khurana, K.-H. Glassmeier, Mass release at Jupiter: substorm-like processes in the Jovian magnetotail. J. Geophys. Res. **110**, A03211 (2005)

E.A. Kronberg, K.-H. Glassmeier, J. Woch, N. Krupp, A. Lagg, M.K. Dougherty, A possible intrinsic mechanism for the quasi-periodic dynamics of the Jovian magnetosphere. J. Geophys. Res. **112**(A5), A05203 (2007). doi:10.1029/2006JA011994

E.A. Kronberg, J. Woch, N. Krupp, A. Lagg, P.W. Daly, A. Korth, Comparison of periodic substorms at Jupiter and Earth. J. Geophys. Res. **113**, A04212 (2008a). doi:10.1029/2007JA012880

E.A. Kronberg, J. Woch, N. Krupp, A. Lagg, Mass release process in the Jovian magnetosphere: statistics on particle burst parameters. J. Geophys. Res. **113**, A10202 (2008b). doi:10.1029/2008JA013332

E.A. Kronberg, J. Woch, N. Krupp, A. Lagg, A summary of observational records on periodicities above the rotational period in the Jovian magnetosphere. Ann. Geophys. **27**, 2565–2573 (2009)

E.A. Kronberg, S. Kasahara, N. Krupp, J. Woch, Field-aligned beams and reconnection in the Jovian magnetotail. Icarus **217**, 55–65 (2012). doi:10.1016/j.icarus.2011.10.011

N. Krupp, J. Woch, A. Lagg, B. Wilken, S. Livi, D.J. Williams, Energetic particle bursts in the predawn Jovian magnetotail. Geophys. Res. Lett. **25**, 1249–1252 (1998)

N. Krupp, A. Lagg, S. Livi, B. Wilken, J. Woch, E.C. Roelof, D.J. Williams, Global flows of energetic ions in Jupiter's equatorial plane: first-order approximation. J. Geophys. Res. **106**, 26017–26032 (2001)

H. Ladreiter, Y. Leblanc, Jovian hectometric radiation—beaming, source extension, and solar wind control. Astron. Astrophys. **226**, 297–310 (1989)

H.R. Lai, H.Y. Wei, C.T. Russell, C.S. Arridge, M.K. Dougherty, Reconnection at the magnetopause of Saturn: perspective from FTE occurrence and magnetosphere size. J. Geophys. Res. **117**, A05222 (2012). doi:10.1029/2011JA017263

224

W.W. Liu, J. Liang, E.F. Donovan, E. Spanswick, If substorm onset triggers tail reconnection, what triggers substorm onset? J. Geophys. Res. **117**, A11220 (2012). doi:10.1029/2012JA018161

P. Louarn, A. Roux, S. Perraut, W. Kurth, D. Gurnett, A study of the largescale dynamics of the Jovian magnetosphere using the Galileo plasma wave experiment. Geophys. Res. Lett. **25**, 2905–2908 (1998)

P. Louarn, A. Roux, S. Perraut, W.S. Kurth, D.A. Gurnett, A study of the Jovian "energetic magnetospheric events" observed by Galileo: role in the radial plasma transport. J. Geophys. Res. **105**, 13073–13088 (2000)

P. Louarn, B. Mauk, D.J. Williams, C. Zimmer, M.G. Kivelson, W.S. Kurth, D.A. Gurnett, A. Roux, A multi-instrument study of a Jovian magnetospheric disturbance. J. Geophys. Res. **106**, 29883 (2001)

P. Louarn et al., Presentation at Magnetosphere of Outer Planets, Athens (2013, in preparation)

A.T.Y. Lui et al., A cross-field current instability for substorm expansions. J. Geophys. Res. **96**, 11389 (1991)

S. Markidis, G. Lapenta, L. Bettarini, M. Goldman, D. Newman, L. Andersson, Kinetic simulations of magnetic reconnection in presence of a background O^+ population. J. Geophys. Res. **116**, A00K16 (2011). doi:10.1029/2011JA016429

A. Masters et al., The importance of plasma β conditions for magnetic reconnection at Saturn's magnetopause. Geophys. Res. Lett. **39**, L08103 (2012). doi:10.1029/2012GL051372

A. Masters, M. Fujimoto, H. Hasegawa, C.T. Russell, A.J. Coates, M.K. Dougherty, Can magnetopause reconnection drive Saturn's magnetosphere? Geophys. Res. Lett. **41**, 1862–1868 (2014). doi:10.1002/2014GL059288

H.J. McAndrews et al., Evidence for reconnection at Saturn's magnetopause. J. Geophys. Res. **113**(A4), A04210 (2008). doi:10.1029/2007JA012581

D.J. McComas, F. Bagenal, Jupiter: a fundamentally different magnetospheric interaction with the solar wind. Geophys. Res. Lett. **34**, L20106 (2007). doi:10.1029/2007GL031078

D.J. McComas et al., Diverse plasma populations and structures in Jupiter's magnetotail. Science **318**, 217–220 (2007)

R.L. McNutt et al., Energetic particles in the Jovian magnetotail. Science **318**, 220–222 (2007)

S.E. Milan, Both solar wind-magnetosphere coupling and ring current intensity control of the size of the auroral oval. Geophys. Res. Lett. **36**, L18101 (2009). doi:10.1029/2009GL039997

D.G. Mitchell, P.C. Brandt, E.C. Roelof, J. Dandouras, S.M. Krimigis, B.H. Mauk, C.P. Paranicas, N. Krupp, D.C. Hamilton, W.S. Kurth, P. Zarka, M.K. Dougherty, E.J. Bunce, D.E. Shemansky, Energetic ion acceleration in Saturn's magnetotail: substorms at Saturn? Geophys. Res. Lett. **32**, L20S01 (2005)

D.G. Mitchell, S.M. Krimigis, C. Paranicas, P.C. Brandt, J.F. Carbary, E.C. Roelof, W.S. Kurth, D.A. Gurnett, J.T. Clarke, J.D. Nichols, J.-C. Gérard, D.C. Grodent, M.K. Dougherty, W.R. Pryor, Recurrent energization of plasma in the midnight-to dawn quadrant of Saturn's magnetosphere, and its relationship to auroral UV and radio emissions. Planet. Space Sci. **57**, 1732–1742 (2009). doi:10.1016/j.pss.2009.04.002

T. Miyoshi, K. Kusano, A global MHD simulation of the Jovian magnetosphere interacting with/without the interplanetary magnetic field. J. Geophys. Res. **106**(A6), 10723–10742 (2001). doi:10.1029/2000JA900153

F.S. Mozer, D. Sundkvist, J.P. McFadden, P.L. Pritchett, I. Roth, Satellite observations of plasma physics near the magnetic field reconnection X line. J. Geophys. Res. **116**(A12), A12224 (2011)

J.D. Nichols, E.J. Bunce, J.T. Clarke, S.W.H. Cowley, J.-C. Gérard, D. Grodent, W.R. Pryor, Response of Jupiter's UV auroras to interplanetary conditions as observed by the Hubble Space Telescope during the Cassini flyby campaign. J. Geophys. Res. **112**, A02203 (2007). doi:10.1029/2006JA012005

A. Nishida, Reconnection in the Jovian magnetosphere. Geophys. Res. Lett. **10**, 451–454 (1983)

T. Ogino, R.J. Walker, M.G. Kivelson, A global magnetohydrodynamic simulation of the Jovian magnetosphere. J. Geophys. Res. **103**(A1), 225–235 (1998). doi:10.1029/97JA02247

V. Olshevsky, G. Lapenta, S. Markidis, Energetics of kinetic reconnection in a three-dimensional null-point cluster. Phys. Rev. Lett. **111**(4), 045002 (2013)

T.G. Onsager, M.F. Thomsen, R.C. Elphic, J.T. Gosling, Model of electron and ion distributions in the plasma sheet boundary layer. J. Geophys. Res. **96**, 20999–21011 (1991). doi:10.1029/91JA01983

L. Pallier, R. Prangé, More about the structure of the high latitude Jovian aurorae. Planet. Space Sci. **49**, 1159 (2001)

L. Pallier, R. Prangé, Detection of the southern counterpart of the Jovian northern polar cusp: shared properties. Geophys. Res. Lett. **31**, L06701 (2004). doi:10.1029/2003GL018041

T.D. Phan, J.F. Drake, M.A. Shay, F.S. Mozer, J.P. Eastwood, Evidence for an elongated (>60 ion skin depths) electron diffusion region during fast reconnection. Phys. Rev. Lett. **99**, 255002 (2007). doi:10.1103/PhysRevLett.99.255002

T.-D. Phan et al., The dependence of magnetic reconnection on plasma β and magnetic shear: evidence from solar wind observations. Astrophys. J. Lett. **719**, L199–L203 (2010). doi:10.1088/2041-8205/719/2/L199

225

R. Prangé, P. Zarka, G.E. Ballester, T.A. Livengood, L. Denis, T.D. Carr, F. Reyes, S.J. Bame, H.W. Moos, Correlated variations of UV and radio emissions during an outstanding Jovian auroral event. J. Geophys. Res. **98**, 18779–18791 (1993). doi:10.1029/93JE01802

R. Prangé et al., An interplanetary shock traced by planetary auroral storms from the Sun to Saturn. Nature **432**, 78–81 (2004)

E. Priest, T. Forbes, *Magnetic Reconnection: MHD Theory and Applications* (Cambridge University Press, Cambridge, 2007)

P.L. Pritchett, F.S. Mozer, The magnetic field reconnection site and dissipation region. Phys. Plasmas **16**, 080702 (2009). doi:10.1063/1.3206947

W.R. Pryor et al., Cassini UVIS observations of Jupiter's auroral variability. Icarus **178**, 312–326 (2005). doi:10.1016/j.icarus.2005.05.021

K.B. Quest, F.V. Coroniti, Linear theory of tearing in a high-β plasma. J. Geophys. Res. **86**(A5), 3299–3305 (1981). doi:10.1029/JA086iA05p03299

A. Radioti, D. Grodent, J.-C. Gérard, B. Bonfond, J.T. Clarke, Auroral polar dawn spots: signatures of internally driven reconnection processes at Jupiter's magnetotail. Geophys. Res. Lett. **35**, L03104 (2008). doi:10.1029/2007GL032460

A. Radioti, D. Grodent, J.-C. Gérard, B. Bonfond, Auroral signatures of flow bursts released during magnetotail reconnection at Jupiter. J. Geophys. Res. **115**, A07214 (2010)

A. Radioti, D. Grodent, J.-C. Gérard, M.F. Vogt, M. Lystrup, B. Bonfond, Nightside reconnection at Jupiter: auroral and magnetic field observations from 26 July 1998. J. Geophys. Res. **116**, A03221 (2011)

A. Radioti, D. Grodent, J.-C. Gérard, B. Bonfond, J. Gustin, W. Pryor, J.M. Jasinski, C.S. Arridge, Auroral signatures of multiple magnetopause reconnection at Saturn. Geophys. Res. Lett. **40**, 4498–4502 (2013). doi:10.1002/grl.50889

I.G. Richardson, S.W.H. Cowley, E.W. Hones Jr., S.J. Bame, Plasmoid-associated energetic ion bursts in the deep geomagnetic tail: properties of plasmoids and the post-plasmoid plasma sheet. J. Geophys. Res. **92**, 9997–10013 (1987). doi:10.1029/JA092iA09p09997

A. Roux, S. Perraut, P. Robert, A. Morane, A. Pedersen, A. Korth, G. Kremser, B. Aparicio, D. Rodgers, R. Pellinen, Plasma sheet instability related to the westward traveling surge. J. Geophys. Res. **96**, 17697–17714 (1991) (ISSN 0148-0227), Oct. 1

H.O. Rucker et al., Saturn kilometric radiation as a monitor for the solar wind? Adv. Space Res. **42**, 40–47 (2008)

A. Runov, V. Angelopoulos, M.I. Sitnov, V.A. Sergeev, J. Bonnell, J.P. McFadden, D. Larson, K.-H. Glassmeier, U. Auster, THEMIS observations of an earthward-propagating dipolarization front. Geophys. Res. Lett. **36**, L14106 (2009). doi:10.1029/2009GL038980

A. Runov, V. Angelopoulos, X.-Z. Zhou, X.-J. Zhang, S. Li, F. Plaschke, J. Bonnell, A THEMIS multicase study of dipolarization fronts in the magnetotail plasma sheet. J. Geophys. Res. **116**, A05216 (2011). doi:10.1029/2010JA016316

C.T. Russell, K.K. Khurana, D.E. Huddleston, M.G. Kivelson, Localized reconnection in the near Jovian magnetotail. Science **280**, 1061–1064 (1998)

C.T. Russell, K.K. Khurana, M.G. Kivelson, D.E. Huddleston, Substorms at Jupiter: Galileo observations of transient reconnection in the near tail. Adv. Space Res. **26**, 1499–1504 (2000)

A. Rymer, D.G. Mitchell, T.W. Hill, E.A. Kronberg, N. Krupp, C.M. Jackman, Saturn's magnetospheric refresh rate. Geophys. Res. Lett. **40**, 2479–2483 (2013). doi:10.1002/grl.50530

D.V. Sarafopoulos, E.T. Sarris, V. Angelopoulos, T. Yamamoto, S. Kokubun, Spatial structure of the plasma sheet boundary layer at distances greater than 180 R_E as derived from energetic particle measurements on GEOTAIL. Ann. Geophys. **15**, 1246–1256 (1997). doi:10.1007/s00585-997-1246-0

M. Scholer, D.N. Baker, G. Gloeckler, B. Klecker, F.M. Ipavich, T. Terasawa, B.T. Tsurutani, A.B. Galvin, Energetic particle beams in the plasma sheet boundary layer following substorm expansion: simultaneous near-Earth and distant tail observations. J. Geophys. Res. **91**, 4277–4286 (1986). doi:10.1029/JA091iA04p04277

L. Scurry, C.T. Russell, Proxy studies of energy transfer to the magnetosphere. J. Geophys. Res. **96**, 9541–9548 (1991). doi:10.1029/91JA00569

V. Sergeev, V. Angelopoulos, S. Apatenkov, J. Bonnell, R. Ergun, R. Nakamura, J. McFadden, D. Larson, A. Runov, Kinetic structure of the sharp injection/dipolarization front in the flow-braking region. Geophys. Res. Lett. **36**, L21105 (2009). doi:10.1029/2009GL040658

M.A. Shay, M. Swisdak, Three species collisionless reconnection: effect of O^+ on magnetotail reconnection. Phys. Rev. Lett. **93**(17), 175001 (2004)

E.J. Smith, L. Davis Jr., D.E. Jones, P.J. Coleman Jr., D.S. Colburn, P. Dyal, C.P. Sonett, A.M.A. Frandsen, The planetary magnetic field and magnetosphere of Jupiter: Pioneer 10. J. Geophys. Res. **79**, 25 (1974)

D.J. Southwood, M.G. Kivelson, A new perspective concerning the influence of the solar wind on the Jovian magnetosphere. J. Geophys. Res. **106**(A4), 6123–6130 (2001)

M. Swisdak, B.N. Rogers, J.F. Drake, M.A. Shay, Diamagnetic suppression of component magnetic reconnection at the magnetopause. J. Geophys. Res. **108**(A5), 1218 (2003). doi:10.1029/2002JA009726

C. Tao, R. Kataoka, H. Fukunishi, Y. Takahashi, T. Yokoyama, Magnetic field variations in the Jovian magnetotail induced by solar wind dynamic pressure enhancements. J. Geophys. Res. **110**, A11208 (2005). doi:10.1029/2004JA010959

M.F. Thomsen, R.J. Wilson, R.L. Tokar, D.B. Reisenfeld, C.M. Jackman, Cassini/CAPS observations of duskside tail dynamics at Saturn. J. Geophys. Res. (2013). doi:10.1002/jgra.50552, in press

A. Vaivads, Y. Khotyaintsev, M. Andre, A. Retino, S.C. Buchert, B.N. Rogers, P. Decreau, G. Paschmann, T.D. Phan, Structure of the magnetic reconnection diffusion region from four-spacecraft observations. Phys. Rev. Lett. **93**, 105001 (2004)

A. Vaivads, A. Retino, M. André, Microphysics of magnetospheric reconnection. Space Sci. Rev. **122**, 19–27 (2006). doi:10.1007/s11214-006-7019-3

V.M. Vasyliunas, Plasma distribution and flow, in *Physics of the Jovian Magnetosphere* (1983), pp. 395–453

V.M. Vasyliunas, Comparing Jupiter and Saturn: dimensionless input rates from plasma sources within the magnetosphere. Ann. Geophys. **26**, 1341–1343 (2008)

V.M. Vasyliunas, L.A. Frank, K.L. Ackerson, W.R. Paterson, Geometry of the plasma sheet in the midnight-to-dawn sector of the Jovian magnetosphere: plasma observations with the Galileo spacecraft. Geophys. Res. Lett. **24**, 869–872 (1997)

M.F. Vogt, M.G. Kivelson, K.K. Khurana, S.P. Joy, R.J. Walker, Reconnection and flows in the Jovian magnetotail as inferred from magnetometer observations. J. Geophys. Res. **115**, A06219 (2010)

M.F. Vogt, M.G. Kivelson, K.K. Khurana, R.J. Walker, B. Bonfond, D. Grodent, A. Radioti, Improved mapping of Jupiter's auroral features to magnetospheric sources. J. Geophys. Res. **116**, A03220 (2011). doi:10.1029/2010JA016148

M.F. Vogt, C.M. Jackman, J.A. Slavin, E.J. Bunce, S.W.H. Cowley, M.G. Kivelson, K.K. Khurana, The structure and statistical properties of plasmoids in Jupiter's magnetotail. J. Geophys. Res. **119**(2), 821–843 (2014). doi:10.1002/2013JA019393

J.H. Waite et al., An auroral flare at Jupiter. Nature **410**, 787–789 (2001). doi:10.1038/35071018

R.J. Walker, C.T. Russell, Flux transfer events at the Jovian magnetopause. J. Geophys. Res. **90**(A8), 7397–7404 (1985)

R.J. Walker, T. Ogino, M.G. Kivelson, Magnetohydrodynamic simulations of the effects of the solar wind on the Jovian magnetosphere. Planet. Space Sci. **49**, 237–245 (2001)

D.J. Williams, R.W. McEntire, S. Jaskulek, B. Wilken, The Galileo energetic particles detector. Space Sci. Rev. **60**, 385–412 (1992)

J. Woch, N. Krupp, A. Lagg, B. Wilken, S. Livi, D.J. Williams, Quasi-periodic modulations of the Jovian magnetotail. Geophys. Res. Lett. **25**, 1253–1256 (1998)

J. Woch, N. Krupp, K.K. Khurana, M.G. Kivelson, A. Roux, S. Perraut, P. Louarn, A. Lagg, D.J. Williams, S. Livi, B. Wilken, Plasma sheet dynamics in the Jovian magnetotail: signatures for substorm-like processes? Geophys. Res. Lett. **26**, 2137–2140 (1999)

J. Woch, N. Krupp, A. Lagg, Particle bursts in the Jovian magnetosphere: evidence for a near-Jupiter neutral line. Geophys. Res. Lett. **29**, 42 (2002)

L. Yin, F.V. Coronoti, P.L. Pritchett, L.A. Frank, L.A. Paterson, Kinetic aspects of the Jovian current sheet. J. Geophys. Res. **115**(A11), 25345 (2000)

P. Zarka, Auroral radio emissions at the outer planets: observations and theories. J. Geophys. Res. **103**(E9), 20159–20194 (1998). doi:10.1029/98JE01323

B. Zieger, K.C. Hansen, T.I. Gombosi, D.L. De Zeeuw, Periodic plasma escape from the mass-loaded Kronian magnetosphere. J. Geophys. Res. **115**, A08208 (2010). doi:10.1029/2009JA014951

G. Zimbardo, Observable implications of tearing-mode instability in Jupiter's nightside magnetosphere. Planet. Space Sci. **41**, 357–361 (1993)

Q.-G. Zong et al., Cluster observations of earthward flowing plasmoid in the tail. Geophys. Res. Lett. **31**, L18803 (2004). doi:10.1029/2004GL020692

DOI 10.1007/978-1-4939-3395-2_7
Reprinted from *Space Science Reviews* Journal, DOI 10.1007/s11214-014-0086-y

Transport of Mass, Momentum and Energy in Planetary Magnetodisc Regions

Nicholas Achilleos · Nicolas André ·
Xochitl Blanco-Cano · Pontus C. Brandt ·
Peter A. Delamere · Robert Winglee

Received: 3 February 2014 / Accepted: 5 August 2014 / Published online: 2 October 2014
© The Author(s) 2014. This article is published with open access at Springerlink.com

Abstract The rapid rotation of the gas giant planets, Jupiter and Saturn, leads to the formation of magnetodisc regions in their magnetospheric environments. In these regions, relatively cold plasma is confined towards the equatorial regions, and the magnetic field generated by the azimuthal (ring) current adds to the planetary dipole, forming radially distended field lines near the equatorial plane. The ensuing force balance in the equatorial magnetodisc is strongly influenced by centrifugal stress and by the thermal pressure of hot ion populations, whose thermal energy is large compared to the magnitude of their centrifugal potential energy. The sources of plasma for the Jovian and Kronian magnetospheres are the respective satellites Io (a volcanic moon) and Enceladus (an icy moon). The plasma produced by these

N. Achilleos (✉)
Department of Physics and Astronomy, Centre for Planetary Sciences, University College London,
London, UK
e-mail: nicholas.achilleos@ucl.ac.uk

N. André
Institut de Recherche en Astrophysique et Planétologie (IRAP)/Centre National de la Recherche
Scientifique (CNRS), Université de Toulouse, Toulouse, France
e-mail: nicolas.andre@irap.omp.eu

X. Blanco-Cano
Instituto de Geofísica, Universidad Nacional Autónoma de México, Ciudad Universitaria, Mexico City,
Mexico
e-mail: xbc@geofisica.unam.mx

P.C. Brandt
Johns Hopkins University Applied Physics Laboratory, Laurel, MD, USA
e-mail: pontus.brandt@jhuapl.edu

P.A. Delamere
Geophysical Institute, University of Alaska Fairbanks, Fairbanks, AK, USA
e-mail: Peter.Delamere@gi.alaska.edu

R. Winglee
Department of Earth and Space Sciences, University of Washington, Seattle, WA, USA
e-mail: winglee@ess.washington.edu

sources is globally transported outwards through the respective magnetosphere, and ultimately lost from the system. One of the most studied mechanisms for this transport is flux tube interchange, a plasma instability which displaces mass but does not displace magnetic flux—an important observational constraint for any transport process. Pressure anisotropy is likely to play a role in the loss of plasma from these magnetospheres. This is especially the case for the Jovian system, which can harbour strong parallel pressures at the equatorial segments of rotating, expanding flux tubes, leading to these regions becoming unstable, blowing open and releasing their plasma. Plasma mass loss is also associated with magnetic reconnection events in the magnetotail regions. In this overview, we summarise some important observational and theoretical concepts associated with the production and transport of plasma in giant planet magnetodiscs. We begin by considering aspects of force balance in these systems, and their coupling with the ionospheres of their parent planets. We then describe the role of the interaction between neutral and ionized species, and how it determines the rate at which plasma mass and momentum are added to the magnetodisc. Following this, we describe the observational properties of plasma injections, and the consequent implications for the nature of global plasma transport and magnetodisc stability. The theory of the flux tube interchange instability is reviewed, and the influences of gravity and magnetic curvature on the instability are described. The interaction between simulated interchange plasma structures and Saturn's moon Titan is discussed, and its relationship to observed periodic phenomena at Saturn is described. Finally, the observation, generation and evolution of plasma waves associated with mass loading in the magnetodisc regions is reviewed.

Keywords Magnetodiscs · Magnetospheres · Plasma · Ion-cyclotron waves · Magnetic reconnection

1 Overview

The rapidly rotating magnetospheres of the 'gas giant' planets, Jupiter and Saturn, combined with the relatively small angles between the magnetic and rotational axes of those worlds, are the principal features which lead to the formation of so-called 'magnetodisc' regions. As the name suggests, the magnetic field in these regions of the magnetosphere may be visualised with dipolar lines of force which have been 'stretched' radially outwards near the equatorial plane. Gledhill (1967) was among the first to show that centrifugal force would confine plasma into a 'disc-like' configuration in a system with a dipole-like, rigidly rotating field.

The most general representation of force balance in a rotating plasma takes into account plasma pressure gradient (including anisotropy), centrifugal force and magnetic ('$\mathbf{J} \times \mathbf{B}$') force. In this general case, we may write the following equations for balance both parallel and perpendicular to the magnetic field $\mathbf{B}$. These forms of the equations, as we shall see, more intuitively illustrate the role of the magnetic field geometry. The field-aligned equation of balance, firstly, may be written:

$$-\frac{dP_{\parallel}}{ds_{\parallel}} + \frac{(P_{\parallel} - P_{\perp})}{B}\frac{dB}{ds_{\parallel}} + N_i m_i r \omega^2 \hat{\rho} \cdot \hat{\mathbf{b}} = 0. \qquad (1)$$

The symbols have the following meanings: field-parallel pressure $P_{\parallel}$, field-perpendicular pressure $P_{\perp}$, magnetic field strength B, arc length along magnetic field line $s_{\parallel}$, ion number density N_i, ion mass m_i, cylindrical radial distance r (measured perpendicular to the axis of rotation), plasma angular velocity ω, unit vectors along the cylindrical radial direction, $\hat{\rho}$, and magnetic field direction $\hat{\mathbf{b}}$. Reading from left to right, the terms in (1) represent the parallel pressure gradient, the mirror force (arising from pressure anisotropy) and the centrifugal volume force (we have neglected the contribution of the electron mass in this term). An isotropic, non-rotating plasma would simply be described by $-\frac{dP_\parallel}{ds_\parallel} = 0$, i.e. uniform pressure all along any field line. The additional forces impose structure on the plasma pressure distribution, and the centrifugal term in particular is effective at confining cold plasma towards the rotational equator (e.g. Caudal 1986; Kivelson and Southwood 2005). Equation (1) is a fluid equation applicable to a quasi-neutral plasma. Equations which separately describe the field-parallel force balance for ions and electrons may also be derived. These would contain terms corresponding to ambipolar electric fields (see, for example, Appendix C in Achilleos et al. 2010; Maurice et al. 1997).

If we now consider force balance in a direction orthogonal to **B**, we may write:

$$\left(\frac{(P_\parallel - P_\perp)}{R_c}\right)\hat{\mathbf{R}}_c - \nabla_\perp(P_\perp) - \nabla_\perp\left(\frac{B^2}{2\mu_o}\right) - \left(\frac{B^2}{\mu_o R_c}\right)\hat{\mathbf{R}}_c + N_i m_i r\omega^2 \hat{\boldsymbol{\rho}}_\perp = 0, \qquad (2)$$

where the additional symbols have the following meanings: field line radius of curvature R_c, unit vector along this radius of curvature $\hat{\mathbf{R}}_c$, component of gradient operator perpendicular to the field $\nabla_\perp$, magnetic vacuum permeability μ_o, component $\hat{\boldsymbol{\rho}}_\perp$ of vector $\hat{\boldsymbol{\rho}}$ in the direction perpendicular to **B**. Reading from left to right, the terms in (2) represent the anisotropy force, perpendicular pressure gradient, magnetic pressure gradient, magnetic curvature force and the centrifugal force component perpendicular to the field. Kivelson and Southwood (2005) pointed out the importance of the pressure anisotropy in the Jovian system, where rotating magnetic flux tubes may become significantly radially distended on the time scale required by ions to travel along them, from polar to equatorial regions. This process may cause the anisotropy term in (2) to grow and violate the force balance conditions (we will return to this aspect in more detail below). When this happens, the corresponding flux tube loses its integrity and 'blows open', thus forming an important channel of mass loss from the magnetosphere. In a broader context, this type of mass loss process, driven by rotation and pressure anisotropy, may also be operating in other astrophysical bodies, such as magnetised accretion discs or proto-stellar discs.

A useful, albeit greatly simplified, representation of force balance in the magnetodisc is obtained if we commence with (2), assume isotropic pressure ($P_\perp = P_\parallel = P$) and apply it at the rotational equator of a system whose magnetic and rotational axes of symmetry are coincident. That is, where the vectors $\hat{\mathbf{R}}_c$ and $\hat{\boldsymbol{\rho}}$ are equal, and both perpendicular to **B**. Under these simplifying assumptions, we may represent the balance between centrifugal force, plasma pressure gradient, and magnetic force along the equatorial, radial direction (r) as follows:

$$N_i m_i r\omega^2 - \frac{dP}{dr} - \frac{d}{dr}\left(\frac{B^2}{2\mu_0}\right) - \frac{B^2}{\mu_0 R_c} = 0. \qquad (3)$$

This equation is appropriate for a simple, rigid plasmadisc which lies entirely in the equatorial plane (e.g. Caudal 1986; Kivelson and Southwood 2005). Achilleos et al. (2010) constructed such isotropic-pressure models for the Jovian and Kronian discs, and noted the following general trends for the various force terms. For average conditions in Saturn's magnetodisc, the total plasma pressure gradient exceeds the centrifugal volume force in the equatorial distance range ~ 8–12 R_S, outside of which centrifugal force is dominant ($R_S = 60268$ km denotes Saturn's 1-bar equatorial radius). They noted, however, that the strong variability in hot plasma pressure at Saturn could easily change this picture. For the average conditions at Saturn, the total plasma beta falls below unity at distances inside ~ 8 R_S. The Jovian disc models (incorporating *Voyager* observations) in this study

Fig. 1 Structure of Jovian magnetodisc model. *Upper panel*: Magnetic field lines (*dark curves*) lying in the magnetic meridian plane, with ρ and Z denoting cylindrical radial and vertical distance, respectively. The superposed *grey curves* are contours of the ratio ξ of the energy density of magnetic field to that of the plasma rotation (see text). *Lower panel*: As for the *upper panel*, but now with contours of the decimal logarithm of the magnetic field magnitude in nano-Tesla. Spacing between adjacent contours is 0.2 units. For clarity, some labels are omitted

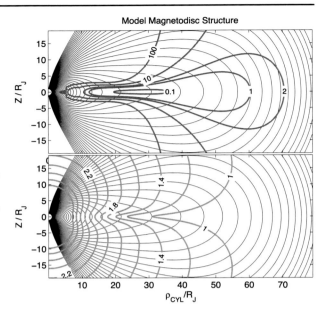

Model Magnetodisc Structure

showed comparable plasma pressure gradient and centrifugal forces in the distance range $\sim$ 20–30 R_J ($R_J = 71492$ km denotes Jupiter's 1-bar equatorial radius), with plasma pressure gradient dominating outside this interval. Total plasma beta falls below unity at distances inside $\sim$ 12 R_J.

If we wish to approximate the region where centrifugal stress alone is large enough to radially displace the planetary dipolar field lines, a simple first approach is to require that the energy density of the field be less than the kinetic energy density of plasma rotation. Using the same symbols as above, we may write:

$$\frac{B^2}{\mu_0} \lesssim N_i m_i r^2 \omega^2, \tag{4}$$

which is equivalent to requiring that the rotational velocity of the plasma, $V_\phi = r\omega$, be trans-Alfvénic.

In Fig. 1, we show in the top panel the magnetic field lines for a cylindrically symmetric model of Jupiter's magnetosphere, based on the force-balance formalism of Caudal (1986). Superposed on the field lines are contours of the quantity $\xi = (B^2/\mu_0)/(N_i m_i r^2 \omega^2)$. The region where this ratio ξ is smaller than unity encloses a disc-like body of centrifugally confined cold plasma, extending between $\sim$ 10–60 R_J (in the present context, 'cold' plasma may be defined as a medium in which the thermal energy of particle motions is small compared to their kinetic energy of bulk rotation). The lower panel, extracted from the same model, shows that the cold plasmadisc has a weaker field strength at its centre than in the neighbouring 'lobe' regions which are relatively devoid of plasma. This property reflects the balance between thermal plasma pressure inside the disc and magnetic pressure outside. Repeating this exercise for the model of Saturn's magnetodisc by Achilleos et al. (2010) reveals a similar structure, with the ratio ξ having a maximum value about half that of the Jovian model, with equatorial value below unity from $\sim$ 8 R_S nearly out to the magnetopause.

This disc-like field structure is supported by a distributed azimuthal current or ring current. On a microscopic scale, the ring current is produced by different ion and electron drift motions, which are associated with the macroscopic forces summarised in (3) (e.g. Caudal 1986; Achilleos et al. 2010; Bunce et al. 2007). One of these drift motions is the inertial drift due to the centrifugal force which arises in a frame of reference that corotates with the cold plasma. The intensity of the corresponding contribution to the ring current will depend on the local angular velocity of the plasma, which is represented as the quantity ω in (3). In situ observations of the plasma angular velocity, ω, were first reported for Jupiter by McNutt et al. (1981) (*Voyager* observations), and for Saturn by Lazarus and McNutt (1983), later augmented by Wilson et al. (2008); Kane et al. (2008). Studies by Kellett et al. (2010, 2011) are among the most detailed analyses of the morphology of Saturn's ring current and the contributions to this current associated with bulk rotation and thermal pressure of the magnetospheric plasma.

For both Jupiter and Saturn, the fact that the plasma rotation rate generally decreases with radial distance is an important observational signature of the interaction, or coupling, between the magnetospheric disc and the planet's ionosphere. Hill (1979) developed a theory for this coupling process in the Jovian system. He described the link between the intensity of field-aligned currents flowing between the ionosphere and the magnetosphere, and the radial profile of plasma angular velocity. In this picture, there are two major factors which determine the dependence of ω on radial distance. The first is related to the radial transport of angular momentum in the plasmadisc. The original source of this momentum is the plasma originating from the Io torus, which, in Hill's theory, is assumed to form a corotating (with the planet) 'inner boundary' for the disc. As the plasma diffuses radially outwards, via processes to be later described, its tendency is to conserve angular momentum, and hence for ω to decrease with increasing radial distance r. In the absence of other forces, the disc plasma would thus exhibit an angular velocity profile $\omega \propto r^{-2}$.

In reality, there is an additional force on the plasma, which arises from the tendency of the most radially distant (equatorial) segments of outward-moving flux tubes to be 'bent back' against the direction of plasma rotation as ω decreases. This action produces a significant, non-zero azimuthal component B_ϕ in the magnetic field ($B_\phi < 0$ north of the magnetodisc current sheet in the equatorial plane, and $B_\phi > 0$ in the south). The corresponding curl of the field is equivalent to a radial current density J_r flowing outwards through the middle magnetosphere. The magnetic force per unit volume exerted on the disc plasma is the second factor which determines the angular velocity profile, and may be written as $-J_r B_z \mathbf{e}_\phi$, where the z direction is orthogonal to the current sheet, $B_z < 0$ (southward-pointing) for both Jupiter and Saturn, and $\mathbf{e}_\phi$ is a unit vector pointing in the local direction of planetary corotation. It follows that the magnetic force acts to accelerate the plasma back towards corotation with the planet. The origin of this force is the angular momentum of the planet itself. This is because the radial current which flows in the magnetospheric current sheet is part of a larger current system which closes through the planet's ionosphere, by means of field-aligned currents that flow between these two regions (Fig. 2). The ionospheric current is driven by ion-neutral collisions. Hence, the angular momentum of the planet's rotating thermosphere is transferred to the magnetospheric plasma by a field-aligned current system.

Taking into account these two mechanisms for angular momentum transport in the Jovian magnetosphere, and assuming a pure dipolar magnetic field with an axis of symmetry aligned with the planet's rotational axis, Hill's (1979) steady-state equation for the change in plasma specific angular momentum ℓ with radial distance is:

$$\frac{d\ell}{dL} = \omega_J R_J^2 \frac{d}{dL}\left[L^2\left(\frac{1-\delta\omega}{\omega_J}\right)\right]$$

$$= 4\pi\left(\frac{\Sigma}{\dot{M}}\right)\delta\omega\, R_J^4 B_J^2 \frac{(1-1/L)^{1/2}}{L^3} \qquad (5)$$

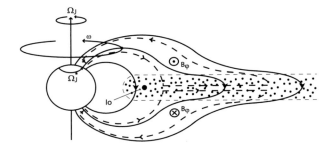

Fig. 2 Figure and caption reprinted from Cowley and Bunce (2001), Copyright (2001), with permission from Elsevier: Sketch of a meridian cross-section through the Jovian magnetosphere, showing the principal features of the inner and middle magnetosphere regions. The *arrowed solid lines* indicate magnetic field lines, which are distended outwards in the middle magnetosphere region by azimuthal currents in the plasma sheet. The plasma sheet plasma originates mainly at Io, which orbits in the inner magnetosphere at ~ 6 R$_J$, liberating $\sim 10^3$ kg s^{-1} of sulphur and oxygen plasma. This plasma is shown by the *dotted* region, which rotates rapidly with the planetary field due to magnetosphere-ionosphere coupling while more slowly diffusing outwards. Three separate angular velocities associated with this coupling are indicated. These are the angular velocity of the planet Ω_J, the angular velocity of a particular shell of field lines ω, and the angular velocity of the neutral upper atmosphere in the Pedersen layer of the ionosphere, Ω_J^*. The latter is expected to lie between ω and Ω_J because of the frictional torque on the atmosphere due to ion-neutral collisions. The oppositely directed frictional torque on the magnetospheric flux tubes is communicated by the current system indicated by the *arrowed dashed lines*, shown here for the case of sub-corotation of the plasma (i.e. $\omega \leq \Omega_J$). This current system bends the field lines out of meridian planes, associated with azimuthal field components B_ϕ as shown

where the symbols have the following meanings: L denotes the equatorial crossing distance of the flux tube in units of R$_J$, the planetary radius; ω_J is Jupiter's angular velocity of rotation; disc plasma along the flux tube extending to L rotates with angular velocity $\omega_J - \delta\omega$; Σ is the height-integrated ionospheric Pedersen conductivity; $\dot{M}$ is the rate at which new plasma mass is added to the disc at its inner boundary, the Io torus; and B$_J$ is the equatorial field strength at the planet's surface.

This equation clearly indicates the important role played by the ionospheric conductance and the mass loading rate of Iogenic plasma. For increasing Σ and decreasing $\dot{M}$, Hill (1979) showed that the 'critical' value of L, beyond which the plasma subcorotates to a significant degree ($\delta\omega/\omega_J \gtrsim 0.3$), would increase. This tells us that the plasma is more efficiently accelerated towards corotation in systems with highly conducting ionospheres and/or low plasma mass loading rates. Mathematically, the value of L at which this level of breakdown in corotation arises is given by:

$$L_0 = \left(\frac{\pi \Sigma B_J^2 R_J^2}{\dot{M}} \right)^{1/4} \tag{6}$$

When we substitute appropriate values of $\Sigma = 0.2$ mho (e.g. Cowley and Bunce 2001), $\dot{M} = 1000$ kg s^{-1} (Bagenal and Delamere 2011), and B$_J = 428000$ nT, we obtain a value $L_0 \sim 28$ R$_J$. Additional variants of Hill's theory have included the effects of non-dipolar fields, global changes in magnetospheric configuration and precipitation-induced enhancement of the ionospheric conductance. (e.g. Cowley and Bunce 2001; Nichols and Cowley 2004; Cowley et al. 2007). Other theoretical studies of the magnetosphere-ionosphere coupling have included more realistic, global-circulation models for the thermosphere-ionosphere (e.g. Achilleos et al. 2001; Smith and Aylward 2009; Yates et al. 2012). The most recent study of this kind (Yates et al. 2012) found that the total energy transferred from the planet's rotation to the atmosphere and magnetosphere varies with magnetodisc size. The total power dissipated in the coupled system changes from ~ 400 TW to ~ 700 TW as the plasmadisc radius expands from 45 R_J to 85 R_J. This prediction is consistent with the theoretical work of Southwood and Kivelson (2001), who demonstrated that the steady-state intensity of the auroral currents should be higher for a more expanded magnetosphere, characterised by a plasmadisc which rotates more slowly.

The addition of newly ionised plasma in the vicinity of the Io torus represents a momentum 'loading' for the system. Freshly created ions initially move at the Keplerian velocity of the neutral molecules from which they formed, and must then be picked up or incorporated into the ambient, corotating plasma flow—this requires an increase in the angular momentum of the newly ionised plasma, which is ultimately provided by the deep layers of the rotating planet. The processes of ion pickup and radial plasma transport both slow the magnetospheric plasma below rigid corotation, extract angular momentum from the planet, and produce magnetosphere-ionosphere coupling currents. However, there is also a process which can change the local angular momentum of the disc plasma, without adding additional plasma mass. This is the phenomenon of charge exchange which may occur between rapidly rotating ions and slower, cold neutrals (e.g. on Keplerian orbits) in a magnetosphere where the neutral-to-ion ratio is considerably high, such as that of Saturn. The transfer of electrons from the cold neutral to the hot ion results in the formation of a cold ion and a hot neutral, the latter now free from the influence of the magnetic field and thus capable of carrying energy away and escaping from the magnetosphere entirely.

Delamere et al. (2007) modelled the energy and particle flows for the Io and Enceladus tori, incorporating the effects of ion pickup and charge exchange (see Sect. 2 for more details). They concluded that the relative importance of charge exchange and radial transport of plasma was sensitive to the ratio of neutral to ion number density in the torus. At Saturn/Enceladus, where this ratio is about ~ 12, the net effect of charge exchange is to carry away ~ 95 % of the mass which is added to the torus per unit time (principally in the form of water group molecules, which may then become ionised). The remaining ~ 5 % of this material is transported radially outwards as plasma. At Jupiter/Io, on the other hand, neutral species in the torus model (mainly sulphur and oxygen) have number densities about ~ 1 % of their corresponding ion products. Under these conditions in the Jovian system, charge exchange is predicted to remove about half of the mass added to the Io torus per unit time. This result demonstrates the more dominant role played by radial plasma transport in the overall mass flow of the Jovian system.

In a related study, Pontius and Hill (2009) developed an equation of motion for the disc plasma at Saturn which included the effects of ion-neutral charge exchange. They then used this theory to 'invert' observations of plasma angular velocity by *Cassini*, and demonstrated that the charge exchange process was the major contributor to decreasing the angular momentum of the plasma within regions $L \lesssim 5$. At $L = 3.4$–3.8 in Pontius and Hill's (2009) modelling study, for example, they show that the charge exchange process, in terms of momentum loading, is *equivalent* to adding ~ 350 $kg\,s^{-1}$ of plasma to the disc. In this same interval of distance, the mass loading rate due to actual, newly ionised plasma (i.e. a genuine increase in plasma mass) is ~ 8 $kg\,s^{-1}$. The same authors estimate a total outward mass flux of ~ 100 $kg\,s^{-1}$ near $L = 7.5$, for an ionospheric height-integrated conductivity of 0.1 mho.

The aforementioned studies suggest, then, that radial transport of plasma plays a major role in the evolution of the Jovian magnetodisc, at least. The question then arises as to the nature of this transport process: How do we move disc plasma in the radial direction without eventually 'breaking' the magnetic flux tubes to which this material is 'frozen in'? An important clue arises from in situ observations of the magnetic field in the vicinity of the plasma tori themselves. On spatial scales which are large compared with Io and Enceladus, the magnetic field near the corresponding tori is always well approximated by a rigid, dipolar configuration, arising mainly from the planet's internal dynamo. Indeed, the model results from Fig. 1 reveal that the $L < 10$ region of Jupiter's magnetosphere is characterised by a magnetic energy density more than ten times the bulk kinetic energy of the rotating plasma. Hence, the Io torus is clearly inside a region where we expect the magnetic field to strongly resist any large-scale 'deformation' due to plasma flow. The radial transport mode required should thus be able to displace plasma mass outwards from its torus of origin, without any attendant displacement of magnetic flux. The most widely proposed transport mechanism which satisfies these requirements is that of flux tube interchange, a process which, in some sense, relies on the development of a small-scale 'texture' in the disc plasma.

In this chapter, we limit our discussion of plasma transport to the interchange process within the magnetodisc region, which arises from a fluid instability. In a more general context, other forms of radial plasma transport may arise when fluctuations of magnetic or electric fields occur. If the fluctuation time scale is larger than a particle bounce period, but small compared to the time scale of azimuthal drift, the first two adiabatic invariants may be conserved, but not the third. Hence particles will radially 'diffuse' (e.g. Brice and Mc-Donough 1973; Walt 1994; Schulz 1979). If field fluctuations occur on adequately small time scales compared to that of bounce motion, the first two adiabatic invariants will not be conserved and particle energy and equatorial pitch angle will also be subject to diffusion. Such rapid fluctuations may be associated with plasma waves, and the interaction of these waves with particles can be the basis for this type of diffusion (e.g. Walt 1994; Thorne 2010).

Sections 4 and 5 discuss more recent, advanced models for the interchange process at both Jupiter and Saturn which incorporate additional effects such as pressure anisotropy and multi-fluid magnetohydrodynamics. However, we may start to introduce the concept here, in a simple way, by appealing to the early description of Gold (1959). He compared the nature of a putative flux-tube-interchange process in the Earth's magnetosphere with that of convection in a gravitating, compressible fluid. In the latter case, consider the temperature gradient $-\nabla T$ in the fluid, whose direction (from higher to lower temperature) opposes gravity. If the magnitude of this gradient is above a critical value, known as the adiabatic temperature gradient, the gas becomes unstable to convection. The critical driver for the process is the fact that, for adequately steep gradient in temperature, the convective mixing of parcels of gas will lead to a lower total energy for the new fluid configuration.

Applying this reasoning to a simple model of interchange, Gold (1959) considered the outward displacement of a single plasma flux tube in a rigid, dipole magnetic field (a compensating inward displacement of a plasma tube at larger radial distance maintains the large-scale, dipole configuration of the field). As the flux tube moves outward, its equatorial cross section intercepts a constant increment in magnetic flux, and its volume V changes in proportion to the quantity L^4. In the limiting case of adiabatic transport, the pressure P of the plasma in the tube would thus change according to $P \propto L^{-4\gamma}$, where γ is the ratio of specific heats for the plasma (e.g. $\gamma = \frac{5}{3}$ for an ideal, monatomic gas). This 'adiabatic gradient' for pressure thus plays an analogous role to the adiabatic temperature gradient for the gravitating fluid.

Hill (1976) extended this idea to a rapidly rotating model magnetosphere, specifically in order to analyse the properties of the Jovian system. He considered the change in centrifugal potential energy of plasma which would result from the exchange in location of two flux tubes embedded in a pure dipole field. For a system where the *flux tube content* M_B (mass of plasma contained in a tube whose cross section encloses unit magnetic flux) decreases

236

monotonically with distance, Hill (1976) supposed that the system would be unstable to such interchange motions when they led to a decrease in the total centrifugal potential, and he showed that this would happen when the following condition was satisfied:

$$\frac{\partial M_B}{\partial r_e} < 0 \tag{7}$$

In other words, a monotonic decrease in flux tube content M_B with equatorial crossing distance r_e encourages interchange motions in a system whose dominant energy arises from bulk rotation. This description is appropriate for the cold, rapidly rotating plasma within the Jovian magnetodisc. It is also important to note that the result of interchange motions is to reduce the spatial gradient in flux tube content, and hence drive the system towards a lower-energy, more stable state. The continual addition of Iogenic plasma to the disc ensures that this stable state is never reached and that interchange continually acts to move newly created plasma radially outwards.

Southwood and Kivelson (1989) further extended the theoretical treatment of interchange motion, and quantified the effects of the interaction between the moving flux tubes and the planetary ionosphere. Specifically, they showed that the timescale for growth of interchange motions increases linearly with ionospheric Pedersen conductance Σ_P. This result reflects the 'damping' effect which arises when a highly conducting ionosphere transmits a retarding force to flux tubes participating in interchange motion. As with the large-scale rotational motions of the disc plasma described above, this conveyance of momentum is achieved by field-aligned currents, which travel along the flanks of the moving tubes and close in the ionosphere. The ionospheric currents are associated with ion-neutral collisions between the ionospheric plasma and the planet's thermosphere. The transfer of momentum due to these collisions is the ultimate source of the retarding force on the moving flux tubes. Southwood and Kivelson (1989) favoured a diffusion-like mechanism to maintain the strong inward gradient of the cold plasma's flux tube content at the outer edge of the Io torus (i.e. tube content strongly decreases with increasing radial distance). Beyond this region, their depiction of the most likely nature of the plasma flow involved interchange motions whose combined effect, over long time scales, was to migrate regions of high flux tube content outwards, and depressed flux tube content inwards. For a temporally variable Io plasma source, the associated flow streamlines would be continually changing in a random manner. The local motion of a flux tube (inward or outward) in this picture is determined according to whether it has an elevated or depressed plasma content compared to the average content of its surroundings.

More than three decades after Gold's analysis (Gold 1959) of the interchange process, the *Galileo* orbiter at Jupiter gave scientists the opportunity to search for observational signatures of this mode of plasma transport. Kivelson et al. (1997b) reported short-duration increases in the magnetic field strength, of typical amplitude 1–2 % of the background field and average duration ~ 26 s. These events were observed over an equatorial distance range ~ 6–7.7 R_J. Although the change in field strength was typically small, the corresponding change in magnetic pressure was often comparable to the background plasma pressure. One particular event at 6.03 R_J was analysed in detail by Kivelson et al. (1997b), and was consistent with an inward-travelling flux tube of depleted plasma density, equal to $\lesssim 50$ % of the background (outside the tube) value. The distance of origin for this depleted tube was estimated to be 7.2 R_J, the location where the flux tube content of the background plasma would be equal to that inside the inward-travelling tube.

The absence of ion-cyclotron wave signatures inside the depleted tube was interpreted as evidence that it had not existed for a long enough period in the near-Io plasma environment for these waves (themselves signatures of ion pickup) to appreciably grow. Indeed,

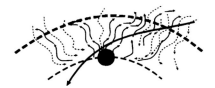

Fig. 3 Schematic of the plasma transport near the Io torus from Kivelson et al. (1997b). *Dashed arcs* denote the orbit of Io, and 7 R_J. Other meandering curves represent averaged, rather than instantaneous, flows. *Solid (dashed) curves* of this type indicate outward (inward) moving flux tubes. Well away from Io, inward and outward flows balance. The vicinity of Io itself (*filled circle*) is a region dominated by mass loading, and thus predominantly outward flow. Inward flow dominates elsewhere, presumably. The inbound Galileo trajectory is shown as a *solid, curved arrow*. The spacecraft crosses both inward- and outward-moving plasma near 7 R_J, but mainly outward-moving material nearer to Io

Huddleston et al. (1997) applied this argument to estimate a radial velocity of travel for the depleted tube around 100 km s^{-1}. A more advanced description of the wave signatures associated with mass loading in the magnetodisc will be given in Sect. 6. Analysis of the same field enhancement event by Thorne et al. (1997), incorporating plasma wave data, indicated that the depleted plasma density could have been as low as ~ 3 % of the background value, thus suggestive of a strong contribution by energetic particles to the plasma pressure. These *Galileo* studies were consistent with a scenario in which heavier flux tubes (i.e. higher plasma content than background) are outward-moving, and outward motions are dominant close to Io itself. They are presumably balanced by inward motions at other longitudes. At larger distances, inward-moving, depleted tubes are observed, as described above. In Fig. 3, we show the schematic diagram from Kivelson et al. (1997b) which illustrates these aspects.

Similar observational signatures of the centrifugally driven interchange process have been observed by *Cassini* at Saturn. Rymer et al. (2009) reported a short-duration field strength enhancement of amplitude ~ 6 % of the background field, corresponding to an electron density about one third of the background value. The particle energy spectrogram acquired by the *Cassini* plasma spectrometer (Young et al. 2004) exhibited the characteristic dispersions associated with the energy-dependent particle drifts that would presumably occur during the inward motion of the depleted tube. The data were consistent with a tube that had moved inward from ~ 10 R_S to 7 R_S at a radial speed of 150–300 km s^{-1}, of the same order as the Jovian observations. Earlier observational studies at Saturn (e.g. Hill et al. 2005; Chen and Hill 2008) analysed what they referred to as plasma injection signatures over an equatorial distance range ~ 6–11 R_S. The azimuthal extent of injection events with electron energies 1–30 keV were estimated to be of the order $\lesssim 1$ R_S, and to survive less than one planetary rotation.

Moving now into the outer reaches of the 'magnetodisc' region and beyond, the nature of plasma transport is observed to change. Generally speaking, the relatively small perturbations to the background plasma which describe the interchange process are replaced in the outer magnetosphere by much larger-scale injections of plasma which is usually more energetic and tenuous than the disc plasma. Mauk et al. (1999, 2002) quantified this transition in the Jovian system through analysis of *Galileo* particle data. They described a change in the 'character' of injections beyond distances of ~ 9 R_J. Specifically, the injections in the more distant magnetosphere are associated with increased fluxes of energetic particles, and more pronounced energy dispersion signatures, even evident in electrons. Mauk et al. (1999) analysed more than 100 events over the distance range 9–27 R_J, which occurred over a wide range in longitude and local time. This property of the Jovian plasma injections highlights the contrast with similar events in the Earth's magnetosphere, whose local time distribution suggests a clustering around the dawn sector, consistent with a physical origin in the terrestrial magnetotail.

Mauk et al. (1999) also found that a typical large-scale plasma injection in their sample moved a few R_J before it became significantly dispersed through differential drift motions. About five injections per day were observed by *Galileo* under quiescent magnetospheric conditions. This rate rose to approximately 30 events per day, following global magnetospheric disturbances or reconfigurations. The typical azimuthal extent of an injection was ~ 30–$40°$, in contrast with the few degrees for the interchange-related events described above. In a related study, Louarn et al. (2001) examined the correlation between *Galileo* plasma data, and radio data relating to regions in the outer Io torus. Their results suggested that relatively young, 'dispersion-free' injections were signatures of an 'unloading' process, which acted quasi-periodically to remove plasma from an active region in the torus. This phenomenon may be related to the 'energetic events' described by Louarn et al. (1998) that trigger sequences of plasma loading and unloading in the magnetodisc (Louarn et al. 2000). The existence of these large-scale, energetic plasma injections in the Jovian system naturally raises the question of whether they and the narrower 'interchange tubes' are in fact different signatures of a unified, global process of plasma transport. At present, we do not have a definitive answer.

The large-scale variety of plasma injections have also been observed in Saturn's magnetosphere. Injections with particle energies in excess of ~ 100 keV have been observed, and inferred to have survived up to several Saturn rotations from their time of formation (Mauk et al. 2005; Paranicas et al. 2007). Plasma injections extending to tens of degrees in azimuth have been remotely observed through energetic neutral atom (ENA) imaging (Krimigis et al. 2005; Carbary et al. 2007). It has been suggested that these may originate in the planet's magnetotail, and rotate more quickly as they move inwards. ENA injections have also been convincingly linked to rotating, arc-like auroral emissions at Saturn, to which they are likely to be magnetically conjugate (Mitchell et al. 2009). Section 3 describes these injections and their analysis in more detail.

After its creation and radial migration through a disc-like region, the final stage of the transport of magnetospheric plasma is, of course, its final release from its parent planetary system. One of the key phenomena which underpins this process is the fact that outward-moving flux tubes cannot maintain their integrity indefinitely. The continual process of internal mass loading (associated with satellite tori) leads to the plasma sheet thickening in the dusk sector, and the strong radial expansion of flux tubes as they rotate into the magnetotail region, now unencumbered by the proximity of the magnetopause (e.g. Kivelson and Southwood 2005).

A well-known scenario for mass loss in this context is the intermittent release of plasma from these highly distended and mass-loaded flux tubes, a process which was first discussed by Vasyliūnas (1983). In this model, the corresponding part of the plasma sheet becomes unstable once these outer flux tubes carry a critical amount of plasma and rotate past the dusk sector into the magnetotail. Without the 'bracing' of a nearby magnetopause boundary to stabilise it, the plasma sheet becomes disrupted and sheds individual parcels of plasma, or plasmoids. A more detailed discussion of this process and magnetic reconnection in the magnetotail region is given in other chapters of this volume. For present purposes, we note that planetary rotation at the gas giant systems is also thought to influence the stability of these heavily loaded flux tubes. Kivelson and Southwood (2005) pointed out that combining concepts from magnetohydrodynamic and kinetic theories were important for understanding the dynamics of plasma embedded in these structures. Specifically, they considered the motion of a heavy (20 proton masses), corotating ion with thermal energy small compared to its rotational energy. Such an ion would gain ~ 20 keV in energy parallel to the field as it moves from 45 to 50 R_J in cylindrical radial distance. If such ions acquire an adequate amount of energy parallel to the field direction, then the plasma at the equatorial portion of the flux tube will become subject to a firehose-type instability, and the magnetic field will not be adequately strong to constrain the rapid, field-aligned motion of the plasma. Consequently, the plasma sheet becomes unstable, distending and thinning, which leads to the loss of plasma from the outer edge of the disc and the return of depleted, closed flux tubes towards the planet.

In summary, then, we have found that the rapid planetary rotation associated with the gas giant systems has a fundamental influence on the transport of magnetospheric plasma, and the communication of energy and momentum between the magnetodiscs and the ionospheres of these worlds. To improve our understanding of both processes, several advancements are required, from both observational and theoretical points of view. We list some of these here:

- Additional surveys of the plasma angular velocity, over comprehensive ranges in distance and in local time.
- More theoretical work coupling global-circulation models of planetary thermospheres with plasma flow models for magnetospheres. Particularly relevant in this context would be the further exploration of time-dependent effects of magnetospheric reconfigurations on the auroral currents, and further work with three-dimensional models, in order to explore local time effects.
- The further development of an observational 'interchange budget', based on the statistics of plasma injection events, as well as the comparison of different injections' spatial scales, dispersion characteristics, and correlations with near-coincident magnetospheric reconfigurations.
- Further refinement of the theory of flux tube interchange, considering, for example, the shapes and scales of the underlying, small 'overturn motions' which drive the global plasma migration. Another major topic to address is the possibly inter-related influence on interchange of: (i) dramatic changes in the rate of internal mass-loading; and (ii) the external driving of the magnetospheric field structure by transient changes in the solar wind dynamic pressure and/or the interplanetary magnetic field.

Finally, we conclude this section with a summary of some of the outstanding questions relating to magnetodisc structure and dynamics (some of the following sections also have summaries related to their relevant topics):

- What is the range of timescales for the *localised* motions, or 'interchange overturns', which support the global radial transport of plasma?
- Do we actually detect interchange signatures with adequate spatial coverage and frequency in order to explain the global rates of plasma transport which are required, on average, at both the Jovian and Kronian magnetospheres?
- How do the larger-scale injection events contribute to global radial transport, and how do they relate to the smaller-scale, centrifugally driven interchange events?
- Does the outward-moving, colder plasma actually extend to very large azimuthal distances between the hotter, more tenuous flux tubes, and is this why its observational signatures are lacking by comparison?
- How do anisotropy forces, for both hot and cold plasma, affect the global structure of the disc and the higher-latitude plasma?
- What are the acceleration mechanisms responsible for establishing the anisotropy in the hot plasma pressure?

Fig. 4 Saturn's E-ring of neutral gas formed near the orbit of Enceladus from water vapor vented from its south pole plumes. (Image credit: NASA/JPL/Space Science Institute)

2 The Neutral-Plasma Interaction

2.1 Introduction

The giant planet magnetodiscs are supplied by mass from satellites orbiting deep within the magnetosphere. At Jupiter, Io revolves around the gas giant with an orbital distance of 6 R_J and supplies the inner magnetosphere with roughly 1 ton per second of neutral sulfur and oxygen from its prodigious volcanic activity (Thomas et al. 2004). At Saturn, Enceladus orbits at a distance of 4 R_S and vents roughly 200 $kg\,s^{-1}$ of water molecules from geysers at its south pole to form Saturn's E-ring (Mauk et al. 2009). The distances to the respective magnetopause boundaries are roughly 75 R_J (Joy et al. 2002) and 24 R_S (Kanani et al. 2010), placing these satellites deep within the inner magnetosphere. The fate of these neutral gases are very different when comparing Jupiter with Saturn. Jupiter's inner magnetosphere is dominated by plasma as most of the neutral gas from Io is ionized (Delamere et al. 2007). Saturn's inner magnetosphere remains neutral dominated due to very low ionization rates (Delamere and Bagenal 2008). This section will compare and contrast the evolution of neutral gas and transport of plasma through the giant planet magnetodiscs.

Escaping neutral gases typically have speeds that are not significantly different from the Keplerian velocity of the parent satellite. For example, the escape velocity from Io is roughly 2.5 $km\,s^{-1}$, while Io's Keplerian velocity is roughly 17 $km\,s^{-1}$. As a result the neutral gases will tend to form a torus of gas about the orbit of the satellite (Fig. 4). However, if ionization occurs then the evolution of the gas clouds is altered significantly.

There are essentially three neutral gas loss mechanisms to consider: 1) photo ionization, 2) electron impact ionization by a thermal background plasma, and 3) charge exchange (Delamere and Bagenal 2003; Fleshman et al. 2010a). Other processes such as electron impact dissociation and photodissociation of molecular species and neutral-neutral collisions can scatter neutral particles resulting in loss to the planet or rings; however, as loss mechanism, these scattering processes are relatively minor (Fleshman et al. 2012; Cassidy and Johnson 2010; Farmer 2009; Johnson et al. 2006). Photoionization timescales at 5 and 9 AU are typically of order of months to years. Compared to the orbital period for Keplerian motion (days) and the spreading of the neutral gas due to small difference in Keplerian speeds (tens of days), a complete neutral torus would be expected to form in the absence of electron impact ionization mechanisms. If, however, electron impact ionization

Fig. 5 An example of Io's partial neutral oxygen torus from Burger (2003)

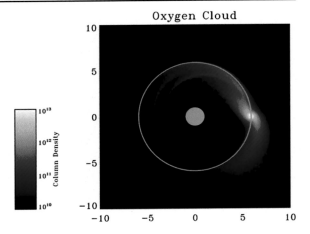

is faster than $\approx$ tens of days, then a partial neutral torus will form due to relatively prompt losses of the neutral gas cloud.

The ionization energy for most neutral species ranges between 10 and 15 eV. For a Maxwellian distribution, significant ionization can still be achieved at lower temperatures via the high energy tail of the distribution. The ionization rates for sulfur (i.e. S to S^+) are sufficient to erode the sulfur neutral clouds extending from Io such that only a partial neutral torus forms. While no global observation of Io's neutral torus has been made (see Brown 1981; Skinner and Durrance 1986 for examples of measurements far from Io inferred from line-of-sight UV observations), model results show that the torus is indeed partial (Smyth and Marconi 2003; Burger 2003) as shown in Fig. 5.

The thermal electron population in the Io plasma torus is observed to be roughly 5 eV (Sittler et al. 2007), while the electron temperature in the Enceladus E-ring region is less then 2 eV (Young et al. 2004; Schippers et al. 2008). The electron impact ionization rates increase by roughly two orders of magnitude between 2 eV and 5 eV. This accounts, in part, for Jupiter's plasma-dominated inner magnetosphere and Saturn's neutral-dominated inner magnetosphere (Delamere et al. 2007). The ubiquitous hot electrons, observed in both systems, can also provide ionization, but as we will discuss below, the hot electrons at Jupiter serve to heat the thermal electrons and increase the ionization state (i.e. average charge of the ions) of the plasma torus. At Saturn, the hot electrons provide ionization at a level that is comparable to photoionization but due to the tenuous plasma, cannot provide significant energy via Coulomb interactions to the thermal electrons.

Additional chemistry must be considered to fully understand the evolution of the neutral clouds and ultimate fate of these gas particles. Charge exchange between ions and neutrals results in high-speed neutrals that can potentially escape from the system completely, or force collisions with the planet or ring systems. This mechanism leads to the permanent loss of neutral material from the system and cannot further participate in the plasma mass loading of the magnetosphere (Cassidy and Johnson 2010; Fleshman et al. 2012). Other reactions involve molecular species that can dissociate via electron impact into smaller molecules or atomic species. Molecular ions likewise can suffer electron impact dissociation, yielding smaller neutral fragments that can likewise escape from the system. Photodissociation for molecular species can also be important (Fleshman et al. 2010a).

In this section we will discuss the mass loading of the inner magnetosphere and the flow of energy and mass through the system.

2.2 Momentum Loading and Neutral Cloud Evolution

Ionization and charge exchange both involve momentum loading of the magnetosphere. The Keplerian speed of neutral particles is significantly less than the local plasma flows that are typically a significant fraction of the corotation speed of the planet. At Jupiter and at Io's orbit, the corotation speed is roughly 74 km s^{-1}, and the difference between the Keplerian speed and the corotation speed is 57 km s^{-1}. This means that new ions will be picked up (from Keplerian speed) to the local flow speed (corotation speed), requiring a transfer of momentum from the planet to the new ions. Mass loading the plasma flow results in a slowing (subcorotation) of the flow and the related magnetic field perturbations will couple (via an Alfvénic perturbation/field-aligned currents) the mass-loaded plasma to the planetary ionosphere. In steady state, the "pickup current" results from the $\mathbf{J} \times \mathbf{B}$ force that balances the momentum loading, specifically when $\dot{M} v_{plasma} = \int \mathbf{J} \times \mathbf{B} dV$, integrated over the mass loaded volume (see Vasyliūnas 2006 for a detailed discussion on the physical origin of the "pickup current").

Alternatively, mass loading and ion pickup can be understood by considering the rest frame of the neutrals. In this frame, the plasma is flowing past the stationary neutrals and has an associated convection electric field, $\mathbf{E} = -\mathbf{v}_{plasma} \times \mathbf{B}$. In this case, the new ion is accelerated by the convection electric field and acquires an average drift velocity equal to the local flow speed and in the flow direction together with a gyromotion with a speed that is also equal to the local flow speed. The combination gives a cycloid motion as illustrated in Fig. 6. The resulting velocity distribution is a two-dimensional ring beam (perpendicular to the local magnetic field) with temperature, $T_\perp = \frac{1}{2} m v_{plasma}^2$. Typically, this temperature is tens to hundreds eV. This unstable distribution can generate electromagnetic ion cyclotron (EMIC) waves which scatter the ring beam toward an equilibrium Maxwellian distribution (Blanco-Cano et al. 2001c; Russell and Huddleston 2000). However, the scattering time scales and the mass loading time scales can be comparable leading to the observed temperature anisotropies ($T_\perp / T_\parallel > 1$) in both Io (Crary et al. 1996) and Enceladus plasma tori (Sittler et al. 2008). Figure 7 illustrates the combined velocity distribution of a ring beam (blue) embedded in a cooler Maxwellian distribution (red).

A critical issue, however, is that the electrons produced during ionization are picked up with an energy that is much less than the ionization potential of the neutral gas (i.e. $\ll 10$ eV). Only through Coulomb coupling to the ions and superthermal electrons will the thermal electrons attain a sufficient energy to ionize. But the effectiveness of the Coulomb coupling is dependent on the plasma density and this energy source can also compete with radiative cooling losses. For a complete discussion of this energy flow, see discussions by Shemansky (1988) and Delamere et al. (2007).

Both ionization and charge exchange momentum load the local plasma flow, but generally only ionization adds mass to the system. Charge exchange reactions can liberate a fast neutral with speeds $\sim v_{plasma}$, in principle, at any point of the cycloid trajectory. These fast neutrals will generally leave the system (i.e. escape from the magnetosphere). However, Fleshman et al. (2012) note that charge exchange cross sections are generally velocity dependent and in the case of symmetric reactions, the cross sections are a strong function of velocity with large increases at low velocity. If the ions move in perfect cycloids, then the reaction rates increase significantly near the cycloid cusps where the relative velocity between the ion and neutral is minimized. The slow moving neutral products will thus linger in the inner magnetosphere and the slow velocity charge exchange reaction can be considered as an important mechanism for spreading the neutral clouds. Other mechanisms for spreading the neutral clouds include dissociation of molecular species due to the residual kinetic energy distribution of the dissociated products (typically 1 km s^{-1}) (Fleshman et al. 2012).

Fig. 6 An illustration of the neutral-plasma interactions near Io. Incident plasma interacting with Io's atmosphere lead to ionization, charge exchange and induced dipole interactions that collectively lead to an atmospheric sputtering process at or near the exobase. (Image credit: F. Bagenal/S. Bartlett)

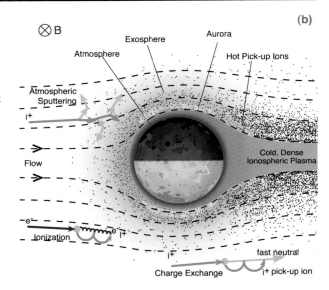

Fig. 7 An illustration of a ring beam distribution (*blue*) combined with a cooler Maxwellian distribution (*red*) near Io

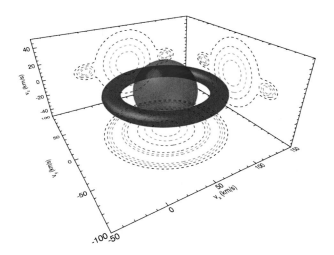

2.3 Satellite-Magnetosphere Interactions

At the root of neutral-plasma interactions at Jupiter is the source of neutral material from Io's atmosphere. Io's atmosphere is created through the direct injection of gases from volcanic vents and through sublimation of the surface frosts originating from volcanic plume fallout. The relative contribution from these processes has generated significant debate (see reviews by McGrath et al. 2004; Lellouch et al. 2007). The atmosphere exhibits day/night asymmetries, supporting the sublimation source, yet the persistence of atmospheric emissions during eclipse supports a direct volcanic source. Io's atmosphere is confined to its equatorial regions and appears to contain multiple components that include a tightly bound SO_2 and SO atmosphere, together with an extended corona of atomic S and O species. The corona is also an expected result of an escaping atmosphere, namely the 1 ton per second of material that escapes from Io to populate the extended neutral clouds.

Various approaches have been adopted to model Io's interaction with Jupiter's magnetosphere. Early models adopted a fully three-dimensional, MHD approach where Io's surface used prescribed boundary conditions (e.g. finite conductivity) and a distributed mass loading rate to represent ionization and charge exchange (Combi et al. 1998; Linker et al. 1998; Kabin et al. 2001; Khurana et al. 2011). Saur et al. (1999) used a two-fluid model to model the flow around a conducting atmosphere, solving for the electric fields and currents associated with the interaction. However, these authors did not self-consistently solve for the magnetic field perturbation, but rather determined the perturbation based on the modeled currents. A hybrid simulation was also attempted by Lipatov and Combi (2006) to improve the treatment of mass loading and the related generation of non-Maxwellian pickup distributions. More recently, Dols et al. (2012) used a MHD approach, coupled with mass loading rates determined from a physical chemistry model of the local Io interaction (Dols et al. 2008). The coupled approach, combined with comparisons with data from the various Galileo flybys, led Dols et al. (2012) to suggest that Io's outer atmosphere/corona is not distributed uniformly in longitude. Instead, it is extended in the wake direction and perhaps even more extended in the anti-Jupiter direction.

There is one possible explanation worth mentioning for a possible asymmetric atmosphere. Using magnetic field perturbations seen in the Galileo magnetometer data, Russell et al. (2003b) concluded that an asymmetric distribution of ion cyclotron wave power could be generated if charge exchange preferentially redistributes the neutral atmosphere to the wakeward and anti-Jupiter directions because ion pickup motion from charge exchange reactions is initially in this direction. Subsequent ionization of these scattered neutrals will lead to the generation of EMIC waves. Crary and Bagenal (2000) used the ion cyclotron waves as a constraint on density and escape rate of the neutral atmosphere. These initial assessments of Io's atmosphere are consistent with the subsequent analysis of Dols et al. (2012).

Roughly 10^{12} W of power is generated by Io's interaction with the plasma torus (Saur et al. 2004; Hess et al. 2010). The escape velocity, v_{esc}, from Io's surface is ~ 2.5 km s^{-1}. If 1 ton of neutral material escapes per second from the atmosphere, then the power associated with neutral escape ($\frac{1}{2}\dot{M}v_{esc}^2 \approx 3 \times 10^9$ W) is only a small fraction (< 1 %) of the interaction power. While most of the power is radiated in Alfvén waves, it is plausible that harnessing just a small fraction of this power can support Io's atmospheric escape. The escape process could involve a variety of different physical mechanisms, all of which might be considered as atmospheric sputtering. These include direct collisional heating of the neutral atmosphere via induced dipole interactions, charge exchange scattering, or even electron impact dissociation of molecular species. The latter mechanism was discussed by Dols et al. (2008) as being potentially a major contributor to the formation of Io's corona and extended neutral clouds because of substantial reaction rates ($\sim$ tons per second) combined with a characteristic kinetic energy distribution of the dissociated products that is comparable with the escape velocity.

Finally, a potentially important aspect of the satellite-magnetosphere interaction is the feedback of field-aligned electrons. Electron beams have been observed near Io by Williams et al. (1996). Dols et al. (2008) and Saur et al. (2002) noted that these beams can interact with Io's atmosphere, providing additional ionization and heating of the atmosphere. The total power attributed to these beams from Frank and Paterson (1999) is roughly 10^9 W, indicating that the beams might contribute to the modification of the atmosphere via the plasma interaction.

A study, similar to that of Dols et al. (2008) for Io, of the Enceladus plume-plasma interaction was conducted by Fleshman et al. (2010b) using a water group physical chemistry model. The tenuous plasma conditions near Enceladus do not result in a strong interaction. Typical mass loading rates, consistent with magnetic field perturbations measured near Enceladus, are small (e.g. Khurana et al. 2007 estimate a plasma mass loading rate of $\lesssim 3$ kg s^{-1} within five Enceladus radii). Field-aligned electron beams are also present and are thought to play a more significant role in the overall chemistry associated with the Enceladus plume (Fleshman et al. 2010b). Simulation studies have investigated the role of charged dust particles in the plasma interaction and Kriegel et al. (2011) concluded that heavy negatively charged dust grains can in fact generate Hall currents in the opposite sense of currents generated by ions and electrons (with the mass of the positive and negative charge carriers reversing roles).

In summary, Io's interaction with its plasma environment results in a fascinating feedback process, modifying the atmosphere that ultimately supplies the plasma torus. Bagenal (1997) showed that most of the mass added to the torus occurs in the escaping neutral cloud rather than through local ionization of the atmosphere. Enceladus, on the other hand, already has an escaping plume of neutral gas and does not require energy input from the magnetosphere to liberate neutral gas from the gravitational well of the satellite.

2.4 Plasma Tori

The neutral-plasma interaction is fundamental for formation of giant planet magnetodiscs. The ultimate fate of escaping neutral gases from satellite sources is a basic problem that affects the flow of mass and energy through magnetodisc systems. Neutral gas can be absorbed by the planet, its rings, or escape from the system at high speed (i.e. greater than the escape velocity for the planet) due to charge exchange reactions. The final mass loss mechanism is via radial plasma transport and eventual loss to the solar wind (primarily down the magnetotail). In other words, the neutral source is either absorbed within the system, scattered outside the system, or lost as plasma to the solar wind. The latter mechanism has significant consequences for magnetospheric dynamics.

Bagenal and Delamere (2011) provided a comprehensive comparative study of mass and energy flow at Jupiter and Saturn. In both cases the transport time scales from satellite to solar wind is roughly tens of days. At Jupiter, roughly half of the neutral source is lost to fast neutral escape, leaving roughly 500 kg s^{-1} of the original 1 ton per second of gas available for plasma transport. At Saturn, Fleshman (2011) showed that only a quarter of the neutral gas survives to the plasma state. The typical neutral source from Enceladus is 200 kg s^{-1}, indicating that only 50 kg s^{-1} is transported as plasma. This order of magnitude difference between Jupiter and Saturn has enormous consequences in terms of magnetosphere-ionosphere (MI) coupling. At Jupiter, the breakdown in corotation of the magnetodisc is dramatically highlighted by the main auroral oval. Saturn's auroral emissions, on the other hand, appear to map to a terrestrial-like boundary to an open polar cap with the internal plasma source not presenting the same level of mass loading to the corotating magnetosphere (Bunce et al. 2008). There has been some debate regarding the relative mass loading of these two systems, with Vasyliūnas (2008) arguing that Saturn is in fact more heavily mass loaded with Saturn's weaker magnetic field playing an important role.

Fundamentally, the plasma mass must be eventually lost to the solar wind flow. Delamere and Bagenal (2013) argued that the scale of the magnetosphere (i.e. distance to the sub-solar magnetopause) must be related to the plasma mass loss rate. At Earth the

Fig. 8 An illustration of the scaling of Jupiter's and Saturn's magnetospheres. The *vertical pink line* approximates the location of the subsolar magnetopause boundary in the absence of an internal plasma source. Jupiter's magnetosphere is substantially more inflated compared with Saturn. (Image credit: F. Bagenal/S. Bartlett)

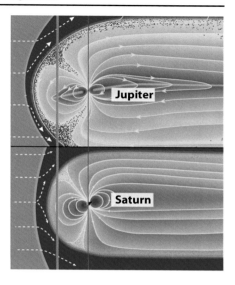

scale of the magnetosphere is determined by balancing the solar wind dynamic pressure with the planetary magnetic field pressure. This is not the case for Jupiter and Saturn where plasma pressure is an essential ingredient in the pressure balance condition. As an alternative, the momentum transfer rates from the solar wind must be able to accommodate this effective mass loading from the magnetosphere. In other words, the amount of solar wind mass per second that can be coupled via an Alfvénic interaction must balance the magnetodisc plasma loss rate. If these effective $\dot{M}$ quantities are not balanced, then the magnetodisc will grow/shrink until momentum balance is once again achieved. In this sense, the neutral-plasma interaction is fundamental to understanding the MI coupling within the magnetosphere, as well as the coupling of the magnetosphere with the solar wind.

2.5 Summary

The interaction between satellite-generated neutral gases and the magnetospheric plasma environment is critical for the formation of giant planet magnetodiscs. In the absence of a source of energy to heat the electrons, ionization of the neutral clouds would not occur. Yet the ionization of the neutral gas leads to a flow of momentum and energy from the planet that in turn heats the plasma. This feedback process can reach a non-linear tipping point where the plasma environment can rapidly erode the neutral clouds. At Saturn, the inner magnetosphere is dominated by neutral gas as electron temperatures near Enceladus are not sufficient to tip the balance in favour of plasma. But at Jupiter, the electron temperatures are sufficient to yield a plasma-dominated inner magnetosphere. The net result is that Saturn's magnetosphere is only slightly inflated by the internal plasma source when compared to Jupiter's highly inflated magnetosphere (Fig. 8). It follows then that Jupiter's magnetodisc is dominated by plasma pressure ($\beta > 1$) while the plasma pressure in Saturn's magnetodisc is comparable to the magnetic field pressure (i.e. $\beta \sim 1$). These very different conditions can significantly affect the solar wind interaction with these magnetospheres.

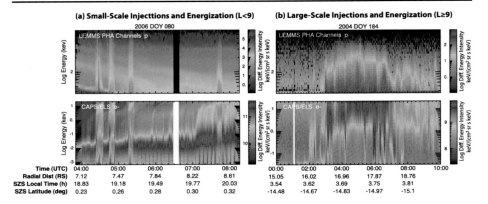

(a) Small-Scale Injecttions and Energization (L<9)   **(b) Large-Scale Injections and Energization (L≥9)**

Time (UTC)	04:00	05:00	06:00	07:00	08:00		00:00	02:00	04:00	06:00	08:00	10:00
Radial Dist (RS)	7.12	7.47	7.84	8.22	8.61		15.05	16.02	16.96	17.87	18.76	
SZS Local Time (h)	18.83	19.18	19.49	19.77	20.03		3.54	3.62	3.69	3.75	3.81	
SZS Latitude (deg)	0.23	0.26	0.28	0.30	0.32		-14.48	-14.67	-14.83	-14.97	-15.1	

Fig. 9 In-situ observations of small- (*left*) and large-scale (*right*) injections at Saturn obtained by Cassini/MIMI (magnetospheric imaging instrument). The *upper panel* shows protons measured by LEMMS (low energy magnetospheric measurement system) and the *lower panel* shows electrons measured by CAPS (Cassini plasma spectrometer). The small-scale injections are observed predominantly inside 9 R_S and have a very narrow spatial extent ($\ll$ 1 R_S). The large-scale injections are observed predominantly beyond 9 R_S and have a relatively broad spatial extent (several R_S). See Fig. 10 for an example of the instantaneous global distribution of large-scale injections

3 Injections and Their Role in Magnetodisc Formation and Stability

3.1 In-Situ Picture

The occurrence of energetic particles, up to several hundreds of keV, is a particularly apparent feature of the Saturnian magnetosphere. In the Jovian magnetosphere, such intensifications of energetic particles are less apparent and occur on top of a "sea" of already enhanced ion intensities, which is most likely due to the lack of a significant radial gradient in phase-space density (Mauk et al. 1999). In Saturn's magnetosphere, the injection of energetic particles seems to stem from two main fundamental mechanisms: (1) small-scale injections observed roughly inside of 9 R_S associated with centrifugally driven interchange (Kennelly et al. 2013; Rymer et al. 2009), and (2) large-scale injections observed roughly beyond 9 R_S, likely associated with planetward propagation due to magnetic field line tension (or buoyancy) set up by reconnection processes in the magnetotail region (Pontius and Hill 1989; Bunce et al. 2005; Mitchell et al. 2009, 2015). Both of these mechanisms are fundamental to the energization and transport of plasma and therefore comprise an important aspect of magnetodisc formation and stability.

Figure 9 illustrates both types of injection. The left-hand column shows the small-scale injections that are associated with $\ll$ 1 R_S particle intensifications. These have been associated with centrifugally driven interchange-type instabilities. The right-hand column provides an example of a large-scale injection observed post-midnight, around about 17 R_S. Its lack of energy dispersion indicates that this is a rather young injection. The injected particle distributions will disperse as they drift due to the sub-corotational electric field (i.e. the convective electric field due to bulk rotation), and magnetic gradient/curvature forces. Such particle dispersions have been observed both in situ (Mauk et al. 2005) and in Energetic Neutral Atom (ENA) images (Brandt et al. 2008) obtained by *Cassini* INCA (Ion Neutral Camera). An initial ion population, evenly distributed (spatially) over finite intervals in radial distance and azimuth, will therefore disperse and form a spiral pattern, with the higher

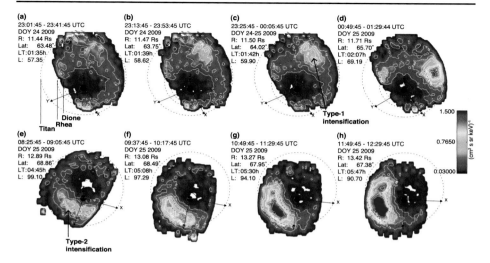

Fig. 10 Hydrogen ENA image sequence obtained from the southern hemisphere by Cassini/INCA in the 24–55 keV range. The sequence shows the first type of large-scale injection with an inward radial propagation from the tail. The injected population then drifts around Saturn due to the sub-corotational electric field, gradient and curvature forces. The drift trajectory is slightly radially inward. As the population reaches the pre-midnight sector it re-intensifies due to yet unknown processes. Position information for the spacecraft is reported in each *panel*. The *axes* shown correspond to the SZS coordinate system, in which the Z axis is parallel to the planet's rotational axis and the planet-Sun direction lies in the XZ plane

energy ions drifting faster than those of lower energy. Although this behaviour has been observed several times, there are other behaviours that remain unexplained.

It is important to remember that the term 'interchange instability' represents a broad range of magnetospheric phenomena and is not exclusively reserved for the small-scale injections, that are most likely centrifugally driven. Large-scale injections can plausibly also be the signature of an interchange-type instability, but they are more likely to be driven by the relatively strong buoyancy (with respect to the ambient plasma) of hot, low-density flux tubes which are generated through magnetic reconnection processes in the magnetotail.

3.2 Global Picture

The relation between the small- and large-scale injections is best observed in remote global images of the energetic ion distributions obtained by INCA. This instrument uses a technique called Energetic Neutral Atom (ENA) imaging, which is based on detecting the direction, energy and species of ENAs produced through charge exchange between singly charged energetic ions and neutral species.

As illustrated in Figs. 10 and 11, there are two main categories of ENA intensifications observed by INCA. The first type ('Type 1') occurs as an inward and slightly azimuthally propagating intensification, and is visible from about 20 R_S in the midnight/post-midnight sector. The second type ('Type 2') is re-brightening of an already existing emission region in the pre-midnight/midnight region. We describe the observations in more detail below.

The hydrogen ENA image sequence shown in Fig. 10 was obtained from the Southern hemisphere on DOY 24–25 2009 in the 24–55 keV energy range. The apparent hard edges of the image are the limitations of the FOV (field of view) of INCA. Figures 10a–10d show

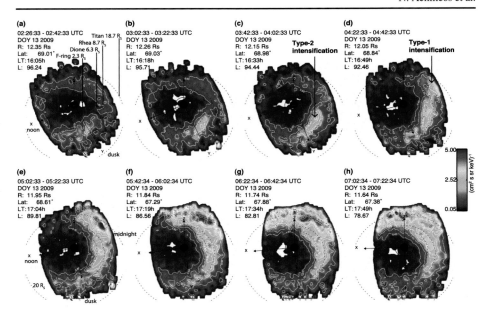

Fig. 11 Hydrogen ENA image sequence obtained from the northern hemisphere by Cassini/INCA in the 24–55 keV range. The sequence displays the occurrence of a Type-2 intensification together with a Type-1. Although tempting to suggest, it is not yet clear if Type-2 intensifications trigger Type-1

a Type 1 intensification becoming first visible around 20 R$_S$ at the upper edge of the FOV, and then propagating inward to the post-midnight sector and intensifying. The intensification could be due to two things: (1) energization of the ions leading to higher ion differential intensities, and/or (2) propagation in to regions of higher neutral densities leading to higher ENA differential intensities. Although this Type 1 intensification shows intriguing fine structure, it is not clear if it is real or is an effect of counting errors and running averages used for the images. Neither can be excluded at this stage.

Figure 10e shows how the Type 1 injection has now drifted around to the dusk sector. There is a decrease of ENA intensities from Figs. 10d to 10e, which is attributed to charge-exchange loss. However, in subsequent images of Figs. 10f to 10h, the intensities are increasing. This kind of (re)intensification is referred to as Type 2. Fine structure is coherent through the observations of the Type 2 intensification and is arguably real, since the image averages are well separated in time, but still display the same general fine structures.

Figure 11 illustrates how a Type 1 intensification can sometimes be preceded by a Type 2 intensification. The sequence shows ENA images in the 24–55 keV range, obtained from the northern hemisphere. Again, a Type 2 intensification of a previous, dimmer population occurs in the pre-midnight sector as shown in Figs. 11a–11c. However, as the Type 2 intensification reaches midnight in Fig. 11d, a Type 1 event is visible at the upper right edge of the FOV of INCA, and propagates rapidly in azimuth (at radial distance about 20 R$_S$), at about twice the rate of rigid planetary corotation—which is consistent with the modeling results by Jia et al. (2012). The reader should be cautioned here that the fine structure of the Type 1 intensification seen in Figs. 11e–11h is likely to be an artefact of the collimator blades of the INCA camera or counting noise.

The Type 1 intensification is most likely the global manifestation of the large-scale injections presented in Fig. 9. This statement is supported by the location and size of the non-dispersive, large-scale injection in Fig. 9. Perhaps, more importantly, both Type 1 intensifications observed by INCA and the large-scale injections observed in-situ appear to be associated with low-frequency extensions (LFE) of the Saturn Kilometric Radiation (Lamy et al. 2013; Mitchell et al. 2015). In addition, modelling by Jia et al. (2012) shows fast plasma flows returning to the inner, post-midnight magnetosphere following reconnection processes in the tail. Therefore, the preliminary conclusion that we draw is that reconnection in the magnetotail creates planetward, fast flows that are driven by the buoyancy of the associated low-density plasma (Pontius and Hill 1989). As the fast plasma flow reaches the inner magnetosphere it leads to plasma energization, either through inward radial transport to regions of higher magnetic field strength, and/or through particle interaction with the thin dipolarization fronts of the plasma flow burst. This type of process is very similar to that believed to be causing the main ion energization in the terrestrial magnetosphere (Runov et al. 2011).

The cause of the Type 2 intensifications is still under investigation, but possible explanations include small-scale plasma injections, based on the radial range where these are observed in-situ. The fact that the Type 1 intensification drifts radially inward to then transform into a Type 2 intensification also supports the idea that small-scale injections are playing a role in transporting plasma radially inward—thus leading to enhanced ENA intensity due to ion energization and the higher neutral densities encountered. Although radial, inward motion is not apparent in the INCA images presented here, Carbary et al. (2008b) demonstrated that a slight tendency for inward propagation in the INCA image sequences was statistically significant. Thomsen et al. (2012) provides further evidence from a range of in-situ measurements that this is indeed a feature of the drifting ion populations. A potentially important clue may also come from the fact that Type-2 intensifications occur more frequently at periods close to that of Saturn's rotation than do the Type-1 intensifications. Brandt et al. (2012) reported four consecutive occurrences of Type-2 ENA intensifications close to the orbit of Dione with a Type-1 intensification only after the fourth Type-2 intensification. As an independent confirmation of the regular nature of the Type-2, Mitchell et al. (2015) reported a close relation between the Type-2 ENA intensifications and narrow-band emissions, although the exact relation between the two remains unclear.

4 The Interchange Instability

4.1 The Interchange Instability Criterion

The structure and dynamics of giant planet magnetospheres are dominated by the presence, within the magnetospheric cavity, of several plasma sources and by the fast planetary rotation. The various plasma populations are trapped by the planetary magnetic field and the relatively cold plasma is confined near the equatorial plane by the action of centrifugal force, thereby giving rise to a thin disk of (sub)corotating plasma. The redistribution of the locally created plasma throughout the magnetospheric system is one of the fundamental dynamical processes occurring in giant planet magnetospheres. This redistribution is achieved through plasma transport perpendicular to magnetic field lines (see Sect. 1). The exact mechanisms responsible for this transport are not yet completely understood, with the transport probably operating through different modes and on different scales, both spatial and temporal. It is, however, widely believed that the outward plasma transport is triggered by the centrifugal instability, a Rayleigh-Taylor type instability in which the centrifugal force plays the role of the gravitational force, and that it proceeds through the interchange of magnetic flux tubes (e.g., Melrose 1967; Ioaniddis and Brice 1971; Hill 1976). Physically, under the action of the centrifugal and magnetic buoyancy forces, mass-loaded flux tubes tend to exchange positions with relatively empty flux tubes, located further out in the magnetosphere.

In terms of *in situ* observations, there is no direct evidence of outward-moving flux tubes laden with cold plasma. However, the outflow velocity in Jupiter's magnetodisc has been inferred by Russell et al. (2000) through the use of various techniques. One of these involved the use of magnetodisc field measurements and stress-balance arguments to estimate a plasma mass density, and then equating the total rate of radial mass outflow to the rate at which plasma mass is added by the Io source. At a distance of 25 R_J, for example, the estimated velocity is 29 km s^{-1}. Considering the density profile of plasma nearer the Io torus (6–9 R_J), Russell et al. (2000) estimated smaller outward radial velocities $\sim$ 9–68 m s^{-1} in this region. These authors also pointed out that magnetic signatures of empty flux tubes in the Io torus region comprised a small fraction, ($\sim$ 0.4 %), of the total time covered by the relevant data. This aspect supports the picture of empty flux tubes which carry magnetic flux inwards at relatively high speeds, in order to compensate for the outward transport of flux via the motion of a body of cold plasma which occupies the majority of the magnetodisc volume (Russell et al. 2000 estimate that the inward-moving tubes have speeds $\sim$ 250 times greater than the outward-moving ones; see also Sect. 1). In this picture, the net radial flux of plasma mass would be outward-directed.

Interchange motions have been extensively discussed in the literature under a variety of simplifying assumptions, sometimes unrealistic and unproven. A review of this abundant literature can be found in the introductory section of Ferrière et al. (1999). Gold (1959) was the first to introduce the concept of interchange of magnetic flux tubes in the magnetospheric context (see also Sect. 1). His so-called strict interchange model assumes a one-to-one interchange between magnetic flux tubes enclosing the same amount of magnetic flux and thus leaving the shape of the field lines unchanged, and the total magnetic energy of the system unperturbed. Cheng (1985) pointed out much later that this model is at odds with the requirement of total pressure balance, and that a realistic flux tube interchange must be accompanied by a corresponding change in field magnitude. The so-called generalized interchange model of Southwood and Kivelson (1987) still assumes that the interchanging flux tubes preserve everywhere the direction of the local magnetic field, but they relax the condition that the energy density of the magnetic field is unperturbed by the interchange. Both models involve approximations for actual interchange motions of plasma elements, which generally entail distortions of the direction of magnetic field lines, preserving the equilibrium of total pressure (plasma plus magnetic pressure).

Newcomb (1961) studied the influence of the gravitational field and of stratification on the three modes of ideal magnetohydrodynamics (the Alfvén mode, the fast and slow modes). He considered a plasma confined in a horizontal magnetic field distribution by a uniform, vertical gravitational field. Newcomb (1961) identified two convective wave modes for this system, whose dispersion relation is influenced by the stratification of the plasma in the vertical direction (for the case of a wave vector nearly perpendicular to the magnetic field). These two modes can be distinguished by their behaviour in the limit of zero parallel wave vector (with respect to the magnetic field). In this limit, the convective wave modes occur as two types, and were given the name of quasi-interchange modes: one of these modes gives rise to plasma motions mainly across the magnetic field lines (hereafter denoted the interchange or type 1 mode), while the other two modes give rise to plasma motions mainly along the magnetic field lines (hereinafter denoted as translation or type 2 modes). These quasi-interchange modes have been studied more comprehensively than in previous attempts by Ferrière et al. (1999) for the case of collisional plasmas, and by Ferrière and André (2003) for the case of collisionless proton-electron plasmas. These studies take into account the effect of both gravitation and magnetic field line curvature. André and Ferrière (2004) derived the local stability criterion of small-scale, low-frequency modes in stratified, rotating plasmas, consisting of multiple ion species and characterised by gyrotropic, bi-Maxwellian distribution functions. They obtained a criterion that consists of the simple superposition of two distinct criteria; the first being related to instabilities triggered by thermal pressure anisotropies, and the second corresponding to instabilities triggered by stratification. All low-frequency modes (Alfvén, fast, slow and mirror modes) were found to be stable over all wave vectors, if and only if:

$$\mathcal{F} \geq 0 \quad \text{and} \quad \mathcal{M} \geq 0 \quad \text{and} \quad \varpi_0^2 - \frac{g_1^2}{\mathcal{M}} \geq 0. \tag{8}$$

The restricted number of parameters appearing in the stability condition (8) are defined as follows. Firstly, the parameter $\mathcal{F}$ is given by

$$\mathcal{F} = V_A^2 + C_\perp^2 - C_\parallel^2, \tag{9}$$

and thus involves the Alfvén speed, V_A and the sound speeds in directions perpendicular and parallel to the magnetic field (respectively $C_\perp$ and $C_\parallel$). The parameter $\mathcal{M}$ is defined as:

$$\mathcal{M} = V_A^2 + \gamma_{\perp\perp} C_\perp^2 - \frac{\gamma_{\perp\parallel}^2 C_\perp^4}{\gamma_{\parallel\parallel} C_\parallel^2}, \tag{10}$$

where the γ coefficients, due to limitations of space, are not given here in their full forms, but may be found in equations (B9)–(B12) of André and Ferrière (2004). These coefficients are complicated functions that contain the relevant information on the composition and velocity distribution of the collisionless plasmas considered. $\mathcal{F}$ and $\mathcal{M}$ have physical meaning, and are respectively linked to the firehose and mirror instability criteria.

The definition of the final parameter in (8) is

$$\varpi_0^2 - \frac{g_1^2}{\mathcal{M}} = \left(\frac{\nabla \rho}{\rho} - \frac{\mathbf{g}}{\gamma_{\parallel\parallel} C_\parallel^2} \right) \cdot \mathbf{g}$$
$$+ \left[\frac{\nabla(2P_M + P_\perp - P_\parallel)}{\rho} - \mathcal{F}\mathbf{c} \right] \cdot \mathbf{c}$$
$$- \frac{([(1 - \frac{\gamma_{\perp\parallel}}{\gamma_{\parallel\parallel}} \frac{C_\perp^2}{C_\parallel^2})\mathbf{g} + \mathcal{F}\mathbf{c}] \cdot \widehat{e}_g)^2}{\mathcal{M}}. \tag{11}$$

This is a threshold parameter with the dimension of frequency squared. In these definitions, ρ denotes the plasma mass density, $\mathbf{g}$ the effective gravity (including a centrifugal component) with unit vector $\widehat{e}_g = \mathbf{g}/g$, $\mathbf{c}$ the magnetic curvature vector, $C_\perp^2 = (P_\perp/\rho)$, $C_\parallel^2 = (P_\parallel/\rho)$ where the corresponding plasma pressures are $P_\perp$ (perpendicular to the magnetic field) and $P_\parallel$ (parallel to the magnetic field). P_M denotes magnetic pressure. All the physical quantities and parameters appearing in these expressions pertain to the equilibrium state, where they are all considered to be invariant along magnetic field lines, with the additional assumption that the effective gravity and magnetic curvature vectors are parallel.

The first two inequalities in (8) are required to guarantee stability against the anisotropy-driven firehose and mirror instabilities, respectively. Together, they constitute the necessary and sufficient condition for the stability of low-frequency modes in a uniform medium. The last inequality represents the effect of stratification in the plasma, and is required to prevent the growth of stratification-driven instabilities. Stratification and rotation alter the physical characteristics of the Alfvén, slow and mirror modes in the limit of nearly field-perpendicular wave vectors, as discussed previously. The types of unstable modes in the different domains of parameter space depend on the four parameters appearing in (8) and on the sign of the generalized Rayleigh-Taylor frequency ω_0, given by:

$$
\omega_0^2 = \left(\frac{\nabla \rho}{\rho} - 2\mathbf{c} \right) \cdot \mathbf{g}
$$
$$
+ \left[\frac{\nabla (2P_M + P_\perp - P_\parallel)}{\rho} - \left(\mathcal{F} - 3C_\parallel^2 \right) \mathbf{c} \right] \cdot \mathbf{c}
$$
$$
- \frac{[\mathbf{g} + (V_A^2 - C_\parallel^2)\mathbf{c}]^2}{V_A^2 + 2C_\perp^2}. \tag{12}
$$

Interestingly, ω_0^2 depends only on fluid parameters and not on the exact composition of the multispecies plasma, whereas the details of the plasma composition enter the overall stability criteria equation (8), such that composition plays a role in determining the plasma's stability against all quasi-interchange modes.

The rest of this section is devoted to a discussion of the stability criterion of the interchange mode of type 1 (quasi-interchange mode) in giant planet magnetospheres. In these environments $\mathcal{F} \geq 0$, which implies that the condition $\omega_0^2 \geq 0$ is necessary and sufficient for the stability of the interchange mode.

4.1.1 Case Including Effective Gravity (Centrifugal Force) Only

If the equilibrium magnetic field is assumed to be straight (i.e. no field line curvature), then the interchange instability criterion reduces to the following:

$$
\mathbf{g} \cdot \left(\frac{\nabla \rho_0}{\rho_0} - \frac{\mathbf{g}}{V_A^2 + 2C_\perp^2} \right) \geq 0. \tag{13}
$$

Although the plasma pressure gradient does not influence the interchange instability criterion, the thermal pressure itself has a stabilizing effect, since it enters the stability criterion through the perpendicular sound speed. In the cold plasma approximation, $C_\perp^2 = 0$ and the stability criterion reads (after integration along magnetic field lines):

$$
\mathbf{g} \cdot \nabla \eta \geq 0, \tag{14}
$$

where $\eta = \int (\rho/B)ds$ is the mass per unit magnetic flux, and one recovers the classical result of the strict interchange model of Gold (1959). This criterion is frequently used in situations where the plasma beta is less than unity. This is typically the case in the innermost magnetospheres of Jupiter and Saturn.

4.1.2 Case Including Curvature Force Only

In the magnetospheric context for rapidly rotating systems, the magnetic curvature force often dominates over that of gravitation (Southwood and Kivelson 1987). In the case of isotropic collisional plasmas, the stability criterion for the interchange mode can be rewritten in terms of the volume of a magnetic flux tube V_0 and the radial gradient of the thermodynamic adiabatic invariant $P_0 V_0^\gamma$, and reads:

$$
-\mathbf{c} \cdot \nabla (P_0 V_0^\gamma) \geq 0, \tag{15}
$$

in which the vector **c** is directed planetward. This expression thus coincides with the interchange stability condition obtained by Southwood and Kivelson (1987). A collisional plasma would therefore be stable against interchange motions provided its thermal pressure decreases outward less rapidly than for the case of adiabatic transport. For completeness, André and Ferrière (2004) have shown that a gyrotropic, collisionless plasma with isotropic thermal pressure at equilibrium is always more stable than the corresponding isotropic, collisional plasma.

4.1.3 Competing Effects of Cold and Hot Plasma Populations

We now turn our attention to the question of whether the local stability criterion against the interchange mode is fulfilled in the Io torus at Jupiter, and in the inner magnetosphere of Saturn. Following the Voyager encounters of Jupiter (specifically observations of the Io torus), it has been proposed that the hot/energetic plasma population plays a critical role in controlling radial plasma transport. For example, Siscoe et al. (1981) suggested that the thermal pressure gradient of the hot plasma in the Io torus could reduce or enhance transport via interchange, depending on whether it is directed in the same or opposite sense to the effective gravity (impoundment mechanism). Mauk et al. (1996, 1998) tested this suggestion by calculating the radial gradient of $P_0 V_0^{\gamma}$ based on the hot/energetic plasma pressures inferred from the *Voyager* and *Galileo* observations. They concluded that the hot plasma distribution indeed impeded radial transport in both cases, and that this impoundment by the hot plasma had probably been more efficient during the Voyager era than during the Galileo era, when the hot plasma appeared significantly depleted (Fig. 12a).

Chen (2003) numerically studied the motion of a mass-loaded magnetic flux tube within empirical models of the Io torus at equilibrium, using a MHD approach. André and Ferrière (2004) also used different equilibrium models of the Io torus in which low- and high-energy plasma components have a specified radial distribution at the Jovian centrifugal equator, similar to the approach of Chen (2003), in order to test the interchange stability criterion. Both studies showed that the relative flux tube content of cold to hot plasma plays an important role in plasma transport for the Jovian system. These studies also showed that most of the torus is close to marginal stability against interchange. This is expected, since the injection of plasma at Io's orbit initially gives rise to unstable density gradients, which trigger plasma transport, which in turn acts to reduce the unstable gradients and restore a configuration closer to marginal stability. In both studies, the Io torus appeared stable against interchange at the time of Voyager when both the contributions of cold and hot plasma were taken into account, in agreement with the conclusions of Mauk et al. (1998). Unstable regions were identified for the Galileo period, in close correlation with the regions where clear observational signatures of the instability were reported (e.g., Kivelson et al. 1997a).

Whereas similar detailed studies of interchange stability remain to be done in the case of Saturn (although see Sect. 5), a few preliminary conclusions may be drawn. Similar to the proposed impoundment mechanism at Jupiter, a large fraction of the inner magnetosphere of Saturn is a 'battleground' between the competing influences of the interchange instability of the cold, internally generated plasma, and the interchange stability of the hotter, energized plasma (Mauk et al. 2009). At Saturn, Sittler et al. (2008) have shown that the main plasma source is at $L = 6$ because that is where the ambient electrons become hot enough to produce fast collisional ionization (Fig. 12b). Outside of the radial location of this peak plasma source at $L = 6$, Saturn's magnetosphere should therefore be unstable to centrifugal interchange in the cold plasma limit—and this picture is

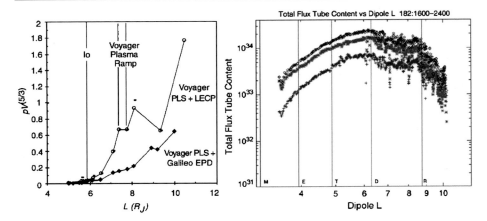

Fig. 12 (a) Radial profile of the thermodynamic invariant for combined *Voyager* PLS, *Voyager* LECP and *Galileo* EPD data obtained at Jupiter. Figure taken from Mauk et al. (1998); (**b**) total flux tube content for water-group ions (in *blue*), protons (in *red*) and total ions (in *black*) versus dipole *L*, from Cassini CAPS observations at Saturn. Figure reprinted from Sittler et al. (2008), Copyright (2007), with permission from Elsevier

indeed in close agreement with Cassini observations of plasma injection events beyond that distance. This simple picture should, however, be contrasted with the analysis of Sergis et al. (2007), which showed that most of the plasma pressure is carried by ions in the energy range 10–150 keV, and that the hot plasma pressure increases radially outward to acquire peak values around $L = 10$. The observational signatures of centrifugal interchange are found to be clustered between 5–6 R_S and 10–11 R_S, exactly where one observes the respective peak values in both cold plasma flux tube content and hot plasma pressure.

4.1.4 Coriolis Influence on the Interchange Instability

The influence of planetary rotation is felt through the centrifugal and Coriolis forces in a rotating frame of reference. Whereas the centrifugal force is included in most studies of the interchange instability, the Coriolis force and its impact are generally not taken into account. Vasyliūnas and Pontius (2007) have recently shown that the presence of the Coriolis force renders the growth rate of the instability complex, adding an imaginary component to this rate. This is equivalent to a real oscillation frequency, with the consequence of making the pattern of fluctuations rotate azimuthally, relative to the corotating frame of reference. The Coriolis force also acts to reduce the magnitude of the real growth rate. The growth rate depends on the ratio of azimuthal to radial wavelength and acquires maximal values for structures elongated in the radial direction. All the Coriolis effects become, however, negligible in the limit of very short azimuthal wavelengths, in agreement with the results of André and Ferrière (2007) that are restricted to that specific case. Although not a generally valid result, neglect of the Coriolis force effects is a reasonable approximation, as long as the fastest-growing instabilities are those with very short azimuthal wavelengths. This result may therefore hold quite well, in general, for practical applications to the magnetospheres of Jupiter and Saturn, although a few additional caveats were mentioned by Vasyliūnas and Pontius (2007).

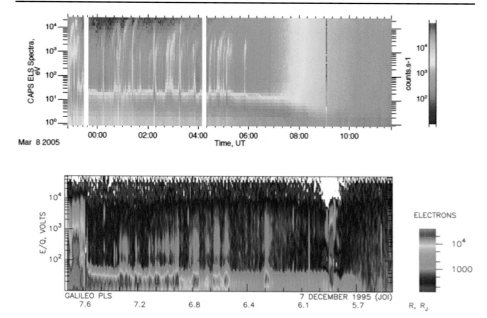

Fig. 13 Energy-time spectrogram of electrons at Saturn (*top*, Cassini CAPS; figure made with the AMDA tool available at CDPP, http://cdpp.cesr.fr) and Jupiter (*bottom*, Galileo PLS; figure taken from Frank and Paterson 2000)

4.2 In Situ Signatures of the Interchange Process

The recent *Galileo* and *Cassini* space missions have amply documented the Jovian and Saturnian magnetospheric systems. In particular, these missions have provided us with new observations and new insights into the mechanisms responsible for the outward plasma transport. There is now considerable observational evidence that centrifugally driven flux tube interchange is at play in the corotation-dominated regions of the Jovian and Saturnian magnetospheres, and that this process contributes to the redistribution of plasma throughout these systems. Signatures of intermittent, short-lived, mass-loaded and relatively empty flux tubes in the Io torus have been detected by the *Galileo* spacecraft, in orbit around Jupiter from December of 1995 to September of 2003 (e.g. Kivelson et al. 1997a; Thorne et al. 1997). Similar signatures have been observed in the E-ring of Saturn by the Cassini spacecraft, in orbit around Saturn since July 2004 (André et al. 2005; Burch et al. 2005, 2007; Hill et al. 2005; Leisner et al. 2005; Mauk et al. 2005; Rymer et al. 2008) and have lent further support to the notion that the centrifugal instability lies at the root of the outward transport.

In both magnetospheres, the reported plasma signatures consisted of discrete hot plasma injection events accompanied by longitudinal dispersion arising from magnetic gradient and curvature drift (Figs. 13, 14). The magnetic signatures of the injections consist of short-duration, sharp-bounded intervals of either enhanced or depressed magnetic pressure, and whose plasma content differs significantly from that of the surrounding medium (Fig. 14). These signatures were consistent with expectations for the interchange of magnetic flux tubes containing dense and cold plasma with those containing tenuous and hot plasma that originate farther out in the magnetosphere. Compared to Jupiter, Saturn's magnetosphere provides a somewhat better 'laboratory' for understanding the observational properties of

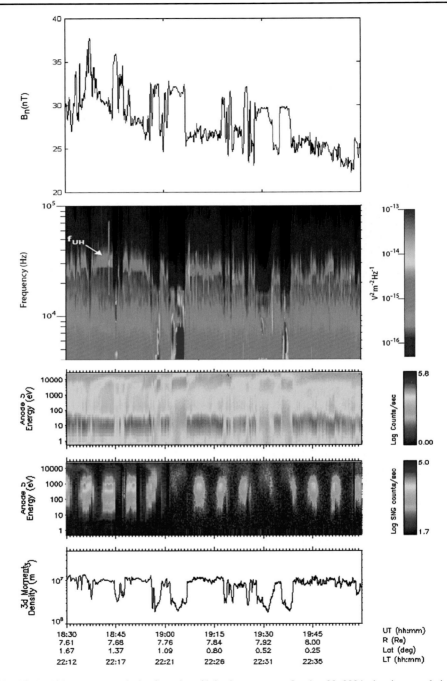

Fig. 14 Multi-instrument analysis of a series of injection events on October 28, 2004, showing correlations between (from *top* to *bottom panels*) enhancements in the magnetic field, the disappearance of the upper hybrid emission band, the depletion of the low-energy plasma population, the coincident appearance of a hot plasma component, and significant density drop-outs. These events occur inside the well-defined boundaries of the interchanging magnetic flux tubes. Figure adapted from André et al. (2007)

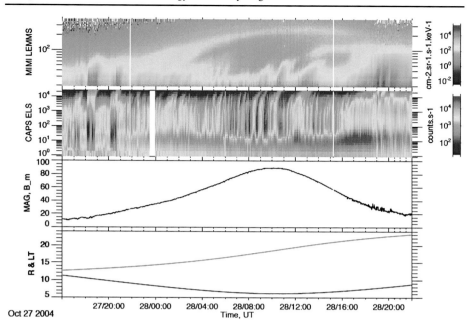

Fig. 15 *Cassini* multi-instrument observations of injection events showing the broad diversity of signatures encountered. From *top* to *bottom*: MIMI LEMMS electron energy-time spectrogram, CAPS ELS electron energy-time spectrogram, MAG magnetic field magnitude, Radial Distance and Local Time (in *red*). Figure made with the AMDA tool available at CDPP, http://cdpp.cesr.fr

the centrifugal interchange instability. This is because, for a given particle energy per unit charge, and a given L value, the gradient/curvature drift is about 25 times faster at Saturn than at Jupiter (Hill et al. 2005), which thus makes Kronian injection events easier to detect.

Due to adiabatic gradient and curvature drifts, the injected plasma indeed exhibits significant longitudinal drift dispersion. On a linear energy-time spectrogram, an injection/dispersion event is revealed as a V-shaped structure, with hot ions (electrons) forming the left (right) legs of the V. The apex of the V marks the original injection longitude, while the width of each leg indicates the width of the injection channel. The slope of the legs of the V is inversely proportional to the elapsed time since the original injection. Hence, by combining the original injection longitude and the elapsed time, the local time of the original injection can also be determined (Hill et al. 2005). Fewer, but spatially larger, events are observed at higher energies by the *Cassini* MIMI instrument (Mauk et al. 2005). These features are long-lived and dominate the appearance of the inner magnetosphere at medium and high energies (Fig. 15); they show a broad diversity of time-dispersed features produced by the combined effects of prograde (ions) and retrograde (electrons) magnetic drifts and the prograde 'ExB' drift (Paranicas et al. 2007). Over 100 injection/dispersion signatures have also been observed at Jupiter by the *Galileo* spacecraft. As at Saturn, the behaviour of the higher-energy particles at Jupiter showed much larger scale injections that are relatively long-lived (Mauk et al. 1999). These injections extended over ~ 9–27 R$_J$ and occurred at all longitudes and local times.

At Saturn, Chen and Hill (2008) and Chen et al. (2010) used *Cassini* plasma spectrometer observations and included 622 events in a statistical study of the smaller-scale injections. Their analysis revealed that the injections are spatially clustered within ~ 5 R$_S$, and also

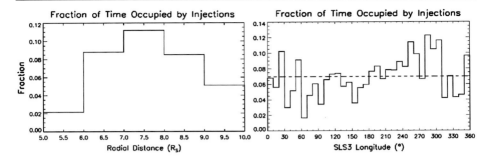

Fig. 16 Plasma inflow statistics at Saturn: (**a**) fraction of observation time occupied by plasma inflow channels within 1 R_S bins, averaged over SLS3 longitudes; (**b**) fraction of observation time occupied by plasma inflow channels within $10°$ SLS3 longitude bins averaged over the radial range 5–10 R_S. The *dashed horizontal line* is the longitudinal average. Figures taken from Chen et al. (2010)

between $\sim$ 10–11 R_S (Fig. 16); their longitudinal widths seldom exceed $\sim$ 1 R_S; their ages since formation seldom exceed the Saturn rotation period; and their spatial distributions in both longitude and local time are essentially random. The combined widths of all injection channels at any given time occupy only a small fraction (about 7 %) of the entire longitude space (Fig. 16). The inflow channels are thus evidently much narrower than the intervening outflow channels. If the total mass fluxes of the incoming hot plasma and outgoing cold plasmas are in balance (on average), then we would expect the ratio of inflow to outflow velocity to be approximately given by $V_i/V_o \sim (M_c/M_h)(\xi/(1-\xi))$, where M denotes typical flux tube content of the plasma, subscripts indicate cold, outflowing (c) or hot, inflowing (h) populations, and the quantity $\xi \sim 7$ % is the fraction of the longitude space occupied in total by the inflowing events. The outflow event speeds are typically $\lesssim$ 1–2 km s^{-1} (Wilson et al. 2008), and so we expect the inflow speed to be at least $\sim$ 13 times greater than this figure. Both inflow and outflow velocities appear to be smaller than the observed azimuthal velocity of the plasma, which is, in turn, lower than the velocity of planetary corotation. This ordering is consistent with theoretical considerations (Hill 2006). This observational finding is not generally predicted by theoretical models of the interchange process, in which the inflow and outflow channels are usually of roughly equal width. This discrepancy will be addressed in the next section.

Many injection events have relatively young ages, as discussed by Burch et al. (2005). The very young events were not included in the analysis of Chen et al. (2010), but a few of them have been studied in detail by Burch et al. (2005) and Rymer et al. (2007). Burch et al. (2005) used low-energy plasma measurements to estimate, for one particular event, an inward injection velocity of 25 km s^{-1} between 10 R_S and 5 R_S, whereas Rymer et al. (2007) derived a radial injection speed of 71 km s^{-1} between 11 R_S and 7 R_S for another event. Whereas the latter estimate is outside the range of velocities inferred by Chen et al. (2010), the former is roughly consistent with their statistical results. More recently, Kennelly et al. (2013) conducted a statistical analysis of very young injection events, using observations from the Cassini Radio and Plasma Wave Science instrument. Whereas their occurrence rate with respect to L shell largely agrees with the distribution published by Chen et al. (2010), they do identify the post-noon (11-17 SLT) and near-midnight (19-3 SLT) regions as preferred sectors for the origin of injection events. At Jupiter, Russell et al. (2005) performed a statistical survey of the plasma-depleted flux tubes observed by Galileo in the Io torus (29 events). These flux tubes occupy 0.32 % of the torus volume outside the orbit of Io. They proposed that the amount of magnetic flux transported by these thin flux tubes could supply the requisite amount of magnetic flux (mass-loaded) transported to the magnetotail provided the inward velocity is ~ 300 times that of the outward plasma motion. This represents a radial velocity of the order of $10 \, \mathrm{km \, s^{-1}}$ in the torus.

4.3 Comparing Simulations and Observations of Interchange

Several numerical models have addressed the large-scale configuration and dynamics of giant plant magnetospheres, including those developed at the University of Michigan (e.g. Hansen et al. 2005; Jia et al. 2012) and at Nagoya/UCLA (e.g. Fukazawa et al. 2007, 2010). However, only a limited number of such codes have focused on the development of the interchange instability itself and how it is influenced by varying internal and external conditions. The rest of this section is devoted to those models which have been specifically used to investigate the interchange process. The main differences between models in this class are related to assumptions concerning rotational/interchange effects, plasma sources, current drivers, solar wind influence, and spatial regions of model validity.

4.3.1 The Rice Convection Model for Jupiter and Saturn

The Rice Convection Model is a multi-fluid physical model which was initially developed at Rice University in order to study solar wind-driven convection in the Earth's magnetosphere and the electromagnetic coupling to the ionosphere (e.g. Wolf 1983). This model was then adapted to simulate the rotation-dominated magnetosphere of Jupiter (Yang et al. 1992, 1994), and recently further extended to investigate the magnetosphere of Saturn (Liu et al. 2010).

The Rice model treats the three-dimensional problem of MHD flow in a planetary magnetosphere as a pair of two-dimensional problems: one in the equatorial plane of the magnetosphere, and the other in the planetary ionosphere. In this approach, the horizontal divergence of the magnetospheric current is balanced in the ionosphere by the analogous divergence of the Pedersen current, the two regions being connected by Birkeland currents flowing along magnetic field lines (Vasyliūnas 1970). In rapidly rotating magnetospheres, the centrifugal drift (or inertial) current is usually one of the major contributions to the total magnetospheric current, whereas, at the Earth, the major currents usually arise from a combination of magnetic gradient drift, magnetic curvature drift and the magnetization current associated with the magnetic moment of the gyrating plasma particles.

Early simulations from the Rice Convection Model applied to the Io torus at Jupiter focused on the influence of the radial width of the initial plasma distribution on the subsequent scale size of the dominant convection cells which emerge in the simulated flow. These simulations produced regularly spaced, long, thin 'fingers' of plasma outflow from the outer edge of the Io torus, interspersed with fingers of inflow from the surrounding magnetosphere (Yang et al. 1994). The plasma outflow in the fingers was suggested to be the main mechanism of plasma transport from the Io torus to the outer magnetosphere through the centrifugal interchange instability. The exponential growth rate of the instability was found to be proportional to the azimuthal wave number of the disturbance, and hence inversely proportional to the azimuthal scale size of the disturbance, in agreement with the analytical linear analysis of Huang and Hill (1991). The instability growth rate for a given wave number scales with the ratio of the total flux tube content divided by the ionospheric Pedersen conductivity.

A narrow, ribbon-like structure near Io's orbit is observed to be embedded within the larger torus. This 'ribbon' is a very dense, narrow annulus of plasma, situated around

5.6–5.8 R_J. In order to understand the stability properties of this ribbon structure, Wu et al. (2007) decreased the size of the simulation region and increased the numerical grid resolution in order to resolve the smallest convection cells. They found that the dominant azimuthal scale of the interchange cells was about half of the radial width of the initial distribution that produced them. The small ribbon-scale structures thus grow faster than the larger torus-scale structures for given values of the flux tube content and ionosphere conductance. This observation emphasizes the importance of stabilizing mechanisms for maintaining the persistent torus structure, and calls for the inclusion in future simulations of ring current impoundment (e.g. Siscoe et al. 1981) and velocity shear (departure from planetary corotation, Pontius et al. 1998).

The relatively large growth rate for the ribbon in the Io torus also calls for the inclusion of the Coriolis and acceleration currents in the Rice Convection Model. This has been recently achieved for Saturn. Liu et al. (2010) have included the effects of the Coriolis force and the pickup current in order to study the plasma convection pattern in the inner magnetosphere of Saturn. To that purpose, a continuously active, distributed inner plasma source has also been added to the simulation. The plasma source model used in that initial simulation had a total mass-loading rate of only 24 kg s^{-1} and an ionospheric Pedersen conductance of 0.3 mho. The simulation confirmed that fast and narrow plasma inflow channels alternate with slower and wider outflow channels, in agreement with the Cassini observations of injection events described above. The Coriolis force was found to be responsible for: (i) the bending of the convection cells in the retrograde direction; (ii) the slowing of their growth; and (iii) the broadening of the outward-moving fingers. In addition, the inclusion of an active plasma source made the inflow sectors much narrower in longitude than the interspersed outflow sectors (Fig. 17). The simulation was then extended to include the effects of finite plasma pressure (i.e. cold plasma with finite temperature). The associated $\mathbf{J} \times \mathbf{B}$ force at Saturn was found to augment the driving provided by the centrifugal force and enhance by 50 % the centrifugal instability growth rate (Liu and Hill 2012). The plasma source model used in the most recent simulation had a total mass-loading rate of 240 kg s^{-1} and the ionospheric Pedersen conductance was increased to 3–6 mho (simulation results not shown here). These incremental adaptations of the Rice Convection Model for Jupiter and Saturn have so far been generally successful in reproducing spacecraft observations of the centrifugal instability in giant planet magnetospheres. A multi-fluid model for Saturn's rapidly rotating magnetosphere is described in Sect. 5.

4.4 Interchange in the Laboratory

Interchange instabilities occur in a variety of natural plasma environments, not only the fast rotating magnetospheres of Jupiter and Saturn just described, but also at the plasmapause and in the F-layer of the ionosphere at Earth. They also arise in laboratory plasmas where magnetic confinement configurations can lead to pressure-driven interchange instability. Although the plasma beta in laboratory experiments is much lower than inside the Jovian or Kronian magnetospheres, observations of the structures resulting from the development of the artificial interchange instability may be relevant to the dynamical processes within giant planet magnetospheres. We briefly illustrate here the potential for such a comparison in the case of Saturn.

Evidence from multiple Cassini instruments exists for some sort of azimuthal asymmetry, moving in the corotational sense, thus causing asymmetric mass and energy flow around the planet (see a review by Carbary and Mitchell 2013). Such an asymmetry has been notably reported in the electron density determined by the Cassini Radio and Plasma Wave Science instrument in the inner magnetosphere (Gurnett et al.

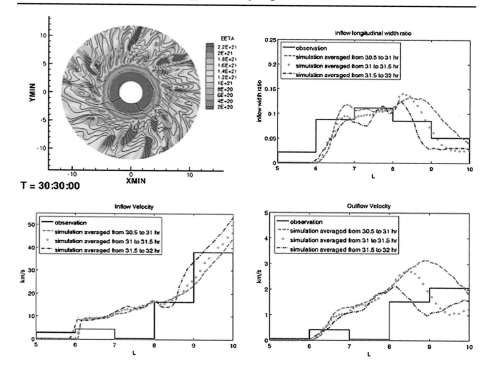

Fig. 17 Simulation results for the evolution of plasma convection in Saturn's inner magnetosphere, with the inclusion of the Coriolis force and an active plasma source. *Top left: Colour* indicates flux tube ion content (ions/Weber) projected on the equatorial plane, in the corotating frame of reference, with equipotential lines (flow streamlines) added at 15 kV intervals. *Top right:* Fraction of the available longitude space occupied by inflow channels, compared with the observations from Chen et al. (2010). *Bottom:* Longitudinally averaged inflow and outflow velocities, compared with observations from Chen et al. (2010). Figures taken from Liu et al. (2010)

2007). One proposed explanation is a corotating convection pattern (Gurnett et al. 2007; Goldreich and Farmer 2007), initially proposed for the Jovian system (Vasyliūnas 1983; Hill et al. 1981). The proposed systematic pattern consists of outflow in an active sector and inflow in the complementary longitude sector (Fig. 18), and the pattern corotates with the planet, although the plasma may not. The resulting two-cell pattern may be generated spontaneously and maintained indefinitely via the centrifugal instability. As originally pointed out by Hill et al. (1981), the flux-tube interchange instability may initiate the plasma outflow, and the proposed corotating convection pattern may represent the steady state toward which the system evolves, i.e. the saturation state of the centrifugal interchange instability.

The two-cell convection pattern has also been observed to extend along a spiral path in the outer magnetosphere of Saturn (Burch et al. 2009). Most previous models of centrifugally driven convection, however, suggest that the convective motions should be dominated by high-order ($m \gg 1$) azimuthal modes that evolve into azimuthally confined fingers, as discussed in the previous section. Gurnett et al. (2007) nevertheless suggested that the lowest-order, $m = 1$ (two-cell) mode should dominate in the Enceladus neutral torus, since this mode produces the longest path length through the source region—thereby giving the largest density increase and the largest growth rate for the instability. How the centrifugal instability maintains a single longitudinal asymmetry ($m = 1$

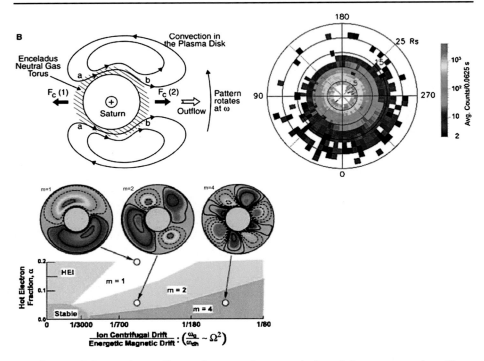

Fig. 18 *Top left*: Proposed two-cell corotating convection pattern in Saturn's inner magnetosphere. Figure from Gurnett et al. (2007), reprinted with permission from AAAS; *Top right*: Average ion counts organized against SLS3 longitudes. Figure from Burch et al. (2009); *Bottom*: Computed variation of the interchange mode structure as the fraction of energetic electrons and the plasma rotation rate (Ω) change. Only the regimes dominated by $m = 1$ and $m = 2$ are observed experimentally. Reprinted with permission from Levitt et al. (2005b). Copyright Physics of Plasmas 2005, AIP Publishing LLC

mode) rather than multiple asymmetries, or how the dominant $m = 1$ mode can persist on decadal time scales are questions which remain unanswered. Additional challenges for such models are presented by the observation of two distinct oscillation periods in Saturn's magnetospheric field and kilometric radiation (e.g. Andrews et al. 2012; Lamy 2011; Gurnett et al. 2009).

Whereas the interchange driven by energetic electron pressure has been studied in detail in artificial plasmas, the excitation of the centrifugal interchange instability in a laboratory plasma confined by a magnetic dipole has been reported by Levitt et al. (2005a). These first experimental observations of the centrifugally driven interchange mode were made in the Collisionless Terrella eXperiment (CTX) (dipole-confined plasma) by inducement of $\mathbf{E} \times \mathbf{B}$ rotation, through application of a radial electric field. Both centrifugally driven interchanges, and interchanges driven by hot electron pressure, were simultaneously observed in the same discharge experiment. The observed mode structures for these instabilities are quasi-coherent in the laboratory frame of reference and are dominated by low azimuthal m numbers and broad radial structures. The dispersive properties of the global, coherent modes are modified by the presence of the hot electron population, which causes long wavelength modes to be more unstable: drift resonant particles induce polarization currents which preferentially stabilize higher m modes (Levitt et al. 2005b). For a relatively low fraction of hot electrons and rapid plasma rotation, the most unstable interchange mode has a shorter wavelength, $m = 4$, and grows with nearly zero real frequency in the rotating frame. As the rotational driver is reduced or as the energetic electron population increases, the azimuthal mode number m decreases and the interchange mode acquires a real frequency in the direction of the electron magnetic drift. The radial phase has a significant shift of $\pi/2$ for the dominant $m = 1$ mode. The corresponding fluctuations presented a broad azimuthal spiral structure, which appeared to rotate rigidly in time. Figure 18 displays the interchange mode structure as a function of the plasma rotation rate and the fraction of energetic electrons.

In this context, energetic electrons also constitute an important population of particles in the magnetospheres of Jupiter and Saturn. For a given ion mass M_i and energetic electron energy E_h, the ratio of the gyrofrequency and drift frequency, denoted ω_g/ω_d, is proportional to $(1/2M_i\Omega^2L^2)/3E_h$, where L is the equatorial radius of the field line and Ω is the angular velocity of the plasma. For the experiments described in Levitt et al. (2005a), $\omega_g/\omega_d \sim 1/100$, whereas the same ratio is $1/55$ at Saturn (and $1/45$ at Jupiter). The mode structure that results from the interaction of the centrifugal instability with energetic particles, and which is observed in these laboratory experiments, bears a striking similarity with some of the Cassini observations reported in Saturn's magnetosphere. The detailed comparison of these two systems, however, requires further studies to be initiated.

5 Magnetodisc-Plasma Interactions

5.1 General Properties

As noted in Sect. 1, the overall position of the plasmadisc and magnetodisc are determined by the plasma velocity, mass density, and the magnetic field. In the outer planets, especially Jupiter and Saturn, the energy and momentum transport is controlled by three influences: (a) solar wind and interplanetary magnetic field (IMF), (b) the planet's rotation and plasma sources, and (c) interactions with the planet's satellites. The influence of these processes is shown in Fig. 19. The interaction of the solar wind with the magnetic field of the planet produces a bow shock and magnetopause that define the outer boundary of the magnetosphere. Along the flanks and in the deep (several tens of planetary radii) tail, solar wind drives plasma flows that are predominantly down-tail. In the middle magnetosphere a two-cell convection pattern, particularly for antiparallel interplanetary magnetic field (IMF), can develop where there are induced sunward plasma flows, some of which are associated with reconnection events. Imposed on these flows is the rotation field set up by the planet and its magnetic field. For the inner planets the rotational effects tend to be small and most dominate out to a few planetary radii, at most. At the outer planets the rotational effects can be much stronger, dominating out to several tens of planetary radii. This effect is strongest within the Jovian magnetosphere, though it is also a strong aspect of the dynamics of the Kronian magnetosphere.

In the presence of a two-cell convection pattern, the convection velocity in the inner to middle magnetosphere tends to be enhanced on the dawn side and reduced on the dusk side. This modulation of the convection speed, as discussed later, has an important effect on the transport of energy and mass within the magnetosphere.

The addition of the solar wind and rotational influences to create the overall, average plasma convection flows is shown in Fig. 20 and illustrates some aspects of the Vasyliūnas convection pattern (Vasyliūnas 1983). Co-rotational flows exist in the inner magnetosphere, while beyond the Alfvén radius in the outer magnetosphere, tailward flows prevail. In between these regimes, there exists a region of transition in the middle magnetosphere with the direction of flow strongly dependent on the prevailing solar wind conditions.

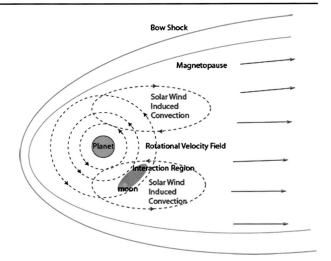

Fig. 19 Schematic of the equatorial velocity fields within the magnetospheres of the outer planets

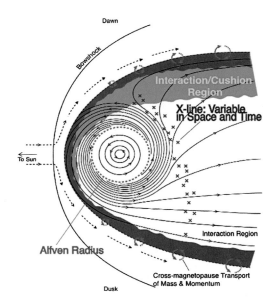

Fig. 20 Schematic of general plasma convection with the magnetospheres of the outer planets (Bagenal and Delamere 2011)

The overall convection pattern in Fig. 20 is not stable, and is subject to an interchange instability where there are alternating longitudinal regions of outward moving cold dense plasma and inward moving hot tenuous plasma (Pontius et al. 1986; Hill et al. 2005, Sects. 3 and 4). Injections of energetic particles into the inner magnetosphere are seen in both the Jovian (Mauk et al. 1997) and Kronian (Mauk et al. 2005) magnetospheres.

Transport of energy and mass is also modified by the presence of moons in the inner magnetosphere. At Jupiter, the strongest interaction is produced by Io, but both Europa and Ganymede have sufficient influence that aurorae can be seen at their magnetic footpoints in the Jovian ionosphere (e.g. Clarke et al. 1998). At Saturn, analogous footpoint emissions from Enceladus have been observed (Pryor et al. 2011). Footpoint emissions associated with Titan have yet to be identified, however the motion of Titan itself has the potential for modulating Saturn's kilometric radiation (SKR)—a phenomenon which is associated with the auroral process (Menietti et al. 2007).

The moons associated with footprint aurorae can generally produce plasma densities that, at least locally, can exceed the density of the ambient plasma in the magnetodisc. Moreover, the plasma tied to these moons initially moves at Keplerian orbits when it is first created through ionisation of neutral species. Since the Keplerian speed tends to be much slower than the convection of the ambient plasma, the moon-magnetosphere interaction produces a 'drag' effect on the ambient plasma. As discussed in Sect. 2 and in the following subsections, there is also potential for global effects developing from these moon-magnetosphere interactions. It is also interesting to note that these interactions can occur in sub-Alfvénic and/or subsonic flows, such that the presence of the moon in question can be 'felt' by the plasma in the region upstream of the moon itself. In the following subsections we provide additional details on these processes as applied to Saturn where there is a wealth of data from the *Cassini* mission (e.g. Blanc et al. 2002).

5.2 Plasmoid Formation at Saturn

Prior to 2011, only a handful of definitive plasmoids had been detected at Saturn by *Cassini* (Jackman et al. 2009; Hill et al. 2008). These reconnection-related events are seen at distances 40–50 R_S from the planet's centre, with a recurrence rate of $\sim$ 5–7 days. During two of these events, *Cassini* was near the centre of Saturn's current sheet and observed a plasmoidal composition dominated by water group ions, strongly suggesting that plasmoid formation is often linked to processes occurring in the inner magnetosphere. A more detailed analysis of the Cassini deep tail orbits in 2006 showed the presence of a total of 34 plasmoids, with an observed occurrence rate of one plasmoid every 2.4 days (Jackman and Arridge 2011).

Another sign that the magnetotail is unstable is the appearance of quasi-periodic flapping of the plasma sheet in this region. This flapping was originally interpreted as the periodic passage of plasmoids by Burch et al. (2009). A far more likely interpretation attributes the flapping to the global magnetospheric oscillations observed in the Saturn system (Jackman et al. 2009; Arridge et al. 2011).

Global MHD modeling has been used to examine the overall shape of the Kronian magnetosphere (Fukazawa et al. 2007) and showed the development of the Kelvin-Helmholtz instability on the flanks during periods where the IMF and planetary equatorial fields are parallel, and the development of magnetic reconnection during periods where these fields are antiparallel. Zieger et al. (2010), using their MHD model, demonstrated that quasi-periodic reconnection, between closed field lines in the tail, could be driven with periods between 20–70 hours, depending on the value of the upstream solar wind's dynamic pressure. Global multi-fluid simulations have been also been performed by Kidder et al. (2012) in order to investigate plasmoid formation. There is scope for improving the inner boundary conditions related to ion density in these multi-fluid calculations, in light of additional plasma moment determinations from the *Cassini* plasma spectrometer experiment (e.g. Thomsen et al. 2010). The multi-fluid approach has the advantage over single-fluid MHD of being able to separately track the dynamics of the different ion species and sources, instead of simply calculating their average properties. The different components in the multi-fluid model include (a) solar wind protons, (b) ionospheric protons, (c) water group ions from Enceladus and (d) heavy ion group that can originate from either Enceladus or Titan. Due to the difference in mass and source, these different ion species can have different trajectories through the magnetosphere, and thereby have very different transport properties to those

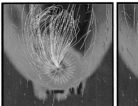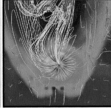

Fig. 21 Development of a plasmoid which includes loop-like structure as seen from a top view (from Kidder et al. 2012)

which would be evident in a one-fluid simulation. The development of a plasmoid in the multi-fluid model, which includes loop-like structures, is shown in Fig. 21. The skewing of the plasmoid towards the dawn sector arises from the relatively fast rotation of Saturn. While external triggering from IMF conditions is possible, the long time scale probably means that this type of triggering is probably not very common. Solar wind pressure pulses, however, can more efficiently produce reconnection and plasmoid formation. The composition of the plasmoid from the model was demonstrated to be primarily water group ions, consistent with the Cassini observations, and thus suggests that there is strong coupling between the inner and middle magnetospheres.

5.3 Interchange Instability

The dynamics of the inner magnetosphere, as noted above and in Sect. 4, are complicated by the potential growth of the interchange instability. The presence of this instability was detected very early in the data from *Cassini* (Burch et al. 2005; André et al. 2005). The first simulations of the interchange instability within a global Saturnian model were developed by Kidder et al. (2009) using the multi-fluid approach. An example of the simulated development of the interchange instability is illustrated in Fig. 22 which shows the evolution of equatorial plasma density. The growth and development of a single, outwelling interchange 'finger' is labeled with an asterisk. The development of the interchange instability is similar to that of the Raleigh-Taylor instability, in which a relatively dense fluid (in this case water group ions) sits over a less dense fluid (in this case solar wind protons), with the free energy being provided by the centrifugal force associated with the rotation of the plasma. The interchange instability starts as small ripples along the plasma density in the inner magnetosphere, where the density of the water group ions (designated CNO^+ here) is relatively high. The ripples start to grow on the dawnside, reaching almost to the magnetopause as they pass through the midday sector. As they move into the dusk sector, components of the interchange structures can break off and move down the tail while, in the inner magnetosphere, they can become truncated, with the potential for regrowth during the next rotation about the planet. As a result, there is much more substantial outward transport of mass from the inner magnetosphere than suggested by the simple picture of Fig. 20.

The corresponding temperature profile is shown in Fig. 23. The plasma in the outer regions is hottest at several tens of keV, while the plasma about the Enceladus torus is only at a few eV. As the water group ions move out, they experience some heating and approach temperatures similar to that of the original magnetospheric plasma. Conversely, between the interchange fingers, hot tenuous magnetospheric plasma is seen to penetrate inwards. Thus the interchange instability is also important for energy transport within the magnetosphere

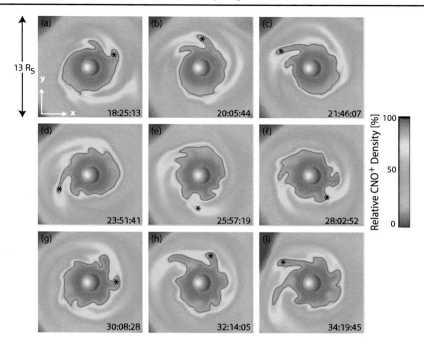

Fig. 22 Equatorial profile of the relative density of the CNO$^+$ ions originating from Saturn. The growth of the interchange instability is illustrated by the *asterisk* which marks the growth of one individual 'finger' (from Kidder et al. 2009)

and the spacecraft cutting through these fingers will see alternative regions of hot, tenuous and cold, dense plasma regions.

A particularly interesting feature seen in Figs. 22 and 23 is that the tip of the interchange fingers lead the finger itself. This effect arises from three-dimensional features of the instability. While much of the forcing is in the equatorial plane, the field lines themselves are tied to the planet at higher latitudes. In the original simulations by Kidder et al. (2009), the plasma density near Enceladus was not fully established by observations and a density nearly an order of magnitude too small was assumed. In this case, the high-latitude components also contributed to the interchange motion and, because of the additional centrifugal acceleration, the tip of the interchange finger leads the motion of the rest of the structure. The relevant point from these initial simulations is that the Kronian magnetosphere is unstable to the interchange instability over quite a wide range of the parameter space.

Subsequent simulations by Snowden et al. (2011a, 2011b) and Winglee et al. (2013) have utilized higher densities for the near-Enceladus plasma. As a result, the previous, forward curvature of the interchange figures is replaced with fingers that tend to have backward curvature (i.e. they lag the bulk finger motion). However, the difference in the velocity field between the dawn and dusk sides (c.f. Fig. 22) still causes the fingers to partially accelerate on the dawn flank and decelerate on the dusk flank, indicative of strong transport occurring where there is strong coupling between rotational and solar wind-related effects. Kidder et al. (2009) also demonstrated that the simulated growth rate of the instability could be enhanced by increases in the solar wind dynamic pressure and/or antiparallel IMF. In both cases, the convective electric field enhances the plasma velocity on the dawnside and thereby brings the plasma closer to the required velocity threshold associated with the

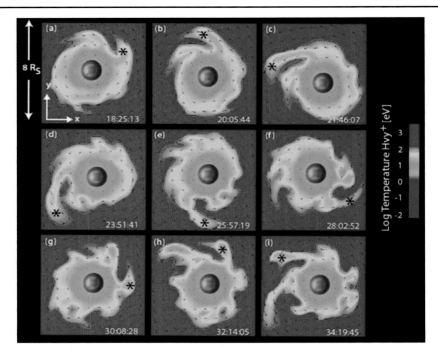

Fig. 23 The temperature profile corresponding to Fig. 22. Threading between the outward moving cold fingers is hot tenuous plasma from the middle magnetosphere (from Kidder et al. 2009)

growth of the instability. These results further emphasize the coupled nature of the problem, and that schematics such as Fig. 20, while providing a useful global picture, may be an oversimplification of the very important processes occurring deeper within the magnetosphere.

5.4 Moon-Magnetosphere Interactions

The simulated interchange fingers described above are sufficiently large that they are able to move beyond the orbit of Titan. The interaction of the Kronian plasma with Titan is known to produce an induced magnetosphere about Titan itself. The characteristics of the induced magnetosphere have been examined using local models, including MHD codes (Nagy et al. 2001; Kopp and Ip 2001; Ma et al. 2004), hybrid codes (Brecht et al. 2000; Ledvina et al. 2004; Simon et al. 2007) and multi-fluid simulations (Snowden et al. 2007). In the latter, the ion tail at Titan was demonstrated to be several Saturn radii in length, which demonstrates that Titan's induced magnetosphere is not an insignificant contributor of plasma to the Kronian magnetosphere in which it is embedded.

Given the potential for feedback between Titan's ion tail and the Kronian magnetosphere, multiscale/multi-fluid simulations that resolve the induced magnetosphere around Titan, as well as the main features of the Kronian magnetosphere, were developed by Winglee et al. (2009). In these initial simulations, time scales of a few tens of hours were considered and, due to the limited time scale, the refinement grid around Titan was held in a fixed position. An example of the results is shown in Fig. 24, where isosurfaces of the plasma density are shown over a region of about 15 R_S square. Prior to the arrival of an interchange finger,

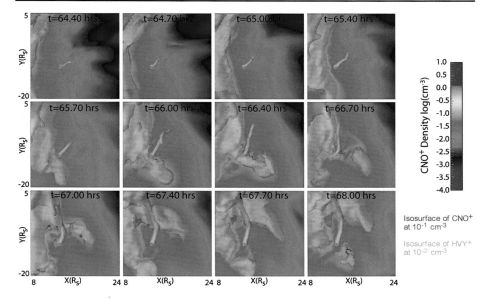

Fig. 24 Titan interacting with an interchange finger, derived from a multiscale global simulation that includes the Kronian magnetosphere and a refinement grid that includes the induced magnetosphere about Titan, with resolution of a few hundred kilometres. Interaction with the ambient plasma causes the flapping of Titan's ion tail and disruption of the near-Titan interchange finger (from Winglee et al. 2009)

Titan's ion tail is seen to lie along the average flow direction expected from the flow patterns in Fig. 20. However, as an interchange finger approaches and passes by Titan, the ion tail direction is substantially altered and moves to point more closely towards the planet. This interaction was also shown to substantially increase the outflow of plasma from Titan by nearly a factor of four. At the same time, there is the back reaction on the properties of the interchange finger, which is seen in the second and third rows of Fig. 24. In particular, the fingers in the vicinity of Titan can be disrupted or broken apart and this effect can occur upstream of the moon. This type of interaction can thus produce modulation of the physical properties of these fingers, as well as of the corresponding plasma mass transport in the Saturn system. In the example in Fig. 24, Titan was in the pre-midnight sector. Simulations that include Titan at pre-noon and post-noon sectors show that this effect of finger disruption is still very strong (Snowden et al. 2011a, 2011b). In the pre-noon sector, Titan's ion tail was sufficiently dense that it could prevent the movement of the magnetopause past Titan during IMF reversals. In the post-noon sector, slow convection of plasma past Titan can lead to the magnetic field being frozen in for periods as long as 15 minutes after IMF field reversals occur. These results again support the fact that a moon with an induced magnetosphere can have important impact on global mass and energy transport within its parent planetary magnetosphere.

5.5 Interaction Between Global and Induced Magnetospheres

The above interaction between the interchange fingers and the induced magnetosphere about Titan has typically been considered only a local interaction, with little consequences for the global dynamics of the Kronian magnetosphere. This paradigm has recently been brought into question by Winglee et al. (2013), who undertook simulations of the interaction for a full rotation of Titan about Saturn. This interaction was shown to have global consequences for the magnetosphere and could make a significant contribution to establishing the periodic signatures seen in many of the observable features of the Kronian magnetosphere (Carbary and Mitchell 2013).

Periodic modulation of Saturn's auroral kilometric radiation (SKR) was observed during the passage of *Voyager 1* and used to determine the rotation period of the planet at 10 h 39 min 24 s ± 7 s (Desch and Kaiser 1981). With the arrival of *Cassini* 24 years later, a slightly shifted period of 10 h 45 min 45 s ± 36 s was determined (Gurnett et al. 2005; Kurth et al. 2008). In addition to a variability of the underlying period, there also appears to be a difference in the periods between the northern hemisphere ($\sim$ 10.6 hrs) and the southern hemisphere at ($\sim$ 10.8 hrs) (Gurnett et al. 2009). The difference in these two rates appeared to converge about seven months after Saturn's equinox in 2009.

The presence of this type of periodicity was a surprise, since the planet's internal magnetic field is highly aligned with its rotational axis. The fact that the period drifts with time suggests that its origin cannot be due to planetary rotation alone. The different periods between the northern and southern hemispheres further complicate any potential explanation of their physical origin. The same quasi-periodic modulation is observed in many of the properties of the Kronian magnetosphere, including the magnetic field (Espinosa and Dougherty 2000; Giampieri et al. 2006), low-energy plasma (Gurnett et al. 2007), energetic charged particles (Carbary et al. 2007), energetic neutral atoms (Paranicas et al. 2005; Carbary et al. 2008a), and the position of the magnetopause (Clarke et al. 2010). Whatever the source of the periodicity may be, it has global consequences for the energy and mass transport within the Kronian system.

The results from the previous section have already demonstrated that the interaction of Titan with an interchange finger causes the flapping of Titan's ion tail and that the interchange finger itself is also modified by the interaction. An important part of this interaction is that Titan's ion tail is seen to be pulled inwards by the interaction with the interchange finger. This means that there is partial reflection of energy and mass that would have otherwise propagated beyond Titan. In so doing, the system has the potential to set up a kind of resonance that favours the growth of modes of the interchange instability that have an apparent frequency near that of the planetary rotation, though the individual fingers themselves rotate with a period much longer than the planetary period. The buildup of this resonance was demonstrated in the long duration simulations of Winglee et al. (2013) and is illustrated in Fig. 25. In this simulation, an equilibrium for the global Kronian magnetosphere is established by running the code for 40 hours in the absence of a simulated Titan. After this period, Titan is then inserted with an initial position at the dusk terminator. The equatorial velocity in the top panels of Fig. 25a–c show that, within about 7 R_S, the flows are approximately rotational. At larger distances the flow pattern is modified by the superposition of a two-cell convection pattern which enhances the plasma velocities on the dawnside and reduces them on the duskside. The Titan obstacle produces a slowdown in the plasma velocities for a few hours in local time about its simulated orbital position. Because conservation of magnetic flux is required there is enhancement of the azimuthal velocity at smaller radial distances. This effect demonstrates that Titan's induced magnetosphere has both local and global consequences for the Kronian system's dynamics.

The simulated interchange instability at earlier times has five to seven large interchange fingers of plasma, similar to the result of Kidder et al. (2009), except that, with higher equatorial plasma density, the interchange fingers are curved in the backward direction (e.g. Fig. 25d). However, at later simulated times in Fig. 25e, f, the number of large interchange fingers is reduced to only three or four. Thus we have a configuration in which a rotationally

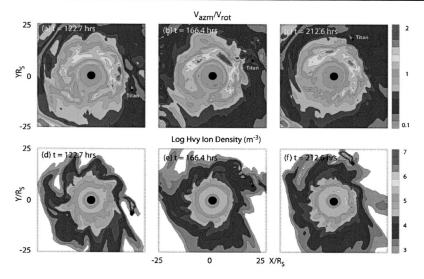

Fig. 25 *Top panels* show the azimuthal velocity in the equatorial plane while the *lower panel* shows the Hvy$^+$ ion density to indicate the positions of the interchange fingers, the position of Titan, and Titan's ion tail. Corotation dominates out to a distance $\sim 7R_S$. This region is then followed by a sector where a two-cell convection pattern is imposed on the rotation field (see text). Titan suppresses the convection velocities over a few hours of local time in its vicinity (from Winglee et al. 2013)

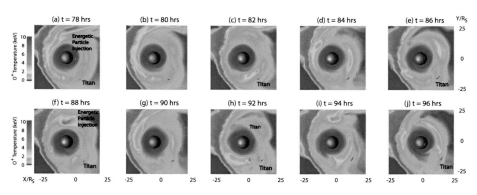

Fig. 26 The equatorial O$^+$ temperature during an energetic particle injection event that is tracked over approximately two Saturn rotations. The panels are arranged so that approximately a full period is shown in each row. Moving down the columns one can see the profiles separated by about eight planetary periods. Similar features in the energetic particles separated by a planetary period are apparent, although the particles themselves are actually different (from Winglee et al. 2013)

driven instability is being modified by an external influence that damps out the higher frequency components. For an observer at a fixed position, this has an interesting consequence which is illustrated in Fig. 26. This figure shows a simulated injection event tracked over two rotation periods. Titan is seen to provide herding of these energetic particles into the inner magnetosphere (top row) and produces some additional heating as they pass by Titan in their second passage around Saturn (bottom row). A second group of energetic particles is launched into the system in near-antiphase with the initial injected population; this effect

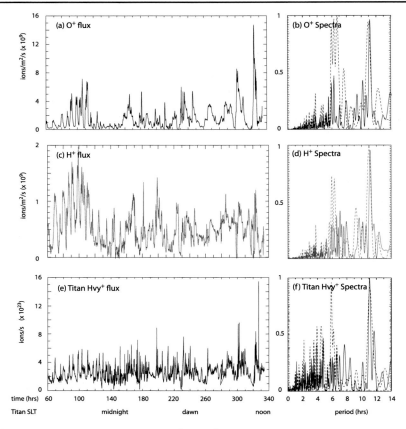

Fig. 27 *Left panels* show calculated fluxes of O^+ and H^+ ions integrated over a 0.4 R_S cube about Titan and the corresponding total flux from Titan. *Right panels* show the calculated power spectrum derived from the full period (*solid line*) and between 60–240 hrs (*dashed line*). Initially the spectrum is dominated by processes at twice the planetary period but, over long timescales, near-planetary periods dominate the fluctuations (from Winglee et al. 2013)

is related to the development of large interchange fingers. As a result, an observer at any fixed azimuthal position will see two distinct spikes in energetic particles moving past them, separated by approximately a planetary period.

Winglee et al. (2013) evaluated Fourier spectra of a variety of physical parameters to determine whether the effects described above could solely produce the observed periods. Figure 27 shows the time history and spectra of the particle fluxes about Titan. The O^+ and H^+ ions originate from the Enceladus torus, while the ions that originate from Titan are designated as Hvy^+ (heavy ions). It is seen that all three ion species are significantly modulated. For the first 40 hours or so the spectrum is dominated by processes at about twice the planetary period. However after this time the spectrum becomes dominated by processes near the planetary period. This change in characteristic period is consistent with the transient buildup of this resonance process and was missing from the early simulations which considered relatively short periods for the interaction. This evolution towards near-planetary periods would also be missing from all global simulations which do not include the Titan-related interaction.

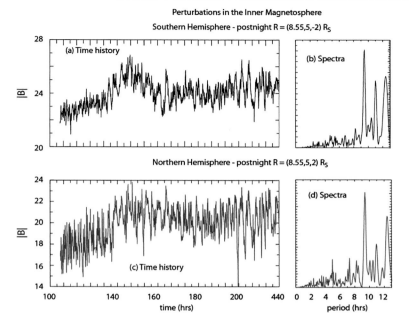

Fig. 28 Comparison of the simulated magnetic perturbations in the northern and southern hemispheres of the Kronian system. The configuration is for southern summer solstice, with Titan being below the center of the magnetodisc. Hence the profiles are not identical, with the northern hemisphere having a slightly shorter period than the southern (from Winglee et al. 2013)

The results in Fig. 27 also show that the particle fluxes moving through the magneto-sphere can have substantial temporal variations. Modulations of nearly an order of magnitude can occur and need to be taken into account when trying to fully describe the mass and energy transport in the Kronian magnetosphere. The above model for the periodicities within the Kronian magnetosphere also sheds light on why there are differences between the northern and southern hemisphere periods. This is illustrated in Fig. 28 which shows the time history and Fourier spectra of representative points in the northern and southern hemi-spheres. The magnetic profiles, while similar, are not identical, and the southern hemisphere generally harbours larger modulations than the northern. This difference occurs because, for the simulated configuration of southern summer solstice, Titan lies below the center of the magnetodisc current sheet. As a result of this difference in proximity, the southern side of the magnetodisc experiences stronger drag and forcing from Titan's induced magneto-sphere. This effect produces a slightly longer period for the southern hemisphere relative to that of the northern hemisphere, qualitatively consistent with the *Cassini* observations cited above. These perturbations are expected to propagate into the magnetosphere where they should be observed simultaneously. However, the simulations have insufficient duration and signal-to-noise ratio to resolve this mixing of the two very close frequencies in the equatorial magnetosphere.

The above results have concentrated mainly on the Kronian magnetosphere due to the wealth of data provided by *Cassini*. The Jovian magnetosphere, with its stronger magnetic field and stronger influences by the Galilean satellites, is expected to produce an equally rich environment in terms of dynamics and characteristic periods, that has yet to be explored in detail. Because of the stronger magnetic field and the high plasma output from Io (Sect. 1),

Fig. 29 The ion pickup geometry at Io. The plasma corotating at 74 $\mathrm{km\,s^{-1}}$ meets Io orbiting at 17 $\mathrm{km\,s^{-1}}$. Pickup ions accelerate in a convective electric field corresponding to 57 $\mathrm{km\,s^{-1}}$. Ions follow cycloidal paths in the inertial frame and form rings in velocity space in the plasma rest frame

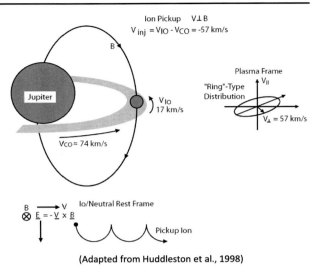

(Adapted from Huddleston et al., 1998)

influence from solar wind forcing on the mass and energy transport within the inner and middle magnetosphere is expected to be less important than for the Kronian system. On the other hand, influences from the Galilean satellites have the potential for being much stronger in terms of moon-magnetosphere interactions. These effects have yet to be explored in detail but, given the physical impact of the effects described above, our understanding of mass and energy transport will change as more data are acquired and improved models developed.

6 Wave Phenomena in Magnetodisc Regions

6.1 Waves in Jupiter's Magnetodisc

The regions of planetary magnetodiscs where mass loading occurs are rich environments for the generation of low-frequency waves. Ion cyclotron and mirror mode waves have been observed in the magnetospheres of both Jupiter and Saturn. In this section, we will discuss the characteristics of these waves. The waves are similar in character to fluctuations found in other mass loading regions, such as those at Venus, Mars, and comets. Waves are important in this context because they can be used as a diagnostic of the local plasma composition, and they tell us where ion pickup takes place. They also participate in wave-particle energy exchange and play an active role in the interaction between some moons and the giant planetary magnetospheres in which these moons are embedded.

In the Jovian system, ion pick up rates are largest near Io (Sects. 1, 2). It is estimated that of the order ~ 500–$1000\ \mathrm{kg\,s^{-1}}$ of neutral gas associated with the Io source becomes ionized through photoionization, electron impact by a thermal background plasma, electron impact by a superthermal electron population (Delamere and Bagenal 2003), and charge exchange. Once ionized, the material is picked up into the plasma torus—that is, incorporated into the ambient flow of the torus plasma (Hill et al. 1983; Frank and Paterson 1999). Figure 29 shows that Jupiter's magnetic field is approximately perpendicular to the corotational flow of the Io torus plasma, and hence ions are injected into the torus with ring-like velocity distributions which, in turn, are able to provide free energy for wave generation. At the orbit of Io, the magnetospheric plasma is rotating around Jupiter with an azimuthal speed of roughly ~ 74 km s^{-1}. Io orbits at ~ 17 km s^{-1}, so the ambient plasma moves at ~ 57 km s^{-1} relative to Io and its ionosphere. Thus, any newborn ion feels an outward electric field (in the reference frame of Io) corresponding to the product of this ~ 57 km s^{-1} relative velocity and the ~ 2000 nT (southward-directed) magnetic field. The newborn ion begins to drift and gyrate about the magnetic field direction, following a cycloidal trajectory in space. In the corotating plasma frame, this motion is circular with a fixed centre and speed of gyration, and the ions thus form a ring-like distribution in velocity space. These ring distributions introduce an effective temperature anisotropy $T_\perp / T_\parallel > 1$ into the plasma and can therefore lead to the growth of ion cyclotron waves and mirror modes.

Galileo observations showed the existence of ion cyclotron waves in the Io torus during most of the spacecraft's encounters with this moon. The corresponding waves have frequencies near the gyrofrequencies of the ion species SO_2^+, SO^+ and S^+. Mirror mode waves were observed only when Galileo crossed Io's wake region. At Europa the mass loading rate representing the pick up of new plasma into the magnetosphere is significantly less than the rate for Io. Galileo observed waves in the Europan wake, with frequencies near the gyrofrequencies of O_2^+ and Cl^+ ions.

By contrast, Cassini data have revealed the existence of ion cyclotron and mirror mode waves in very extended regions of Saturn's magnetodisc. Thus, at Saturn's magnetosphere, pickup processes occur not only near the moons, but over wider regions in which water group ions are produced at all local times over distance intervals covering several R_S.

6.1.1 Ion Cyclotron Waves

Figure 30 shows one example of ion cyclotron waves observed near Io and mirror mode waves observed in the moon's wake. Ion cyclotron waves were observed in the torus during six flybys near Io. The waves had large amplitudes of $\lesssim 100$ nT, or up to ~ 5 % of the background field ($\delta B/B \lesssim 0.05$). These waves also exhibited a left-handed, near-circular polarization, with wave power primarily near the cyclotron frequencies of SO_2^+ and SO^+ (Russell and Kivelson 2000; Blanco-Cano et al. 2001a, 2001b; Russell et al. 2003a, 2003b). Wave power became stronger closer to the moon, and the intensity ratio of the SO_2^+ to SO^+ waves also varied with distance from Io. In some regions, additional cyclotron waves associated with the S^+ ion were observed.

Figure 31 shows Galileo trajectories for five crossings of the Io wake region. On the first orbit, on December 7, 1995 (flyby I0), Galileo flew 900 km downstream of Io at closest approach, revealing strong magnetic fluctuations with periods ~ 2–3 s, corresponding to the gyrofrequencies of SO_2^+ and SO^+ (Kivelson et al. 1996). The waves were observed inbound and outbound (sinusoidal signatures), extending over distances of ~ 20 R_I (Io radii) and ~ 7 R_I, respectively. Inbound, the SO_2^+-related waves had larger amplitudes than those associated with SO^+. In contrast, the outbound portion showed intervals with stronger waves near the SO^+ gyrofrequency, and a region with a burst near the S^+ gyrofrequency. SO_2^+ waves reached amplitudes of around 100 nT peak-to-peak, which is consistent with a mass loading rate near Io of about 300 kg s^{-1} (Huddleston et al. 1998). On this orbit, Galileo crossed Io's wake and mirror mode waves were observed at the wake edges (square signatures in Fig. 31; see also Fig. 30, and section below).

On October 11, 1999 (flyby I24) Galileo crossed the upstream nose of the interaction region (Fig. 31a (left panel)), where a weak, short burst of SO^+ ion cyclotron waves was observed. Later, on Io's downstream side, stronger waves were detected. Power spectra for this flyby (Fig. 32) show two major differences with respect to the waves of orbit I0. Firstly,

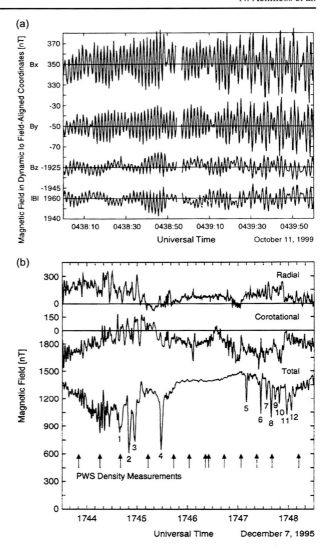

Fig. 30 Galileo measurements of (**a**) ion cyclotron waves (adapted from Kivelson et al. 1996), and (**b**) mirror modes (adapted from Russell et al. 1999)

instead of a broad peak, there are two separate narrow peaks at the SO_2^+ and SO^+ gyrofrequencies (hereafter referred to as $\Omega(SO_2^+)$ and $\Omega(SO^+)$). Secondly, the waves at frequencies near $\Omega(SO^+)$ were stronger and more persistent than those with frequencies closer to $\Omega(SO_2^+)$. In some intervals, a third peak in the spectra appeared near the S^+ gyrofrequency, denoted $\Omega(S^+)$. Because there is a background, thermalized component of S^+ ions in the torus, the presence of waves with frequencies near $\Omega(S^+)$ indicates that there is a strong S^+ ring distribution which is able to generate the relevant instability for wave growth, and presumably overcome the wave damping due to the thermalized S^+ population.

On November 26, 1999 (I25), there was only a short interval of *Galileo* magnetometer data after closest approach to Io. Figure 32c shows two peaks in power, with similar amplitudes at frequencies near $\Omega(SO_2^+)$ and $\Omega(SO^+)$. On February 22, 2000 (I27), *Galileo* again moved from upstream to downstream (Fig. 31a), and ion cyclotron waves appeared

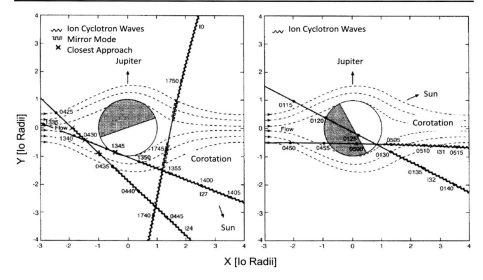

Fig. 31 Galileo trajectories during five encounters with Io. The regions where ion cyclotron waves and mirror modes were observed are indicated by sinusoidal and 'square wave' lines, see *legend* at *top* of both *panels*. The coordinate system keeps the flow velocity to the right and the convective electric field associated with the corotation of the plasma is projected on the Y-direction, outward from Jupiter, according to the instantaneous model field at Io. (Adapted/reprinted from Russell et al. 2003b, Copyright (2003), with permission from Elsevier)

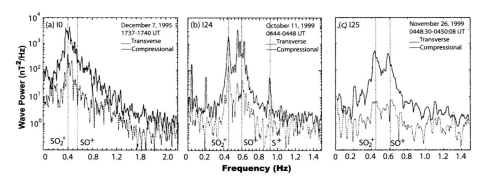

Fig. 32 Transverse and compressional wave power for three intervals with ion cyclotron waves in the Io torus

as Galileo entered the downstream side. Power spectra show peaks near all of the above frequencies ($\Omega(SO_2^+)$, $\Omega(SO^+)$ and $\Omega(S^+)$). In contrast to I24, in some intervals, the amplitude of the peak near $\Omega(S^+)$ is comparable to the $\Omega(SO^+)$ peak.

The last two Io orbits with magnetic field data took place on August 6, 2001 (I31), and October 16, 2001 (I32) (Fig. 31b (right panel)). Both flybys crossed the polar region above or below the moon, so that field lines connected Io and the spacecraft during the relevant time intervals. The I31 orbit was almost parallel to Io's wake, and almost parallel to the corotating flow of the ambient plasma. The field lines through Galileo connected it to the centre of the wake, a region of local minimum in field strength. Wave properties along I31 are quite different to other passes (Russell et al. 2003a). Power spectra show broad peaks at frequencies below all of the gyrofrequencies $\Omega(\mathrm{SO}_2^+)$, $\Omega(\mathrm{SO}^+)$ and $\Omega(\mathrm{S}^+)$. The waves exhibit the properties of the ion cyclotron mode in the frequency range 0.3–1.1 Hz. It follows that the waves are generated along field lines where the strength of the magnetic field varies, such that the waves resonate with ions over a broad range of frequencies as they propagate to the higher latitudes where they are observed. For SO_2^+ waves with frequency 0.3 Hz, a local field of 1259 nT would be required. Fields this low were indeed encountered on the I0 flyby through the wake.

Finally, orbit I32 had a geometry intermediate between I24 and I27, and wave power peaked near the $\Omega(\mathrm{SO}_2^+)$, comparable to the spectra of flyby I0. During orbits I0, I24, I25, I27, and I32, the waves observed were transverse, left-handed, and elliptically polarized. Their angle of propagation was variable. In some intervals the waves were almost parallel-propagating (i.e. along the field direction), and almost circularly polarized, while in others they propagate at angles $\lesssim 40°$ with respect to the ambient magnetic field (Russell et al. 2001). In contrast, during orbit I31, compressive waves, propagating at large angles to the magnetic field, were also present (Russell et al. 2003a).

Dynamic spectra summarizing wave properties for all flybys are given in Blanco-Cano et al. (2001a) and Russell et al. (2003b). In summary, ion cyclotron waves have generally been observed only downstream from Io, except on the I24 pass when there was a short burst of wave activity upstream of the moon. Their characteristics change along each flyby, and from one orbit to the next. Figure 31 shows that the I0 trajectory crosses both the I24 and I27 trajectories. However, where they cross, the spectra are quite different on the different passes, suggesting that the physical conditions in the Io atmosphere are quite variable in time. Another interesting point is that ion cyclotron waves can arise far beyond the region in which the flow and the atmosphere should be interacting. The waves are generated by ion ring distributions via the cyclotron instability. They are observed with frequencies mostly close to $\Omega(\mathrm{SO}_2^+)$ and $\Omega(\mathrm{SO}^+)$. This is because, for other ionic species (O^+, S^+, S_2^{++}) there is a thermalized distribution in the background torus whose presence acts to damp the waves and suppress the instability. On the other hand, because SO_2^+ and SO^+ dissociate within tens of minutes, they only exist as ring distributions with no thermalized component.

The existence of S^+ waves was unexpected because we expect S^+ background ions to damp any waves generated by S^+ pickup. Thus, a dense S^+ ring distribution is indicated to exist in some regions, with a sufficient population for overcoming wave damping. The observed wave variability shows that the Io torus is not uniform, and that ion pickup composition changes with time and spatial location. Since pickup ions originate from Io's atmosphere, observed wave variations suggest that this atmosphere is changing spatially as well as temporally (Russell and Kivelson 2000; Russell et al. 2003b).

6.1.2 Origin of Ion Cyclotron Waves: Dispersion Analysis and Hybrid Simulations

It is well known that SO_2 and SO are the main constituents of Io's atmosphere (e.g., Wong and Smyth 2000). The waves associated with the ions of these molecules are generated by ion ring distributions via the cyclotron resonant instability, due to the aforementioned temperature anisotropy in the source plasma. The detection of waves with frequencies near the gyro frequency values $\Omega(\mathrm{SO}_2^+)$ and $\Omega(\mathrm{SO}^+)$ indicates ongoing ionisation and pickup of ions from the atmosphere. However, SO_2^+ and SO^+ are minor components of the torus plasma (e.g. Frank et al. 1996; Frank and Paterson 2001), since they dissociate within tens

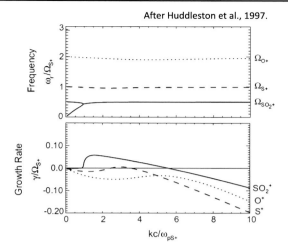

Fig. 33 Wave frequency and growth as a function of wave number for ion cyclotron waves in a multispecies plasma with O^+, S^+ and SO_2^+ ring distributions, along with O^+ and S^+ thermalized background components (after Huddleston et al. 1997). Dispersion curves are for parallel propagating waves

of minutes. The primary constituents of the background plasma torus are O^+ and S^+. However, only weak waves near the $\Omega(S^+)$ were observed. The immediate question that arises is: Why would these minority constituents generate waves while the majority plasma species did not? In a multi-species plasma, multiple ion cyclotron modes are possible, each with a growth rate dependent upon the anisotropy and free energy provided by the specific ion species. To understand the generation of the observed waves, kinetic dispersion analyses for a plasma that resembles the Io torus have been performed. Warnecke et al. (1997) and Huddleston et al. (1997) showed that the presence of the dense O^+/S^+ thermalized background population damps the waves generated by picked-up O^+ and S^+ ions, while, on the other hand, SO_2^+ and SO^+ waves can grow easily because these ions exist only as ring distributions, without a damping, thermal component.

Figure 33 shows frequency and growth rate for waves propagating parallel to the magnetic field in a plasma which harbours SO_2^+, O^+ and S^+ ring distributions, along with O^+ and S^+ thermalized populations. It is possible to see that the O^+ and S^+ wave modes are damped, with negative values of the growth rate parameter γ, while the SO_2^+ waves remain undamped, and have positive γ value for a range of k values. Blanco-Cano et al. (2001a, 2001b) demonstrated that ion cyclotron waves associated with the species SO_2^+, SO^+ and S^+ can grow simultaneously under certain conditions, and that S^+ pickup ions can generate waves, provided the density of this component is adequate ($\gtrsim 10~\%$ of total density) to overcome damping by the thermal background of S^+ ions. Linear kinetic analysis also shows that, while wave growth at parallel propagation is the most rapid, growth at oblique angles can also be significant (Blanco-Cano et al. 2001b), in agreement with the observations described herein.

More recently, hybrid simulations have been used to study wave generation and nonlinear evolution near Io (Cowee et al. 2006, 2008; Cowee and Gary 2012). These studies have considered continuous ion injection, and are useful for investigating wave-ion energy exchange and the scattering of ion velocity distributions. These works have been successful in reproducing wave spectra observed near Io, and have yielded important insight regarding wave interaction with the various pickup ion populations.

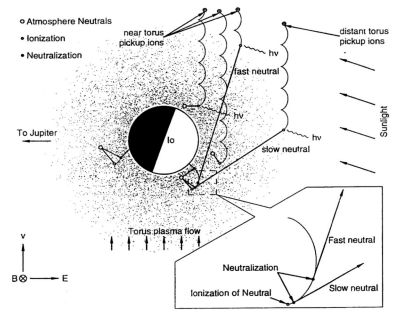

Fig. 34 Mechanism for forming a thin, mass-loading disc that enables the production of pickup ions far to the side of the moon, but not in the upstream direction. Ions are first formed near Io in its moderately dense atmosphere, where charge exchange occurs frequently enough that freshly accelerated ions can be converted to fast neutrals. The fast neutrals then travel for significant distances across the magnetic field before they again become ionized and thus release their free energy in the form of ion cyclotron waves (after Wang et al. 2001)

6.1.3 Morphology of the Mass Loading Region in the Magnetodisc

The regions of ion cyclotron wave generation seen by *Galileo*, and the established absence of this phenomenon in the *Voyager* data from 1979, 10 R_I beneath Io, constrain the dimensions of the mass loading region where ions are injected into the torus. This region covers a plane extending ~ 20 R_I in the anti-Jupiter direction, and only ~ 7 R_I toward Jupiter (see Fig. 31). In the flow direction the region extends downstream at least ~ 10 R_I, but it does not extend upstream of Io. This 'fan-shaped disc' downstream from Io can be produced by the multi-step mechanism proposed by Wang et al. (2001), in which ions are accelerated in the exosphere of Io by the ambient corotational electric field, followed by neutralization and transport across field lines to regions much further from Io, where they can be reionized and picked up into ring distributions, thus generating waves (see Fig. 34). This mechanism limits wave growth to the downstream region, as opposed to growth in a more isotropic source region that would be expected if sputtering were the main mechanism producing Io's neutral torus. Pickup ions produced in this way have their highest fluxes near the wake axis and extend downstream of Io. Observed wave amplitudes are consistent with this distribution of pickup ions, decreasing with distance from the wake axis (Russell et al. 2001).

The fact that ion cyclotron waves appear downstream from Io is also in agreement with recent findings by Dols et al. (2012), who found that Io's atmosphere has longitudinal asymmetries, with a limited radial extension upstream, and a significantly larger scale on the anti-Jovian downstream side. Using a MHD model of the flow and magnetic perturbations

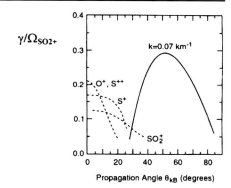

Fig. 35 Growth rate for ion cyclotron waves and the mirror mode in a plasma with multiple species. The growth of the mirror mode (*solid line*) benefits from the combined temperature anisotropy of the various ion distributions. Growth rates of the ion cyclotron waves (*dashed lines*) are smaller (adapted from Huddleston et al. 1999)

around the moon in conjunction with a multi-species chemistry model, they predicted the formation of a very extended corona of SO_2 and SO, mainly downstream from Io.

6.1.4 Mirror Mode Waves in Io's Wake Region

As mentioned earlier, during flyby I0, *Galileo* also observed mirror mode waves on the edges of Io's wake region (see Fig. 30) in addition to the ion cyclotron fluctuations. The waves were highly compressional, with wave normals at large angles ($\sim 54°-85°$) to the local magnetic field. Their properties are discussed in Russell et al. (1999). The mirror mode is typically excited in high-beta plasmas when there is a significant pressure anisotropy (and the condition $P_\perp/P_\parallel > 1 + 1/\beta_\perp$ is satisfied), such as that created by pickup ion ring distributions. Maximum growth for the mirror mode occurs at oblique angles, in contrast to the ion cyclotron instability, with maximum growth corresponding to parallel (field-aligned) propagation. The near-stagnant wake region of Io contains a multi-species, anisotropic plasma, a reduced magnetic field, and thus a relatively high plasma β. Plasma data showed that, at the wake edges, pickup ion densities increased rapidly (Frank et al. 1996), hence providing a large pressure anisotropy. Using kinetic dispersion analysis, Huddleston et al. (1999) showed that in these regions the mirror mode can become dominant.

Figure 35 shows the growth rate of the mirror mode and of O^+, S_2^+, S^+ and SO_2^+ ion cyclotron waves for a plasma which is physically similar to that at the wake edges. While ion cyclotron waves grow due to the anisotropy in the distributions of each species, the mirror mode is fed by the combined anisotropy of all ions present in the plasma and can thus be the dominant plasma perturbation.

6.1.5 Waves Near Europa

At Europa, the plasma mass loaded into the magnetosphere per unit time is less than that from the Io source. Interaction of the exosphere with the Jovian magnetosphere produces pickup ion rates of the order of a few $kg\,s^{-1}$ (Saur et al. 1998; Khurana et al. 2002) in contrast to ~ 1000 $kg\,s^{-1}$ at Io. Evidence for ion pickup and wave signatures was found on three of eleven passes of *Galileo* near Europa (Volwerk et al. 2001; Volwerk and Khurana 2010). In contrast to Io, most of the waves were observed in Europa's wake and have been identified as ion cyclotron fluctuations with frequencies near the values $\Omega(O_2^+)$, $\Omega(Cl^+)$, $\Omega(Na^+)$, $\Omega(Ca^+)$, $\Omega(K^+)$ and $\Omega(H_2O^+)$ (using the gyrofrequency nomenclature of previous sections). The waves have both right- and left-hand polarizations. Wave power distributions suggest that the ion pickup rate is larger when Europa is near the center of the Jovian current sheet than when the moon is outside it. In a recent work, Volwerk and Khurana (2010) found evidence of H_2O^+ cyclotron waves upstream of Europa, observed during the *Galileo* E26 pass.

6.2 Waves at Saturn

At Saturn, magnetosphere neutral sources, whether from moons or rings, are weaker than at Jupiter. Nonetheless, Saturn has abundant ion-cyclotron waves, and mirror mode waves have also been observed. In contrast to Jupiter, waves at Saturn appear not only near the moons, but also exist in a extended region covering distances $\sim$ 4–8 R_S, at all local times, regardless of whether or not the spacecraft was near a moon (Leisner et al. 2005, 2006; Russell et al. 2006).

Ion cyclotron waves were observed at Saturn during the *Pioneer 11* and *Voyager 1* passes through the system (Smith and Tsurutani 1983; Barbosa 1993). Due to the limited spatial coverage of those flybys, the full radial extent of these waves was not seen until the *Cassini* spacecraft arrived at Saturn on July 1, 2004 (GMT). In addition to the ion cyclotron waves observed near the fundamental gyrofrequency of pickup ions, there were regions where harmonic ion cyclotron modes were found. Beyond $\sim$ 6 R_S, mirror mode waves and interchanging flux tubes are present and replace the ion cyclotron waves up to regions $\sim$ 8 R_S distant from the planet (Russell et al. 2006).

6.2.1 Ion Cyclotron Wave Properties

Ion cyclotron waves in Saturn's magnetosphere are generated by water-group ions (O^+, OH^+, H_2O^+). The source of these ions has been identified as Saturn's 'extended neutral cloud', which is primarily sourced by the moon, Enceladus. Enceladus ejects an estimated 10^{27} H_2O molecules/s into the surrounding space through fissures in its surface (e.g., Waite et al. 2006; Tokar et al. 2006). Modelling of these ejected neutrals has shown that they do, indeed, form a neutral cloud which extends over the radial range where the ion cyclotron waves are seen (Johnson et al. 2006). Thus, the source of waves at Saturn is more extended and not strictly limited to the vicinity of a moon (or that moon's torus), as it is at Jupiter.

As in the case of the Jovian magnetosphere, ions in the Kronian system are picked up into ring distributions able to generate waves via the ion cyclotron instability. Such anisotropic ($T_\perp > T_\parallel$) ring-type distributions have been observed near Enceladus (Tokar et al. 2006) and out to radial distance 5 R_S (Tokar et al. 2008). Unsurprisingly, wave properties varied between the Cassini passes, but in general the waves displayed left-handed, near-circular to elliptical polarizations, propagating at angles within $\sim$ 20° of the background magnetic field. The wave amplitudes varied, but were typically in the range 0.5–2 nT, with peak amplitudes occurring beyond the orbit of Enceladus at around 4–5 R_S.

Figure 36a shows power spectra of *Cassini* observations made at different dates, but at a similar location. It is clear that the largest peak occurred near the gyrofrequencies of water group ions ($\Omega(O^+)$, $\Omega(OH^+)$, $\Omega(H_2O^+)$), and that the amplitude of this peak varied for the two dates. A secondary peak appears near the cyclotron frequency of ions with mass $\sim$ 35 amu, which are thought to be O_2^+. This peak is more prominent on the observation of December 15, 2004. As shown in Fig. 36b, a clear beating structure was seen in the wave time series, indicative of wave generation by these multiple ion species with markedly different masses.

Using many Cassini orbits, Martens et al. (2008) showed that molecular oxygen is a minor constituent of the magnetospheric plasma. Due to the low observed densities of this

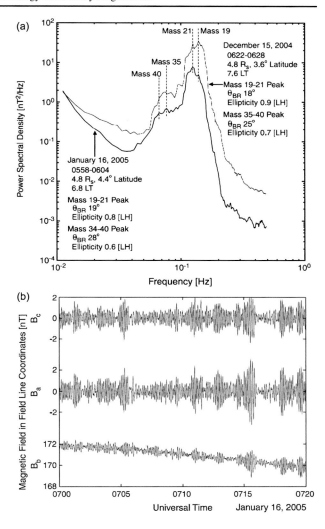

Fig. 36 (**a**) Power spectrum of ion cyclotron waves at Saturn magnetodisc. These two observations were made when *Cassini* was at a similar location. (**b**) Wave series showing three components of the magnetic field. (From Leisner et al. 2006)

molecule, it is difficult to determine the source, or sources, of the corresponding ions. The two most probable sources are fast neutrals escaping Saturn's main rings and Saturn's dusty E ring. The work by Leisner et al. (2006) supports the conclusion that these ions are a consistent, but minor, plasma species in the magnetosphere. There were a few intervals where the dominant waves had a frequency near the value $\Omega(O_2^+)$ (Rodríguez-Martínez et al. 2010). The variation of peak amplitudes indicates that the pickup ion composition changes across the extent of the neutral cloud.

The ion cyclotron waves were observed over a large region of space so the assumption of a homogeneous plasma may not be appropriate. The pickup geometry in the extended neutral cloud region is roughly perpendicular ($\alpha \sim 90°$, with α the angle between Saturn's magnetic field and $\mathbf{v}$, the velocity of injection of the ions), but the pickup injection velocity also changes with radial distance. Figure 37 shows how the Keplerian and corotational velocities vary with distance from the planet. The pickup velocity may be represented as $V_{\text{pickup}} = V_K - V_c$, where V_K is the Keplerian velocity with which the neutral cloud is orbiting Saturn,

Fig. 37 Pickup velocities as a function of radial distance. (From Russell et al. 2006)

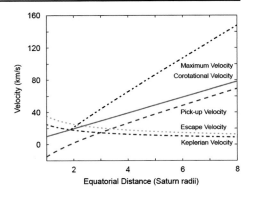

Fig. 38 Wave frequency and growth rate for ion cyclotron waves in Saturn's magnetodisc. Curves are labelled with ion pickup speed (*left panels*), given in kilometres per second. On *right panels*, curves correspond to different ring ion density, given in ions per cubic centimetre (from Leisner et al. 2006)

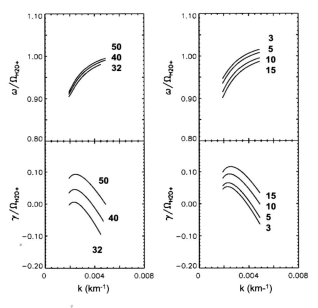

and V_c is the local corotation velocity of the ambient plasma (which, of course, changes with radial distance). At a radial distance of ~ 4 R_S, the nominal pickup velocity is ~ 25 km s^{-1}, while at ~ 8 R_S, it is ~ 70 km s^{-1}. Therefore, we may expect that a fixed density of pickup ions at ~ 4 R_S will be less energetic than those at ~ 8 R_S, and will thus generate lower-amplitude waves. There are no ion-cyclotron waves inside ~ 3.8 R_S; ions picked up at these small radial distances will have such small V_{pickup} values that they will not become unstable to wave generation.

In this context, kinetic dispersion analysis has been very useful to determine how wave growth varies with ring properties. Figure 38 shows the values of wave frequency and growth rate as function of pickup ion injection velocity and density. Ring distributions have more energy to give to the waves as these two values increase, and this is reflected in enhanced values of growth rate γ. *Cassini* data also has revealed the presence of weak waves at twice the cyclotron frequency of the water group ions (Rodriguez-Martinez et al. 2008). These waves propagate at angles $\gtrsim 20°$ to the background field, and are mainly compressional. Their power is around half that of the fundamental mode. They have been observed on

    Springer

multiple Cassini passes and are present only when the background field strength B_o exceeds ~ 150 nT.

High-inclination passes by *Cassini* allowed the study of the vertical structure of the ion cyclotron wave 'belt'. Both the water group and O_2^+ cyclotron waves were found to propagate from the equator up to a vertical distance of ~ 0.4 R_S, and wave amplitude was found to increase off the equator, peaking at a distance of ~ 0.2 R_S and then rapidly falling off above this limit (Leisner et al. 2011). Why the production region is limited to within ~ 0.04 R_S of the magnetic equator, and why the waves rapidly damp out beyond ~ 0.2 R_S are aspects which have not been explained.

6.2.2 Mirror Modes

The ion ring distributions produced by pickup in the E-ring torus are unstable to both ion-cyclotron waves and mirror-mode waves. For low plasma beta and moderate temperature anisotropy, the growth rate of the ion-cyclotron waves generally exceeds the growth rate of mirror-mode waves. As described above, ion cyclotron waves have been observed in Saturn's magnetosphere out to ~ 5 R_S. Beyond ~ 7 R_S, mirror mode waves dominate the spectra. As shown in Fig. 39, there is an intermediate region where ion cyclotron waves are dominant, but weaker peaks in the spectra suggest the existence of compressive, mirror-mode waves. The simultaneous detection of ion cyclotron and mirror modes is an interesting effect, since linear theory predicts only the mode with the highest growth rate should be observed. At larger distances from Saturn, the mirror mode becomes dominant (see Fig. 39), with strong depths in magnetic field amplitude, and the ion cyclotron waves disappear.

While both instabilities grow from the same pressure anisotropy, the mirror mode dominance at large distances may be due to the fact that the ion cyclotron waves can propagate along the background field out of the wave growth region, while the mirror mode waves remain within it and convect with the bulk motion of the plasma. It is also possible that larger values of plasma beta occur in regions where the mirror mode waves are observed, enhancing the growth of this instability. More work is needed to solve this puzzle.

6.2.3 The Region Near Titan

Titan's dense atmosphere is a strong mass-loading source and plasma observations have clearly identified pickup ion populations near this moon (e.g. Hartle et al. 2006; Szego et al. 2005). However, no significant wave power has been identified close to the moon, counter to intuition. In the few examples were some wave growth has been observed near Titan, the waves are mostly left-hand, elliptically polarized (ellipticity ~ 0.6) and propagate at a large angle ($\sim 60°$) to the background field. However, the observed wave power is only slightly above the noise level and wave characteristics do not coincide with those of ion cyclotron waves (i.e. near-circular polarization and almost field-aligned propagation).

Linear theory and hybrid simulation results predict that the ion cyclotron ring instability may be unstable at Titan for ideal conditions (i.e. $\alpha = 90°$) but that waves driven by heavy pickup ions (e.g. those with mass-to-charge values $m/q = 16$) may simply take too long to grow to observable amplitudes before they are convected by the flowing background plasma into the Titan wake (Cowee et al. 2010). Appreciable growth of waves by the lighter ion species (e.g. $m/q \lesssim 2$) could be possible on the timescale of plasma convection, however these waves are expected to be strongly damped by the thermalized background plasma. Because Titan is in the outer magnetosphere, the pickup angle and pickup velocity can vary

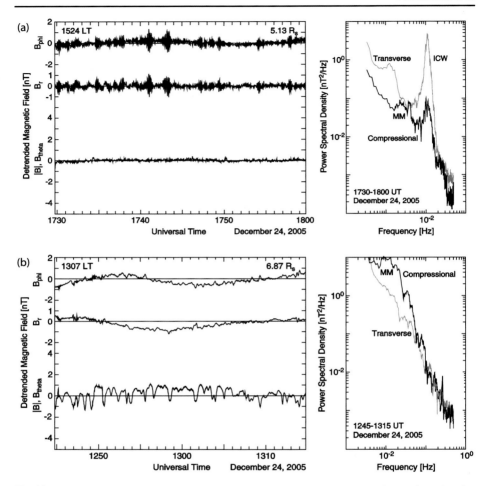

Fig. 39 (a) Ion cyclotron waves and mirror modes: time series and power spectrum in a region where ion cyclotron waves dominate. (b) Time series and power spectrum in a region where the mirror mode is dominant (from Russell et al. 2006)

considerably, especially since the moon may quasi-periodically encounter the centre of the plasma sheet, and it may also, infrequently, emerge into the solar wind. Thus, the pickup velocity may vary between sub-Alfvénic and super-Alfvénic values. Further analysis of the observations is needed, in order to determine under what specific physical conditions ion cyclotron waves can exist near Titan.

6.3 Conclusions and Future Directions

Ion cyclotron waves can be generated by pickup ions in Jupiter's and Saturn's magnetodiscs. At Jupiter, the waves appear in a localized region near Io and have frequencies near the gyrofrequencies of the ions SO_2^+, SO^+ and S^+. Wave properties are not uniform which indicates that ion composition is not homogeneous and varies due to volcanic activity at Io. Mirror mode waves have been observed only in Io's wake region. Observations made by

Galileo showed the existence of ion cyclotron waves in Europa's wake. In contrast to the Jovian observations, ion cyclotron waves in the Saturnian magnetodisc appear in larger extensions within the extended neutral cloud. Waves are generated by water group ions picked up from Enceladus and the E ring. In some regions, O_2^+ cyclotron waves are dominant, indicating variations in pickup ion composition. Ion cyclotron waves extend ± 0.3 R_S out of the equator. Mirror mode waves have also been observed in extended regions of the Saturnian magnetodisc. Pickup ions have been observed near Titan, but no well-defined ion cyclotron waves have been identified there. Future studies concentrating on ion distributions, when available, are needed to understand wave evolution and particle distribution thermalization. Additional simulation studies are also needed to understand the dynamics of waves propagating far from the equatorial disc, the variations in wave propagation properties, and the conditions that favour the mirror mode to be dominant in extended regions of the magnetodisc. *Galileo* and *Cassini* data have been very valuable in showing the complexity of these two magnetodiscs. More observations are needed to gain insight regarding the extension of the waves in the region between Io and Jupiter and, in the case of Saturn, to have a better understanding of the wave belt which lies out of the equator.

Acknowledgements   NA wishes to acknowledge the support of the ISSI and the STFC UCL Astrophysics Consolidated Grant ST/J001511/1; he also wishes to acknowledge useful discussions with staff at the Japanese Aerospace Exploration Agency (JAXA) during a Visiting Professorship. PB was supported by NASA Grants NNX10AF17G and NNX12AG81G. Nicolas André wishes to acknowledge the support of CNES for Cassini data analysis, and CDPP for the use of the AMDA (Automated Multi-Dataset Analysis) tool available at URL http://cdpp.cesr.fr. XBC acknowledges support of the ISSI and UNAM-DGAPA-PAPIIT grants IN110511 and IN105014. The authors acknowledge the support of EUROPLANET RI project (Grant agreement no.: 228319) funded by EU; and also the support of the International Space Science Institute (Bern). The Editor thanks the work of two anonymous referees.

References

N. Achilleos, S. Miller, R. Prangé, G. Millward, M.K. Dougherty, A dynamical model of Jupiter's auroral electrojet. New J. Phys. **3**, 3 (2001). doi:10.1088/1367-2630/3/1/303

N. Achilleos, P. Guio, C.S. Arridge, A model of force balance in Saturn's magnetodisc. Mon. Not. R. Astron. Soc. **401**, 2349–2371 (2010). doi:10.1111/j.1365-2966.2009.15865.x

N. André, K.M. Ferrière, Low-frequency waves and instabilities in stratified, gyrotropic, multicomponent plasmas: Theory and application to plasma transport in the Io torus. J. Geophys. Res. **109**, 12225 (2004). doi:10.1029/2004JA010599

N. André, K.M. Ferrière, Comments on Vasyliunas' and Pontius' studies of the effects of the planetary ionosphere and of the Coriolis force on the interchange instability. J. Geophys. Res. **112**, 10203 (2007). doi:10.1029/2006JA011732

N. André, M.K. Dougherty, C.T. Russell, J.S. Leisner, K.K. Khurana, Dynamics of the Saturnian inner magnetosphere: First inferences from the Cassini magnetometers about small-scale plasma transport in the magnetosphere. Geophys. Res. Lett. **32**, 14 (2005). doi:10.1029/2005GL022643

N. André, A.M. Persoon, J. Goldstein, J.L. Burch, P. Louarn, G.R. Lewis, A.J. Coates, W.S. Kurth, E.C. Sittler, M.F. Thomsen, F.J. Crary, M.K. Dougherty, D.A. Gurnett, D.T. Young, Magnetic signatures of plasma-depleted flux tubes in the Saturnian inner magnetosphere. Geophys. Res. Lett. **34**, 14108 (2007). doi:10.1029/2007GL030374

D.J. Andrews, S.W.H. Cowley, M.K. Dougherty, L. Lamy, G. Provan, D.J. Southwood, Planetary period oscillations in Saturn's magnetosphere: Evolution of magnetic oscillation properties from southern summer to post-equinox. J. Geophys. Res. **117**, 4224 (2012). doi:10.1029/2011JA017444

C.S. Arridge, N. André, K.K. Khurana, C.T. Russell, S.W.H. Cowley, G. Provan, D.J. Andrews, C.M. Jackman, A.J. Coates, E.C. Sittler, M.K. Dougherty, D.T. Young, Periodic motion of Saturn's nightside plasma sheet. J. Geophys. Res. **116**, 11205 (2011). doi:10.1029/2011JA016827

F. Bagenal, Ionization source near Io from Galileo wake data. Geophys. Res. Lett. **24**, 2111 (1997). doi:10.1029/97GL02052

F. Bagenal, P.A. Delamere, Flow of mass and energy in the magnetospheres of Jupiter and Saturn. J. Geophys. Res. **116**, 5209 (2011). doi:10.1029/2010JA016294

D.D. Barbosa, Theory and observations of electromagnetic ion cyclotron waves in Saturn's inner magnetosphere. J. Geophys. Res. **98**, 9345–9350 (1993). doi:10.1029/93JA00476

M. Blanc, S. Bolton, J. Bradley, M. Burton, T.E. Cravens, I. Dandouras, M.K. Dougherty, M.C. Festou, J. Feynman, R.E. Johnson, T.G. Gombosi, W.S. Kurth, P.C. Liewer, B.H. Mauk, S. Maurice, D. Mitchell, F.M. Neubauer, J.D. Richardson, D.E. Shemansky, E.C. Sittler, B.T. Tsurutani, P. Zarka, L.W. Esposito, E. Grün, D.A. Gurnett, A.J. Kliore, S.M. Krimigis, D. Southwood, J.H. Waite, D.T. Young, Magnetospheric and plasma science with Cassini-Huygens. Space Sci. Rev. **104**, 253–346 (2002). doi:10.1023/A:1023605110711

X. Blanco-Cano, C.T. Russell, R.J. Strangeway, The Io mass-loading disk: Wave dispersion analysis. J. Geophys. Res. **106**, 26261–26276 (2001a). doi:10.1029/2001JA900090

X. Blanco-Cano, C.T. Russell, R.J. Strangeway, M.G. Kivelson, K.K. Khurana, Galileo observations of ion cyclotron waves in the Io torus. Adv. Space Res. **28**, 1469–1474 (2001b). doi:10.1016/S0273-1177(01)00548-8

X. Blanco-Cano, C.T. Russell, D.E. Huddleston, R.J. Strangeway, Ion cyclotron waves near Io. Planet. Space Sci. **49**, 1125–1136 (2001c). doi:10.1016/S0032-0633(01)00020-4

P.C. Brandt, C.P. Paranicas, J.F. Carbary, D.G. Mitchell, B.H. Mauk, S.M. Krimigis, Understanding the global evolution of Saturn's ring current. Geophys. Res. Lett. **35**, 17101 (2008). doi:10.1029/2008GL034969

P.C. Brandt, D.G. Mitchell, D.A. Gurnett, A.M. Persoon, N.A. Tsyganenko, Saturn's periodic magnetosphere: the relation between periodic hot plasma injections, a rotating partial ring current, global magnetic field distortions, plasmapause motion, and radio emissions, in *EGU General Assembly Conference Abstracts*, ed. by A. Abbasi, N. Giesen. EGU General Assembly Conference Abstracts, vol. 14 (2012), p. 12906

S.H. Brecht, J.G. Luhmann, D.J. Larson, Simulation of the Saturnian magnetospheric interaction with Titan. J. Geophys. Res. **105**, 13119–13130 (2000). doi:10.1029/1999JA900490

N. Brice, T.R. McDonough, Jupiter's radiation belts. Icarus **18**, 206–219 (1973). doi:10.1016/0019-1035(73)90204-2

R.A. Brown, The Jupiter hot plasma torus: Observed electron temperature and energy flows. Astrophys. J. **244**, 1072 (1981)

E.J. Bunce, S.W.H. Cowley, D.M. Wright, A.J. Coates, M.K. Dougherty, N. Krupp, W.S. Kurth, A.M. Rymer, In situ observations of a solar wind compression-induced hot plasma injection in Saturn's tail. Geophys. Res. Lett. **32**, 20 (2005). doi:10.1029/2005GL022888

E.J. Bunce, S.W.H. Cowley, I.I. Alexeev, C.S. Arridge, M.K. Dougherty, J.D. Nichols, C.T. Russell, Cassini observations of the variation of Saturn's ring current parameters with system size. J. Geophys. Res. **112**(A11), 10202 (2007). doi:10.1029/2007JA012275

E.J. Bunce, C.S. Arridge, J.T. Clarke, A.J. Coates, S.W.H. Cowley, M.K. Dougherty, J.-C. GéRard, D. Grodent, K.C. Hansen, J.D. Nichols, D.J. Southwood, D.L. Talboys, Origin of Saturn's aurora: Simultaneous observations by Cassini and the Hubble Space Telescope. J. Geophys. Res. **113**, 9209 (2008). doi:10.1029/2008JA013257

J.L. Burch, J. Goldstein, T.W. Hill, D.T. Young, F.J. Crary, A.J. Coates, N. André, W.S. Kurth, E.C. Sittler, Properties of local plasma injections in Saturn's magnetosphere. Geophys. Res. Lett. **32**, 14 (2005). doi:10.1029/2005GL022611

J.L. Burch, J. Goldstein, W.S. Lewis, D.T. Young, A.J. Coates, M.K. Dougherty, N. André, Tethys and Dione as sources of outward-flowing plasma in Saturn's magnetosphere. Nature **447**, 833–835 (2007). doi:10.1038/nature05906

J.L. Burch, A.D. DeJong, J. Goldstein, D.T. Young, Periodicity in Saturn's magnetosphere: Plasma cam. Geophys. Res. Lett. **36**, 14203 (2009). doi:10.1029/2009GL039043

M.H. Burger, Io's neutral clouds: From the atmosphere to the plasma torus, PhD thesis, University of Colorado at Boulder, 2003

J.F. Carbary, D.G. Mitchell, Periodicities in Saturn's magnetosphere. Rev. Geophys. **51**, 1–30 (2013). doi:10.1002/rog.20006

J.F. Carbary, D.G. Mitchell, S.M. Krimigis, D.C. Hamilton, N. Krupp, Charged particle periodicities in Saturn's outer magnetosphere. J. Geophys. Res. **112**, 6246 (2007). doi:10.1029/2007JA012351

J.F. Carbary, D.G. Mitchell, P. Brandt, C. Paranicas, S.M. Krimigis, ENA periodicities at Saturn. Geophys. Res. Lett. **35**, 7102 (2008a). doi:10.1029/2008GL033230

J.F. Carbary, D.G. Mitchell, P. Brandt, E.C. Roelof, S.M. Krimigis, Track analysis of energetic neutral atom blobs at Saturn. J. Geophys. Res. **113**, 1209 (2008b). doi:10.1029/2007JA012708

T.A. Cassidy, R.E. Johnson, Collisional spreading of Enceladus's neutral cloud. Icarus **209**, 696–703 (2010). doi:10.1016/j.icarus.2010.04.010

G. Caudal, A self-consistent model of Jupiter's magnetodisc including the effects of centrifugal force and pressure. J. Geophys. Res. **91**, 4201–4221 (1986). doi:10.1029/JA091iA04p04201

C.X. Chen, Numerical simulation of the Io-torus-driven radial plasma transport. J. Geophys. Res. **108**, 1376 (2003). doi:10.1029/2002JA009460

Y. Chen, T.W. Hill, Statistical analysis of injection/dispersion events in Saturn's inner magnetosphere. J. Geophys. Res. **113**, 7215 (2008). doi:10.1029/2008JA013166

Y. Chen, T.W. Hill, A.M. Rymer, R.J. Wilson, Rate of radial transport of plasma in Saturn's inner magnetosphere. J. Geophys. Res. **115**, 10211 (2010). doi:10.1029/2010JA015412

A.F. Cheng, Magnetospheric interchange instability. J. Geophys. Res. **90**, 9900–9904 (1985). doi:10.1029/JA090iA10p09900

J.T. Clarke, L. Ben Jaffel, J.-C. Gérard, Hubble Space Telescope imaging of Jupiter's UV aurora during the Galileo orbiter mission. J. Geophys. Res. **103**, 20217–20236 (1998). doi:10.1029/98JE01130

K.E. Clarke, D.J. Andrews, C.S. Arridge, A.J. Coates, S.W.H. Cowley, Magnetopause oscillations near the planetary period at Saturn: Occurrence, phase, and amplitude. J. Geophys. Res. **115**, 8209 (2010). doi:10.1029/2009JA014745

M.R. Combi, K. Kabin, T.I. Gombosi, D.L. DeZeeuw, Io's plasma environment during the Galileo flyby: Global three-dimensional MHD modeling with adaptive mesh refinement. J. Geophys. Res. **103**, 9071 (1998)

M.M. Cowee, S.P. Gary, Electromagnetic ion cyclotron wave generation by planetary pickup ions: One-dimensional hybrid simulations at sub-Alfvénic pickup velocities. J. Geophys. Res. **117**, 6215 (2012). doi:10.1029/2012JA017568

M.M. Cowee, R.J. Strangeway, C.T. Russell, D. Winske, One-dimensional hybrid simulations of planetary ion pickup: Techniques and verification. J. Geophys. Res. **111**, 12213 (2006). doi:10.1029/2006JA011996

M.M. Cowee, C.T. Russell, R.J. Strangeway, One-dimensional hybrid simulations of planetary ion pickup: Effects of variable plasma and pickup conditions. J. Geophys. Res. **113**, 8220 (2008). doi:10.1029/2008JA013066

M.M. Cowee, S.P. Gary, H.Y. Wei, R.L. Tokar, C.T. Russell, An explanation for the lack of ion cyclotron wave generation by pickup ions at Titan: 1-D hybrid simulation results. J. Geophys. Res. **115**, 10224 (2010). doi:10.1029/2010JA015769

S.W.H. Cowley, E.J. Bunce, Origin of the main auroral oval in Jupiter's coupled magnetosphere-ionosphere system. Planet. Space Sci. **49**, 1067–1088 (2001). doi:10.1016/S0032-0633(00)00167-7

S.W.H. Cowley, J.D. Nichols, D.J. Andrews, Modulation of Jupiter's plasma flow, polar currents, and auroral precipitation by solar wind-induced compressions and expansions of the magnetosphere: a simple theoretical model. Ann. Geophys. **25**, 1433–1463 (2007). doi:10.5194/angeo-25-1433-2007

F.J. Crary, F. Bagenal, Ion cyclotron waves, pickup ions, and Io's neutral exosphere. J. Geophys. Res. **105**, 25379–27066 (2000). doi:10.1029/2000JA000055

F.J. Crary, F. Bagenal, J.A. Ansher, D.A. Gurnett, W.S. Kurth, Anisotropy and proton density in the Io plasma torus derived from whistler wave dispersion. J. Geophys. Res. **101**, 2699–2706 (1996). doi:10.1029/95JA02212

P.A. Delamere, F. Bagenal, Modeling variability of plasma conditions in the Io torus. J. Geophys. Res. **108**, 1276 (2003)

P.A. Delamere, F. Bagenal, Longitudinal plasma density variations at Saturn caused by hot electrons. Geophys. Res. Lett. **35**, 3107 (2008). doi:10.1029/2007GL031095

P.A. Delamere, F. Bagenal, Jupiter and Saturn: Colossal comets? (2013 in preparation)

P.A. Delamere, F. Bagenal, V. Dols, L.C. Ray, Saturn's neutral torus versus Jupiter's plasma torus. Geophys. Res. Lett. **34**, 9105 (2007). doi:10.1029/2007GL029437

M.D. Desch, M.L. Kaiser, Voyager measurement of the rotation period of Saturn's magnetic field. Geophys. Res. Lett. **8**, 253–256 (1981). doi:10.1029/GL008i003p00253

V. Dols, P.A. Delamere, F. Bagenal, A multispecies chemistry model of Io's local interaction with the Plasma Torus. J. Geophys. Res. **113**, 9208 (2008). doi:10.1029/2007JA012805

V. Dols, P.A. Delamere, F. Bagenal, W.S. Kurth, W.R. Paterson, Asymmetry of Io's outer atmosphere: Constraints from five Galileo flybys. J. Geophys. Res., Planets **117**, 10010 (2012). doi:10.1029/2012JE004076

S.A. Espinosa, M.K. Dougherty, Periodic perturbations in Saturn's magnetic field. Geophys. Res. Lett. **27**, 2785–2788 (2000). doi:10.1029/2000GL000048

A.J. Farmer, Saturn in hot water: Viscous evolution of the Enceladus torus. Icarus **202**, 280–286 (2009). doi:10.1016/j.icarus.2009.02.031

K.M. Ferrière, N. André, A mixed magnetohydrodynamic-kinetic theory of low-frequency waves and instabilities in stratified, gyrotropic, two-component plasmas. J. Geophys. Res. **108**, 1308 (2003). doi:10.1029/2003JA009883

K.M. Ferrière, C. Zimmer, M. Blanc, Magnetohydrodynamic waves and gravitational/centrifugal instability in rotating systems. J. Geophys. Res. **104**, 17335–17356 (1999). doi:10.1029/1999JA900167

B.L. Fleshman, The roles of dissociation and velocity-dependent charge exchange in Saturn's extended neutral clouds, in *Magnetospheres of the Outer Planets* (2011)

B.L. Fleshman, P.A. Delamere, F. Bagenal, A sensitivity study of the Enceladus torus. J. Geophys. Res., Planets **115**, 4007 (2010a). doi:10.1029/2009JE003372

B.L. Fleshman, P.A. Delamere, F. Bagenal, Modeling the Enceladus plume-plasma interaction. Geophys. Res. Lett. **37**, 3202 (2010b). doi:10.1029/2009GL041613

B.L. Fleshman, P.A. Delamere, F. Bagenal, T. Cassidy, The roles of charge exchange and dissociation in spreading Saturn's neutral clouds. J. Geophys. Res., Planets **117**, 5007 (2012). doi:10.1029/2011JE003996

L.A. Frank, W.R. Paterson, Intense electron beams observed at Io with the Galileo spacecraft. J. Geophys. Res. **104**, 28657 (1999)

L.A. Frank, W.R. Paterson, Production of hydrogen ions at Io. J. Geophys. Res. **104**, 10345–10354 (1999). doi:10.1029/1999JA900052

L.A. Frank, W.R. Paterson, Observations of plasmas in the Io torus with the Galileo spacecraft. J. Geophys. Res. **105**, 16017–16034 (2000). doi:10.1029/1999JA000250

L.A. Frank, W.R. Paterson, Passage through Io's ionospheric plasmas by the Galileo spacecraft. J. Geophys. Res. **106**(A11), 26209–26224 (2001). doi:10.1029/2000JA002503

L.A. Frank, W.R. Paterson, K.L. Ackerson, V.M. Vasyliunas, F.V. Coroniti, S.J. Bolton, Plasma observations at Io with the Galileo spacecraft. Science **274**(5286), 394–395 (1996). doi:10.1126/science.274.5286.394

K. Fukazawa, T. Ogino, R.J. Walker, Vortex-associated reconnection for northward IMF in the Kronian magnetosphere. Geophys. Res. Lett. **34**, 23201 (2007)

K. Fukazawa, T. Ogino, R.J. Walker, A simulation study of dynamics in the distant Jovian magnetotail. J. Geophys. Res. **115**, 9219 (2010). doi:10.1029/2009JA015228

G. Giampieri, M.K. Dougherty, E.J. Smith, C.T. Russell, A regular period for Saturn's magnetic field that may track its internal rotation. Nature **441**, 62–64 (2006). doi:10.1038/nature04750

J.A. Gledhill, Magnetosphere of Jupiter. Nature **155** (1967). doi:10.1038/214155a0

T. Gold, Motions in the magnetosphere of the earth. J. Geophys. Res. **64**, 1219–1224 (1959)

P. Goldreich, A.J. Farmer, Spontaneous axisymmetry breaking of the external magnetic field at Saturn. J. Geophys. Res. **112**, 5225 (2007). doi:10.1029/2006JA012163

D.A. Gurnett, W.S. Kurth, G.B. Hospodarsky, A.M. Persoon, T.F. Averkamp, B. Cecconi, A. Lecacheux, P. Zarka, P. Canu, N. Cornilleau-Wehrlin, P. Galopeau, A. Roux, C. Harvey, P. Louarn, R. Bostrom, G. Gustafsson, J.-E. Wahlund, M.D. Desch, W.M. Farrell, M.L. Kaiser, K. Goetz, P.J. Kellogg, G. Fischer, H.-P. Ladreiter, H. Rucker, H. Alleyne, A. Pedersen, Radio and plasma wave observations at Saturn from Cassini's approach and first orbit. Science **307**, 1255–1259 (2005). doi:10.1126/science.1105356

D.A. Gurnett, A.M. Persoon, W.S. Kurth, J.B. Groene, T.F. Averkamp, M.K. Dougherty, D.J. Southwood, The variable rotation period of the inner region of Saturn's plasma disk. Science **316**, 442 (2007). doi:10.1126/science.1138562

D.A. Gurnett, A.M. Persoon, J.B. Groene, A.J. Kopf, G.B. Hospodarsky, W.S. Kurth, A north-south difference in the rotation rate of auroral hiss at Saturn: Comparison to Saturn's kilometric radio emission. Geophys. Res. Lett. **36**, 21108 (2009). doi:10.1029/2009GL040774

K.C. Hansen, A.J. Ridley, G.B. Hospodarsky, N. Achilleos, M.K. Dougherty, T.I. Gombosi, G. Tóth, Global MHD simulations of Saturn's magnetosphere at the time of Cassini approach. Geophys. Res. Lett. **32**, 20 (2005). doi:10.1029/2005GL022835

R.E. Hartle, E.C. Sittler, F.M. Neubauer, R.E. Johnson, H.T. Smith, F. Crary, D.J. McComas, D.T. Young, A.J. Coates, D. Simpson, S. Bolton, D. Reisenfeld, K. Szego, J.J. Berthelier, A. Rymer, J. Vilppola, J.T. Steinberg, N. Andre, Initial interpretation of Titan plasma interaction as observed by the Cassini plasma spectrometer: Comparisons with Voyager 1. Planet. Space Sci. **54**, 1211–1224 (2006). doi:10.1016/j.pss.2006.05.029

S.L.G. Hess, P. Delamere, V. Dols, B. Bonfond, D. Swift, Power transmission and particle acceleration along the Io flux tube. J. Geophys. Res. **115**, 06205 (2010). doi:10.1029/2009JA014928

T.W. Hill, Interchange stability of a rapidly rotating magnetosphere. Planet. Space Sci. **24**, 1151–1154 (1976)

T.W. Hill, Inertial limit on corotation. J. Geophys. Res. **25**, 6554–6558 (1979)

T.W. Hill, Effect of the acceleration current on the centrifugal interchange instability. J. Geophys. Res. **111**, A03214 (2006). doi:10.1029/2005JA011338

T.W. Hill, A.J. Dessler, L.J. Maher, Corotating magnetospheric convection. J. Geophys. Res. **86**, 9020–9028 (1981). doi:10.1029/JA086iA11p09020

T.W. Hill, A.M. Rymer, J.L. Burch, F.J. Crary, D.T. Young, M.F. Thomsen, D. Delapp, N. André, A.J. Coates, G.R. Lewis, Evidence for rotationally driven plasma transport in Saturn's magnetosphere. Geophys. Res. Lett. **32**, 14 (2005). doi:10.1029/2005GL022620

⧠ Springer

T.W. Hill, M.F. Thomsen, M.G. Henderson, R.L. Tokar, A.J. Coates, H.J. McAndrews, G.R. Lewis, D.G. Mitchell, C.M. Jackman, C.T. Russell, M.K. Dougherty, F.J. Crary, D.T. Young, Plasmoids in Saturn's magnetotail. J. Geophys. Res. **113**, 1214 (2008). doi:10.1029/2007JA012626

T.W. Hill, A.J. Dessler, C.K. Goertz, Magnetospheric models, ed. by A.J. Dessler 1983, pp. 353–394

T.S. Huang, T.W. Hill, Drift wave instability in the Io plasma torus. J. Geophys. Res. **96**, 14075 (1991). doi:10.1029/91JA01170

D.E. Huddleston, R.J. Strangeway, J. Warnecke, C.T. Russell, M.G. Kivelson, F. Bagenal, Ion cyclotron waves in the Io torus during the Galileo encounter: Warm plasma dispersion analysis. Geophys. Res. Lett. **24**, 2143 (1997). doi:10.1029/97GL01203

D.E. Huddleston, R.J. Strangeway, J. Warnecke, C.T. Russell, M.G. Kivelson, Ion cyclotron waves in the Io torus: Wave dispersion, free energy analysis, and SO_2^+ source rate estimates. J. Geophys. Res. **103**, 19887–19900 (1998). doi:10.1029/97JE03557

D.E. Huddleston, R.J. Strangeway, X. Blanco-Cano, C.T. Russell, M.G. Kivelson, K.K. Khurana, Mirror-mode structures at the Galileo-Io flyby: Instability criterion and dispersion analysis. J. Geophys. Res. **104**, 17479–17490 (1999). doi:10.1029/1999JA900195

G. Ioaniddis, N. Brice, Plasma densities in the Jovian magnetosphere: plasma slingshot or Maxwell demon? Icarus **14**, 360–373 (1971). doi:10.1016/0019-1035(71)90007-8

C.M. Jackman, C.S. Arridge, Solar cycle effects on the dynamics of Jupiter's and Saturn's magnetospheres. Solar Phys. **274**(1–2), 481–502 (2011). doi:10.1007/s11207-011-9748-z

C.M. Jackman, L. Lamy, M.P. Freeman, P. Zarka, B. Cecconi, W.S. Kurth, S.W.H. Cowley, M.K. Dougherty, On the character and distribution of lower-frequency radio emissions at Saturn and their relationship to substorm-like events. J. Geophys. Res. **114**, 8211 (2009). doi:10.1029/2008JA013997

X. Jia, K.C. Hansen, T.I. Gombosi, M.G. Kivelson, G. Tóth, D.L. DeZeeuw, A.J. Ridley, Magnetospheric configuration and dynamics of Saturn's magnetosphere: A global MHD simulation. J. Geophys. Res. **117**, 5225 (2012). doi:10.1029/2012JA017575

R.E. Johnson, H.T. Smith, O.J. Tucker, M. Liu, M.H. Burger, E.C. Sittler, R.L. Tokar, The Enceladus and OH Tori at Saturn. Astrophys. J. Lett. **644**, 137–139 (2006). doi:10.1086/505750

S.P. Joy, M.G. Kivelson, R.J. Walker, K.K. Khurana, C.T. Russell, T. Ogino, Probabilistic models of the Jovian magnetopause and bow shock locations. J. Geophys. Res. **107**, 1309 (2002). doi:10.1029/2001JA009146

K. Kabin, M.R. Combi, T.I. Gombosi, D.L. DeZeeuw, K.C. Hansen, K.G. Powell, Io's magnetospheric interaction: an MHD model with day-night asymmetry. Planet. Space Sci. **49**, 337–344 (2001)

S.J. Kanani, C.S. Arridge, G.H. Jones, A.N. Fazakerley, H.J. McAndrews, N. Sergis, S.M. Krimigis, M.K. Dougherty, A.J. Coates, D.T. Young, K.C. Hansen, N. Krupp, A new form of Saturn's magnetopause using a dynamic pressure balance model, based on in situ, multi-instrument Cassini measurements. J. Geophys. Res. **115**, 6207 (2010). doi:10.1029/2009JA014262

M. Kane, D.G. Mitchell, J.F. Carbary, S.M. Krimigis, F.J. Crary, Plasma convection in Saturn's outer magnetosphere determined from ions detected by the Cassini INCA experiment. Geophys. Res. Lett. **35**, 4102 (2008). doi:10.1029/2007GL032342

S. Kellett, C.S. Arridge, E.J. Bunce, A.J. Coates, S.W.H. Cowley, M.K. Dougherty, A.M. Persoon, N. Sergis, R.J. Wilson, Nature of the ring current in Saturn's dayside magnetosphere. J. Geophys. Res. **115**, 8201 (2010). doi:10.1029/2009JA015146

S. Kellett, C.S. Arridge, E.J. Bunce, A.J. Coates, S.W.H. Cowley, M.K. Dougherty, A.M. Persoon, N. Sergis, R.J. Wilson, Saturn's ring current: Local time dependence and temporal variability. J. Geophys. Res. **116**, 5220 (2011). doi:10.1029/2010JA016216

T.J. Kennelly, J.S. Leisner, G.B. Hospodarsky, D.A. Gurnett, Ordering of injection events within Saturnian SLS longitude and local time. J. Geophys. Res. **118**, 832–838 (2013). doi:10.1002/jgra.50152

K. Khurana, M. Kivelson, M. Volwerk, The interactions of Europa and Callisto with the magnetosphere of Jupiter, in *34th COSPAR Scientific Assembly*. COSPAR Meeting, vol. 34 (2002)

K.K. Khurana, M.K. Dougherty, C.T. Russell, J.S. Leisner, Mass loading of Saturn's magnetosphere near Enceladus. J. Geophys. Res. **112**, 8203 (2007). doi:10.1029/2006JA012110

K.K. Khurana, X. Jia, M.G. Kivelson, F. Nimmo, G. Schubert, C.T. Russell, Evidence of a global magma ocean in Io's interior. Science **332**, 1186 (2011). doi:10.1126/science.1201425

A. Kidder, R.M. Winglee, E.M. Harnett, Regulation of the centrifugal interchange cycle in Saturn's inner magnetosphere. J. Geophys. Res. **114**, 2205 (2009). doi:10.1029/2008JA013100

A. Kidder, C.S. Paty, R.M. Winglee, E.M. Harnett, External triggering of plasmoid development at Saturn. J. Geophys. Res. **117**, 7206 (2012). doi:10.1029/2012JA017625

M.G. Kivelson, D.J. Southwood, Dynamical consequences of two modes of centrifugal instability in Jupiter's outer magnetosphere. J. Geophys. Res. **110**, 12209 (2005)

M.G. Kivelson, K.K. Khurana, R.J. Walker, J. Warnecke, C.T. Russell, J.A. Linker, D.J. Southwood, C. Polanskey, Io's interaction with the plasma torus: Galileo magnetometer report. Science **274**, 396–398 (1996). doi:10.1126/science.274.5286.396

M.G. Kivelson, K.K. Khurana, C.T. Russell, R.J. Walker, P.J. Coleman, F.V. Coroniti, J. Green, S. Joy, R.L. McPherron, C. Polanskey, D.J. Southwood, L. Bennett, J. Warnecke, D.E. Huddleston, Galileo at Jupiter—changing states of the magnetosphere and first looks at Io and Ganymede. Adv. Space Res. **20**, 193–204 (1997a). doi:10.1016/S0273-1177(97)00533-4

M.G. Kivelson, K.K. Khurana, C.T. Russell, R.J. Walker, Intermittent short-duration magnetic field anomalies in the Io torus: Evidence for plasma interchange? Geophys. Res. Lett. **24**, 2127 (1997b). doi:10.1029/97GL02202

A. Kopp, W.-H. Ip, Asymmetric mass loading effect at Titan's ionosphere. J. Geophys. Res. **106**, 8323–8332 (2001). doi:10.1029/2000JA900140

H. Kriegel, S. Simon, U. Motschmann, J. Saur, F.M. Neubauer, A.M. Persoon, M.K. Dougherty, D.A. Gurnett, Influence of negatively charged plume grains on the structure of Enceladus' Alfvén wings: Hybrid simulations versus Cassini Magnetometer data. J. Geophys. Res. **116**, 10223 (2011). doi:10.1029/2011JA016842

S.M. Krimigis, D.G. Mitchell, D.C. Hamilton, N. Krupp, S. Livi, E.C. Roelof, J. Dandouras, T.P. Armstrong, B.H. Mauk, C. Paranicas, P.C. Brandt, S. Bolton, A.F. Cheng, T. Choo, G. Gloeckler, J. Hayes, K.C. Hsieh, W.-H. Ip, S. Jaskulek, E.P. Keath, E. Kirsch, M. Kusterer, A. Lagg, L.J. Lanzerotti, D. LaVallee, J. Manweiler, R.W. McEntire, W. Rasmuss, J. Saur, F.S. Turner, D.J. Williams, J. Woch, Dynamics of Saturn's magnetosphere from MIMI during Cassini's orbital insertion. Science **307**, 1270–1273 (2005). doi:10.1126/science.1105978

W.S. Kurth, T.F. Averkamp, et al., An update to a Saturnian longitude system based on kilometric radio emissions. J. Geophys. Res. **113**, 05222 (2008)

L. Lamy, Variability of southern and northern periodicities of Saturn Kilometric Radiation, in *Planetary, Solar and Heliospheric Radio Emissions (PRE VII)* (2011), pp. 38–50

L. Lamy, R. Prangé, W. Pryor, J. Gustin, S.V. Badman, H. Melin, T. Stallard, D.G. Mitchell, P.C. Brandt, Multispectral simultaneous diagnosis of Saturn's aurorae throughout a planetary rotation. J. Geophys. Res. **118** (2013). doi:10.1002/jgra.50404.

A.J. Lazarus, R.L. McNutt Jr., Low-energy plasma ion observations in Saturn's magnetosphere. J. Geophys. Res. **88**, 8831–8846 (1983). doi:10.1029/JA088iA11p08831

S.A. Ledvina, S.H. Brecht, J.G. Luhmann, Ion distributions of 14 amu pickup ions associated with Titan's plasma interaction. Geophys. Res. Lett. **31**, 17 (2004). doi:10.1029/2004GL019861

J.S. Leisner, C.T. Russell, K.K. Khurana, M.K. Dougherty, N. André, Warm flux tubes in the E-ring plasma torus: Initial Cassini magnetometer observations. Geophys. Res. Lett. **32**, 14 (2005). doi:10.1029/2005GL022652

J.S. Leisner, C.T. Russell, M.K. Dougherty, X. Blanco-Cano, R.J. Strangeway, C. Bertucci, Ion cyclotron waves in Saturn's E ring: Initial Cassini observations. Geophys. Res. Lett. **33**, 11101 (2006). doi:10.1029/2005GL024875

J.S. Leisner, C.T. Russell, H.Y. Wei, M.K. Dougherty, Probing Saturn's ion cyclotron waves on high-inclination orbits: Lessons for wave generation. J. Geophys. Res. **116**, 9235 (2011). doi:10.1029/2011JA016555

E. Lellouch, M.A. McGrath, K.L. Jessup, Io's atmosphere, ed. by R.M.C. Lopes, J.R. Spencer. *Io After Galileo. A New View of Jupiter's Volcanic Moon.* Springer Praxis Books/Geophysical Sciences (Springer, Berlin, 2007), pp. 231–264. doi:10.1007/978-3-540-48841-5_10

B. Levitt, D. Maslovsky, M.E. Mauel, Observation of centrifugally driven interchange instabilities in a plasma confined by a magnetic dipole. Phys. Rev. Lett. **94**, 175002 (2005a). doi:10.1103/PhysRevLett.94.175002

B. Levitt, D. Maslovsky, M.E. Mauel, J. Waksman, Excitation of the centrifugally driven interchange instability in a plasma confined by a magnetic dipolea). Phys. Plasmas **12**(5), 055703 (2005b). doi:10.1063/1.1888685

J.A. Linker, K.K. Khurana, M.G. Kivelson, R.J. Walker, MHD simulations of Io's interaction with the plasma torus. J. Geophys. Res. **103**, 19867 (1998)

A.S. Lipatov, M.R. Combi, Effects of kinetic processes in shaping Io's global plasma environment: A 3D hybrid model. Icarus **180**, 412–427 (2006). doi:10.1016/j.icarus.2005.08.012

X. Liu, T.W. Hill, Effects of finite plasma pressure on centrifugally driven convection in Saturn's inner magnetosphere. J. Geophys. Res. **117**, 7216 (2012). doi:10.1029/2012JA017827

X. Liu, T.W. Hill, R.A. Wolf, S. Sazykin, R.W. Spiro, H. Wu, Numerical simulation of plasma transport in Saturn's inner magnetosphere using the Rice Convection Model. J. Geophys. Res. **115**, 12254 (2010). doi:10.1029/2010JA015859

P. Louarn, A. Roux, S. Perraut, W. Kurth, D. Gurnett, A study of the large-scale dynamics of the Jovian magnetosphere using the Galileo Plasma Wave Experiment. Geophys. Res. Lett. **25**, 2905–2908 (1998). doi:10.1029/98GL01774

P. Louarn, A. Roux, S. Perraut, W.S. Kurth, D.A. Gurnett, A study of the Jovian "energetic magnetospheric events" observed by Galileo: role in the radial plasma transport. J. Geophys. Res. **105**, 13073–13088 (2000). doi:10.1029/1999JA900478

P. Louarn, B.H. Mauk, M.G. Kivelson, W.S. Kurth, A. Roux, C. Zimmer, D.A. Gurnett, D.J. Williams, A multi-instrument study of a Jovian magnetospheric disturbance. J. Geophys. Res. **106**, 29883–29898 (2001). doi:10.1029/2001JA900067

Y.-J. Ma, A.F. Nagy, T.E. Cravens, I.V. Sokolov, J. Clark, K.C. Hansen, 3-D global MHD model prediction for the first close flyby of Titan by Cassini. Geophys. Res. Lett. **31**, 22803 (2004). doi:10.1029/2004GL021215

H.R. Martens, D.B. Reisenfeld, J.D. Williams, R.E. Johnson, H.T. Smith, Observations of molecular oxygen ions in Saturn's inner magnetosphere. Geophys. Res. Lett. **35**, 20103 (2008). doi:10.1029/2008GL035433

B.H. Mauk, S.A. Gary, M. Kane, E.P. Keath, S.M. Krimigis, T.P. Armstrong, Hot plasma parameters of Jupiter's inner magnetosphere. J. Geophys. Res. **101**, 7685–7696 (1996). doi:10.1029/96JA00006

B.H. Mauk, D.J. Williams, R.W. McEntire, Energy-time dispersed charged particle signatures of dynamic injections in Jupiter's inner magnetosphere. Geophys. Res. Lett. **24**, 2949–2952 (1997). doi:10.1029/97GL03026

B.H. Mauk, R.W. McEntire, D.J. Williams, A. Lagg, E.C. Roelof, S.M. Krimigis, T.P. Armstrong, T.A. Fritz, L.J. Lanzerotti, J.G. Roederer, B. Wilken, Galileo-measured depletion of near-Io hot ring current plasmas since the Voyager epoch. J. Geophys. Res. **103**, 4715 (1998). doi:10.1029/97JA02343

B.H. Mauk, D.J. Williams, R.W. McEntire, K.K. Khurana, J.G. Roederer, Storm-like dynamics of Jupiter's inner and middle magnetosphere. J. Geophys. Res. **104**, 22759–22778 (1999). doi:10.1029/1999JA900097

B.H. Mauk, J.T. Clarke, D. Grodent, J.H. Waite, C.P. Paranicas, D.J. Williams, Transient aurora on Jupiter from injections of magnetospheric electrons. Nature **415**, 1003–1005 (2002)

B.H. Mauk, J. Saur, D.G. Mitchell, E.C. Roelof, P.C. Brandt, T.P. Armstrong, D.C. Hamilton, S.M. Krimigis, N. Krupp, S.A. Livi, J.W. Manweiler, C.P. Paranicas, Energetic particle injections in Saturn's magnetosphere. Geophys. Res. Lett. **32**, 14 (2005). doi:10.1029/2005GL022485

B.H. Mauk, D.C. Hamilton, T.W. Hill, G.B. Hospodarsky, R.E. Johnson, C. Paranicas, E. Roussos, C.T. Russell, D.E. Shemansky, E.C. Sittler, R.M. Thorne, Fundamental plasma processes in Saturn's magnetosphere, in *Saturn from Cassini-Huygens*, ed. by M.K. Dougherty, L.W. Esposito, S.M. Krimigis (Springer, Berlin, 2009), p. 281. doi:10.1007/978-1-4020-9217-6_11

S. Maurice, M. Blanc, R. Prangé, E.C. Sittler, The magnetic-field-aligned polarization electric field and its effects on particle distribution in the magnetospheres of Jupiter and Saturn. Planet. Space Sci. **45**, 1449–1465 (1997)

M.A. McGrath, E. Lellouch, D.F. Strobel, P.D. Feldman, R.E. Johnson, Satellite atmospheres, in *Jupiter: The Planet, Satellites and Magnetosphere* (2004), pp. 457–483

R.L. McNutt, J.W. Belcher, H.S. Bridge, Positive ion observations in the middle magnetosphere of Jupiter. J. Geophys. Res. **86**, 8319–8342 (1981). doi:10.1029/JA086iA10p08319

R.L. Melrose, Rotational effects on the distribution of thermal plasma in the magnetosphere of Jupiter. Planet. Space Sci. **15**, 381–393 (1967). doi:10.1016/0032-0633(67)90202-4

J.D. Menietti, J.B. Groene, T.F. Averkamp, G.B. Hospodarsky, W.S. Kurth, D.A. Gurnett, P. Zarka, Influence of Saturnian moons on Saturn kilometric radiation. J. Geophys. Res. **112**, 8211 (2007). doi:10.1029/2007JA012331

D.G. Mitchell, S.M. Krimigis, C. Paranicas, P.C. Brandt, J.F. Carbary, E.C. Roelof, W.S. Kurth, D.A. Gurnett, J.T. Clarke, J.D. Nichols, J.-C. Gérard, D.C. Grodent, M.K. Dougherty, W.R. Pryor, Recurrent energization of plasma in the midnight-to-dawn quadrant of Saturn's magnetosphere, and its relationship to auroral UV and radio emissions. Planet. Space Sci. **57**, 1732–1742 (2009). doi:10.1016/j.pss.2009.04.002

D.G. Mitchell, P.C. Brandt, J.F. Carbary, W.S. Kurth, S.M. Krimigis, C. Paranicas, N. Krupp, D.C. Hamilton, B.H. Mauk, G.B. Hospodarsky, M.K. Dougherty, W.R. Pryor, Injection, interchange and reconnection: Energetic particle observations in Saturn's magnetotail, in *Magnetotails in the Solar System*. AGU Geophysical Monograph Series (2015)

A.F. Nagy, Y. Liu, K.C. Hansen, K. Kabin, T.I. Gombosi, M.R. Combi, D.L. DeZeeuw, K.G. Powell, A.J. Kliore, The interaction between the magnetosphere of Saturn and Titan's ionosphere. J. Geophys. Res. **106**, 6151–6160 (2001). doi:10.1029/2000JA000183

W.A. Newcomb, Convective instability induced by gravity in a plasma with a frozen-in magnetic field. Phys. Fluids **4**, 391–396 (1961). doi:10.1063/1.1706342

J. Nichols, S. Cowley, Magnetosphere-ionosphere coupling currents in Jupiter's middle magnetosphere: effect of precipitation-induced enhancement of the ionospheric Pedersen conductivity. Ann. Geophys. **22**, 1799–1827 (2004). doi:10.5194/angeo-22-1799-2004

C. Paranicas, D.G. Mitchell, E.C. Roelof, P.C. Brandt, D.J. Williams, S.M. Krimigis, B.H. Mauk, Periodic intensity variations in global ENA images of Saturn. Geophys. Res. Lett. **32**, 21101 (2005). doi:10.1029/2005GL023656

C. Paranicas, D.G. Mitchell, E.C. Roelof, B.H. Mauk, S.M. Krimigis, P.C. Brandt, M. Kusterer, F.S. Turner, J. Vandegriff, N. Krupp, Energetic electrons injected into Saturn's neutral gas cloud. Geophys. Res. Lett. **34**, 2109 (2007). doi:10.1029/2006GL028676

D.H. Pontius Jr., T.W. Hill, Rotation driven plasma transport—the coupling of macroscopic motion and microdiffusion. J. Geophys. Res. **94**, 15041–15053 (1989). doi:10.1029/JA094iA11p15041

D.H. Pontius Jr., T.W. Hill, M.E. Rassbach, Steady state plasma transport in a corotation-dominated magnetosphere. Geophys. Res. Lett. **13**, 1097–1100 (1986). doi:10.1029/GL013i011p01097

D.H. Pontius, T.W. Hill, Plasma mass loading from the extended neutral gas torus of Enceladus as inferred from the observed plasma corotation lag. Geophys. Res. Lett. **36**, 23103 (2009). doi:10.1029/2009GL041030

D.H. Pontius, R.A. Wolf, T.W. Hill, R.W. Spiro, Y.S. Yang, W.H. Smyth, Velocity shear impoundment of the Io plasma torus. J. Geophys. Res. **103**, 19935–19946 (1998). doi:10.1029/98JE00538

W.R. Pryor, A.M. Rymer, D.G. Mitchell, T.W. Hill, D.T. Young, J. Saur, G.H. Jones, S. Jacobsen, S.W.H. Cowley, B.H. Mauk, A.J. Coates, J. Gustin, D. Grodent, J.-C. Gérard, L. Lamy, J.D. Nichols, S.M. Krimigis, L.W. Esposito, M.K. Dougherty, A.J. Jouchoux, A.I.F. Stewart, W.E. McClintock, G.M. Holsclaw, J.M. Ajello, J.E. Colwell, A.R. Hendrix, F.J. Crary, J.T. Clarke, X. Zhou, The auroral footprint of Enceladus on Saturn. Nature **472**, 331–333 (2011). doi:10.1038/nature09928

M. Rodriguez-Martinez, X. Blanco-Cano, C. Russell, J.S. Leisner, M.M. Cowee, M.K. Dougherty, Harmonic growth of ion cyclotron waves in Saturn's Magnetosphere, in *37th COSPAR Scientific Assembly*. COSPAR Meeting, vol. 37 (2008), p. 2638

M. Rodríguez-Martínez, X. Blanco-Cano, C.T. Russell, J.S. Leisner, R.J. Wilson, M.K. Dougherty, Harmonic growth of ion-cyclotron waves in Saturn's magnetosphere. J. Geophys. Res. **115**, 9207 (2010). doi:10.1029/2009JA015000

A. Runov, V. Angelopoulos, X.-Z. Zhou, X.-J. Zhang, S. Li, F. Plaschke, J. Bonnell, A THEMIS multi-case study of dipolarization fronts in the magnetotail plasma sheet. J. Geophys. Res. **116**, 5216 (2011). doi:10.1029/2010JA016316

C.T. Russell, D.E. Huddleston, Ion-cyclotron waves at Io. Adv. Space Res. **26**, 1505–1511 (2000)

C.T. Russell, M.G. Kivelson, Detection of SO in Io's exosphere. Science **287**, 1998–1999 (2000). doi:10.1126/science.287.5460.1998

C.T. Russell, D.E. Huddleston, R.J. Strangeway, X. Blanco-Cano, M.G. Kivelson, K.K. Khurana, L.A. Frank, W. Paterson, D.A. Gurnett, W.S. Kurth, Mirror-mode structures at the Galileo-Io flyby: Observations. J. Geophys. Res. **104**, 17471–17478 (1999). doi:10.1029/1999JA900202

C.T. Russell, M.G. Kivelson, K.K. Khurana, D.E. Huddleston, Circulation and dynamics in the Jovian magnetosphere. Adv. Space Res. **26**, 1671–1676 (2000). doi:10.1016/S0273-1177(00)00115-0

C.T. Russell, Y.L. Wang, X. Blanco-Cano, R.J. Strangeway, The Io mass-loading disk: Constraints provided by ion cyclotron wave observations. J. Geophys. Res. **106**, 26233–26242 (2001). doi:10.1029/2001JA900029

C.T. Russell, X. Blanco-Cano, M.G. Kivelson, Ion cyclotron waves in Io's wake region. Planet. Space Sci. **51**, 233–238 (2003a). doi:10.1016/S0032-0633(02)00198-8

C.T. Russell, X. Blanco-Cano, Y.L. Wang, M.G. Kivelson, Ion cyclotron waves at Io: implications for the temporal variation of Io's atmosphere. Planet. Space Sci. **51**, 937–944 (2003b). doi:10.1016/j.pss.2003.05.005

C.T. Russell, M.G. Kivelson, K.K. Khurana, Statistics of depleted flux tubes in the Jovian magnetosphere. Planet. Space Sci. **53**, 937–943 (2005). doi:10.1016/j.pss.2005.04.007

C.T. Russell, J.S. Leisner, C.S. Arridge, M.K. Dougherty, X. Blanco-Cano, Nature of magnetic fluctuations in Saturn's middle magnetosphere. J. Geophys. Res. **111**, 12205 (2006). doi:10.1029/2006JA011921

A.M. Rymer, B.H. Mauk, T.W. Hill, C. Paranicas, N. André, E.C. Sittler, D.G. Mitchell, H.T. Smith, R.E. Johnson, A.J. Coates, D.T. Young, S.J. Bolton, M.F. Thomsen, M.K. Dougherty, Electron sources in Saturn's magnetosphere. J. Geophys. Res. **112**, 2201 (2007). doi:10.1029/2006JA012017

A.M. Rymer, B.H. Mauk, T.W. Hill, C. Paranicas, D.G. Mitchell, A.J. Coates, D.T. Young, Electron circulation in Saturn's magnetosphere. J. Geophys. Res. **113**, 1201 (2008). doi:10.1029/2007JA012589

A.M. Rymer, B.H. Mauk, T.W. Hill, N. André, D.G. Mitchell, C. Paranicas, D.T. Young, H.T. Smith, A.M. Persoon, J.D. Menietti, G.B. Hospodarsky, A.J. Coates, M.K. Dougherty, Cassini evidence for rapid interchange transport at Saturn. Planet. Space Sci. **57**, 1779–1784 (2009). doi:10.1016/j.pss.2009.04.010

J. Saur, D.F. Strobel, F.M. Neubauer, Interaction of the Jovian magnetosphere with Europa: Constraints on the neutral atmosphere. J. Geophys. Res. **103**, 19947–19962 (1998). doi:10.1029/97JE03556

J. Saur, F.M. Neubauer, D.F. Strobel, M.E. Summers, Three-dimensional plasma simulation of Io's interaction with the Io plasma torus: Asymmetric plasma flow. J. Geophys. Res. **104**, 25105–25126 (1999). doi:10.1029/1999JA900304

J. Saur, F.M. Neubauer, D.F. Strobel, M.E. Summers, Interpretation of Galileo's Io plasma and field observations: I0, I24, and I27 flybys and close polar passes. J. Geophys. Res. **107**, 1422 (2002). doi:10.1029/2001JA005067

J. Saur, F.M. Neubauer, J.E.P. Connerney, P. Zarka, M.G. Kivelson, Plasma interaction of Io with its plasma torus, in *Jupiter: The Planet, Satellites and Magnetosphere* (2004), pp. 537–560

P. Schippers, M. Blanc, N. André, I. Dandouras, G.R. Lewis, L.K. Gilbert, A.M. Persoon, N. Krupp, D.A. Gurnett, A.J. Coates, S.M. Krimigis, D.T. Young, M.K. Dougherty, Multi-instrument analysis of electron populations in Saturn's magnetosphere. J. Geophys. Res. **113**, 7208 (2008). doi:10.1029/2008JA013098

M. Schulz, Jupiter's radiation belts. Space Sci. Rev. **23**, 277–318 (1979). doi:10.1007/BF00173813

N. Sergis, S.M. Krimigis, D.G. Mitchell, D.C. Hamilton, N. Krupp, B.M. Mauk, E.C. Roelof, M. Dougherty, Ring current at Saturn: Energetic particle pressure in Saturn's equatorial magnetosphere measured with Cassini/MIMI. Geophys. Res. Lett. **34**, 09102 (2007). doi:10.1029/2006GL029223

D.E. Shemansky, Energy branching in the Io plasma torus: The failure of neutral cloud theory. J. Geophys. Res. **93**, 1773 (1988)

S. Simon, G. Kleindienst, A. Boesswetter, T. Bagdonat, U. Motschmann, K.-H. Glassmeier, J. Schuele, C. Bertucci, M.K. Dougherty, Hybrid simulation of Titan's magnetic field signature during the Cassini T9 flyby. Geophys. Res. Lett. **34**, 24 (2007). doi:10.1029/2007GL029967

G.L. Siscoe, A. Eviatar, R.M. Thorne, J.D. Richardson, F. Bagenal, J.D. Sullivan, Ring current impoundment of the Io plasma torus. J. Geophys. Res. **86**, 8480–8484 (1981). doi:10.1029/JA086iA10p08480

E.C. Sittler, M. Thomsen, R.E. Johnson, R.E. Hartle, M. Burger, D. Chornay, M.D. Shappirio, D. Simpson, H.T. Smith, A.J. Coates, A.M. Rymer, D.J. McComas, D.T. Young, D. Reisenfeld, M. Dougherty, N. Andre, Erratum to "Cassini observations of Saturn's inner plasmasphere: Saturn orbit insertion results". [Planetary and Space Science 54 (2006) 1197–1210]. Planet. Space Sci. **55**, 2218–2220 (2007). doi:10.1016/j.pss.2006.11.022

E.C. Sittler, N. Andre, M. Blanc, M. Burger, R.E. Johnson, A. Coates, A. Rymer, D. Reisenfeld, M.F. Thomsen, A. Persoon, M. Dougherty, H.T. Smith, R.A. Baragiola, R.E. Hartle, D. Chornay, M.D. Shappirio, D. Simpson, D.J. McComas, D.T. Young, Ion and neutral sources and sinks within Saturn's inner magnetosphere: Cassini results. Planet. Space Sci. **56**, 3–18 (2008). doi:10.1016/j.pss.2007.06.006

T.E. Skinner, S.T. Durrance, Neutral oxygen and sulfur densities in the Io torus. Astrophys. J. **310**, 966–971 (1986). doi:10.1086/164747

C.G.A. Smith, A.D. Aylward, Coupled rotational dynamics of Jupiter's thermosphere and magnetosphere. Ann. Geophys. **27**, 199–230 (2009)

E.J. Smith, B.T. Tsurutani, Saturn's magnetosphere—observations of ion cyclotron waves near the Dione L shell. J. Geophys. Res. **88**, 7831–7836 (1983). doi:10.1029/JA088iA10p07831

W.H. Smyth, M.L. Marconi, Nature of the Iogenic plasma source in Jupiter's magnetosphere I. Circumplanetary distribution. Icarus **166**(1), 85–106 (2003)

D. Snowden, R. Winglee, C. Bertucci, M. Dougherty, Three-dimensional multifluid simulation of the plasma interaction at Titan. J. Geophys. Res. **112**, 12221 (2007). doi:10.1029/2007JA012393

D. Snowden, R. Winglee, A. Kidder, Titan at the edge: 1. Titan's interaction with Saturn's magnetosphere in the prenoon sector. J. Geophys. Res. **116**, 8229 (2011a). doi:10.1029/2011JA016435

D. Snowden, R. Winglee, A. Kidder, Titan at the edge: 2. A global simulation of Titan exiting and reentering Saturn's magnetosphere at 13:16 Saturn local time. J. Geophys. Res. **116**, 8230 (2011b). doi:10.1029/2011JA016436

D.J. Southwood, M.G. Kivelson, Magnetospheric interchange instability. J. Geophys. Res. **92**, 109–116 (1987). doi:10.1029/JA092iA01p00109

D.J. Southwood, M.G. Kivelson, Magnetospheric interchange motions. J. Geophys. Res. **94**, 299–308 (1989)

D.J. Southwood, M.G. Kivelson, A new perspective concerning the influence of the solar wind on the Jovian magnetosphere. J. Geophys. Res. **106**, 6123–6130 (2001). doi:10.1029/2000JA000236

K. Szego, Z. Bebesi, G. Erdos, L. Foldy, F. Crary, D.J. McComas, D.T. Young, S. Bolton, A.J. Coates, A.M. Rymer, R.E. Hartle, E.C. Sittler, D. Reisenfeld, J.J. Bethelier, R.E. Johnson, H.T. Smith, T.W. Hill, J. Vilppola, J. Steinberg, N. Andre, The global plasma environment of Titan as observed by Cassini Plasma Spectrometer during the first two close encounters with Titan. Geophys. Res. Lett. **32**, 20 (2005). doi:10.1029/2005GL022646

N. Thomas, F. Bagenal, T.W. Hill, J.K. Wilson, The Io neutral clouds and plasma torus, in *Jupiter. The Planet, Satellites and Magnetosphere*, ed. by F. Bagenal, T.E. Dowling, W.B. McKinnon (2004), pp. 561–591

M.F. Thomsen, D.B. Reisenfeld, D.M. Delapp, R.L. Tokar, D.T. Young, F.J. Crary, E.C. Sittler, M.A. Mc-Graw, J.D. Williams, Survey of ion plasma parameters in Saturn's magnetosphere. J. Geophys. Res. **115**, 10220 (2010). doi:10.1029/2010JA015267

M.F. Thomsen, E. Roussos, M. Andriopoulou, P. Kollmann, C.S. Arridge, C.P. Paranicas, D.A. Gurnett, R.L. Powell, R.L. Tokar, D.T. Young, Saturn's inner magnetospheric convection pattern: Further evidence. J. Geophys. Res. **117**, 9208 (2012). doi:10.1029/2011JA017482

R.M. Thorne, Radiation belt dynamics: The importance of wave-particle interactions. Geophys. Res. Lett. **37**, 22107 (2010). doi:10.1029/2010GL044990

R.M. Thorne, T.P. Armstrong, S. Stone, D.J. Williams, R.W. McEntire, S.J. Bolton, D.A. Gurnett, M.G. Kivelson, Galileo evidence for rapid interchange transport in the Io torus. Geophys. Res. Lett. **24**, 2131 (1997). doi:10.1029/97GL01788

R.L. Tokar, R.E. Johnson, T.W. Hill, D.H. Pontius, W.S. Kurth, F.J. Crary, D.T. Young, M.F. Thomsen, D.B. Reisenfeld, A.J. Coates, G.R. Lewis, E.C. Sittler, D.A. Gurnett, The interaction of the atmosphere of Enceladus with Saturn's plasma. Science **311**, 1409–1412 (2006). doi:10.1126/science.1121061

R.L. Tokar, R.J. Wilson, R.E. Johnson, M.G. Henderson, M.F. Thomsen, M.M. Cowee, E.C. Sittler, D.T. Young, F.J. Crary, H.J. McAndrews, H.T. Smith, Cassini detection of water-group pick-up ions in the Enceladus torus. Geophys. Res. Lett. **35**, 14202 (2008). doi:10.1029/2008GL034749

V.M. Vasyliūnas, Mathematical models of magnetospheric convection and its coupling to the ionosphere, in *Particles and Field in the Magnetosphere*, ed. by B.M. McCormack, A. Renzini. Astrophysics and Space Science Library, vol. 17 (1970), p. 60

V.M. Vasyliūnas, Plasma distribution and flow, in *Physics of the Jovian Magnetosphere*, ed. by A.J. Dessler (Cambridge University Press, New York, 1983), pp. 395–453. ISBN 0521520061 (paperback)

V.M. Vasyliūnas, Physical origin of pickup currents, in *European Planetary Science Congress* (2006)

V.M. Vasyliūnas, Comparing Jupiter and Saturn: dimensionless input rates from plasma sources within the magnetosphere. Ann. Geophys. **26**, 1341–1343 (2008)

V.M. Vasyliūnas, D.H. Pontius, Rotationally driven interchange instability: Reply to André and Ferrière. J. Geophys. Res. **112**(A10), A10204 (2007). doi:10.1029/2007JA012457

M. Volwerk, K.K. Khurana, Ion pick-up near the icy Galilean satellites, in *American Institute of Physics Conference Series*, ed. by J. Le Roux, G.P. Zank, A.J. Coates, V. Florinski. American Institute of Physics Conference Series, vol. 1302 (2010), pp. 263–269. doi:10.1063/1.3529982

M. Volwerk, M.G. Kivelson, K.K. Khurana, Wave activity in Europa's wake: Implications for ion pickup. J. Geophys. Res. **106**, 26033–26048 (2001). doi:10.1029/2000JA000347

J.H. Waite, M.R. Combi, W.-H. Ip, T.E. Cravens, R.L. McNutt, W. Kasprzak, R. Yelle, J. Luhmann, H. Niemann, D. Gell, B. Magee, G. Fletcher, J. Lunine, W.-L. Tseng, Cassini ion and neutral mass spectrometer: Enceladus plume composition and structure. Science **311**, 1419–1422 (2006). doi:10.1126/science.1121290

M. Walt, *Introduction to Geomagnetically Trapped Radiation*. Cambridge Atmospheric and Space Science Series, vol. 10 (1994)

Y. Wang, C.T. Russell, J. Raeder, The Io mass-loading disk: Model calculations. J. Geophys. Res. **106**, 26243–26260 (2001). doi:10.1029/2001JA900062

J. Warnecke, M.G. Kivelson, K.K. Khurana, D.E. Huddleston, C.T. Russell, Ion cyclotron waves observed at Galileo's Io encounter: Implications for neutral cloud distribution and plasma composition. Geophys. Res. Lett. **24**, 2139 (1997). doi:10.1029/97GL01129

D.J. Williams, B. Mauk, R.E. McEntire, E.C. Roelof, S.M. Krimigis, T.P. Armstrong, B. Wilken, J.G. Roederer, T.A. Fritz, L.J. Lanzerotti, Energetic electron beams measured at Io. Bull. Am. Astron. Soc. **28**, 1055 (1996)

R.J. Wilson, R.L. Tokar, M.G. Henderson, T.W. Hill, M.F. Thomsen, D.H. Pontius, Cassini plasma spectrometer thermal ion measurements in Saturn's inner magnetosphere. J. Geophys. Res. **113**, 12218 (2008). doi:10.1029/2008JA013486

R.M. Winglee, D. Snowden, A. Kidder, Modification of Titan's ion tail and the Kronian magnetosphere: Coupled magnetospheric simulations. J. Geophys. Res. **114**, 5215 (2009). doi:10.1029/2008JA013343

R.M. Winglee, A. Kidder, E. Harnett, N. Ifland, C. Paty, D. Snowden, Generation of periodic signatures at Saturn through Titan's interaction with the centrifugal interchange instability. J. Geophys. Res. **118**, 4253–4269 (2013). doi:10.1002/jgra.50397

R.A. Wolf, Computer model of inner magnetospheric convection, in *Solar-Terrestrial Physics: Principles and Theoretical Foundations*, ed. by R.L. Carovillano, J.M. Forbes. Astrophysics and Space Science Library, vol. 104 (1983), p. 342

M.C. Wong, W.H. Smyth, Model calculations for Io's atmosphere at Eastern and Western elongations. Icarus **146**, 60–74 (2000). doi:10.1006/icar.2000.6362

H. Wu, T.W. Hill, R.A. Wolf, R.W. Spiro, Numerical simulation of fine structure in the Io plasma torus produced by the centrifugal interchange instability. J. Geophys. Res. **112**, 2206 (2007). doi:10.1029/2006JA012032

Y.S. Yang, R.A. Wolf, R.W. Spiro, A.J. Dessler, Numerical simulation of plasma transport driven by the Io torus. Geophys. Res. Lett. **19**, 957–960 (1992). doi:10.1029/92GL01031

Y.S. Yang, R.A. Wolf, R.W. Spiro, T.W. Hill, A.J. Dessler, Numerical simulation of torus-driven plasma transport in the Jovian magnetosphere. J. Geophys. Res. **99**, 8755–8770 (1994). doi:10.1029/94JA00142

J.N. Yates, N. Achilleos, P. Guio, Influence of upstream solar wind on thermospheric flows at Jupiter. Planet. Space Sci. **61**, 15–31 (2012). doi:10.1016/j.pss.2011.08.007

D.T. Young, J.J. Berthelier, M. Blanc, J.L. Burch, A.J. Coates, R. Goldstein, M. Grande, T.W. Hill, R.E. Johnson, V. Kelha, D.J. McComas, E.C. Sittler, K.R. Svenes, K. Szegö, P. Tanskanen, K. Ahola, D. Anderson, S. Bakshi, R.A. Baragiola, B.L. Barraclough, R.K. Black, S. Bolton, T. Booker, R. Bowman, P. Casey, F.J. Crary, D. Delapp, G. Dirks, N. Eaker, H. Funsten, J.D. Furman, J.T. Gosling, H. Hannula, C. Holmlund, H. Huomo, J.M. Illiano, P. Jensen, M.A. Johnson, D.R. Linder, T. Luntama, S. Maurice, K.P. McCabe, K. Mursula, B.T. Narheim, J.E. Nordholt, A. Preece, J. Rudzki, A. Ruitberg, K. Smith, S. Szalai, M.F. Thomsen, K. Viherkanto, J. Vilppola, T. Vollmer, T.E. Wahl, M. Wüest, T. Ylikorpi, C. Zinsmeyer, Cassini plasma spectrometer investigation. Space Sci. Rev. **114**, 1–4 (2004). doi:10.1007/s11214-004-1406-4

B. Zieger, K.C. Hansen, T.I. Gombosi, D.L. De Zeeuw, Periodic plasma escape from the mass-loaded Kronian magnetosphere. J. Geophys. Res. **115**, 8208 (2010). doi:10.1029/2009JA014951

DOI 10.1007/978-1-4939-3395-2_8
Reprinted from *Space Science Reviews* Journal, DOI 10.1007/s11214-015-0145-z

Sources of Local Time Asymmetries in Magnetodiscs

C.S. Arridge[1,2,3] · M. Kane[4] · N. Sergis[5] ·
K.K. Khurana[6] · C.M. Jackman[7]

Received: 27 September 2014 / Accepted: 11 March 2015 / Published online: 8 April 2015
© Springer Science+Business Media Dordrecht 2015

Abstract The rapidly rotating magnetospheres at Jupiter and Saturn contain a near-equatorial thin current sheet over most local times known as the magnetodisc, resembling a wrapped-up magnetotail. The Pioneer, Voyager, Ulysses, Galileo, Cassini and New Horizons spacecraft at Jupiter and Saturn have provided extensive datasets from which to observationally identify local time asymmetries in these magnetodiscs. Imaging in the infrared and ultraviolet from ground- and space-based instruments have also revealed the presence of local time asymmetries in the aurora which therefore must map to local time asymmetries in the magnetosphere. Asymmetries are found in (i) the configuration of the magnetic field and magnetospheric currents, where a thicker disc is found in the noon and dusk sectors; (ii) plasma flows where the plasma flow has local time-dependent radial components; (iii) a thicker plasma sheet in the dusk sector. Many of these features are also reproduced in global MHD simulations. Several models have been developed to interpret these various observations and typically fall into two groups: ones which invoke coupling with the solar wind (via reconnection or viscous processes) and ones which invoke internal rotational processes operating inside an asymmetrical external boundary. In this paper we review these observational in situ findings, review the models which seek to explain them, and highlight open questions and directions for future work.

Keywords Saturn · Jupiter · Magnetodisc · Local time asymmetry · Plasma flows · Current sheet · Solar wind interaction · Reconnection · Kelvin–Helmholtz

✉ C.S. Arridge
c.arridge@lancaster.ac.uk

[1] Mullard Space Science Laboratory, University College London, Holmbury St. Mary, Dorking, Surrey, RH5 6NT, UK

[2] The Centre for Planetary Sciences at UCL/Birkbeck, Gower Street, London, WC1E 6BT, UK

[3] Department of Physics, Lancaster University, Lancaster, LA1 4YB, UK

[4] Harford Research Institute, 1411 Saratoga Dr, Bel Air, MD 21014, USA

[5] Office of Space Research and Technology, Academy of Athens, Athens, Greece

[6] Institute of Geophysics and Planetary Physics, University of California, Los Angeles, CA, USA

[7] Department of Physics and Astronomy, University of Southampton, Southampton, UK

1 Introduction

The terrestrial magnetotail contains a thin current sheet separating the tail into two lobes with the current running from dawn to dusk, closing on the magnetopause. The solar wind provides the confining stress which produces a noon-midnight asymmetry in the terrestrial magnetosphere. One of the most remarkable magnetospheric findings made at Jupiter by Pioneer 10/11 and Voyager 1/2 in the 1970s was of a disc-like configuration of the magnetic field and plasma, now known as the magnetodisc (e.g., Goertz 1979, and references therein). This resembled Earth's magnetotail current sheet except it was found at all local times (midnight-to-noon sector) sampled by these spacecraft. Subsequent missions, such as Ulysses and Galileo, have studied this magnetodisc in more detail and at a wider range of local times. Further analysis of these data has also revealed the presence of local-time asymmetries. The disc is not an azimuthally uniform current sheet, but is stronger and weaker in different local time sectors. At Saturn, such a structure was not detected on the dayside at Saturn in Pioneer 11 and Voyager 1/2 data, but some evidence of such a thin current sheet was found on the dawn flank (Smith et al. 1980). Observations made by Cassini at Saturn have subsequently found evidence for this disc-like configuration in the noon, dawn and midnight sectors (Arridge et al. 2008b).

The magnetic field in the magnetodiscs of Jupiter and Saturn has both poloidal (B_r, B_θ in spherical polar coordinates) and toroidal (B_φ) components associated with azimuthal and radial currents through the current sheet. The azimuthal currents are associated with radial stress balance in the magnetodisc, between magnetic tension and centrifugal, thermal and magnetic pressure gradient and pressure anisotropy forces. The radial currents are associated with azimuthal stress balance where ion-neutral collisions in the planet's atmosphere exert an azimuthal torque which is transmitted via field-aligned currents to the equatorial plasma, and attempts to maintain the outflowing equatorial plasma in corotation with the planet. Hence, the toroidal component of the field is implicated in the outward transport of plasma and angular momentum conservation, although the toroidal component can weaken in response to solar wind compression (e.g., Hanlon et al. 2004). The presence of a toroidal component, which reverses in direction about the centre of the current sheet and hence does not take the form of a guide field as in the case of the Earth's magnetotail, produces a swept-back or bent-back configuration in the magnetic field, similar to an Archimedean spiral, but where the degree of bend-back is a function of local-time (e.g., Bunce et al. 2003). The geometry of the current sheet at both Jupiter and Saturn is not simply planar and collocated with the dipole magnetic equators of these planets. The location of the surface is deformed by the solar wind, internal disturbances and other periodic mechanisms. As such, the current sheet is time-dependent and three-dimensional.

The observed diamagnetic depression in the current sheet is evidence of the presence of significant particle pressure at both Jupiter and Saturn. Due to rapid azimuthal motion of the magnetodisc, centrifugal forces affect the meridional distribution of the particles. The centrifugal scale height for particles of mass, m, and thermal energy, $k_B T$, rotating at an angular velocity, ω, in a dipole field is proportional to $\sqrt{k_B T / m\omega^2}$ (Hill and Michel 1976). Hence, heavy cold magnetospheric ions are centrifugally confined to the centre of the plasma sheet, although this picture is modified in the presence of multiple ion and electron populations and ambipolar electric fields (e.g., Maurice et al. 1997). Since the thermal energy of energetic particles is much larger than "thermal" populations they have much larger scale heights and are effectively free to fill the field lines. However, due to the disc-like geometry, the sheet has a finite width normal to the current sheet, even for the energetic particles. The sheet thickness has an observed local-time asymmetry at both Jupiter and Saturn. Particle flows have significant local-time asymmetries at Jupiter (Krupp et al. 2001) and Saturn (Kane et al. 2014).

In a steady state, the magnetic stress associated with the stretched-out field lines in the magnetodisc is balanced by mechanical stresses in the plasma and energetic particles in the disc. The tension force associated with stretched magnetodisc field lines points inwards and mechanical stresses include centrifugal stress (outwards), pressure gradient forces (typically outwards), and pressure anisotropy (inward or outward). The principal mechanical stress at Jupiter appears to be the anisotropy force of the energetic heavy ions where $P_\parallel > P_\perp$, by a factor of 1.1–1.2, at least on the nightside (Paranicas et al. 1991) although there is some evidence of increasing importance of centrifugal forces at larger distances (Arridge 2011). At Saturn, pressure gradient forces are important in the middle magnetosphere in the region where the field makes the transition to a disc-like configuration, while anisotropy and centrifugal forces are seen to be important in the inner magnetosphere (Sergis et al. 2010; Kellett et al. 2010). In the magnetodisc there is evidence that the centrifugal force may be the dominant mechanical stress (Arridge et al. 2007), but a full particle analysis similar to Sergis et al. (2010) has not yet been carried out.

In this review we consider these aspects of the magnetodisc and examine the evidence for local-time asymmetries. Specifically, we look at local-time asymmetries in the geometry of the magnetic field, currents, and how divergences in radial and azimuthal currents are closed; the geometry of the current sheet with local-time; and plasma flows and particle populations. Comparisons with the results from global MHD simulations will be mentioned where appropriate. We close this review by considering the physical mechanisms underlying these asymmetries and posing outstanding questions.

2 Field Geometry, Current Sheets and Current Closure

2.1 Magnetic Field Geometry at Jupiter

Figure 1 shows magnetometer data from four different local time sectors in the Jovian magnetosphere as measured by the Galileo spacecraft. The data are presented in a spherical polar coordinate system where the radial direction ($\mathbf{e}_r$) points away from the planet along a line connecting the spacecraft and the planet, the polar direction ($\mathbf{e}_\theta$) points in the direction of increasing colatitude with respect to the north rotational pole, and the azimuthal direction ($\mathbf{e}_\varphi$) is in a prograde direction around the planet (i.e. points in the local direction of planetary corotation). In the dawn sector the B_φ and B_r components regularly change sign as the magnetodisc current sheet moves up and down over the spacecraft with each rotation of Jupiter. The centre of the current sheet is located close to the region where B_φ and B_r change sign but is not exactly located at that point due to the difference between the coordinate frame in which the magnetic field is measured and the local coordinate frame of the current sheet (e.g., Vasyliūnas 1983; Khurana 2001; Jackman et al. 2009). Regions where B_φ and B_r approach asymptotic values and fluctuations in the field reach a minimum are the lobe-type regions adjacent to the current sheet. Such a morphology is also seen in the midnight sector. On the dawn sector B_φ is large (~50 % of the B_r component) and out of phase with B_r such that B_φ and B_r have opposite signs producing a "swept-back" configuration to the magnetic field. In the midnight sector there is still an anti-phase relationship, but B_φ is somewhat smaller in magnitude. The B_θ component of the field is approximately normal to the current sheet and is small but positive, indicating closed field lines in a thin current sheet geometry. B_θ varies very little as

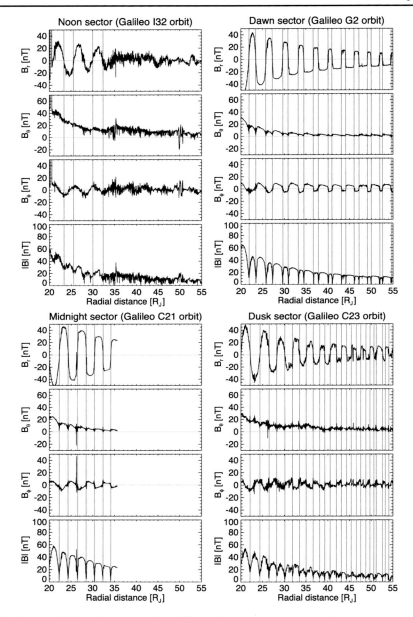

Fig. 1 Magnetic field configuration in four different local time sectors in the Jovian magnetosphere, as measured by the Galileo spacecraft. The data are presented in spherical polar (radial-theta-phi) coordinates, plus the field magnitude, as a function of radial distance. *Vertical grey lines* indicate current sheet crossings or contact with the current sheet

the spacecraft moves relative to the current sheet. From the divergenceless of B, the polar gradient of B_θ is related to the radial and azimuthal gradients of B_r and B_φ respectively. Hence, the fact that B_θ is effectively constant with the motion of the current sheet indicates fairly weak radial and azimuthal gradients in the disc. Small oscillations in the magnitude

Fig. 2 Equatorial projection of current sheet (B_ρ and B_φ components) magnetic field unit vectors to indicate the sweep-forward and sweep-back of the current sheet as a function of local time. Average magnetopause and bow shock locations are indicated. From Khurana (2001)

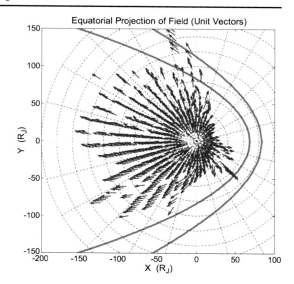

of B_θ are associated with motions of the current sheet and are also seen in data from Saturn (e.g., Jackman et al. 2009). The field magnitude is seen to dip at current sheet crossings and is associated with the diamagnetic effect of the plasma in the magnetodisc.

In the dusk sector B_θ is larger in magnitude (but generally still positive) and the noise in the B_r component indicates that the spacecraft never entirely leaves the plasma sheet. This is indicative of the presence of a thicker plasma/current sheet in the dusk sector. We also notice additional oscillations in the B_r component showing that the current sheet is somewhat "sloppy" and is being deformed by additional processes. B_φ has a structure that includes both swept-back and swept-forward (B_φ and B_r have the same sign) configurations. In the noon sector the field shows similar evidence of a thick current/plasma sheet but which, for this pass, terminates around 32 R_J from Jupiter. After the end of the current sheet the spacecraft is in a region of fluctuating magnetic fields and very little evidence of the periodicity imposed by the rotation of the Jovian dipole. This region is known as the "cushion" region and is a persistent feature of the noon magnetosphere at Jupiter (Went et al. 2011). Hence, the magnetic field data demonstrate a considerable local-time asymmetry in the magnetodisc field.

Figure 2 presents the tangential (B_φ and B_r) component of the field as unit vectors projected onto the equatorial plane as a function of local time. One can clearly see the strongly swept-back configuration on the dawn flank and in the post-midnight sector. But in the pre-noon through to pre-midnight sectors the field often points away from the planet or is weakly swept-forward/swept-back, and hence the sign of B_φ has a distinct local time asymmetry. The source of this asymmetry has been the source of controversy and research for more than 30 years (Vasyliūnas 1983). The presence of B_φ and its role in azimuthal stress balance in accelerating plasma back up to corotation is one interpretation. Another interpretation (see Vasyliūnas 1983, and references therein) is that B_φ is associated with the solar wind via either normal or tangential stress. These two interpretations predict opposite sweep-back configurations on the dusk and dawn sectors—with B_φ reversing sign across the noon-midnight meridian. One can see from Fig. 2 that this is not observed and the field still has a swept-back configuration in the pre-midnight sector.

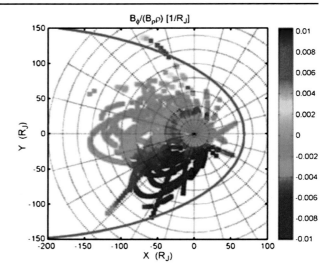

Fig. 3 The observed ratio $B_\varphi/\rho B_\rho$ in the Jovian magnetosphere calculated from all the available spacecraft data (Pioneer, Voyager, Ulysses, Galileo). The calculated ratios were binned into $10 \times 10\ R_J^2$ bins. From Khurana and Schwarzl (2005)

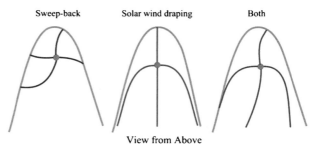

Fig. 4 Sketches of the spiral field configuration for an outflow-type interaction (*left*), solar wind interaction (*middle*), and a combined interaction (*right*) as inferred from the data and Figs. 2, 3. From Khurana (2001)

The quantity $B_\varphi/\rho B_\rho$ or $B_\varphi/r B_r$, in cylindrical polar or spherical polar coordinates respectively, is often used to quantify the sweep-back in the field and this is plotted for Jupiter in Fig. 3, from Khurana and Schwarzl (2005). Within 10 R_J of Jupiter this ratio is close to zero, but becomes increasingly negative at larger distances over most local time sectors of the magnetosphere. On the dawn flank the ratio reaches a plateau whereas on the dusk sector beyond $\sim$60 R_J the sign reverses indicating a swept-forward configuration, as could be seen in Fig. 2. Therefore, the simple picture of outflow, angular momentum conservation and magnetosphere-ionosphere coupling to enforce corotation is not valid for the whole of the Jovian magnetosphere. Figure 4 illustrates the spiral configuration of the field lines for the outflow picture, with swept-back configuration at all local times, the solar wind influence with sweep-back at dawn and sweep-forward at dusk, and the combination of the two (Khurana 2001). This latter configuration best matches the observational data in Figs. 2 and 3 and suggests a role for the solar wind in modifying the outflow picture to introduce local time asymmetries (see Sect. 5 for further discussion).

2.2 Jovian Radial and Azimuthal Currents

Bunce and Cowley (2001a) presented the first extensive study of local-time asymmetries in the radial fields at Jupiter using data from Voyager, Pioneer and Ulysses, leading to a description of the divergence of the azimuthal currents. Within a radial distance of $\sim$20 R_J

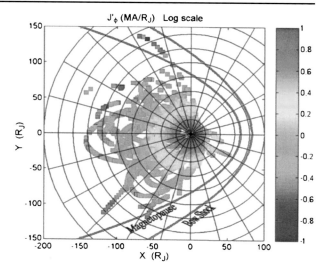

Fig. 5 Jovian azimuthal currents integrated vertically through the current sheet and inferred from magnetometer data. From Khurana (2001)

the radial field, and hence the azimuthal currents, were approximately symmetrical in local time. However, beyond that distance the current sheet field fell with radial distance more rapidly on the dayside compared to the nightside. This leads to current sheet fields that are systematically weaker on the dayside compared to the nightside at the same radial distance. However, because they only analysed the radial field and hence calculated the azimuthal component of the divergence of **j** they could not determine if the azimuthal currents were diverted into radial or field-aligned currents. Bunce and Cowley (2001b) took the next step and analysed the azimuthal field, and hence the radial currents, and therefore calculated both the radial and azimuthal terms in $\nabla \cdot \mathbf{j}$ and could estimate the field-aligned current. They found significant field-aligned currents were required and that these currents flowed into the ionosphere in the pre-noon sector, and out of the ionosphere into the magnetodisc in the post-noon sector.

Khurana (2001) also analysed similar data, but also included data from Galileo to comprehensively survey the radial and azimuthal currents as a function of local time and radial distance. Figures 5 and 6 from Khurana (2001) show the calculated azimuthal and radial currents. Between 10 and 50 R_J the azimuthal currents were much stronger on the nightside at 144 MA compared with 88 MA on the dayside. Stronger radial currents were found on the dawn flank and weaker currents at dusk and noon. These asymmetries in the current sheet can be matched with the strong current sheets found at dawn and midnight in the magnetometer data in Fig. 1, and the strong sweep-back found on the dawn flank in Fig. 2.

Khurana (2001) also calculated the divergence and Fig. 7 shows the divergence of the perpendicular current (azimuthal and radial terms), where negative (blue) indicates current is drawn out of the magnetodisc into field-aligned currents, while positive (red) indicates that current is fed into the magnetodisc via field-aligned currents. Except in the dusk sector, $\nabla \cdot \mathbf{j}_\perp$ is positive between 10 and 30 R_J, indicating current is being fed into the magnetodisc from the ionosphere. These currents are associated with the radial current system associated with corotation enforcement. Torque exerted on the plasma by ion-neutral collisions in the ionosphere is transmitted to the equatorial plasma via field-aligned currents which feed a radial current system flowing outwards from Jupiter through the magnetodisc. These currents exert an azimuthal $\mathbf{j} \times \mathbf{B}$ force to enforce (partial) corotation. The upward field-aligned currents at the inner edge of the magnetodisc (positive divergence inside 340 R_J) naturally

Fig. 6 Jovian radial currents integrated vertically through the current sheet and inferred from magnetometer data. From Khurana (2001)

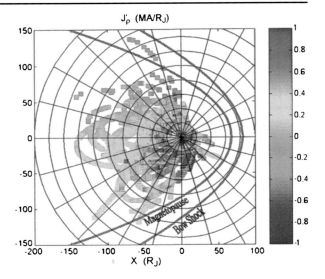

Fig. 7 The divergence of the height integrated current in the magnetodisc. From Khurana (2001)

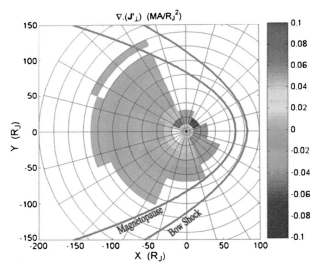

require downward electron precipitation into the ionosphere and this is associated with the Jovian main auroral oval. Beyond 30 R_J no current was added to the magnetodisc. In the noon sector of the magnetosphere, the divergence was negative or close to zero indicating current is drawn out of the magnetodisc. From the divergence of the azimuthal current, there is a strong noon-midnight asymmetry in the azimuthal current associated with a more dipolar dayside and more stretched current sheet at midnight and dawn. This is similar to a partial ring current at Earth, except that it is opposite in sense due to the fact that Earth's ring current flows in the opposite direction. Therefore, this noon-midnight asymmetry in the magnetodisc is fed and emptied by region-2-like field-aligned currents, although how this asymmetry is maintained in the presence of nearly corotational flow is unknown. Khurana (2001) concluded that this revealed the presence of a solar wind influence, since the region 1 and region 2 currents at Earth are associated with solar wind convection. This analysis did not identify where the radial currents close in the outer magnetosphere—so did not find the return field-aligned currents matching the positive divergence inside 30 R_J. It may be that part of this current system closes on the magnetopause, at least over some local time sectors. However, some evidence for closure currents was found in the dusk sector by Kivelson et al. (2002) where a field-aligned current sheet carrying 6 MA was found 2° poleward of the main auroral oval.

2.3 Saturn's Magnetodisc

At Saturn, a magnetodisc was not found in the Pioneer or Voyager datasets but thin current sheets were found on the dawn flank (Smith et al. 1980) suggesting either a disc with a significant local time asymmetry (a spatial asymmetry) or time-dependence driven by the solar wind. Evidence for a magnetodisc configuration in the midnight, dawn and noon local time sectors was identified in Cassini magnetometer data (Arridge et al. 2008b). Figure 8 shows Cassini magnetometer data from the noon, dawn and midnight sectors of the magnetosphere. Because of Saturn's extremely small dipole tilt (e.g., Cao et al. 2011) the magnetic equator and current sheet does not move over the spacecraft in a periodic manner as at Jupiter. However, current sheet oscillations (e.g., Arridge et al. 2011) can be seen, particularly in the midnight sector. These are thought to be produced by magnetic perturbations introduced by a rotating system of field-aligned currents (e.g., Southwood and Cowley 2014). In each sector of the magnetosphere B_φ and B_r have opposite signs indicating a swept-back configuration. In the noon sector the B_θ component is larger and so indicates a more dipolar configuration however the field strength is higher than that of a dipole indicating the importance of magnetodisc currents. Two passes of Cassini are shown for the noon sector: the red trace is for a compressed magnetosphere with a magnetopause subsolar distance of 16.8 R_S and the black trace for a more expanded magnetosphere with a subsolar position of 24.8 R_S (Arridge et al. 2008b). In the compressed case the B_θ component is stronger and the B_r component weaker indicating a weaker current sheet and a quasi-dipolar (non-magnetodisc) configuration. In the expanded case the current sheet is more pronounced and the field thus takes the form of a magnetodisc. The transition to the magnetodisc at Jupiter is near a radial distance of 20 R_J compared to a magnetopause distance of $\gtrsim 45$ R_J, however at Saturn the transition region is at $\gtrsim 16$ R_S compared to a magnetopause distance of $\gtrsim 15$ R_S. Hence, Arridge et al. (2008b) argued that the formation of Saturn's magnetodisc at noon was highly sensitive to the confining effect of the solar wind. The suppression of the dayside magnetodisc under compressed magnetospheric conditions must require similar divergences in the azimuthal current, similar to that found by Khurana (2001) for the Jovian magnetosphere, but further work is required in this area.

Giampieri and Dougherty (2004) quantitatively studied the magnetic field of Saturn's magnetodisc using a simple axisymmetric current sheet model to fit the magnetic field data from Pioneer 11 and Voyagers 1 and 2. They found that a single current sheet model did not fit both the inbound and outbound segments of each flyby but could not distinguish between a local time asymmetry or a temporal effect associated with variable solar wind conditions. The inbound trajectories were all near the noon sector, whereas the outbound trajectories were near dawn or in the post-midnight sector. They found systematic differences in the total current between the inbound and outbound segments of each flyby, typically 10 MA for the outbound (near dawn/post-midnight) leg and 5–8 MA for the inbound (near noon) leg. However, they did not have sufficient data to claim a local time asymmetry.

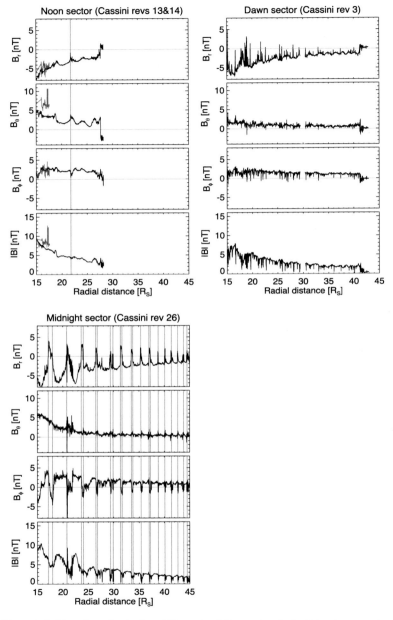

Fig. 8 Magnetic field configuration in three different local time sectors in the Saturnian magnetosphere, as measured by the Cassini spacecraft. The data are presented in spherical polar (radial-theta-phi) coordinates, plus the field magnitude, as a function of radial distance. *Vertical grey lines* indicate current sheet crossings or contact with the current sheet. Two profiles are shown in the noon sector to match two different upstream solar wind pressures corresponding to an expanded system (rev. 13, *black trace*) and compressed system (rev. 14, *red trace*). Gaps in the time series indicate data gaps or calibration periods. One calibration period (identified by sharp increases in the field strength) has been included in the rev. 14 (*red*) trace to show the overall behaviour of the time series

Fig. 9 Radial profiles of the azimuthal current density associated with various mechanical stresses and the magnetic field separated into four local time sectors: midnight to dawn (*blue*), dawn to noon (*orange*), noon to dusk (*red*) and dusk to midnight (*green*). The *solid lines* show means in 0.25 R_S-wide radial bins and the shaded grey regions show the associated standard error. Panel (**a**) shows the inertia ($j_\varphi > 0$, associated with balancing an outward mechanical force) and pressure anisotropy ($j_\varphi < 0$, associated with balancing an inward mechanical force) currents; (**b**) the sum of inertia and pressure anisotropy currents; (**c**) pressure gradient currents; (**d**) the total azimuthal current density (typically, $j_\varphi > 0$, associated with balancing a largely outward mechanical force and thus an inward $\mathbf{j} \times \mathbf{B}$ force); (**e**) B_θ component with a model of Saturn's internal field subtracted. From Kellett et al. (2011)

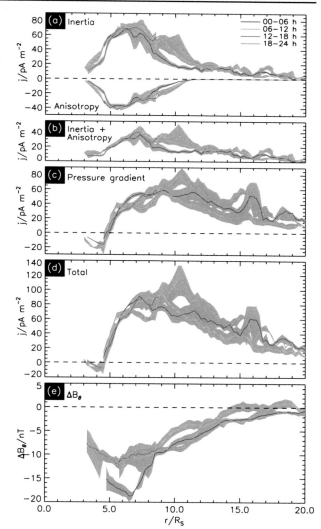

2.4 Stress Balance and Local Time Asymmetries in Saturn's Magnetodisc

Recently, Kellett et al. (2011) have investigated local time asymmetries in the ring current region inside the inner edge of the Saturnian magnetodisc. This work extended the results of Kellett et al. (2010) who computed mechanical stresses associated with the ring current: computing pressure gradient forces of energetic particles and plasma, the centrifugal force, and pressure anisotropy in the plasma. From these stresses and the B_θ component of the magnetic field they were able to calculate the azimuthal component of the volume current density. Figure 9 shows these results for four different local time sectors (only to 10 R_S for the noon-to-dusk sector due to the trajectory of the spacecraft). These currents were found to not vary greatly with local time, but were generally larger in the dusk to midnight sector (green and blue traces), and fell gradually from midnight through to noon, with a factor of 1.5 difference between the nightside and dayside. From Fig. 9c we can see that

Fig. 10 Sketch of the cross-tail configuration at Jupiter drawn from the perspective of an observer in the tail looking towards Jupiter

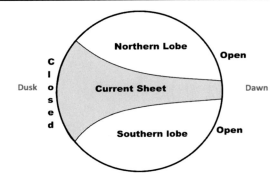

these changes are mainly driven by changes in the perpendicular pressure gradients. In the magnetodisc proper—beyond about 15 R_S—only a preliminary analysis of stress balance has been carried out. Arridge et al. (2007) demonstrated the importance of centrifugal stress, but such a detailed analysis as that carried out by Kellett et al. (2010, 2011) has not been carried out in the magnetodisc at Jupiter or Saturn.

Local time asymmetries have also been detected across the tail on the nightside at Jupiter (Kivelson et al. 2002). The magnetic pressure in the tail lobes was found to be smaller by 20–40 % near dusk compared to midnight between 25 and at least 60 R_J. This implies that the field lines are much less stretched in the dusk sector compared to the dawn sector, in accordance with the Khurana (2001) results, and therefore that the magnetotail does not have cross-tail symmetry. A proposed cross-tail geometry is sketched in Fig. 10 showing a thicker current sheet on the dusk flank and is also consistent with an observed thicker plasma sheet on the dusk flank (Krupp et al. 1999, 2001). No local time dependences in the magnetotail field at Saturn have yet been reported, but at the time of writing the Cassini trajectory does not provide good coverage of the magnetotail in the dusk-midnight sector.

3 Plasma and Current Sheet Geometry

The magnetodisc and its associated plasma sheet are not simply planar structures rotating in the dipole equator of their parent planet. Similar to the terrestrial tail current sheet (e.g., Tsyganenko and Fairfield 2004) the sheet adopts a complex three-dimensional and time-dependant shape due to the influences of the rotation of the dipole axis around the planet, magnetospheric oscillations, and solar wind forcing. Local time dependences manifest themselves due to solar wind distortions in the location of the magnetodisc, pushing the current sheet out of the equatorial plane, and in the timing of current sheet crossings seen by an observer near the equatorial plane due to local time asymmetries in the configuration of the magnetic field, the plasma distribution, and plasma flows.

The prime meridian of these current sheet oscillations is the longitude at which the current sheet elevation is a maximum. Figure 11 shows the inferred prime meridian of Jovian current sheet crossings projected onto the X–Y plane of JSO coordinate system, as calculated from all available current sheet crossing pairs (north to south and south to north). The prime meridian was averaged into 10×10 R_J^2 bins to smooth natural variation in the data. Near Jupiter, the prime meridian lies close to that expected from rigid rotation of the magnetodisc lying in the dipole equator. At larger radial distances the current sheet is delayed from that expected of rigid rotation. The delay is also a function of local time with a larger delay on the flanks compared to the noon-midnight meridian.

 🌀 Springer

Fig. 11 Prime meridian of the Jovian current sheet (in degrees) in the X–Y plane in Jovicentric Solar Orbital (JSO) coordinates where X points to the Sun, Y is opposite the direction of Jupiter's orbital motion around the Sun, and Z completes the right-handed set. From Khurana and Schwarzl (2005)

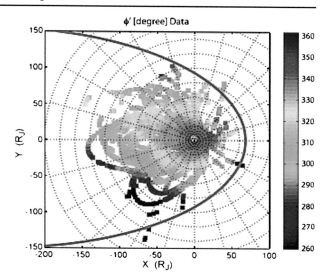

This delay in the arrival of current sheet crossings is interpreted in terms of a combination of delays introduced by MHD outflow, wave travel time and the geometry of the field (Northrop et al. 1974). Equation (1) describes the longitudinal delay, δ, in the arrival of current sheet crossings with cylindrical radial distance, ρ, expressed as an incremental delay $d\delta/d\rho$, where Ω_J is the angular velocity of Jupiter, Ω_i is the angular velocity of Jupiter's ionosphere, V_A is the Alfvén speed, and u_ρ is the mass outflow rate.

$$\frac{d\delta}{d\rho} = \frac{B_\varphi}{\rho B_\rho} - \frac{\Omega_J - \Omega_i}{u_\rho + V_A B_\rho / B} \tag{1}$$

The first term on the right hand side is related to the geometry of swept-back field lines, and the second term is associated with wave delay in subcorotating and partially outflowing plasma. As shown in Fig. 3, $B_\varphi/\rho B_\rho$ can be calculated from data (Khurana 2001; Khurana and Schwarzl 2005) and has a significant local time dependency. Using the data in Fig. 11 with the bend-back data in Fig. 3 the second term on the right-hand side can be calculated. The delay from both terms are of a similar order of magnitude but where the bend-back term has a smaller local time dependence than the second term. The local time dependence is particularly strong between the noon-midnight meridian and the dawn/dusk flanks.

It is also found that the north to south crossings are delayed more than the south to north crossings, and this is interpreted in terms of the solar wind flow distorting the current sheet such that it asymptotically becomes parallel to the solar wind flow (Khurana and Givelson 1989; Steffl et al. 2012). Such hinging is also clearly seen in Earth's magnetotail (e.g., Tsyganenko and Fairfield 2004). The evidence for such hinging is clear on the nightside of Jupiter. Evidence for such a global hinging of the current sheet is also found in the Saturnian current sheet (Arridge et al. 2008a, 2011) but where evidence for solar wind warping was also found at noon and dawn local times. Thus the current sheet takes on a bowl-shaped profile, distorted above the rotational equator at all measured local times during southern hemisphere summer. Evidence for dayside warping has also subsequently been found in the terrestrial magnetosphere (Tsyganenko and Andreeva 2014) and may also be present at Jupiter. Even though there is no measurable dipole tilt at Saturn, current sheet crossings are still observed as part of a system of global magnetospheric oscillations (see, e.g., Carbary

Fig. 12 Signatures of plasma sheet crossings as observed by the Galileo spacecraft. From Waldrop et al. (2005)

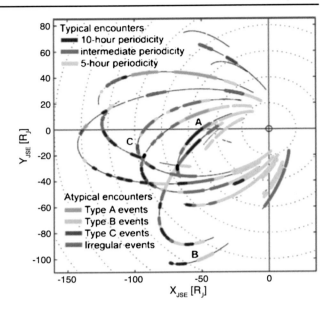

and Mitchell (2013) for a recent review). The current sheet is delayed as a function of radial distance but no evidence for a local time asymmetry in the delay (e.g., the wave speed does not appear to vary with local time) has been found. However, the prime meridian of the current sheet oscillations does appear to have a variation in local time which is connected to a local time asymmetry in the oscillations of magnetosphere (Arridge et al. 2011).

A variety of plasma sheet crossing signatures are observed in particles and fields data at both Jupiter and Saturn. Waldrop et al. (2005) studied Jovian magnetodisc crossing signatures in Galileo magnetometer and energetic particle data. Figure 12 shows a summary of their findings where different plasma sheet crossing morphologies are represented by different colour codes. Depending on the location of the spacecraft relative to the mean location of the magnetodisc either 10 h, 5 h or some intermediate periodicity is recorded. When the spacecraft is displaced from the magnetodisc's mean location by an amount similar to the oscillation amplitude 10 h period signatures are observed as the magnetodisc moves to meet the spacecraft and then moves back away again. If the spacecraft is located exactly at the mean location of the magnetodisc then two crossings (north to south, followed by south to north) will be recorded during each rotation of Jupiter, producing a 5 h periodicity. Between these two extremes an intermediate periodicity will be produced. Most of these clear current sheet crossing morphologies are found in the midnight to pre-noon sector of the magnetosphere. In the dusk sector many anomalous crossings are observed, some of which show no periodicity over many Jupiter rotations. Waldrop et al. (2005) identified three types of anomalous encounter, as labelled in Fig. 12. Type A was associated with a depressed field strength and high particle fluxes, Type B events had little or no field periodicity but intense particle fluxes, and Type C was associated with disruption of the periodic motion of the sheet, after which the magnetodisc's mean location was found to be shifted to a new location. Hence, Type C events are indicative of the result of a solar wind compression reaching the magnetosphere. Only a single anomalous encounter was found on the dawn flank, and was found to be Type B, but this may have been an example of an encounter with the cushion region. Waldrop et al. (2005) noted that the erratic nature of plasma sheet encounters in

        Springer

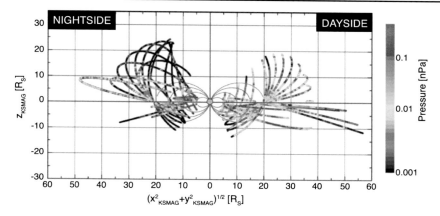

Fig. 13 Energetic particle pressure at Saturn, projected into the noon-midnight meridional plane and showing a thicker plasma sheet on the dayside compared to the nightside. From Krimigis et al. (2007)

the dusk sector perhaps indicated that the dusk sector was particularly sensitive to changes in global magnetospheric configuration, perhaps induced by solar wind activity. However, Galileo explored the dusk sector as solar activity was rising during solar cycle 23, so the predominance of anomalous activity in the dusk sector may be the result of biases in the trajectory and so this requires further work to confirm. Similar disruptions to the current sheet have also been noted at Saturn (André et al., private communication) but insufficient statistics are available to determine a dawn-dusk local time asymmetry.

The plasma sheet at Saturn is noted to be asymmetrical in local time. Figure 13 shows particle pressures as a function of distance from Saturn in the equatorial plane, and distance above/below the equatorial plane on the dayside and nightside (collapsed onto the noon-midnight meridian) (Krimigis et al. 2007). From this it can be seen that the plasma sheet appears thinner on the nightside compared to the dayside, where it is observed up to $\pm 45°$ latitude. Krimigis et al. (2007) noted that the plasma sheet appeared to thin gradually towards the nightside from noon. The data were not sorted according to the subsolar distance to the magnetopause, so it is not known if the energetic particle pressure follows the magnetodisc geometry on the dayside during an expanded magnetosphere, as identified in magnetometer data by Arridge et al. (2008b). The highly inclined orbits of Cassini during 2007 and 2009 provided an opportunity to study the plasma sheet thickness during north-south transects of the magnetodisc in the nightside (2007) and dusk (2009) sectors (Sergis et al. 2011). The plasma sheet was found to exhibit an energy-dependant vertical structure with a thicker plasma sheet in energetic particles, and a current sheet that was thinner than the thermal electrons. Intense dynamical behaviour was found in the 2009 passes in the dusk sector, therefore perhaps similar to the dusk-side magnetodisc encounters reported for the Jovian system by Waldrop et al. (2005), but this observation may be influenced by trajectory biases since the 2009 examples were close to equinox.

4 Plasma Flows and Particle Populations

4.1 Energetic Particles and Plasma

As highlighted in Fig. 13, evidence for local time asymmetries in energetic particles at Saturn were found by Krimigis et al. (2007). Even though their study did not focus on the

Fig. 14 (a) Local time distribution of the suprathermal ($E > 3$ keV) pressure measured by Cassini/MIMI near the magnetic equatorial plane of the Saturnian magnetosphere for radial distances between 6 R$_S$ and 15 R$_S$, in the period between July 2004 and July 2012. (b) Same as panel (a) but focused between 8 R$_S$ and 11 R$_S$, where the radial dependence of the suprathermal pressure is moderate. A simple sinusoidal function is over-plotted to describe the local time dependence

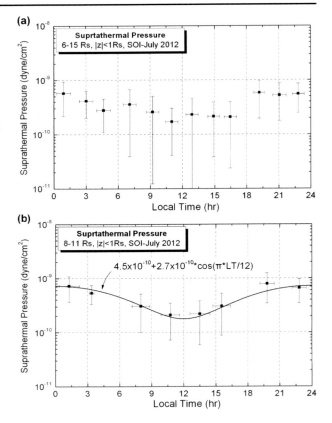

detailed local time distribution of the energetic particle pressure, a clear day-night asymmetry was revealed when the pressure data were projected in the cylindrical ρ–z plane. The dayside structure of the energetic ion plasma sheet is characterized by flux tubes filled with energetic ions to latitudes as high as 45°, extending in range to the dayside magnetopause. In contrast, the night side plasma sheet is much narrower as the field lines become strongly stretched there. Nevertheless, the energetic particles still extend to higher latitudes compared to the thermal plasma, as later shown by Sergis et al. (2009) using more extended sets of thermal plasma and energetic particle data. As shown in Fig. 9, Kellett et al. (2011) provided radial profiles of the thermal and suprathermal ion pressure components and the azimuthal ring current density, separately for 4 h local time sectors, supplying a slightly more detailed binning in local time. Their results indicated modest local time variation of the current density, with the pressure gradient current being stronger in the dusk-to-midnight sector, declining modestly by factors of ~2 or less in the midnight to dawn and dawn-to-noon sectors. Pass-to-pass temporal variability by factors of ~2–3 is also present in the outer region, particularly in the dawn to noon sector, probably reflecting both hot plasma injection events as well as solar wind-induced variations. Figure 14 depicts the local time distribution of the energetic ion pressure ($E > 3$ keV) as measured by Cassini during the first seven years of the mission (July 2004–July 2011) near the equatorial plane of the Saturnian magnetosphere ($|z| < 1$ R$_S$). Panel (a) includes radial distances between 6 and 15 R$_S$, while panel (b) focuses on ranges between 8–11 R$_S$, where the radial dependence of the suprathermal pressure is moderate. With a local time binning of 2 h, a clear day-night asymmetry is revealed with maximum pressure peaking near midnight and a minimum at noon. Although some local time sectors are not sufficiently covered, the pressure distribution appears nearly sinusoidal in local time, with a maximum-to-minimum pressure ratio of around five. This can be described by a simple function of the form:

$$P(\varphi) = \left[4.5 + 2.7 \cos\left(\frac{\pi\varphi}{12}\right) \right] \times 10^{-10} \qquad (2)$$

where P is the suprathermal ($E > 3$ keV) particle pressure in dyne cm^{-2} and φ is the Saturnian local time measured in hours from local midnight (LT $= 0$) to noon (LT $= 12$). We should mention, however, that the distribution we described and approximated here concerns the average hot plasma conditions and is the statistical outcome of long term measurements, the cumulative result of the short scale dynamic activity seen systematically during each Cassini pass.

A similar behaviour was also revealed by the analysis of long term ENA emission measurements, obtained by MIMI/INCA (Carbary et al. 2008a). Contrary to the rest of the particle sensors onboard Cassini, data obtained by INCA are not spatially limited along the spacecraft trajectory. INCA is a wide field-of-view camera that can capture the activity of energetic ions through their charge exchange interaction. During Cassini's high latitude orbits, INCA was capable of viewing a large part of the equatorial magnetosphere. Figure 15 summarises the morphology of the energetic hydrogen (panel a) and oxygen (panel b) atoms during the 120-day period when Cassini was above 40° north latitude. The colour-coded map shows that the ENA emissions originate in a quasi-toroidal region between ∼5 and ∼20 R$_S$. The ENA emissions are not azimuthally uniform around this torus but exhibit a peak near midnight (23.6 h) for hydrogen and at 21.6 h for oxygen, more easily observed in the intensity line plots of Fig. 14c. The maximum to minimum intensity ratio is ∼2.3 for the hydrogen and ∼3.5 for the more variable oxygen. Energetic particle injection events, measured in-situ by MIMI/LEMMS, also follow a similar distribution in local time, as shown by Müller et al. (2010) where a clear day-night asymmetry is observed, with a factor of 2.5 more injections on the nightside compared to the dayside.

These independent long term statistical analyses summarise the local time asymmetries in the energetic particle activity at Saturn as measured by Cassini. In Fig. 16 we compare the discussed local time distributions and we attempt to fit each of them with the same type of sinusoidal function we used to describe the local time dependence of the suprathermal pressure in Fig. 14. The comparison reveals a remarkable similarity to the average long term behaviour of the hot plasma, in terms of its local time distribution. Suprathermal pressure, ENA emission and hot plasma injections, measured independently, in-situ or remotely, with different sensors, seem to follow the same simple pattern in their local time distribution. In particular, the remote ENA imaging of the Saturnian magnetosphere from high/low latitudes during off-equatorial orbits of Cassini has illustrated the systematic plasma energisation that takes place in the night side (midnight to dawn local time sector), often in the form of hot plasma injections, possibly a result of dipolarisation, where the changing magnetic field accelerates plasma over a wide region (e.g., Mitchell et al. 2009). In the presence of sufficient neutral gas densities, these energetic particles appear as a discrete corotating population in ENA imaging. The large amplitude in the variations of the hot particle pressure also reflects its sensitivity to the heavy ion (W^+) distribution, which is linked to the presence of rotating blobs of hot plasma.

As discussed in Sect. 2, the Saturnian plasma sheet is displaced above the rotational equator, during southern hemisphere summer (Krimigis et al. 2007; Arridge et al. 2008a; Carbary et al. 2008b; Sergis et al. 2011), essentially over all local times until equinox in

Fig. 15 (**a**) Bin averaged image for 20–50 keV hydrogen neutrals obtained when Cassini was above 40 °N latitude during the first 120 days of 2007. Intensities were projected and averaged into 2×2 R$_S$ bins in the equatorial plane of Saturn. In this coordinate system x is toward the Sun, z is along the spin axis of Saturn, and y is toward dusk. *Small black crosses* indicate peaks in 36 radial profiles spaced at 10° intervals in azimuth. A *solid line* indicates a circular fit to these points; the *encircled cross* shows the centre of the circle. The *large cross with triangle at centre* shows the centroid peak intensity on the nightside. (**b**) Same as panel (a) but for oxygen neutrals. (**c**) Local time variation of intensities of radial profile peaks in the neutral H (*squares*) and neutral O (*circles*). The *continuous lines* represent third-order harmonic fits to the peaks. From Carbary et al. (2008a)

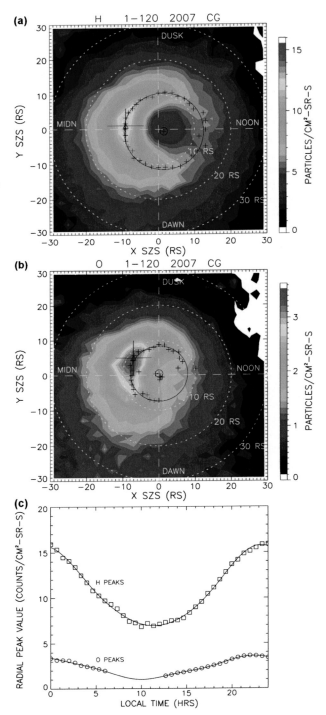

Fig. 16 Summary of the local time dependence of the energetic particle population in the Saturnian magnetosphere, as revealed by (**a**) long term measurements of suprathermal pressure; (**b**) ENA intensities; (**c**) weighted local time distribution of injection sites observed by Cassini between 3 R_S and 13 R_S (from Müller et al. 2010). In all distributions we have included the same type of sinusoidal function that was used to fit the pressure (*red lines*). Notice that the particle injection distribution in panel (c) is shifted in local time by 3 h

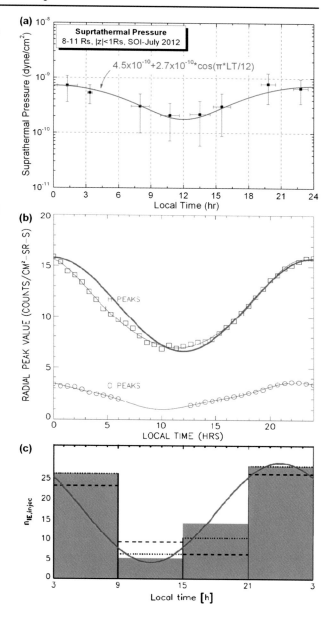

August 2009. On top of its seasonal displacement, the Saturnian plasma sheet is also periodically oscillating (flapping) vertically to the equatorial plane with a period close to the planetary rotation (e.g., Arridge et al. 2011). The question that naturally emerges is in what degree the results we reach from long term measurements obtained near the equatorial plane of the magnetosphere are affected by the warping and flapping of the plasma sheet. According to Arridge et al. (2011), the center of the plasma sheet is expected to be displaced from the equatorial plane by less than 2 R_S for radial distances below 15 R_S (their Fig. 6),

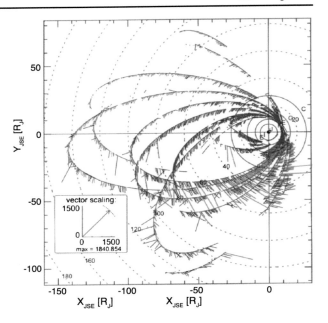

Fig. 17 Sulphur ion flow velocities projected onto the Jovian equatorial plane. The colour scale indicates the fraction of subcorotation where $0 < v/v_{cor} < 0.2$ (*cyan*), $0.2 < v/v_{cor} < 0.8$ (*blue*), $0.8 < v/v_{cor} < 1.2$ (*green*—near corotation), and $v/v_{cor} > 1.2$ (*red*—super-corotation). The vector scale is in units of km/s. From Krupp et al. (2001)

which is the region of maximum particle pressure and plasma energisation. The expected vertical displacement computed in this study can be directly compared to the plasma sheet scale heights for different particle populations. A typical suprathermal pressure scale height of 2.5 R_S (Sergis et al. 2011), indicates that studying the plasma sheet at the rotational (and magnetic) equatorial plane would impose an error not exceeding a factor of 2 (which naturally becomes less for the dayside and near-equinox conditions). This uncertainty is in part reflected in the pressure error bars in Fig. 14 and does not affect considerably our conclusions regarding the local time dependence of the energetic particle properties, as it is usually overwhelmed by the intense dynamics of the system, manifested particularly in the keV energy range. At greater distances the selection of data from the plasma sheet should be based on the orientation of the magnetic field and not be purely geometrical.

4.2 Plasma Flows and Convection

Excluding the magnetotail, the motion of plasma within the magnetospheres of Jupiter and Saturn is very different from that of the surrounding solar wind stream. At Jupiter and Saturn there is essentially a quasi-stationary boundary structure encapsulating a rapidly rotating magnetodisc that interacts with the solar wind at the boundaries between the two regimes. The magnetodiscs also respond to the internal forces that maintain their rotation around their planets and the addition of plasma mass from inner sources. Details of these interactions and forces affect the local time asymmetries observed in the flows and convection patterns in these magnetodiscs.

To understand the local time differences in plasma flows in the magnetodisc, knowledge of the local time coverage provided by the Galileo and Cassini missions is essential and thus the focus of our discussion. These missions provide more information in the nightside region, where local time asymmetries manifest themselves. Figure 17 shows plasma flow observations (Krupp et al. 2001) derived from energetic ion anisotropies measured by the Galileo Energetic Particles Detector (EPD, Williams et al. 1992). To generate the first-order

Fig. 18 Water group ion flow velocities projected onto Saturn's equatorial plane from measurements below ±5° latitude and with above average densities. Only every third point is shown for clarity. From Thomsen et al. (2010)

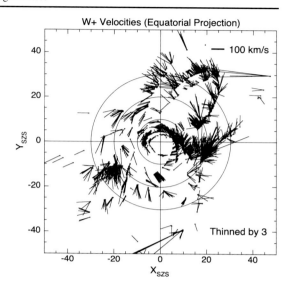

anisotropies, which usually arise from bulk flow of the plasma, a spherical harmonic expansion was fitted to the data. Two adjacent time of flight channels that discriminate mass were used to generate the spectral slope for sulphur, oxygen, and hydrogen. The harmonic analysis was then performed for each species, and when combined with the spectral slope the first order anisotropy may yield a bulk velocity. The observations show a clear asymmetry with near corotation on the dawn flank, and subcorotation in the dusk/pre-midnight sector. There are occasional radial components to the flow but the dominant flow direction is azimuthal. The data from the Galileo Plasma (PLS) instrument have also recently been reanalysed (Bagenal et al. 2011) and flow profiles have been generated. In their results, which are generally accurate in the inner ($\lesssim$10 R$_J$) to middle ($\sim$10–40 R$_J$) magnetosphere, show that no significant dawn-dusk asymmetry in the azimuthal speed is observed. Examination of the asymmetry from Krupp et al. (2001) does show a decrease in the inner regions, so that the two results may be consistent. In any case, one may expect flow asymmetries to become more important at greater distances from Jupiter since the internal forces weaken with distance and external influences increase with distance. Accordingly, anti-corotational, anti-sunward flows were found on the inbound leg of the Ulysses flyby in the pre-noon sector of the magnetosphere (Cowley et al. 1996).

There are both significant similarities and differences between the convection pattern at Jupiter and Saturn. Both systems have rapidly rotating magnetodiscs that, when scaled to the planetary radius and the standoff distance, are similarly sized (Kane et al. 2014). Both have similar interaction regions in the outermost regions of their magnetodiscs (Went et al. 2011), and are subject to the same internal stresses. Thus, one might expect the local time asymmetries to be similar. But, in fact, there are significant differences. Figure 18 shows a plot of plasma flow at Saturn derived from thermal plasma ion moments by Thomsen et al. (2010). Inside of 20 R$_S$ at all local times plasma flows are dominantly in the corotational direction, although significantly subcorotating. Beyond 20 R$_S$ there are departures from local time symmetry. In the dusk sector the plasma flows were aligned with the magnetopause, while in the pre-midnight to dawn sector the flows typically had a tailward and dawnward component (also reported by McAndrews et al. (2009) using a forward modelling technique, instead of numerical integration, to derive ion moments). No evidence was found for return

Fig. 19 Radial plasma speeds at Saturn between 10 and 60 R_S determined from Cassini MIMI/INCA data. The *red points* indicate ~26 minute measurements and the *black points* are means over 1 h local time bins with the error bars reflecting the standard deviation. Pre-midnight the radial speeds are centred about 60 km s^{-1} but after midnight they decline to be consistent with zero after 0200 local time. From Kane et al. (2014)

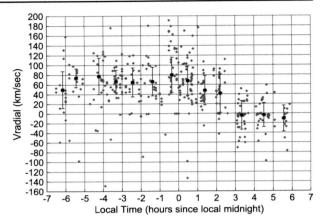

flow from more distant reconnection sites. These data were restricted to moments where the density was above average and hence Thomsen et al. (2010) interpreted these observations as indicative of heavily loaded flux tubes containing dense water group plasma (hence of internal magnetospheric origin) which were unable to return to the dayside (since the flows were dawnward and tailward).

In order to remove the limitations on density and to consider both inflow and outflow equally, Thomsen et al. (2014) have used CAPS moments analysis, selecting for data where the detector field of view spans the corotation direction and inflow/outflow sectors approximately equally, so there is no sampling bias for inflow/outflow. Their results confirm minimal local time dependence for the azimuthal speed. The radial component, however, does exhibit a local time anomaly at larger (>25 R_S) radial distances. For data analysed in the pre-dawn sector, there is evidence for significant inflows in addition to the usual outflows at larger distances. Local time coverage is somewhat limited at larger distances in this study but some evidence was also seen for radial outflow in the dusk-pre-midnight sector. In the inner magnetosphere a number of studies have suggested the presence of a circulation pattern, fixed in local time but superimposed on corotation, where plasma flows outward at dawn and inward at dusk such that plasma drift paths were offset towards noon (Andriopoulou et al. 2012; Thomsen et al. 2012; Wilson et al. 2013). Although originally envisaged as the result of a noon-to-midnight electric field, Wilson et al. (2013) have shown evidence that the electric field is offset towards the post-midnight sector field oriented towards 0100–0200 local time. However, further work is required to confirm the presence of this convection pattern. The origin of this circulation pattern and its connection with the rest of the magnetosphere is unknown.

Kane et al. (2014) have used the intensity spectrum and anisotropies in hot hydrogen ions from the Cassini INCA instrument to determine azimuthal and radial flow speeds at Saturn. In their analysis, data from five time-of-flight channels were used, and data were restricted to those times when the Cassini spacecraft was spinning to increase the field of view to nearly the full-sky. The data in each channel are binned into a 16 × 16 pixel array; thus there is a wealth of information available from which convection speeds are determined. Their analysis allowed a radial and azimuthal speed to be determined in ~26 minute intervals. Their resulting azimuthal speeds were generally consistent with those calculated by Thomsen et al. (2010), with minimal local time asymmetry (in apparent contrast with the case at Jupiter between the Galileo EPD and PLS results). Good local time coverage in the outer (>20 R_S) regions allowed a pattern in the radial flow speed to be detected. Figure 19

Springer

shows radial speeds as a function of local time with net radial outward flows before midnight (consistent with Thomsen et al. 2014) and inward or near zero in the post-midnight sector beyond 0200 LT.

One possible conclusion based on the radial speed local time dependence determined by Kane et al. (2014) is that this is an artifact generated by external forcing and possibly internal mass unloading. Their interpretation of the artifact is that plasma on approach to the dawn magnetopause is constrained by that boundary and possibly affected by unloading of mass in the pre-dawn sector near the magnetopause boundary. Their findings would imply that plasma they are measuring is being funnelled back into the return flow to continue around the dayside. One might expect the azimuthal speed to be affected as in the Krupp et al. (2001) analysis, though measurements indicate otherwise. At greater distance in the pre-dawn tail region, they propose that plasma cannot make the turn and is entrained into the magnetosheath flow and exhausted down the tail. Some events, although limited, show anti-sunward flow in this distant region near but within the dawn magnetopause.

5 Discussion

5.1 Observations Showing Local Time Asymmetries

In Sects. 2–4 we have presented and discussed the observational evidence for local time asymmetries in the magnetodisc. At both Jupiter and Saturn the magnetodisc magnetic field was more highly stretched in the midnight and dawn sectors but less so in the noon sector. At Jupiter the field is also much less stretched in the dusk sector, with some evidence for this at Saturn. In addition, the Saturnian magnetodisc can be prevented from forming in the noon sector if the solar wind dynamic pressure is sufficiently high. At Jupiter the magnetic field forms a spiral shape such that it is swept back over much of the magnetosphere but with a region near dusk where the field is almost entirely meridional, or weakly swept-forward or swept-back.

In energetic particles the plasma sheet is much thicker on the dusk flank at both Jupiter and Saturn. The magnetospheric plasma flows in the magnetodisc display local time asymmetries at Saturn, where plasma flows out at dusk and in at dawn, except for a layer adjacent to the magnetopause where the plasma flows anti-sunward (opposite to corotation). At Jupiter there is some discrepancy between energetic particle and plasma data on the presence of local time asymmetries in plasma flows, where EPD observations show a local time asymmetry, but PLS observations do not appear to show such a feature. Evidence for anti-sunward flows in the pre-noon sector at Jupiter has also been presented. Local-time asymmetries in the inner/middle magnetosphere at both planets has also been presented in the form of Io torus brightness variations (Barbosa and Kivelson 1983) and flows, plasma temperature and microsignature shifts (Andriopoulou et al. 2012; Thomsen et al. 2012; Wilson et al. 2013). The current sheet and plasma sheet itself does not rotate as a rigid structure, but experiences warping due to forces applied by the solar wind and delays introduced due to a combination of outflows, magnetic field structure and plasma distribution.

Finally, evidence for local time asymmetries can be found in the aurorae at both Jupiter and Saturn—see for example, Badman et al. (this volume) and Delamere et al. (this volume).

Many of these observational asymmetries are also found in MHD simulations, for example weaker azimuthal currents on the dayside and flow stagnation in the dusk sector (Walker and Ogino 2003), flows reaching the dawn sector and outflowing anti-sunward along the

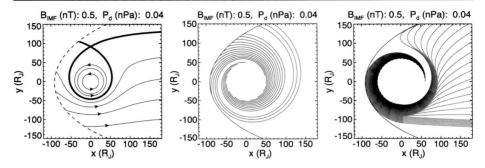

Fig. 20 (*Left*) Superposition of potentials associated with corotation and solar wind-driven convection. The *bold curve* shows the Alfvén layer separating the corotation-dominated region from the solar wind-dominated region. (*Middle*) Superposition of flows associated with corotation and outflow. (*Right*) Superposition of flows associated with corotation, outflow and anti-sunward solar wind flow tailward of a hypothetical X-line. From Delamere and Bagenal (2010)

dawn flank of the magnetotail (Fukazawa et al. 2006), asymmetries in $B_\varphi / \rho B_\rho$ (Fukazawa et al. 2010), and magnetotail flapping at Saturn (Jia and Kivelson 2012).

5.2 Models of Local Time Asymmetries

Theoretical models to understand local time asymmetries in the magnetospheres of Jupiter and Saturn either invoke (a) solar wind-driving via either reconnection (Cowley et al. 2005; Dungey 1961) or viscous (Delamere and Bagenal 2010, 2013; Axford and Hines 1961) interactions; or (b) rotational effects within a rotating disc confined inside an asymmetrical obstacle (e.g., Kivelson and Southwood 2005). The flow structure within the magnetosphere can be represented by the superposition of a corotational flow with solar wind-driven flows, equivalently by the superposition of electric fields associated with corotation and a solar wind convection electric field. The potential associated with corotation is proportional to the field strength leading Brice and Ioannidis (1970) to point out that the Jovian field should push the region of corotation out beyond the magnetopause producing a rotation-dominated magnetosphere. Delamere and Bagenal (2010) reproduced this calculation using observed azimuthal flow speeds and realistic field strengths and this can be seen in Fig. 20. Even for realistic azimuthal flow speeds (i.e., not full corotation) the region where rotation dominates extends over a large fraction of the magnetosphere, with a region near dusk that is dominated by solar wind-driven convection. As an alternative to flows driven by a solar wind convection electric field, Delamere and Bagenal (2010) considered the superposition of corotational flow and slow MHD outflow (centre panel, Fig. 20) producing flow streamlines that take circular paths around the planet before a final "lap of honour" where plasma elements beyond 60 R_J impact the dawn-to-noon magnetopause. Delamere and Bagenal found that they could replicate the canonical corotation plus solar wind convection flow pattern by adding flows due to a tail X-line (right panel, Fig. 20). Because these patterns do not show the inflowing flux tubes, which conserve magnetic flux, these patterns are strictly speaking not flow streamlines but are streamlines of momentum flux. Empty flux tubes that return via the dayside are thought to form the "cushion region" which is seen as a turbulent region of dipolar field at Jupiter (Kivelson and Southwood 2005; Went et al. 2011). Little evidence for this cushion region is found at Saturn, suggesting that the region is much thinner and perhaps a result of the lack of asymmetry in the magnetodisc during expanded magnetospheric conditions (Went et al. 2011).

Fig. 21 Theoretical and data-based models of flows in the magnetosphere for Saturn but also applicable to Jupiter. (*Left*) From Cowley et al. (2005) and similar to the theoretical flow pattern proposed by Vasyliūnas (1983) but with the addition of a Dungey cycle tail X-line in the post-midnight sector. (*Right*) Flow pattern inspired by hot plasma observations adding a region of anti-sunward flow in the post-midnight-to-dawn sector (Kane et al. 2014). In this sketch the *purple region* indicates a low-latitude boundary layer, *red arrows* indicate plasma diffusion into the magnetosheath via the Kelvin–Helmholtz instability, *black* and *red crosses* indicate small scale reconnection at the magnetopause and in the tail respectively

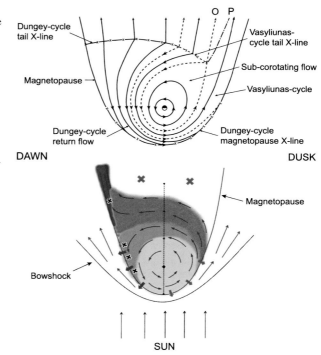

5.2.1 Solar Wind Driving as a Generator of Local Time Asymmetries

Figure 21 illustrates two conceptual pictures of flow patterns at Saturn, but which are also be applicable to Jupiter. Both show rotational flow in the inner and middle magnetosphere but local time asymmetries and X-lines at larger distances. The left-hand sketch in Fig. 21 is based on a similar diagram from Vasyliūnas (1983). Mass is continually added in the inner magnetosphere via ionisation of neutrals which originate in natural satellites and rings. Centrifugal forces drive centrifugal interchange motions producing outward transport. At large distances centrifugal forces must be balanced by magnetic stress and at some distance the required force cannot be supplied by the magnetic field and the field will stretch and eventually reconnect in the manner described by Vasyliūnas (1983), and shown in Fig. 21, now known as the "Vasyliūnas cycle".

Cowley et al. (2003, 2005) added a Dungey cycle tail X-line to this picture to close magnetic flux opened at the dayside. Thus, Fig. 21 shows three regions: (i) a sub-corotating region extending out to 10 s of R_J where the plasma flows around the planet, (ii) a sub-corotating region beyond this where the planet's ionosphere is still in control of the equatorial plasma, but where plasma is lost along the dusk flank as the field lines stretch into the tail and reconnect on the dusk flank. The return flow from this Vasyliūnas cycle through down and the morning sector is adjacent to the inner sub-corotating "core" region. Finally, the Dungey-cycle return flow in the outer magnetosphere originates in an X-line that is confined to the dawn flank of the magnetotail, because plasma on the dusk flank is dominated by outflowing, internally produced heavy ion plasma. This produces a sunward return flow channel adjacent to the dawn magnetopause (e.g., Badman and Cowley 2007). Of course, anti-sunward flow will be found tailward of the Dungey-cycle X-line.

This model is in agreement with a number of observational facts, such as the divergence of the magnetodisc currents in the dawn-noon sector (Bunce and Cowley 2001b; Khurana 2001), and the flow pattern presented by Krupp et al. (2001). Although Krupp et al. (2001) did not find evidence for the X-lines in Fig. 21, it is entirely possible that the spacecraft coverage was sunward of the X-line. However, Vogt et al. (2010) studied Galileo magnetometer data and found evidence for a statistical X-line in the tail which was a strong function of local time, from a radial distance of 90 R_J near 0500, to 100 R_J near midnight, and 140 R_J near 2200 pre-midnight. This is somewhat opposite to the orientation sketched in Fig. 21 but agrees in local-time extent. This is also compatible with the X-line derived by Woch et al. (2002) using Galileo energetic particle data.

Khurana (2001) also reached similar conclusions based on the analysis of the configuration of the field, magnetospheric currents, and plasma flows, but this has recently been extended by adding a distant tail neutral line to Vasyliūnas' original sketch, shown in Fig. 21 as modified by Cowley et al. (2003, 2005) (Khurana, private communication). In this model, dawn-dusk asymmetries in the plasma flows are produced by the return flow from the distant neutral line slowing the flow at dusk and the return flow from the near-tail neutral line accelerating it in the dawn sector. Bursty plasma flows observed in the dawn sector (Krupp et al. 2001) are the result of outflows from the near-Jupiter neutral line. Asymmetries in the equatorial field strength are associated with stagnation of the plasma flow in the dusk sector and asymmetries in the current sheet and magnetodisc field are produced by differences in plasma flow speed and cross-tail asymmetries in open flux. Finally, the partial ring current centred at midnight is a manifestation of solar wind-driven convection.

These works rely on magnetopause reconnection to drive flows via the Dungey cycle. However, the presence of a significant Dungey cycle has been challenged by some, based on the low efficiency of magnetic reconnection at giant planets (e.g., Desroche et al. 2012, 2013; Masters et al. 2012; and references therein) and also the size of the magnetosphere compared with typical flow speeds (McComas and Bagenal 2007, 2008; Cowley et al. 2008). Although there is evidence for reconnection at the magnetopauses of both Jupiter and Saturn, direct observations of dayside reconnection are relatively rare. It is not clear if this is the result of inadequate coverage of the magnetopause surface combined with only periodic driving of the system during the passage of corotating interaction regions, or if this reflects generally low reconnection rates.

The right-hand sketch in Fig. 21 from Kane et al. (2014) shows a model which does not require large-scale convection cycles as a result of dayside reconnection, and which was inspired by flow observations. In common with the Vasyliūnas (1983) and Cowley et al. (2003, 2005) models, the inner and middle magnetosphere contains plasma which flows around the planet. When it approaches the dawn magnetopause it must "squeeze" through along the dawn flank. In the more distant nightside the plasma is entrained in a viscous interaction at the dawn magnetopause and flows down tail forming a low latitude boundary layer. Sporadic reconnection in the tail will pinch off blobs of plasma that move tailward or dawnward across the tail. Hence, in this model the viscous solar wind interaction is a crucial element.

Another such model is also presented in Fig. 22 from Delamere and Bagenal (2010) where the viscous solar wind interaction plays a critical role. Within 60 R_J the magnetosphere is sub-corotating and outflowing forming spiral plasma streamlines. Beyond 60 R_J the streamlines can reach sink regions at the dawn-to-noon magnetopause and plasma can also be lost via plasmoids in the tail. Kelvin–Helmholtz vortices along the magnetopause provides a viscous interaction between the magnetosphere and the solar wind where mass and momentum can be intermittently interchanged. In this model this viscous interaction region forms the cushion region. An extended version of this model draws a comparison between the magnetospheres of Jupiter and Saturn and comets, where the viscous interaction

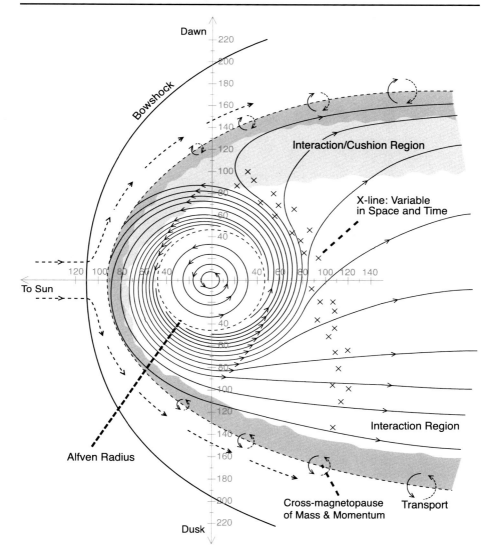

Fig. 22 Sketch of the dynamics of the Jovian magnetosphere from Delamere and Bagenal (2010)

provides a mechanism for draping of the interplanetary magnetic field (IMF) around the magnetosphere, becoming entrained in the magnetosphere (Delamere and Bagenal 2013). This model also draws on the findings of MHD simulations (Jia et al. 2012) that show regions of closed flux along the dawn and dusk flanks of Saturn's magnetopause where the viscous interaction is operating.

5.2.2 Rotational Dynamics Inside an Asymmetrical Boundary as a Generator of Local Time Asymmetries

Kivelson and Southwood (2005) examined the behaviour of the magnetodisc as a function of local time by considering the disc to be confined in an asymmetrical boundary. Figure 23

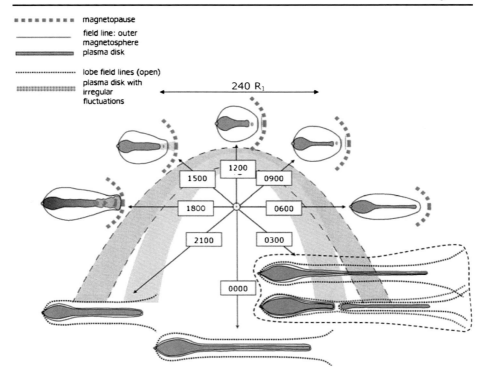

Fig. 23 Sketch showing the consequences of thickening and destabilisation of the plasma sheet in the dusk sector. Kivelson and Southwood (2005)

shows a schematic of the disc as a function of time as the disc rotates through different sectors. In the dawn sector the disc is thin and highly stretched out but as it feels the influence of the magnetopause the disc starts to thicken and hence the currents weaken. However, the outer edge of the plasma sheet becomes marginally unstable to the firehose instability and the outer edge starts to shed material. This process continues through the noon sector. Once in the post-noon sector one might expect the plasma sheet to thin again as it expands with the magnetopause. However, from observations the opposite is observed, with a thick current and plasma sheet in the dusk sector. In the Kivelson and Southwood (2005) picture this is due to the plasma sheet becoming explosively firehose unstable and the sheet breaks up in the dusk-to-pre-midnight sector. In the post-midnight sector the plasma sheet once again has a thin configuration and reconnection can occur in the stretched current sheet, producing plasmoids with the return flow populating the cushion region. In support of this mechanism, Vogt et al. (2014) have simulated the development of large $P_{\parallel} > P_{\perp}$ pressure anisotropies as the magnetodisc rotates through various local time sectors, where the field geometry periodically changes producing a non-adiabatic heating of the plasma sheet ions.

5.3 Open Questions

A wide range of observational data demonstrates the clear presence of local time asymmetries in the magnetodiscs of both Jupiter and Saturn. These asymmetries are found in magnetic fields, particle distributions, plasma flows and auroral data. A key physical question that remains to be solved is whether local time asymmetries require solar wind-driven convection (via viscous interaction or reconnection) or if we can explain these asymmetries in terms of internal/rotational processes operating in an asymmetrical cavity formed by the normal stress of the solar wind. Some models call for an entirely different approach to the solar wind-magnetosphere interaction at giant planets (e.g., Delamere and Bagenal 2010; McComas and Bagenal 2007) whilst others interpret this in terms of reconnection and solar wind-driven convection (e.g., Khurana 2001; Cowley et al. 2003, 2005). In reality, multiple elements of these models might be operating at Jupiter and Saturn and so further work is required to understand in what limit these different models are applicable.

Additional data in appropriate local time sectors and at large distances with new data and sufficient data return, are required to solve this important problem in giant planet magnetospheres. Future key observations to test these ideas more fully, include searching for more detailed evidence of anti-sunward Dungey cycle flow and Dungey cycle and Vasyliūnas cycle return flows on the dawn flank via the measurement of plasma distributions, flows and composition (including mass resolution to separate H_2^+ and He^{++} at Saturn for example). A point of potential disagreement between some of these models and data is the degree to which plasma escapes the dusk region. Some data suggest radial plasma outflow at dusk (Thomsen et al. 2014) and some data are not consistent with strong radial outflow (Kane et al. 2014). More complete coverage of the dusk magnetotail sector should clarify the existence or otherwise of radial outflow, and may also produce evidence, or otherwise, for more distant X-lines.

There are also other outstanding questions that are important to clarify in order to understand local time asymmetries in magnetodiscs:

- Are there local time asymmetries in the plasma flow at Jupiter? If not how do we reconcile the results from Galileo PLS (Bagenal et al. 2011) and EPD (Krupp et al. 2001)?
- How asymmetrical is the Jovian magnetodisc when the solar wind pressure is very low?
- Is there solar wind hinging in the dayside magnetodisc at Jupiter?
- What are the physical contributions to wave delays in current sheet motion and why are these asymmetrical in local time?
- The magnetodisc can be thought of as wrapped-around magnetotail (Piddington 1969) so have spacecraft at the giant planets ever entered the magnetotail proper—i.e., enter a region not controlled by the ionosphere (e.g., Vasyliūnas 1994)?
- Might we expect to see auroral spots, equatorward of the dusk main emission at Saturn, associated with the dipolarisation of the dayside during magnetospheric compression?
- What is the pressure anisotropy in the magnetodisc at large distances, to test the stability of the plasma sheet (e.g., Kivelson and Southwood 2005)?

Further progress can also be made in the development of new analysis techniques, for understanding spatial and temporal variability with a single spacecraft and no upstream monitor; for example: how can we separate local time asymmetries from temporal variability? Can we practically sort particle data using the magnetic field to order the particles by distance from the centre of the current sheet rather than simply geometrically sorting by distance from the centre of a model current sheet? Missions such as Juno and JUICE, and the ongoing Cassini mission will enhance our understanding of the Jovian and Saturnian magnetospheres.

Acknowledgements   The authors acknowledge the support of EUROPLANET RI project (Grant agreement No.: 228319) funded by EU; and also the support of the International Space Science Institute (Bern). The authors thank M.G. Kivelson for access to Galileo magnetometer data and M.K. Dougherty for access to Cassini magnetometer data, accessed via the NASA Planetary Data System. CSA was supported by a Royal Society Research Fellowship. CMJ was supported by a Science and Technology Facilities Council Ernest Rutherford Fellowship.

References

M. Andriopoulou, E. Roussos, N. Krupp, C. Paranicas, M. Thomsen, S. Krimigis, M.K. Dougherty, K.-H. Glassmeier, A noon-to-midnight electric field and nightside dynamics in Saturn's inner magnetosphere, using microsignature observations. Icarus **220**, 503–513 (2012). doi:10.1016/j.icarus. 2012.05.010

C.S. Arridge, Large-scale structure in the magnetospheres of Jupiter and Saturn, in *The Dynamic Magnetosphere*, ed. by W. Lui, M. Fujimoto. IAGA Special Sopron Book Series, vol. 3 (Springer, Berlin, 2011). ISBN 978-94-007-0500-5

C.S. Arridge, C.T. Russell, K.K. Khurana, N. Achilleos, N. André, A.M. Rymer, M.K. Dougherty, A.J. Coates, Mass of Saturn's magnetodisc: Cassini observations. Geophys. Res. Lett. **34**, L09108 (2007). doi:10.1029/2006GL028921

C.S. Arridge, K.K. Khurana, C.T. Russell, D.J. Southwood, N. Achilleos, M.K. Dougherty, A.J. Coates, H.K. Leinweber, Warping of Saturn's magnetospheric and magnetotail current sheets. J. Geophys. Res. **113**, A08217 (2008a). doi:10.1029/2007JA012963

C.S. Arridge, C.T. Russell, K.K. Khurana, N. Achilleos, S.W.H. Cowley, M.K. Dougherty, D.J. Southwood, E.J. Bunce, Saturn's magnetodisc current sheet. J. Geophys. Res. **113**, A04214 (2008b). doi:10.1029/ 2007JA012540

C.S. Arridge, N. André, K.K. Khurana, C.T. Russell, S.W.H. Cowley, G. Provan, D.J. Andrews, C.M. Jackman, A.J. Coates, E.C. Sittler Jr., M.K. Dougherty, D.T. Young, Periodic motion of Saturn's nightside plasma sheet. J. Geophys. Res. **116**, A11205 (2011). doi:10.1029/2011JA016827

W.I. Axford, C.O. Hines, A unifying theory of high-latitude geophysical phenomena and geomagnetic storms. Can. J. Phys. **39**, 1433 (1961). doi:10.1139/p61-172

S.V. Badman, S.W.H. Cowley, Significance of Dungey-cycle flows in Jupiter's and Saturn's magnetospheres, and their identification on closed equatorial field lines. Ann. Geophys. **25**, 941–951 (2007)

F. Bagenal, R.J. Wilson, J.D. Richardson, W.R. Paterson, Jupiter's plasmasheet: Voyager and Galileo observations, abstract SM11A-2001 presented at 2011 Fall Meeting, AGU, San Francisco, Calif., 5–9 Dec. (2011)

D.D. Barbosa, M.G. Kivelson, Dawn-dusk electric field asymmetry of the Io plasma torus. Geophys. Res. Lett. **10**, 210–213 (1983). doi:10.1029/GL010i003p00210

N.M. Brice, G.A. Ioannidis, The magnetospheres of Jupiter and Earth. Icarus **13**, 173 (1970). doi:10.1016/ 0019-1035(70)90048-5

E.J. Bunce, S.W.H. Cowley, Local time asymmetry of the equatorial current sheet in Jupiter's magnetosphere. Planet. Space Sci. **49**, 261–274 (2001a)

E.J. Bunce, S.W.H. Cowley, Divergence of the equatorial current in the dawn sector of Jupiter's magnetosphere: analysis of Pioneer and Voyager magnetic field data. Planet. Space Sci. **49**, 1089–1113 (2001b)

E.J. Bunce, S.W.H. Cowley, J.A. Wild, Azimuthal magnetic fields in Saturn's magnetosphere: effects associated with plasma sub-corotation and the magnetopause-tail current system. Ann. Geophys. **21**, 1709–1722 (2003). doi:10.5194/angeo-21-1709-2003

H. Cao, C.T. Russell, U.R. Christensen, M.K. Dougherty, M.E. Burton, Saturn's very axisymmetric magnetic field: no detectable secular variation or tilt. Earth Planet. Sci. Lett. **304**(1–2), 22–28 (2011). doi:10.1016/j.epsl.2011.02.035

J.F. Carbary, D.G. Mitchell, Periodicities in Saturn's magnetosphere. Rev. Geophys. **51**(1), 1–30 (2013). doi:10.1002/rog.20006

J.F. Carbary, D.G. Mitchell, P. Brandt, E.C. Roelof, S.M. Krimigis, Statistical morphology of ENA emissions at Saturn. J. Geophys. Res. **113**, A05210 (2008a). doi:10.1029/2007JA012873

J.F. Carbary, D.G. Mitchell, C. Paranicas, E.C. Roelof, S.M. Krimigis, Direct observation of warping in the plasma sheet of Saturn. Geophys. Res. Lett. **35**, L24201 (2008b). doi:10.1029/2008GL035970

S.W.H. Cowley, A. Balogh, M.K. Dougherty, M.W. Dunlop, T.M. Edwards, R.J. Forsyth, R.J. Hynds, N.F. Laxton, K. Staines, Plasma flow in the Jovian magnetosphere and related magnetic effects: Ulysses observations. J. Geophys. Res. **101**(A7), 15197–15210 (1996)

S.W.H. Cowley, E.J. Bunce, T.S. Stallard, S. Miller, Jupiter's polar ionospheric flows: theoretical interpretation. Geophys. Res. Lett. **30**(5), 1220 (2003). doi:10.1029/2002GL016030

S.W.H. Cowley, S.V. Badman, E.J. Bunce, J.T. Clarke, J.-C. Gérard, D. Grodent, C.M. Jackman, S.E. Milan, T.K. Yeoman, Reconnection in a rotation-dominated magnetosphere and its relation to Saturn's auroral dynamics. J. Geophys. **110**, A02201 (2005). doi:10.1029/2004JA010796

S.W.H. Cowley, S.V. Badman, S.M. Imber, S.E. Milan, Comment on "Jupiter: a fundamentally different magnetospheric interaction with the solar wind" by D.J. McComas and F. Bagenal. Geophys. Res. Lett. **35**, L10101 (2008). doi:10.1029/2007GL032645

P.A. Delamere, F. Bagenal, Solar wind interaction with Jupiter's magnetosphere. J. Geophys. Res. **115**, A10201 (2010). doi:10.1029/2010JA015347

P.A. Delamere, F. Bagenal, Magnetotail structure of the giant magnetospheres: implications of the viscous interaction with the solar wind. J. Geophys. Res. **118**, 7045–7053 (2013). doi:10.1002/2013JA019179

M. Desroche, F. Bagenal, P.A. Delamere, N. Erkaev, Conditions at the expanded Jovian magnetopause and implications for the solar wind interaction. J. Geophys. Res. **117**, A07202 (2012). doi:10.1029/2012JA017621

M. Desroche, F. Bagenal, P.A. Delamere, N. Erkaev, Conditions at the magnetopause of Saturn and implications for the solar wind interaction. J. Geophys. Res. Space Phys. **118**(6), 3087–3095 (2013). doi:10.1002/jgra.50294

J.W. Dungey, Interplanetary magnetic field and the auroral zones. Phys. Rev. Lett. **6**, 47 (1961). doi:10.1103/PhysRevLett.6.47

K. Fukazawa, T. Ogino, R.J. Walker, Configuration and dynamics of the Jovian magnetosphere. J. Geophys. Res. **111**, A10207 (2006). doi:10.1029/2006JA011874

K. Fukazawa, T. Ogino, R.J. Walker, A simulation study of dynamics in the distant Jovian magnetotail. J. Geophys. Res. **115**, A09219 (2010). doi:10.1029/2009JA015228

G. Giampieri, M.K. Dougherty, Modelling of the ring current in Saturn's magnetosphere. Ann. Geophys. **22**, 653–659 (2004). doi:10.5194/angeo-22-653-2004

C.K. Goertz, The Jovian magnetodisk. Space Sci. Rev. **23**, 319–343 (1979)

P.G. Hanlon, M.K. Dougherty, N. Krupp, K.C. Hansen, F.J. Crary, D.T. Young, G. Tóth, Dual spacecraft observations of a compression event in the Jovian magnetosphere: signatures of externally triggered supercorotation? J. Geophys. Res. **109**, A09S09 (2004). doi:10.1029/2003JA010116

T.W. Hill, F.C. Michel, Heavy ions from the Galilean satellites and the centrifugal distortion of the Jovian magnetosphere. J. Geophys. Res. **81**, 4561–4565 (1976). doi:10.1029/JA081i025p04561

C.M. Jackman, C.S. Arridge, H.J. McAndrews, M.G. Henderson, R.J. Wilson, Northward field excursions in Saturn's magnetotail and their relationship to magnetospheric periodicities. Geophys. Res. Lett. **36**, L16101 (2009). doi:10.1029/2009GL039149

X. Jia, M.G. Kivelson, Driving Saturn's magnetospheric periodicities from the upper atmosphere/ionosphere: magnetotail response to dual sources. J. Geophys. Res. **117**, A11219 (2012). doi:10.1029/2012JA018183

X. Jia, K.C. Hansen, T.I. Gombis, M.G. Kivelson, G. Tóth, D.L. DeZeeuw, A.J. Ridley, Magnetospheric configuration and dynamics of Saturn's magnetosphere: a global MHD simulation. J. Geophys. Res. **117**, A05225 (2012). doi:10.1029/2012JA017575

M. Kane, D.G. Mitchell, J.F. Carbary, S.M. Krimigis, Plasma convection in the nightside magnetosphere of Saturn determined from energetic ion anisotropies. Planet. Space Sci. **91**, 1–13 (2014)

S. Kellett, C.S. Arridge, E.J. Bunce, A.J. Coates, S.W.H. Cowley, M.K. Dougherty, A.M. Persoon, N. Sergis, R.J. Wilson, Nature of the ring current in Saturn's dayside magnetosphere. J. Geophys. Res. **115**, A08201 (2010). doi:10.1029/2009JA015146

S. Kellett, C.S. Arridge, E.J. Bunce, A.J. Coates, S.W.H. Cowley, M.K. Dougherty, A.M. Persoon, N. Sergis, R.J. Wilson, Saturn's ring current: local time dependence and temporal variability. J. Geophys. Res. **116**, A05220 (2011). doi:10.1029/2010JA016216

K.K. Khurana, Influence of solar wind on Jupiter's magnetosphere deduced from currents in the equatorial plane. J. Geophys. Res. **106**(A11), 25999–26016 (2001)

K.K. Khurana, M.G. Givelson, On Jovian plasma sheet structure. J. Geophys. Res. **94**(A9), 11791–11803 (1989). doi:10.1029/JA094iA09p11791

K.K. Khurana, H.K. Schwarzl, Global structure of Jupiter's magnetospheric current sheet. J. Geophys. Res. **110**, A07227 (2005). doi:10.1029/2004JA010757

M.G. Kivelson, D.J. Southwood, Dynamical consequences of two modes of centrifugal instability in Jupiter's outer magnetosphere. J. Geophys. Res. **110**, A12209 (2005). doi:10.1029/2005JA011176

M.G. Kivelson, K.K. Khurana, R.J. Walker, Sheared magnetic field structure in Jupiter's dusk magnetosphere: implications for return currents. J. Geophys. Res. **107**(A6), 1116 (2002). doi:10.1029/2001JA000251

S.M. Krimigis, N. Sergis, D.G. Mitchell, D.C. Hamilton, N. Krupp, A dynamic, rotating ring current around Saturn. Nature **450**, 1050–1053 (2007). doi:10.1038/nature06425

N. Krupp, M.K. Dougherty, J. Woch, R. Seidel, E. Keppler, Energetic particles in the duskside Jovian magnetosphere. J. Geophys. Res. **104**, 14867 (1999)

N. Krupp, A. Lagg, S. Livi, B. Wilken, J. Woch, Global flows of energetic ions in Jupiter's equatorial plane: first-order approximation. J. Geophys. Res. **106**(A11), 26017–26032 (2001)

A. Masters, J.P. Eastwood, M. Swisdak, M.F. Thomsen, C.T. Russell, N. Sergis, F.J. Crary, M.K. Dougherty, A.J. Coates, S.M. Krimigis, The importance of plasma β conditions for magnetic reconnection at Saturn's magnetopause. Geophys. Res. Lett. **39**, L08103 (2012). doi:10.1029/2012GL051372

S. Maurice, M. Blanc, R. Prangé, E.C. Sittler Jr., The magnetic-field-aligned polarization electric field and its effects on particle distribution in the magnetospheres of Jupiter and Saturn. Planet. Space Sci. **45**(11), 1449–1465 (1997)

H.J. McAndrews, M.F. Thomsen, C.S. Arridge, C.M. Jackman, R.J. Wilson, M.G. Henderson, R.L. Tokar, K.K. Khurana, E.C. Sittler, A.J. Coates, M.K. Dougherty, Plasma in Saturn's nightside magnetosphere and the implications for global circulation. Planet. Space Sci. **57**(14–15), 1714–1722 (2009). doi:10.1016/j.pss.2009.03.003

D.J. McComas, F. Bagenal, Jupiter: a fundamentally different magnetospheric interaction with the solar wind. Geophys. Res. Lett. **34**, L20106 (2007). doi:10.1029/2007GL031078

D.J. McComas, F. Bagenal, Reply to comment by S.W.H. Cowley et al. on "Jupiter: a fundamentally different magnetospheric interaction with the solar wind". Geophys. Res. Lett. **35**, L10103 (2008). doi:10.1029/2008GL034351

D.G. Mitchell, S.M. Krimigis, C. Paranicas, P.C. Brandt, J.F. Carbary, E.C. Roelof, W.S. Kurth, D.A. Gurnett, J.T. Clarke, J.D. Nichols, J.-C. Gérard, D.C. Grodent, M.K. Dougherty, W.R. Pryor, Recurrent energization of plasma in the midnight-to-down quadrant of Saturn's magnetosphere, and its relationship to auroral UV and radio emissions. Planet. Space Sci. **57**(14–15), 1732–1742 (2009). doi:10.1016/j.pss.2009.04.002

A.L. Müller, J. Saur, N. Krupp, E. Roussos, B.H. Mauk, A.M. Rymer, D.G. Mitchell, S.M. Krimigis, Azimuthal plasma flow in the Kronian magnetosphere. J. Geophys. Res. **115**, A08203 (2010). doi:10.1029/2009JA015122

T.G. Northrop, C.K. Goertz, M.F. Thomsen, The magnetosphere of Jupiter as observed with Pioneer 10: 2. Non-rigid rotation of the magnetodisc. J. Geophys. Res. **79**, 3579 (1974)

C.P. Paranicas, B.H. Mauk, S.M. Krimigis, Pressure anisotropy and radial stress balance in the Jovian neutral sheet. J. Geophys. Res. **96**(A12), 21135–21140 (1991). doi:10.1029/91JA01647

J.H. Piddington, *Cosmic Electrodynamics* (Wiley-Interscience, New York, 1969)

N. Sergis, S.M. Krimigis, D.G. Mitchell, D.C. Hamilton, N. Krupp, B.H. Mauk, E.C. Roelof, M.K. Dougherty, Energetic particle pressure in Saturn's magnetosphere measured with the magnetospheric imaging instrument on Cassini. J. Geophys. Res. **114**, A02214 (2009). doi:10.1029/2008JA013774

N. Sergis, S.M. Krimigis, E.C. Roelof, C.S. Arridge, A.M. Rymer, D.G. Mitchell, D.C. Hamilton, N. Krupp, M.F. Thomsen, M.K. Dougherty, A.J. Coates, D.T. Young, Particle pressure, inertial force, and ring current density profiles in the magnetosphere of Saturn, based on Cassini measurements. Geophys. Res. Lett. **37**, L02102 (2010). doi:10.1029/2009GL041920

N. Sergis, C.S. Arridge, S.M. Krimigis, D.G. Mitchell, A.M. Rymer, D.C. Hamilton, N. Krupp, M.K. Dougherty, A.J. Coates, Dynamics and seasonal variations in Saturn's magnetospheric plasma sheet, as measured by Cassini. J. Geophys. Res. **116**, A04203 (2011). doi:10.1029/2010JA016180

E.J. Smith, L. Davis, D.E. Jones, P.J. Coleman, D.S. Colburn, P. Dyal, C.P. Sonett, Saturn's magnetosphere and its interaction with the solar wind. J. Geophys. Res. **85**, 5655–5674 (1980). doi:10.1029/JA085iA11p05655

D.J. Southwood, S.W.H. Cowley, The origin of Saturn's magnetic periodicities: northern and southern current systems. J. Geophys. Res. Space Phys. **119**(3), 1563–1571 (2014). doi:10.1002/2013JA019632

A.J. Steffl, A.B. Shinn, G.R. Gladstone, J.W. Parker, K.D. Retherford, D.C. Slater, M.H. Versteeg, S.A. Stern, MeV electrons detected by the Alice UV spectrograph during the *New Horizons* flyby of Jupiter. J. Geophys. Res. **117**, A10222 (2012). doi:10.1029/2012JA017869

M.F. Thomsen, D.B. Reisenfeld, D.M. Delapp, R.L. Tokar, D.T. Young, F.J. Crary, E.C. Sittler, M.A. McGraw, J.D. Williams, Survey of ion plasma parameters in Saturn's magnetosphere. J. Geophys. Res. **115**, A10220 (2010). doi:10.1029/2010JA015267

M.F. Thomsen, E. Roussos, M. Andriopoulou, P. Kollmann, C.S. Arridge, C.P. Paranicas, D.A. Gurnett, R.L. Powell, R.L. Tokar, D.T. Young, Saturn's inner magnetospheric convection pattern; further evidence. J. Geophys. Res. **117**, A09208 (2012). doi:10.1029/2011JA017482

M.F. Thomsen, C.M. Jackman, R.L. Tokar, R.J. Wilson, Plasma flows in Saturn's nightside magnetosphere. J. Geophys. Res. Space Phys. **119**, 4521–4535 (2014). doi:10.1002/2014JA019912

N.A. Tsyganenko, V.A. Andreeva, On the "bowl-shaped" deformation of planetary equatorial current sheets. Geophys. Res. Lett. **41**(4), 1079–1084 (2014). doi:10.1002/2014GL059295

N.A. Tsyganenko, D.H. Fairfield, Global shape of the magnetotail current sheet as derived from Geotail and Polar data. J. Geophys. Res. **109**, A03218 (2004)

V.M. Vasyliūnas, Plasma distribution and flow, in *Physics of the Jovian Magnetosphere*, ed. by A.J. Desler (Cambridge University Press, Cambridge, 1983)

V.M. Vasyliūnas, Role of the plasma acceleration time in the dynamics of the Jovian magnetosphere. Geophys. Res. Lett. **21**(6), 401–404 (1994)

M.F. Vogt, M.G. Kivelson, K.K. Khurana, S.P. Joy, R.J. Walker, Reconnection and flows in the Jovian magnetotail as inferred from magnetometer observations. J. Geophys. Res. **115**, A06219 (2010). doi:10.1029/2009JA015098

M.F. Vogt, M.G. Kivelson, K.K. Khurana, R.J. Walker, M. Ashour-Abdalla, E.J. Bunce, Simulating the effect of centrifugal forces in Jupiter's magnetosphere. J. Geophys. Res. Space Phys. **119**, 1925–1950 (2014). doi:10.1002/2013JA019381

L.S. Waldrop, T.A. Fritz, M.G. Kivelson, K. Khurana, N. Krupp, A. Lagg, Jovian plasma sheet morphology: particle and field observations by the Galileo spacecraft. Planet. Space Sci. **53**, 681–692 (2005). doi:10.1016/j.pss.2004.11.003

R.J. Walker, T. Ogino, A simulation study of currents in the Jovian magnetosphere. Planet. Space Sci. **51**, 295–307 (2003)

D.R. Went, M.G. Kivelson, N. Achilleos, C.S. Arridge, M.K. Dougherty, Outer magnetospheric structure: Jupiter and Saturn compared. J. Geophys. Res. **116**, A04224 (2011). doi:10.1029/2010JA016045

D.J. Williams, R.W. McEntire, S. Jaskulek, B. Wilken, The Galileo energetic particles detector. Space Sci. Rev. **60**, 385–412 (1992)

R.J. Wilson, F. Bagenal, P.A. Delamere, M. Desroche, B.L. Fleshman, V. Dols, Evidence from radial velocity measurements of a global electric field in Saturn's inner magnetosphere. J. Geophys. Res. Space Phys. **118**, 2122–2132 (2013). doi:10.1002/jgra.50251

J. Woch, N. Krupp, A. Lagg, Particle bursts in the Jovian magnetosphere: evidence for a near-Jupiter neutral line. Geophys. Res. Lett. **29**(7), 1138 (2002). doi:10.1029/2001GL014080